Dictionary of Zoo Biology and Animal Management

The back cover of the catalogue of Bostock and Wombwell's Royal No. 1 Menagerie (1917 edition), illustrating a visit to Windsor Castle in 1847. (Courtesy of Chetham's Library, Manchester.)

Dictionary of Zoo Biology and Animal Management

A guide to terminology used in zoo biology, animal welfare, wildlife conservation and livestock production

Paul A. Rees

Senior Lecturer, School of Environment and Life Sciences, University of Salford, UK

A John Wiley & Sons, Ltd., Publication

This edition first published 2013 © 2013 by John Wiley & Sons, Ltd.

Wiley-Blackwell is an imprint of John Wiley & Sons, formed by the merger of Wiley's global Scientific, Technical and Medical business with Blackwell Publishing.

Registered office: John Wiley & Sons, Ltd, The Atrium, Southern Gate, Chichester, West Sussex, PO19 8SQ, UK

Editorial offices: 9600 Garsington Road, Oxford, OX4 2DQ, UK
 The Atrium, Southern Gate, Chichester, West Sussex, PO19 8SQ, UK
 111 River Street, Hoboken, NJ 07030-5774, USA

For details of our global editorial offices, for customer services and for information about how to apply for permission to reuse the copyright material in this book please see our website at www.wiley.com/wiley-blackwell.
The right of the author to be identified as the author of this work has been asserted in accordance with the UK Copyright, Designs and Patents Act 1988.

Designations used by companies to distinguish their products are often claimed as trademarks. All brand names and product names used in this book are trade names, service marks, trademarks or registered trademarks of their respective owners. The publisher is not associated with any product or vendor mentioned in this book. This publication is designed to provide accurate and authoritative information in regard to the subject matter covered. It is sold on the understanding that the publisher is not engaged in rendering professional services. If professional advice or other expert assistance is required, the services of a competent professional should be sought.

Library of Congress Cataloging-in-Publication Data
Rees, Paul A.
  Dictionary of zoo biology and animal management / Paul A. Rees.
      pages cm
  Includes bibliographical references and index.
  ISBN 978-0-470-67148-1 (cloth) – ISBN 978-0-470-67147-4 (pbk.)   1. Zoology–Terminology.   2. Wildlife conservation–Terminology.   3. Animal welfare–Terminology.   4. Livestock–Terminology.   I. Title.
  QL10.R44 2013
  590.3–dc23
                        2012031386

A catalogue record for this book is available from the British Library.

Wiley also publishes its books in a variety of electronic formats. Some content that appears in print may not be available in electronic books.

Cover image: Male western lowland gorilla (*Gorilla gorilla gorilla*)
Cover design by Simon Levy Associates

Set in 8/10 pt Meridien by Toppan Best-set Premedia Limited

1   2013

Contents

For Katy, Clara, Mum and Dad

'There's nothing more interesting than an orangutan'.

Terry Maple (in a lecture given at a symposium held at Detroit Zoo 6 – 7 August 2011: *From Good Care to Great Welfare – Advancing Zoo Animal Welfare Science and Policy*).

An entry from the journal of Princess Victoria, aged 17 years, following a visit to the Zoological Gardens in Regent's Park, London on Saturday 18 June 1836, the year before her coronation. (An extract from Lord Esher's typescripts of Queen Victoria's Journals, volume 2, page 19, The Royal Archives.)

'At 10 we went with Lehzen to the Zoological Gardens to see the Giraffes. There are four; they are young and not full grown, though already very tall, and are in very good health. There is a Frenchman and 3 Nubians with them. The Frenchman, from having been a long time in Africa, has adopted the costume of a native and wears a long beard. The natives are of a dark mahogany complexion, and wear red caps like the Turks. Two wore long, loose linen robes, which reached to above their ancles; the youngest, who was only 18 years old, had his face tatooed and wore a blue calico dress. The Giraffes have beautiful large black eyes, and are very good tempered.'

Preface

This book is intended, in part, as a companion to my previous work, *An Introduction to Zoo Biology and Management* (Wiley-Blackwell). It focuses on zoo animals, farm animals and companion animals and covers a wide range of topics from animal anatomy to zoo history. I hope it will be useful as a source of information for students studying a wide range of courses concerned with the management of animals, and professionals working in these areas.

In addition to entries directly concerned with the management of animals, I have also included terms which readers may encounter in the course of their studies or words which will help them to appreciate the context in which animal management occurs. For example, I have included definitions of legal terms and entries describing particular laws (including international treaties) which are relevant to agriculture, conservation, companion animals, hunting, sport and zoos. I have also included a number of ecological terms which, although they might be better placed in a dictionary of ecology, are important in understanding interactions within natural populations and in describing the habitats that zoos often attempt to simulate in naturalistic exhibits, or to which they may refer in the interpretation of their exhibits.

Zoo biology is a relatively new discipline and I have placed considerable emphasis on terms used in this field. The academic journal *Zoo Biology* only came into existence as recently as 1982, although the concept of a science devoted to the study of animals living in zoos can be traced back to the work of Heini Hediger. He wrote a number of important books including *Studies of the Psychology and Behaviour of Captive Animals in Zoos and Circuses* (1955), *Wild Animals in Captivity: an Outline of the Biology of Zoological Gardens* (1964) and *Psychology and Behaviour of Animals in Zoos and Circuses* (1969). I have attempted to include a very wide range of terms that anyone studying zoo biology or animal management, or a related subject, might encounter. In total there are over 5000 entries, including terms used in:

- Agriculture
- Animal behaviour
- Animal husbandry
- Animal welfare
- Biochemistry
- Cell biology
- Companion animal studies
- Ecology
- Experimental design
- Histology
- Law
- Nutrition
- Parasitology
- Physiology
- Reproduction
- Statistics
- Veterinary science
- Wildlife conservation
- Zoo biology

Some of the terms defined in this book may have additional meanings in common usage and other specific meanings in other disciplines which have been omitted. For example, 'saltation' has a meaning in relation to genetics and evolution, and is also a method of locomotion. Both of these definitions are of interest in the context of this book. However, the term also has a meaning in geology which is unlikely to be of interest to readers of this work, so this has been omitted. I have attempted to give only brief definitions of terms likely to be explained elsewhere (e.g. in a biology dictionary) and more detailed accounts of terms which are the proper concern of a specialised dictionary of zoo biology and animal management.

A number of references are made to the laws of various countries. By its very nature, law changes over time and the reader should not assume that any legal references necessarily relate to current law. In some cases I have made specific reference to laws which are no longer in force because of their historical importance. In other cases I have referred to legal definitions to illustrate the differences between the meaning of words between legal jurisdictions and the differences between the scientific and legal meanings of a particular term.

Several years ago my daughter and an ex-student both independently gave me the same small poster with a picture of an elephant and a speech bubble declaring '*Those of you who think you know everything are annoying to those of us that do.*' I prefer to think that these gifts were selected because of my fondness for elephants rather than the sentiment expressed by the text. There is nothing more humbling than writing a dictionary and finding that you have to look up the meaning of most of the words you want to include. However, I remember reading J. Z. Young's *Life of Vertebrates* many years ago as a young undergraduate and being surprised that he did not already know everything he had written in his book. In his preface he wrote: '*A glance through the book will show that I have not been successful in producing anything very novel . . . However, I have very much enjoyed the attempt, which has provided the stimulus to try to find out many things I had always wanted to know.*' Young's book was just about vertebrates. This book is about many subjects and it has been impossible for me to include all of the things I should like to have included. A colleague once advised me never to buy a tool that purports to serve more than one function, as it would perform none of them properly. In writing a book that attempts to explain words used in such a very wide range of disciplines I hope I have not made this mistake.

Paul A. Rees *BSc(Hons) LLM PhD CertEd*
University of Salford

Acknowledgements

A number of people have assisted with the supply of suitable photographs for this book. I am grateful to Radosław Fedacyński, Andrzej Fedaczyński and Jakub Kotowicz of the Rehabilitation Centre for Protected Animals in Przemysl, Poland, for allowing me to reproduce the photograph in Fig. S11 and to Jane Muskett who allowed me access to the archives of Belle Vue Zoo – at Chetham's Library in Manchester – which include the material photographed for Fig. B7 and the frontispiece. I am also grateful to Shaun Williams, for allowing me to photograph a number of the facilities at Reaseheath College (Fig. F2, I2, L2, S9), and Louise Bell at Blackpool Zoo for access to the materials photographed for Fig. T4. My daughter Clara kindly took the image used for Fig. T6, and my colleague Sandra Hickman kindly allowed me to reproduce the image used for Fig. C6. The images used in Figs. C5 and J3 were obtained from the Library of Congress Prints and Photographs Division, Washington DC. Extracts of the law of the United Kingdom are reproduced under the Open Government Licence for public sector information.

At Wiley-Blackwell I am indebted to Ward Cooper (Senior Commissioning Editor, Ecology, Conservation and Evolution) for believing that this project was worthwhile, Kelvin Matthews (Project Editor) for his assistance and encouragement throughout the production process, and Kenneth Chow (Production Editor). This book could not have been produced without the efforts of Ruth Swan (Toppan Best-Set Premedia Limited), Brenda Sibbald and Joanna Brocklesby MRCVS, and I thank them for their patience and help during the production process.

I am grateful to many of my colleagues and students who have unwittingly given me ideas for terms to include in this dictionary. Finally, I must thank my wife Katy, for pretending not to notice that I have been too busy with this book to learn to use our new washing machine.

How to use this book

Items are listed alphabetically by the first letter of the term described (headword), except where the first word is usually 'the' or 'a'. So, for example, 'The Royal Society for the Prevention of Cruelty to Animals' is listed under 'R' not 'T'. Plurals are included in the headwords and indicated with *pl.* and adjectives with *adj.* where these are not obvious. English spellings have been used for the headwords with American English spellings added after the first headword and cross-referenced. In some cases the main entry for a term appears under its abbreviation, especially where this is widely used, e.g. RNA. A list of acronyms and abbreviations is provided at the back of the book. Under each headword references to other entries are indicated in **BOLD AND SMALL CAPITALS**. I have attempted to provide cross-references within each entry which are essential to the understanding of the text therein or which may be of particular interest. However I have not cross-referenced every single mention of common terms such as 'blood' or 'zoo' as this would have meant that some entries consisted almost entirely of cross-references. Terms which the reader may wish to consult for comparison are indicated in each entry after the word '*compare*' and additional entries that may be of interest are indicated after the words '*see also*'. The titles of legislation are followed by the name of the country where the law applies except where this is part or all of the United Kingdom, e.g. **Marine Turtles Conservation Act of 2002 (USA), Wildlife and Countryside Act 1981**.

A note on classification

This work is not primarily concerned with the classification of animals but I have included short entries on a wide range of taxa, particularly within the mammals, birds and fishes. The classifications used were based on those adopted by the following publications:

Nelson, J.S. (1994) *Fishes of the World*. Wiley-Interscience, New York.

Nowak, R.M. (1999) *Walker's Mammals of the World* (6th ed). Johns Hopkins University Press, Baltimore and London.

Peters, J.L., Mayr, E., Greenway, J.C. *et al.* (1931–1987) *Checklist of Birds of the World* (volumes 1–16). Museum of Comparative Zoology, Cambridge, Mass.

Fig. A1 A is for aardvark (*Orycteropus afer*).

Dictionary of Zoo Biology and Animal Management: A guide to terminology used in zoo biology, animal welfare, wildlife conservation and livestock production, First Edition. Paul A. Rees.

A *See* ADENINE (A)

aardvark (*Orycteropus afer*) Traditionally the animal that represents the letter A in the alphabet. It is the only extant member of the mammalian family Orycteropodidae. Adults are the size of a small pig, with little body hair (Fig. A1). The aardvark is NOC-TURNAL and lives in underground burrows. It possesses large ears, a long snout and a long thin tongue which it uses for collecting insects. Its limbs are specialised for digging (*see also* FOSSORIAL). Aardvarks occur in Africa south of the Sahara.

AAZK *See* AMERICAN ASSOCIATION OF ZOO KEEPERS (AAZK). *See also* KEEPER ASSOCIATION

AAZPA American Association of Zoological Parks and Aquariums, now the ASSOCIATION OF ZOOS AND AQUARIUMS (AZA). *See also* ZOO ORGANISATION

Abandonment of Animals Act 1960 An Act in Britain that made it a criminal offence to abandon an animal, or permit it to be abandoned, in circumstances likely to cause the animal any unnecessary SUFFERING. The Act was repealed and effectively replaced in England and Wales by the ANIMAL WELFARE ACT 2006 and in Scotland by the Animal Health and Welfare (Scotland) Act 2006.

ABC species The species that visitors expect to see in a traditional zoo that are often used to illustrate letters of the alphabet, e.g. AARDVARK, bear, camel, deer, elephant, ZEBRA etc. *See also* AMBASSADOR SPECIES

abdomen
1. The part of the body between the thorax and the pelvis in vertebrates, bounded by the diaphragm in mammals but not other classes. It contains the viscera (e.g. most of the organs of the digestive system, kidneys etc.).
2. The part of the trunk of the body which is posterior to the thorax in arthropods.

abdominal skinfold A method of measuring OBESITY by assessing the amount of excess fat under the skin around the ABDOMEN (1). The method has been used in chimpanzees. *See also* BODY CONDITION SCORE, BODY MASS INDEX (BMI)

abductor muscle A muscle that moves a limb or other structure away from the centre line (midline) of the body. *Compare* ADDUCTOR MUSCLE (1)

abiotic Without life.

abiotic environment *See* PHYSICAL ENVIRONMENT

abiotic stress Stress caused by non-biological factors in the environment, e.g. heat and cold.

abnormal behaviour A general term for behaviour which is not part of the usual (normal) repertoire of a species, especially that which is exhibited in the wild. May occur as a result of a pathological condition, including ANXIETY and STRESS, and is sometimes observed in captive animals. *See also* APATHY, STEREOTYPIC BEHAVIOUR

abomasal ulcer An ulcer in the ABOMASUM. A common disease in cattle, especially milk-fed calves.

Causes loss of appetite, poor growth and, in extreme cases, death from bleeding.

abomasum In RUMINANTS, the fourth (and last) stomach. It is a 'secretory stomach' the lining of which produces hydrochloric acid and PROTEOLYTIC ENZYMES, and is therefore equivalent to the stomach of other mammals.

aboral Located on the side of the body opposite the mouth, especially in relation to ECHINODERM anatomy. *Compare* ORAL

abortion, miscarriage The natural or intentional termination of a pregnancy by the removal or expulsion of the EMBRYO or FOETUS. Spontaneous abortion (miscarriage) may result from a problem that arises during the development of the embryo or foetus (e.g. infection, UMBILICAL CORD TORSION, congenital defects) or abortion may be induced by a veterinary surgeon, e.g. to preserve the health of the pregnant female.

abrasion
1. A scraped area of the skin or a MUCOUS MEMBRANE which has been caused by an injury or irritation.
2. Pathological wearing away of the teeth by grinding.

abscess A swollen, inflamed area of the body containing a collection of PUS formed by the disintegration of tissue. May be caused by the presence of disease (e.g. bacteria, parasites) or a foreign object (e.g. a bullet wound, wood splinter). Depending upon the cause and condition, may be treated with antibiotics and may need to be surgically drained.

absorption
1. The uptake by cells of water by (OSMOSIS) and/or solutes (e.g. by DIFFUSION or ACTIVE TRANSPORT).
2. In the digestive system, the passage of water and solutes through the gut wall into the blood system.

ABWAK *See* ASSOCIATION OF BRITISH AND IRISH WILD ANIMAL KEEPERS (ABWAK). *See also* KEEPER ASSOCIATION

abyssal Relating to the depths of the oceans. The abyssal zone is located between approximately 2000 and 6000 m below the surface of the sea, and is in perpetual darkness, cold and nutrient poor.

academic journal A periodical publication in which original scientific and other academic work is published after a process of PEER REVIEW (Table A1). *See also* ACADEMIC PAPER

academic paper A paper (article) published in an ACADEMIC JOURNAL, which describes the results of an original scientific study, discusses a scientific problem, reviews research published on a particular subject, or is an account of some other academic endeavour.

acariasis A skin condition caused by mites and ticks.

acaricide, acaridicide A substance which kills mites and ticks.

accelerometer An electronic device that detects movements by measuring acceleration. A tri-axial

Table A1 Selected academic journals which publish original scientific studies of interest in zoo biology and animal management.

Acta Primatologica	Journal of Environmental Education
Acta Zoologica	Journal of Equine Veterinary Science
African Journal of Ecology	Journal of Experimental Psychology
American Journal of Primatology	Journal of Experimental Zoology
American Naturalist	Journal of Field Ornithology
Animal	Journal of Herpetology
Animal Behaviour	Journal of International Wildlife Law and Policy
Animal Conservation	Journal of Mammalogy
Animal Genetics	Journal of Medical Primatology
Animal Production Science	Journal of Parasitology
Animal Welfare	Journal of Reproduction and Fertility
Anthrozoös	Journal of the Bombay Natural History Society
Applied Animal Behaviour Science	Journal of Theoretical Biology
Aquatic Conservation: Marine and Freshwater Ecosystems	Journal of Tourism Studies
Auk	Journal of Veterinary Behavior
Avian Diseases	Journal of Veterinary Epidemiology
Behaviour	Journal of Veterinary Science
Biodiversity and Conservation	Journal of Wildlife Diseases
Biological Conservation	Journal of Wildlife Management
BioScience	Journal of Wildlife Rehabilitation
Bird Study	Journal of Zoo and Aquarium Research
British Poultry Science	Journal of Zoo and Wildlife Medicine
Canadian Journal of Animal Science	Journal of Zoology
Canadian Journal of Zoology	Laboratory Animal Science
Companion Animal	Laboratory Primate Newsletter
Consciousness and Cognition	Leisure Studies
Conservation	Mammal Review
Conservation Biology	Molecular Ecology
Conservation Genetics	Museum Studies
Conservation Letters	Nature
Copeia	Oryx – The International Journal of Conservation
Current Trends in Audience Research	Parks and Recreation
Ecology and Evolution	PLOS Biology
Environment and Behavior	Primate Report
Equine Veterinary Journal	Reproductive Biology and Endocrinology
Ethology	Restoration Ecology
Fisheries Research	Science
Folia Primatologica	Scientific American
Folia Zoologica	Sexuality, Reproduction and Menopause
Frontiers in Zoology	Social Cognitive and Affective Neuroscience
Herpetological Journal	Society and Animals Journal
Human Dimensions of Wildlife	South African Journal of Wildlife Research
Ibis	The Bulletin of Zoological Nomenclature
Insect Conservation and Diversity	The Veterinary Journal
International Journal of Fisheries and Aquaculture	Trends in Ecology and Evolution
International Journal of Livestock Production	Trends in Neurosciences
International Journal of Primatology	Tropical Animal Health and Production
International Journal of Zoonoses	Veterinary Ophthalmology
International Zoo Educators Journal	Veterinary Pathology
International Zoo Yearbook	Veterinary Quarterly
Journal of Agricultural Science	Veterinary Record
Journal of Animal Ecology	Visitor Behavior
Journal of Animal Physiology and Animal Nutrition	Visitor Studies
Journal of Animal Production	Wildlife Society Bulletin
Journal of Animal Science	Zoo Biology
Journal of Applied Animal Welfare Science	Zoologica
Journal of Aquatic Animal Health	Zoological Journal of the Linnean Society
Journal of Biological Macromolecules	Zoonoses and Public Health
Journal of Comparative Psychology	Zoos' Print Journal
Journal of Dairy Science	

accelerometer can detect movement in three dimensions. May be used to detect movements of animals and sometimes built into a collar incorporating a **DATA LOGGER** or wireless transmitter. Accelerometers have been used to study swimming behaviour in sharks, walking in zoo elephants, 'flying' behaviour in colugos (**DERMOPTERA**: flying lemurs), and movements of cattle on farms.

acceptable daily intake The daily amount of a substance (e.g. a nutrient, vitamin, food additive or pollutant) that an animal may safely consume throughout its life.

accidental An animal which occurs in a particular location by accident. Often applied to birds blown off course so that they appear in an area which is outside their normal range.

acclimation *See* **ACCLIMATISATION**

acclimatisation, acclimation A reversible adjustment made by an organism to the local environmental conditions. Usually occurs in nature in response to seasonal climate changes. Also occurs in fish when introduced to a new tank. *See also* **NEW TANK SYNDROME**

accredited herd In the UK, a herd of animals that has been registered as being free from certain diseases, e.g. **BRUCELLOSIS,** Johne's disease, **LEPTOSPIROSIS**.

accredited vet *See* **APPROVED VET**

accredited zoo A **ZOO** which is accredited by a zoo organisation, e.g. AZA-accredited zoo. Accreditation is conditional upon the zoo conforming to a number of standards in relation to, for example, **ANIMAL WELFARE**, **ENCLOSURE DESIGN**, provision of **ENVIRONMENTAL ENRICHMENT** etc.

acetylcholine (ACh) A **NEUROTRANSMITTER** which creates an **ACTION POTENTIAL** across the membrane of a neurone, thereby propagating a nerve impulse.

acid A chemical which produces hydrogen ions when dissolved in water. Acid solutions have a **pH** below 7.0.

acidity The relative concentration of hydrogen ions in a solution. Acid solutions have a high concentration of hydrogen ions and a **pH** of less than 7.0.

acidosis A condition in which there is a high proportion of acid waste such as urea in the blood. This may be the result of a number of conditions, for example **DIABETES MELLITUS**, kidney disease, respiratory acidosis (caused by **HYPOVENTILATION**), or lactic acidosis (when oxygen levels in the blood fall).

Acipenseriformes An order of fishes: sturgeons and their allies.

acoelomate Possessing no body cavity. *Compare* **COELOMATE**

acoustic signalling The use of sound as a method of communication in animals. Sound allows animals to produce a great variety of signals by varying frequency, pitch, loudness and temporal pattern. *See also* **ALARM CALL**

acoustic tag A small electronic device that emits radio signals whose presence is detected by listening stations. These tags may be attached to aquatic animals (e.g. manta rays, sharks) and used to record their movements.

acoustic-lateralis system, lateral line system A system for sensing the external environment in fish and some amphibians which consists of the **INNER EAR** region (which responds to sound and gravity) and lateral line organs in the skin (which respond to changes in water pressure and displacement).

acrosome A **LYSOSOME** in a sperm that contains enzymes capable of digesting a path through the covering of an egg during the process of **FERTILISATION**.

Act of Congress A statute enacted by the Congress of the United States of America (or other legislature called a congress), e.g. **ASIAN ELEPHANT CONSERVATION ACT OF 1997 (USA)**, **WILD BIRD CONSERVATION ACT OF 1992 (USA)**.

Act of Parliament Primary legislation of the UK and some other jurisdictions with a legislature called a parliament (especially countries within the Commonwealth). Many Acts are concerned with the welfare or conservation of animals, e.g. the **ANIMAL WELFARE ACT 2006** (England and Wales), **ZOO LICENSING ACT 1981** (Great Britain), **HEALTH OF ANIMALS ACT 1990 (CANADA)**.

ACTH *See* **ADRENOCORTICOTROPHIC HORMONE (ACTH)**

actin *See* **MUSCLE**

actinomycosis This disease occurs mainly in cattle and is one of several conditions known as '**LUMPY JAW**'. It is caused by an anaerobic bacterium, *Actinomyces bovis*, which probably only becomes pathogenic by invading tissues through a wound. It commonly occurs when the permanent cheek teeth are erupting. Typically, lesions occur on the cheeks, pharynx and the jaws. Swelling in bone and other tissue may cause interference with mastication, swallowing or breathing depending on the location of the lesion. Antibiotics are rarely an effective treatment. Other *Actinomyces* species can cause infections in dogs, pigs, sheep, horses, reptiles and humans.

Actinopterygii A subclass of bony fishes: ray-finned fishes.

action potential The condition in which the inside of an **AXON** is positively charged and the outside is negatively charged which occurs when the nerve membrane is depolarised. This is the reverse of the situation when the axon is exhibiting the **RESTING POTENTIAL** (i.e. the electrical polarity is reversed). A nerve impulse is the result of the movement of this depolarised area along the axon. A similar process occurs in muscle cells when excited.

activated charcoal A highly porous form of carbon which has been processed to produce a very large surface area. Formed by heating wood and other materials in the absence of oxygen. The large surface area makes the material suitable for the removal of unwanted chemicals by adsorption. Used in some filters in aquariums.

active site The region of an ENZYME that binds to the substrate when it catalyses a chemical reaction and gives the enzyme its specificity.

active sleep The part of the sleep cycle in mammals and birds during which RAPID EYE MOVEMENTS (REM) or ear movements occur and during which a characteristic ELECTROENCEPHALOGRAPH (EEG) is produced along with signs of DREAMING. Thermoregulatory mechanisms in animals do not respond to thermal stress during active sleep. This may be problematic for small mammals as their body temperature is influenced by ambient temperature much more than is the case in large mammals.

active transport The movement of molecules or ions across a biological membrane, against a chemical or electrochemical gradient, with the use of energy. The process uses 'pumps' made of PROTEINS which cross the CELL MEMBRANE.

activity budget, behaviour budget A description of the amount of time an individual animal spends on various activities during the day (e.g. feeding, sleeping, resting, walking, etc.) as defined by an ETHOGRAM. Usually expressed as a percentage (or proportion) of the total amount of time the animal is observed (Table A2). Data is often collected by instantaneous SCAN SAMPLING.

acupressure A treatment similar to ACUPUNCTURE but which involves the gentle massage of acupuncture points and channels.

acupuncture An ancient Chinese treatment that involves placing needles into special locations on the body to treat the pain associated with a wide variety of illnesses including ARTHRITIS, HIP DYSPLASIA and SPONDYLOSIS.

acute condition A disease or disorder which appears rapidly, lasts a short time, has distinct signs, and which may require short-term treatment and care. It may or may not be severe. *Compare* CHRONIC CONDITION

ad lib *See* AD LIBITUM

Table A2 Activity budget: mean proportion of time active coyotes (*Canis latrans*) kept in outdoor pens spent exhibiting different behaviours (based on Shivik *et al.*, 2009).

Behaviour	Proportion of time
Resting	0.58
Locomotion	0.21
Standing	0.16
Foraging	0.03
Social	0.02
Eating	0.01

ad lib **feeder** An animal feeder designed in such a way that the animal may obtain food whenever desired.

ad lib **sampling, opportunistic sampling** In the context of studying behaviour, *ad libitum* sampling refers to opportunistic observations which are made at the convenience of the recorder or when the opportunity arises. This is especially important for some relatively rare behaviours (e.g. mating) which might be missed during other types of sampling, such as instantaneous SCAN SAMPLING. *Compare* ONE-ZERO SAMPLING

ad libitum, ad lib Latin for 'at one's pleasure'.

Adamson, George (1906–1989) and Joy (1910–1980) A game warden and his artist wife who lived in Kenya and became famous as a result of their successful rehabilitation of an orphaned lion cub (*Elsa*) who was returned to the wild, mated with a wild lion, and produced her own cubs. Joy Adamson wrote a book about Elsa entitled *BORN FREE* which became a bestseller and was made into a film of the same name. Bill Travers and Virginia McKenna played the Adamsons in the film and as a result of their experiences working with lions established the BORN FREE FOUNDATION and ZOO CHECK.

adaptation
1. A beneficial character possessed by an organism as a result of EVOLUTION, e.g. the ability to survive for long periods without water.
2. An adjustment made by a SENSE ORGAN to changes in the strength of stimulation; a reduction in the sensitivity of a sensory receptor due to use.
3. A behavioural change caused by LEARNING which allows the animal to adjust to a variety of environmental changes.
4. A physiological change which allows the animal to adjust to a change in climate, food quality etc. *See also* ACCLIMATISATION

adaptive heterothermy A physiological adaptation to heat stress found in some mammals that inhabit arid areas, e.g. camels, ARABIAN ORYX (*ORYX LEUCORYX*). These animals allow their core body temperature to increase during the heat of the day, reducing evaporative losses by storing body heat. This excess heat is then lost to the environment at night.

adaptive radiation The evolutionary process which causes diversification from a single ancestral type which results in descendant POPULATIONS occupying increasing numbers of ECOLOGICAL NICHES.

adductor muscle
1. A muscle that moves a limb or other structure towards the centre line of the body. *Compare* ABDUCTOR MUSCLE
2. A muscle that closes the valves in a bivalve mollusc.

adenine (A) A NUCLEOTIDE base which pairs with T (thymine) in DNA and is also found in ADENOSINE TRIPHOSPHATE (ATP).

adenohypophysis *See* **ANTERIOR PITUITARY**

adenosine diphosphate (ADP) *See* **ADENOSINE TRIPHOSPHATE (ATP)**

adenosine triphosphate (ATP) The energy 'currency' of the cell; a chemical which moves energy around the cell by releasing an inorganic phosphate ion (as a result of hydrolysis) and becoming adenosine diphosphate (ADP). ADP may combine with an inorganic phosphate ion (P_i), using energy, to form ATP.

$$ATP \leftrightarrow ADP + P_i + energy \ (34 \ kJ/mol)$$

adipose Relating to cells or tissue where fat is deposited. *See also* **ADIPOSE TISSUE**

adipose fin A fatty, fin-like lobe located behind the **DORSAL** fin of some male **SALMONID** fishes.

adipose tissue A type of **CONNECTIVE TISSUE** which contains fat.
1. Brown adipose tissue (brown fat) appears to be concerned with the release of heat in neonate mammals and occurs around the neck and between the scapulae in these and hibernating mammals.
2. White adipose tissue occurs widely in animal bodies.

adoption scheme A method of fund-raising which involves visitors paying an annual subscription for the right to 'adopt' an animal kept in a zoo. The adopter may receive an adoption certificate, a photograph and updates on the animal's activities at regular intervals. The adopter's name is often displayed in a prominent place near the animal's enclosure. Similar schemes are used by wildlife conservation **NON-GOVERNMENTAL ORGANISATIONS (NGOs)** and animal **SANCTUARIES** to raise funds for conservation or animal welfare.

ADP *See* **ADENOSINE TRIPHOSPHATE (ATP)**

adrenal cortex The outer region of the adrenal gland which produces a number of hormones including **CORTISOL** and other glucocorticoids, aldosterone (which promotes water retention by the kidneys) and some sex (mainly male) hormones.

adrenal gland A hormone-secreting gland located near each kidney in most vertebrates. Consists of an inner **ADRENAL MEDULLA** and an outer **ADRENAL CORTEX**.

adrenal medulla The central region of the **ADRENAL GLAND** which secretes **ADRENALINE** (epinephrine) in response to **STRESS**. It also secretes a little **NORADRENALINE** (norepinephrine).

adrenaline, adrenalin, epinephrine A hormone secreted by the **ADRENAL MEDULLA** and to some extent by sympathetic nerve endings. It is secreted in response to fear, excitement and anger. Adrenaline increases heart rate and blood pressure, and diverts blood from the intestines and towards the muscles, preparing the body for 'flight or fright'.

adrenocorticotrophic hormone (ACTH) A hormone secreted by the **PITUITARY GLAND** which stimulates the release of a number of glucocorticoid hormones from the **ADRENAL CORTEX**, especially **CORTISOL**. These stimulate the conversion of amino acids into glucose to provide energy.

advertisement A visual, oral or chemical display used by an individual to indicate the possession of a territory by a male, the fertility state of a female, or for some other reason. *See also* **ADVERTISEMENT CALL**

advertisement call A **VOCALISATION** used by animals, especially male frogs, during rivalry and **COURTSHIP**.

aerial census *See* **AERIAL SURVEY**

aerial gunning The use of aircraft to pursue and shoot animals. *See also* **AIRBORNE HUNTING ACT OF 1971 (USA)**

aerial photography The process of producing a film or digital image of the ground taken from an aircraft. May be used to count and record the location of wild animals (especially colonial birds or large grassland mammals), and to map vegetation types. *See also* **REMOTE SENSING, SATELLITE IMAGE**

aerial survey, aerial census A survey undertaken from the air, usually from a light aircraft. Often involves flying along invisible **TRANSECT** lines and counting animals within a strip of land of known length and width, sometimes utilising **AERIAL PHOTOGRAPHY**. Often used to count large mammals in **SAVANNA** habitats. *See also* **ELTRINGHAM**

aerobic
1. Of a chemical process which requires free oxygen, e.g. **AEROBIC RESPIRATION**.
2. Of an organism which requires oxygen to survive.

aerobic respiration The biochemical process by which cells extract the energy from sugars using oxygen. This begins with the splitting of glucose into two three-carbon sugars in the process of **GLYCOLYSIS**, thereby producing **ADENOSINE TRIPHOSPHATE (ATP)**. Each of these sugars enters the **KREBS CYCLE** and produces additional molecules of ATP. The NAD (an electron carrier) produced during glycolysis and the Krebs cycle generates further ATP via an **ELECTRON TRANSPORT SYSTEM** within the **MITOCHONDRIA**. The entire process produces 36–38 molecules of ATP from a single molecule of glucose, along with carbon dioxide and water.

$$C_6H_{12}O_6 + 6O_2 \rightarrow 6CO_2 + 6H_2O + energy$$

Compare **ANAEROBIC RESPIRATION, FERMENTATION**

aerosol delivery system *See* **METERED DOSE INHALER (MDI), NEBULISER**

aestivation, estivation A dormant condition exhibited by some animals (e.g. lungfish) which allows them to avoid excessive heat during the summer or a dry period. Commonly occurs in desert species. *See also* **HIBERNATION**

aetiology, etiology The study of the cause or origin of a phenomenon, especially a disease or an **ABNORMAL BEHAVIOUR**, e.g. the aetiology of **STEREOTYPIC BEHAVIOUR** in a particular animal.

AEWA *See* **AGREEMENT ON THE CONSERVATION OF AFRICAN–EURASIAN MIGRATORY WATERBIRDS 1995 (AEWA)**

affective states Emotional states, e.g. happy, sad, excited, calm, aroused, alert.

afferent Conduction towards. For example, an afferent blood vessel carries blood towards the heart, an afferent nerve conducts impulses from the **SENSORY RECEPTORS** towards the **CENTRAL NERVOUS SYSTEM (CNS)**. *Compare* **EFFERENT**

affiliative behaviour A form of social behaviour which involves an animal's tendency to approach, interact with and remain near a **CONSPECIFIC**. Behaviour between animals which promotes group cohesion, e.g. **GROOMING**, touching, positive gestures.

affiliative exhibit An exhibit which encourages **AFFILIATIVE BEHAVIOUR** among and between people and other animals in the arrangement of activities, space and features of the design, in collaboration with management practices. For example, food, shelter, shade, water and **ENVIRONMENTAL ENRICHMENT** features are provided throughout the exhibit to reduce confrontation and competition between individuals; focal points are provided for collaborative activities, e.g. an artificial termite mound for apes, located near public viewing areas; visual access is provided between holding and isolation areas so social contact can be maintained between animals.

afforestation The process or practice of planting trees (or tree seeds), to create new forest from open land, often to replace that which has been removed by **DEFORESTATION**.

AfiFarm Management software used for dairy farming and herd management which analyses information from individual cows using pedometers and milk meters.

Africa USA The first drive-through **SAFARI PARK** in the USA, which opened in Florida in 1953 and closed in 1961. Visitors travelled around the artificially created African landscape in a 'Jungle Train'. *See also* **LAND TRAIN**

African Elephant Conservation Act of 1989 (USA) A US law whose purpose is to assist in the conservation and protection of the African elephant by financially supporting the conservation programmes of African countries and the **CITES** Secretariat, and to restrict trade in ivory.

Afrotropical region *See* **FAUNAL REGIONS**

aftershaft A second feather growing from the main shaft at the base of the vane. Important in insulation in some taxa, e.g. grouse and quail.

agar plate A Petri dish containing agar as a growth medium for microbes. Used to culture microbes and to test for the presence of infection.

age class A category into which individual animals in a population are grouped based on their age. For example, we could count all animals that are 1 year old and place them in a single class, and all animals that are 2 years old and allocate them to the next class. Alternatively we could group together animals aged 0 to less than 4 years into one class, and then those between 4 and up to 8 years in the next class and so on. A population of animals must be separated into age classes in order to construct a **LIFE TABLE** or **AGE PYRAMID**.

age pyramid A diagrammatic representation of the **AGE STRUCTURE** of a population that uses horizontal bars to represent the number of males and females in each **AGE CLASS**. Useful for comparing the age structures of different populations and for predicting future changes in the size and structure of a population.

age structure The relative numbers of animals of different ages present in a population. *See also* **AGE PYRAMID, LIFE TABLE**

age-specific mortality rate The death rate of a specific age class within a population (e.g. individuals that are 1 year old), especially in the context of a **LIFE TABLE** (Table L1).

agglutination The process of sticking together, especially in bacteria and blood cells. *See* **BLOOD GROUP**

aggregation A group of animals which has formed as a result of each being individually attracted to the same place, e.g. birds attracted to a food source, woodlice attracted to a location with high humidity. *See also* **GREGARIOUSNESS**

aggression *See* **AGONISTIC BEHAVIOUR**

aggressive mimicry The resemblance of a predator (the mimic) to a harmless animal or object (the model) that is attractive to a third organism (the dupe) on which the mimic preys, e.g. the angel fish possesses a dorsal spine that mimics a small organism which acts as a lure for its prey. *Compare* **AUTOMIMICRY, BATESIAN MIMICRY, MÜLLERIAN MIMICRY**

Agnatha A superclass of vertebrates which are fish-like but jawless and possess sucker-like mouths: lampreys, hagfishes and allies.

agonistic behaviour The complex of agonistic behaviours which may occur when two individuals of the same species encounter one another: Aggression, competition, threat, **APPEASEMENT BEHAVIOUR, RECONCILIATION**, avoidance, retreat/flight, offensive attacks, defensive fighting. This often includes species-specific **DISPLAYS**. *See also* **REDIRECTED BEHAVIOUR**.

Agreement on the Conservation of African–Eurasian Migratory Waterbirds 1995 (AEWA) An international agreement requiring parties to engage in a wide range of conservation actions for waterbirds including species and habitat conservation, management of human activities, research and monitoring, education and the provision of information. Also known as the African–Eurasian Waterbird Agreement (AEWA).

Agreement on the Conservation of Bats in Europe 1991 (EUROBATS) A regional agreement on the protection of bats concluded under the auspices of the CONVENTION ON THE CONSERVATION OF MIGRATORY SPECIES OF WILD ANIMALS 1979 (CMS).

Agreement on the Conservation of Cetaceans of the Black Sea, Mediterranean Sea and contiguous Atlantic Area 1996 (ACCOBAMS) A regional agreement on the protection of CETACEANS concluded under the auspices of the CONVENTION ON THE CONSERVATION OF MIGRATORY SPECIES OF WILD ANIMALS 1979 (CMS). It covers all Odontoceti and Mysticeti. Its purpose is to prohibit and, where possible, eliminate any deliberate taking of cetaceans and to create and maintain a network of specially protected areas to conserve cetaceans, and to promote education, research and the management of human–cetacean interactions.

Agreement on the Conservation of Polar Bears 1973 An agreement between Canada, Denmark (for Greenland), Norway, the Union of Soviet Socialist Republics (now Russia) and the USA which prohibits the taking of polar bears, except for scientific or conservation purposes, to protect other living resources, or by indigenous peoples according to their traditional rights. It requires the parties to protect polar bear habitat (especially denning and feeding sites, and migratory routes), conduct research and cooperate in the management and conservation of migrating populations.

Agreement on the Conservation of Small Cetaceans of the Baltic, North East Atlantic, Irish and North Seas 1991 (ASCOBANS) A regional agreement on the protection of small CETACEANS concluded under the auspices of the CONVENTION ON THE CONSERVATION OF MIGRATORY SPECIES OF WILD ANIMALS 1979 (CMS). Originally called the Agreement on the Conservation of Small Cetaceans of the Baltic and North Seas but extended and renamed in 2008. It covers all toothed whales (Odontoceti), except the sperm whale (*Physeter macrocephalus*) and requires parties to prohibit the taking and killing of small cetaceans, control marine pollution, conduct research, take measures to reduce and collect data on bycatches, and educate the public.

Agricultural Revolution A series of changes that took place in the agricultural practices used in England and later across Western Europe between 1700 and 1850 (and possibly earlier), including the intensification of land use, the enclosure of land and the development of scientific animal breeding.

agricultural show An event at which farmers exhibit their animals to the public and compete for prizes awarded in recognition of the quality of their livestock. May also include exhibitions of pet animals, agricultural machinery, countryside skills (e.g. SHEEP DOG TRIALS) etc. In the UK often organised as a county show, e.g. Cheshire Show, Great Yorkshire Show (Fig. A2).

Agriculture and Agri-Food Canada (AAFC) The department of the Canadian government responsible for agriculture and food (formerly the Department of Agriculture). It provides information, research, policies and programmes to achieve environmentally sustainable and competitive agricultural products.

agri-environment scheme A payment system which rewards farmers for environmentally sensitive land management.

agronomy The scientific study of the use of plants for food, fuel, fibre and other purposes; the study of soil science and crop production.

AI *See* ARTIFICIAL INSEMINATION (AI)

air bladder *See* SWIM BLADDER

air flow The movement or exchange of air. Adequate air flow is important in regulating the environment by providing VENTILATION (3) within a cage, vivarium, animal house, cow shed or other space or building containing animals. It prevents the build-up of noxious gases and helps to control temperature and humidity. Air flow may be achieved naturally as a result of the design of an enclosed space or 'forced' using fans. Ventilation is important in preventing the spread of airborne diseases. Air flow should be proportionate to the density of animals, i.e. the higher the density the more air exchanges necessary per unit time. *See also* VENTILATION SHUTDOWN

air sac One of several air-filled sacs which form part of the respiratory system of birds and assist in creating a one-way air flow resulting in VENTILATION (1) of the lungs.

air stone A block of porous material which produces a column or curtain of fine air bubbles that aerate aquarium water when it is connected to a pump via a plastic tube (Fig. A3).

Airborne Hunting Act of 1971 (USA) An Act in the USA which bans the shooting, harassing, capturing or killing of any bird, fish or other animal from aircraft, except for legitimate wildlife management purposes.

airplane wing *See* ANGEL WING

air-stripper *See* PROTEIN SKIMMER

airway The route through which oxygen reaches the LUNGS, i.e. via the nose or mouth and TRACHEA (1).

AKAA Animal Keepers' Association of Africa. *See also* KEEPER ASSOCIATION

alarm A device for generating a sound during an emergency. Some zoos use alarms with different sounds for emergencies in different sections of the zoo. This ensures that staff respond appropriately. *See also* ALARM CALL, ALARM PHEROMONE, ALARM RESPONSE, DETECTOR BEAM ACTIVATED SURVEILLANCE SYSTEM

alarm call A VOCALISATION made by an animal (particularly a social animal, especially birds and mammals) to indicate the presence of danger to others of the same species (particularly the same social group). In some species, for example prairie dogs (*Cynomys* spp.) and some primates, distinctive

Fig. A2 Agricultural show. Two Prizewinning British Blue cattle at the Cheshire Show in the UK.

Fig. A3 Air stone.

calls are used to indicate different types of predator, e.g. snake, bird of prey. *See* **ALARM RESPONSE, SENTRY**

alarm pheromone A short-lived chemical signal released by an animal to indicate the presence of danger to conspecifics, e.g. by ants, bees and other insects.

alarm response A response made to a sign of danger by an individual animal to warn others. It may be visual, olfactory or auditory. An animal may draw the attention of a predator to itself when giving an **ALARM CALL** to assist others. This type of **ALTRU-**

ISM occurs when the animal giving the warning is closely related to those it is trying to warn. *See also* **ALARM PHEROMONE**

albinism A condition in which the pigment **MELANIN** fails to develop in the hair, skin and iris. Individuals have pale skin, white hair and pink pupils. Albinism in mammals is inherited via an autosomal recessive gene. Albinos are homozygous for the gene. *Compare* **LEUCISM, MELANISM**

albino An animal that exhibits **ALBINISM**.

Albuliformes An order of fishes: bonefishes and their allies.

albumen Egg white. The fluid contained in an **AMNI-OTIC EGG** that contains large quantities of **ALBUMIN**.

albumin One of a group of soluble proteins found in blood, **ALBUMEN** and many other tissues.

aldosterone *See* **ADRENAL CORTEX**

alertness A state of readiness, in an animal, to detect environmental changes. *See also* **VIGILANCE BEHAVIOUR**

alevin A newly hatched fish, especially a salmon or trout.

***Alex* the African grey parrot** An African grey parrot (*Psittacus erithacus*) who was the subject of a 30-year study by Irene Pepperberg. She taught him to recognise words, colours, objects and some numbers and could communicate with him using words.

alfalfa *See* LUCERNE

algal bloom An increase in the number of algae in a body of water. May be seasonal or caused by EUTROPHICATION as a result of pollution, especially fertilisers. *See also* BLUE-GREEN ALGAE

alien species
1. A species which occurs in a location which is not part of its normal geographical range. Often established as a result of accidental or intentional introduction by humans, e.g. FERAL CATS, goats, rats.
2. In the US, Executive Order 13112 of February 3, 1999 defines alien species as '. . . *with respect to a particular ecosystem, any species, including its seeds, eggs, spores, or other biological material capable of propagating that species, that is not native to that ecosystem*'.

alimentary canal The gut.

alkalage A fodder produced by preserving wholecrop (grain, stem and leaves) using a process of ammoniation, which involves spreading an additive pellet containing urea and urease. This is a means of storing cereal crops as forage which helps to retain the nutrient qualities of the fodder. *See also* SILAGE

alkali A hydroxide of any of a number of metallic elements, e.g. sodium hydroxide, that dissolves in water to produce an alkaline solution: a solution with a **pH** above 7.0 as a result of an excess of hydroxyl ions, which neutralises acids to form salts.

alkaline Possessing the properties of an ALKALI.

allantois A cavity, formed by a membrane, which stores metabolic wastes and assists with gaseous exchange through the egg shell in birds and reptiles. Also forms part of the PLACENTA in mammals.

allele, allelomorph Any of a number of different forms of a gene that may exist at a specific LOCUS on a chromosome, each of which produces a different variety of the same trait. In diploid organisms, each individual normally possesses two alleles (one from each parent), for each gene in each SOMATIC CELL. *See also* DOMINANT ALLELE, RECESSIVE ALLELE

allele fixation *See* FIXATION

allele frequency *See* GENE FREQUENCY

allele retention The expected proportion of founder X's alleles that have survived to generation *t*. This is a measure of the extent to which a population has been able to retain the alleles originally present in the founder individuals. Conservation management should attempt to maximise allele retention.

allelomorph *See* ALLELE

Allen's rule A rule in biology which states that endotherms that have evolved in warmer climates have longer appendages or extremities than related forms that have evolved in colder climates. The rule may apply to different, related species or to individuals of the same species that originate from different

parts of its geographical distribution. The rule is named after Joel Asaph Allen who proposed it in 1877. Long, thin appendages (e.g. ears or legs) have a high SURFACE AREA:VOLUME RATIO, resulting in high heat loss. Animals that live in cold climates need to conserve heat so should evolve short appendages. The reverse is true in warm climates. The red fox (*Vulpes vulpes*) of temperate latitudes has relatively small ears, whereas the fennec (*Fennecus zerda*) and bat-eared fox (*Otocyon magalotis*), that both inhabit areas of Africa with a warmer climate, both have very large ears (Fig. A4). *See also* BERGMANN'S RULE

allergen Any foreign substance (usually a protein) which induces an ALLERGIC REACTION in the body of an animal who is hypersensitive to it.

allergic reaction An IMMUNE REACTION with no purpose which occurs when the body responds to a non-threatening foreign substance (usually a protein). Results in the release of HISTAMINE which causes allergic signs including INFLAMMATION. *See also* ANAPHYLACTIC SHOCK, ANTIHISTAMINE

allergy An immune response to an ANTIGEN which is otherwise harmless. *See also* ALLERGIC REACTION

alligator farm A facility where alligators are bred commercially for their meat, skins and other products. Some are open to the public as visitor attractions, particularly in the southern United States.

allocarer *See* ALLOPARENT

allogrooming Social GROOMING. The grooming of one individual by another within a social group. May have hygienic and SIGNALLING functions, e.g. appeasement. Occurs widely in primate societies (Fig. A5). *See also* ALLOPREENING, APPEASEMENT BEHAVIOUR

allometry The study of biological scaling: the differential effect of changing the linear, area and volume dimensions of organisms, and the impact this has on their evolution and ecology. Allometric relationships may be studied during the growth of a single organism; between different organisms within a single species; or between individuals which belong to different species. For example, the allometric relationship between brain size and body size is such that animals with bigger bodies tend to have bigger brains.

allomother *See* ALLOPARENT

alloparent, allocarer A 'false' or 'other' parent or mother (allomother). An animal who assists in providing PARENTAL CARE to an individual who is not its offspring and may not even be related, e.g. cow elephants (allomothers) may take care of the offspring of others. *See also* ALTRUISM, SURROGATE (1)

allopatric Having non-overlapping distributions. Two species are said to be allopatric if their distributions do not overlap in time and space. Often refers to a mode of evolution caused by the geographical separation of members of a population which then

Fig. A4 Allen's rule. Bat-eared fox (*Otocyon megalotis*), left; red fox (*Vulpes vulpes*), right.

Fig. A5 Allogrooming. Hamadryas baboons (*Papio hamadryas*).

diverge into separate species. *Compare* **PARAPAT-RIC, SYMPATRIC**

allopatric speciation The formation of new species as a result of the geographical separation of populations of the ancestral species such that each evolves independently, adapting to local conditions and eventually becoming genetically isolated from other populations and unable to interbreed. This may occur when habitats become fragmented or when major dispersal barriers such as mountains or oceans separate populations. *Compare* **PARAPATRIC SPE-CIATION, SYMPATRIC SPECIATION**

allopreening Mutual **PREENING** that occurs between birds. May have a hygienic function in removing **ECTOPARASITES** and also a social function in helping to maintain social bonds. It may also replace aggressive behaviour in some species. *See also* **ALLOGROOMING**

all-terrain vehicle (ATV) *See* **QUAD BIKE**

alpaca farming The rearing of alpacas (*Vicugna pacos*) for their fleeces, most of which are processed into yarn for knitting or weaving. Alpacas are **CAMELIDS** used as pack animals in South America and also bred in North America, Australia, New Zealand, South Africa, China, the UK and throughout Europe. Some alpaca farm owners also offer 'alpaca walking' and 'alpaca trekking' experiences. *See also* **BRITISH ALPACA SOCIETY (BAS)**

alpha (α) diversity The biological diversity within a particular area or **ECOSYSTEM** (usually expressed as the number of species in that ecosystem, i.e. species richness). *See also* **BETA (β) DIVERSITY, BIODIVERSITY, GAMMA (γ) DIVERSITY**

alpha male *See* **ALPHA STATUS**

alpha predator *See* **APEX PREDATOR**

alpha status The status afforded to the most dominant or important individual in a group of animals, to which all other individuals are subordinate. The alpha male is the most dominant male animal in a DOMINANCE HIERARCHY. *See also* BETA STATUS

alpha-tocopherol *See* VITAMIN E

alternating tripod gait A walking pattern found in insects in which three alternating legs support the body at any one time.

alternative hypothesis (H₁) *See* NULL HYPOTHESIS (H₀)

altricial Referring to a species (especially a mammal or bird) in which very young animals are helpless and incapable of caring for themselves at birth or when they hatch, e.g. canids, rodents. In relation to birds, a young bird which is incapable of moving around on its own soon after hatching. Altricial chicks possess little or no down, hatch with their eyes closed and are incapable of leaving the nest. They are fed by their parents. All PASSERINES produce altricial chicks. *Compare* PRECOCIAL

altruism The act of unselfishly helping or showing concern for another. An animal may suffer some detriment as a result of engaging in an altruistic act. *See also* ALARM CALL, INCLUSIVE FITNESS, INDIRECT FITNESS, PROSOCIAL BEHAVIOUR, RECIPROCAL ALTRUISM

altruistic behaviour *See* ALTRUISM

alula (alulae *pl.*) The first digit on a bird's wing.

alveolus (alveoli *pl.*)
1. A minute air sac in the lungs over which GASEOUS EXCHANGE occurs. These sacs have the effect of increasing the total surface area of the LUNGS substantially.
2. A milk reservoir in a MAMMARY GLAND.

ambassador species A species kept by a zoo which may have little conservation value, e.g. common zebra (*Equus burchelli*). Visitors may expect to see such species and they may help to generate interest in conservation in general. *See also* ABC SPECIES, EDUCATION OUTREACH ANIMAL

ambient temperature The surrounding air temperature.

ambivalent behaviour A behaviour which is typical of conflict situations and in which an animal appears to be trying to perform two incompatible activities at the same time, e.g. pecking at food while moving away from it. This is the result of motivational conflict and the behaviour may have evolved into a stereotypical display as a result of RITUALISATION. *See also* STEREOTYPICAL BEHAVIOUR

ambush behaviour A form of hunting whereby a predator lurks (often camouflaged) in a place which prey are likely to frequent, rather than actively hunting.

ambush predator *See* AMBUSH BEHAVIOUR

amebiasis *See* AMOEBIASIS

amenorrhoea The absence or suppression of MENSTRUATION (OVULATION). Lactational amenorrhoea is the suppression of menstruation when the mother is producing milk and prevents her from becoming pregnant while nursing a young infant. *See also* MENSTRUAL CYCLE

amensalism An ecological relationship between two organisms of different species in which one is adversely affected and the other is unaffected.

American Anti-vivisection Society (AAVS) The first non-profit animal advocacy and educational organisation in the USA (founded in 1883) dedicated to ending experimentation on animals in research, testing and education. It also opposes other forms of cruelty to animals. *See also* VIVISECTION

American Association of Zoo Keepers (AAZK) A professional organisation for zoo keepers and aquarists, which supports keeper education and education of the public in conservation.

American Association of Zoo Veterinarians (AAZV) An association of veterinary surgeons whose aim is to advance programmes for preventive medicine, husbandry and scientific research in the field of veterinary medicine dealing with captive and free-ranging wild animals. It disseminates research by publishing the *Journal of Zoo and Wildlife Medicine*.

American Association of Zoos and Aquariums *See* ASSOCIATION OF ZOOS AND AQUARIUMS (AZA)

American Birding Association (ABA) An organisation which represents the North American BIRDING community. It supports birders through publications, conferences, workshops, tours, partnerships and networks.

American Farm Bureau An NGO in the USA that promotes the wellbeing of farm and ranch communities and acts on behalf of these communities in dealings with the government and others at local, county, state, national and international levels. Each state has its own State Farm Bureau. *See also WYOMING FARM BUREAU FEDERATION V. BABBITT* (1997)

American Horse Council An organisation that represents all aspects of the horse industry in the United States.

American Humane Association An organisation that exists to protect children, pets and farm animals from abuse and neglect in the United States. It was founded in 1877.

American Kennel Club An organisation, founded in 1884, which registers dog breeds and whose aim is to advance the study, breeding, exhibiting, running and maintenance of purebred dogs. *See also* AMERICAN RARE BREED ASSOCIATION (ARBA), KENNEL CLUB

American Livestock Breeds Conservancy (ALBC) A non-profit membership organisation, founded in 1977, which works to protect over 150 breeds of livestock and poultry from extinction, including asses, cattle, goats, horses, sheep, pigs, rabbits, chickens, ducks, geese and turkeys. *See also* RARE BREEDS CONSERVATION SOCIETY OF NEW ZEALAND, RARE BREEDS SURVIVAL TRUST (RBST)

American Museum of Natural History A major natural history museum in New York which was founded in 1869.

American Rare Breed Association (ARBA) An organisation which provides a registration system and show venues across the USA for 'rare breeds' of dogs which are not recognised by the **AMERICAN KENNEL CLUB**.

American Society for the Prevention of Cruelty to Animals (ASPCA) The first humane organisation in the Western Hemisphere. It was founded by Henry Bergh in 1866. Its aim is to prevent cruelty to animals and it works to rescue animals from abuse, and pass humane laws. It operates animal shelters and animal adoption schemes.

American Society of Animal Science An organisation which fosters the discovery, sharing and application of scientific knowledge regarding the responsible use of animals to enhance human life and well-being. It publishes the *Journal of Animal Science*.

American Stud Book The registry maintained by the **JOCKEY CLUB (USA)** for all **THOROUGH-BREDS** foaled in the United States, Puerto Rico and Canada and for all thoroughbreds imported into the United States, Puerto Rico and Canada from countries that have a studbook approved by the Jockey Club and the **INTERNATIONAL STUD BOOK COMMITTEE**.

American Veterinary Medical Association A non-profit organisation founded in 1863 to represent the interests of veterinary surgeons in the United States. It works to advance veterinary medicine, including its relationship to public health, biological science and agriculture.

Amiiformes An order of fishes: bowfin.

amino acid A subunit of a protein which contains an amine group and an acidic carboxyl group, along with a side chain that varies between different amino acids. There are 22 different types of standard amino acid.

amnion A fluid-filled cavity formed by a membrane which encloses the embryo of birds, reptiles and mammals.

amniote
1. Possessing an **AMNION**.
2. A vertebrate (mammal, bird or reptile) in which the embryo possesses an **AMNION**.

amniotic egg The egg of an **AMNIOTE (2)**.

amniotic fluid Fluid contained within the amnion formed in reptiles, birds and some mammals. Provides a buffer against mechanical damage and helps to stabilise temperature, especially in placental mammals.

Amoeba A genus of sarcodine **PROTOZOANS**, members of which consist of a single irregular-shaped cell which moves and feeds using pseudopodia (cytoplasmic extensions). Some cause disease. *See also* **AMOEBIASIS**

amoebiasis, amebiasis An infection caused by the **AMOEBA** *Entamoeba histolytica*.

Amphibia A class of chordates; **POIKILOTHERMIC**, mostly terrestrial **TETRAPOD** vertebrates. Most species return to water to lay eggs which develop into tadpoles, but some are **VIVIPAROUS**. Fertilisation is internal or external but there is no **INTROMITTANT ORGAN**. The skin is soft, naked and glandular (being rich in **MUCUS** glands) and used for gaseous exchange. Some species possess poison glands in the skin.

amphibian Member of the class **AMPHIBIA**.

Amphibian Ark (AArk) An organisation which was established as a joint effort between the **WORLD ASSOCIATION OF ZOOS AND AQUARIUMS (WAZA)**, the IUCN/SSC **CONSERVATION BREEDING SPECIALIST GROUP (CBSG)** and the IUCN/SSC Amphibian Specialist Group (ASG), and other partners around the world, aimed at ensuring the global survival of amphibians. Since 2006, AArk has been assisting the *ex-situ* conservation community to address the captive components of the Amphibian Conservation Action Plan of the **INTERNATIONAL UNION FOR THE CONSERVATION OF NATURE AND NATURAL RESOURCES (IUCN)**. This involves taking species at immediate risk of extinction into captivity in order to establish captive-survival assurance colonies. The survival of many amphibian species is threatened by **CHYTRIDIOMYCOSIS**.

amplexus The mating embrace in frogs in which the male grasps the female from behind and both sexes release gametes (Fig. A6).

ampulla (ampullae *pl.*)
1. A bulge at the end of each semi-circular canal in the **INNER EAR** containing hair cells that detect the acceleration of the head.
2. A reservoir in the water vascular system of **ECHINODERMS**.
3. A low-frequency electroreceptor found in some fishes.

amputation The removal of a limb, usually as part of a surgical procedure.

Fig. A6 Amplexus in glass frogs (Centrolenidae).

amylase An enzyme found in saliva in some taxa, which breaks starch down into simple sugars. It is also released into the small intestine from the pancreas.

anabolic
1. Relating to **ANABOLISM**.
2. Describing a substance which promotes the growth of body tissue, e.g. **ANABOLIC STEROID**.

anabolic steroid A drug that simulates the effect of male sex hormones and increases tissue growth, especially in **SKELETAL MUSCLES**. Sometimes used to enhance performance in racehorses and greyhounds.

anabolism **METABOLISM** in which complex organic molecules are synthesised from simpler ones, storing energy, e.g. carbohydrate anabolism involves the conversion of glucose to glycogen; protein anabolism involves the creation of complex proteins from amino acids. *Compare* **CATABOLISM**.

anachoresis The avoidance of predators by living in a crevice, hole or other retreat. Species which exhibit this behaviour are called anachoretes. Some live entirely in a burrow (e.g. some polychaete worms), while others emerge at night (e.g. rabbits, badgers).

anadromous Migrating from the oceans to freshwater for **SPAWNING**. Anadromous fishes are those that are born in freshwater, spend most of their lives at sea, and then return to freshwater streams and rivers to spawn, e.g. salmon, trout, lampreys. *Compare* **CATADROMOUS, DIADROMOUS**

anaemia, anemia An abnormal reduction in the amount of **HAEMOGLOBIN** in red blood cells resulting in reduced oxygen-carrying capacity. This causes fatigue and breathlessness. It may be caused by blood loss, iron deficiency, red cell destruction or an inability to produce a sufficient quantity of red cells (as in pernicious anaemia caused by vitamin B_{12} deficiency).

anaerobic Relating to conditions in which no oxygen is present, e.g. **ANAEROBIC RESPIRATION**. *See also* **ANOXIC**

anaerobic respiration The cellular process by which energy is released from food molecules (e.g. **GLUCOSE**) in the absence of oxygen. **ADENOSINE TRIPHOSPHATE (ATP)** is generated from an **ELECTRON TRANSPORT SYSTEM**. *See also* **LACTIC ACID FERMENTATION**

anaesthesia, anesthesia A loss of sensation which may affect the whole body, and involve a loss of consciousness (general anaesthesia), or a localised area (local anaesthesia). *See also* **ANALGESIA**

anaesthesia induction chamber A container used for anaesthetising small animals which usually consists of a sealed transparent plastic box with a gas inlet and a waste gas outlet.

anaesthesia vaporiser A device attached to an **ANAESTHETIC MACHINE** which delivers a specific concentration of a volatile anaesthetic agent, which is liquid at room temperature but vaporises easily.

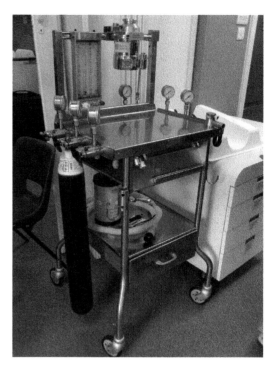

Fig. A7 Anaesthetic machine.

anaesthesiologist *See* **ANAESTHETIST**
anaesthesiology *See* **ANAESTHETICS**
anaesthetic, anesthetic
1. A drug that causes a temporary loss of sensation (**ANAESTHESIA**). *See also* **ANALGESIC (1)**
2. Relating to or inducing a loss of sensation.

anaesthetic machine A device used to support the administration of anaesthesia (Fig. A7). It generally consists of a ventilator, a gas delivery system (for oxygen, air and anaesthetic), including a vaporiser, flow meters and monitors to measure and record the patient's vital signs, e.g. **BLOOD PRESSURE (BP)**, **HEART RATE, OXYGEN SATURATION** etc.

anaesthetics, anaesthesiology The study and application of anaesthetics. *See also* **ANAESTHETIC (1)**

anaesthetist, anaesthesiologist Someone who is expert in the use of anaesthetics. *See also* **ANAESTHETIC (1)**

anal glands Paired sacs located either side of the anus in many mammals including most carnivores. Their secretions contain chemicals which allow individual animals to identify other particular **CONSPECIFICS**. *See also* **SCENT GLAND**

analgesia Pain relief without loss of consciousness. *See also* **ANAESTHESIA**

analgesic
1. A drug that provides pain relief. *See also* **ANAESTHETIC (1)**
2. Having the effect of providing pain relief.

analogous

1. An anatomical structure in one type of animal is said to be analogous with a structure in another type of animal when both have the same function and when they are not **HOMOLOGOUS (1)**, i.e. they have a different evolutionary origin such as the wings of insects and those of birds. *See also* **CONVERGENT EVOLUTION**

2. The term may be used to describe structures with a similar function whether or not these structures are **HOMOLOGOUS (1)**.

analysis of variance *See* **ANOVA**

anaphylactic shock A fall in **BLOOD PRESSURE (BP)** caused by an extreme **IMMUNE REACTION**.

anapsid A vertebrate, especially a reptile, which does not posses a temporal opening in its skull.

anatomical terms of location Standard terms of anatomical location used within **ZOOLOGY** (Fig. A8).

ancient forest *See* **PRIMARY FOREST**

ancient woodland *See* **PRIMARY FOREST**

androgen A **STEROID** hormone which has masculinising effects, e.g. **TESTOSTERONE**.

anemia *See* **ANAEMIA**

anesthetic *See* **ANAESTHETIC**

anestrous *See* **ANOESTROUS**

anestrus *See* **ANOESTRUS**

angel wing, airplane wing A deformity of the scapulae in which they bow outward as a result of the pull of the scapular muscles (*see* **PECTORAL GIRDLE**). Caused by a condition called osteodystrophia fibrosa. Occurs particularly in kittens.

Anguilliformes An order of fishes: eels.

animal

1. An organism which belongs to the animal kingdom (Animalia) and is characterised by being **MOTILE** (in some stage of its life cycle), multicellular, made of **EUKARYOTIC** cells which are almost always **DIPLOID**, and usually arranged into **TISSUES**, **HETEROTROPHIC**, possessing cells without cell walls, usually reproducing sexually, and with an embryo which has a **BLASTULA** stage. In law, animals are divided into **DOMESTIC ANIMALS**, **CAPTIVE ANIMALS** and **WILD ANIMALS**. In addition, the general term 'animal' may have a specific meaning within a particular piece of legislation which is different from the zoological meaning.

2. In English law, in the **PERFORMING ANIMALS (REGULATION) ACT 1925**, the **PET ANIMALS ACT 1951** and the **ANIMAL WELFARE ACT 2006**, '*animal*' means a **VERTEBRATE**.

Fig. A8 Anatomical terms of location. Barbary sheep (*Ammotragus lervia*).

3. In the **Zoo Licensing Act 1981**, s21(1), '"*animals" means animals of the classes* **Mammalia**, **Aves**, **Reptilia**, **Amphibia**, **Pisces** *and* **Insecta** *and any other multi cellular organism that is not a plant or a fungus . . . ,*' (i.e. the legal definition is essentially the same as the zoological definition).

4. In the **Protection of Animals Act 1911** the expression 'animal' means any **domestic animal (2)** or **captive animal (2)**.

5. Under the Animal Boarding Establishments Act 1963 (s.5(2)), '"*animal" means any dog or cat*'.

6. In the USA, the **Animal Welfare Act of 1966 (USA)**, (USC § 2132 (g)), '*The term "animal" means any live or dead dog, cat, monkey (nonhuman primate mammal), guinea pig, hamster, rabbit, or such other warm-blooded animal, as the Secretary may determine is being used, or is intended for use, for research, testing, experimentation, or exhibition purposes, or as a pet; but such term excludes (1) birds, rats of the genus Rattus, and mice of the genus Mus, bred for use in research, (2) horses not used for research purposes, and (3) other farm animals, such as, but not limited to livestock or poultry, used or intended for use as food or fiber, or livestock or poultry used or intended for use for improving animal nutrition, breeding, management, or production efficiency, or for improving the quality of food or fiber. With respect to a dog, the term means all dogs including those used for hunting, security, or breeding purposes;*'

7. In New York State, under the Agriculture and Markets Law § 350 (1) "Animal," . . . includes '*every living creature except a human being.*'

animal £ The amount of money the public is prepared to donate to (spend on) charities which support animals. Much of this is given to organisations concerned with animal cruelty, animal **sanctuaries** etc., rather than those that support wildlife **conservation**.

animal actors Animals which are used in the entertainment industry. They are usually taken from their parents at an early age and trained to perform in films, television programmes and advertisements. In many cases social animals, such as apes, are forced to live alone. When they become too large and difficult to handle many in the USA live out the rest of their lives in **roadside zoos**. *See also* **animals in film**, *Cheetah, Flipper, Lassie, Skippy the Bush Kangaroo*

animal advocate A person who acts for or speaks on behalf of animals, usually in relation to animal welfare issues. *See also* **legal personality**, *locus standi*

Animal Aid An animal rights group founded in the UK in 1977. It campaigns peacefully against animal abuse and promotes a cruelty-free lifestyle.

Animal and Plant Health Inspection Service (APHIS) The Agency of the **United States Department of Agriculture (USDA)** responsible for protecting and promoting agricultural health, the regulation of genetically engineered organisms, and carrying out wildlife damage management activities. It administers the **Animal Welfare Act of 1966 (USA)** and inspects zoos.

animal ark *See* **ark (2), poultry ark**

Animal Behavior Society (ABS) A society whose purpose is to promote and encourage the biological study of animal behaviour, including studies at all levels of organisation using both descriptive and experimental methods under natural and controlled conditions. It encourages research studies and the dissemination of knowledge about animal behaviour through publications, educational programmes and activities. *See also* **Animal Behaviour**

Animal Behaviour An academic journal which publishes original work on animal behaviour. Published by the **Association for the Study of Animal Behaviour (ASAB)** in collaboration with the **Animal Behavior Society (ABS)**.

Animal Boarding Establishments Act 1963 An Act which regulates animal boarding establishments (for dogs or cats) in Great Britain by a system of licensing and inspections.

Animal By-Products (Enforcement) (England) Regulations 2011 A statutory instrument which regulates the collection, transportation, storage, handling, processing and disposal of animal by-products in England, in compliance with EU law.

animal carer Alternative name for an animal **keeper (1)**.

animal cognition The study of the mental life of animals. *See* **cognition**

Animal Conservation A scientific journal which publishes research on the conservation of species and their habitats, published by the **Zoological Society of London (ZSL)**.

animal cruelty Treating an animal in such a way as to cause pain or suffering either intentionally or by **neglect**. Apart from **cruelty** caused by neglect many instances of gratuitous cruelty towards animals have been recorded: a kitten killed by being heated in a microwave oven, a dog dragged behind a car, a deer killed by dogs, a hamster attached to a firework rocket, a cat boil-washed in a washing machine, a puppy kicked to death. *See also* **animal cruelty laws, Royal Society for the Prevention of Cruelty to Animals (RSPCA)**

animal cruelty and violence towards humans In the USA the Federal Bureau of Investigation (FBI) has recognised animal cruelty as an indicator of violence against people since the 1970s, including children, partners and elders. In a study of serial killers the FBI found that most had killed or tortured animals as children.

animal cruelty as a component of domestic violence There is a strong association between domestic violence between people and cruelty to animals. In America the **Humane Society of the United States (HSUS)** estimates that nearly 1 million animals a year are abused or killed in connection with domestic violence. Ascione (1998) found that 71% of battered women who had pets reported that their partner had either threatened or had actually hurt or killed one of their pets, and actual pet abuse was reported by 57% of the women.

animal cruelty laws Legislation aimed at the prevention of **CRUELTY** to animals developed first in the UK in the early 19th century (*see* **MARTIN'S ACT**). Most countries have laws which protect **COMPANION ANIMALS**, animals used in experiments, **FARM ANIMALS** and animals living in zoos. In more recent times this has been extended to some wild animals (e.g. **WILD MAMMALS (PROTECTION) ACT 1996**). In the USA each state has its own animal cruelty laws. *See also* **ANIMAL CRUELTY, ANIMAL WELFARE ACT 1999 (NZ), ANIMAL WELFARE ACT 2006, ANIMAL WELFARE ACT OF 1966 (USA), CRUELTY TO ANIMALS ACT 1835, CRUELTY TO ANIMALS ACT 1876, PROTECTION OF ANIMALS ACT 1911**

animal dealer A person who buys (or captures) and then sells animals. In the past zoos relied on such dealers to provide their animals, e.g. the **WORLD'S ZOOLOGICAL TRADING COMPANY LTD**. *See also* **GAME AUCTION, HAGENBECK**

Animal Defenders International (ADI) An organisation whose aim is to create awareness of, and campaign against, cruelty to animals, including **ANIMAL TESTING** and circuses. It works to alleviate suffering, and conserve and protect animals and the environment.

animal density *See* **STOCKING DENSITY**

animal economics *See* **BEHAVIOURAL ECONOMIC THEORY**

animal enclosure
1. A space which is enclosed in order to contain animals.
2. In India, under the **RECOGNITION OF ZOO RULES 2009 (INDIA)**, '*"Enclosure" means any accommodation provided for zoo animals.*'
See also **ANIMAL EXHIBIT, ENCLOSURE DESIGN, ENCLOSURE SIZE**

Animal Equity UK An animal rights organisation based in the UK.

animal exhibit An **ANIMAL ENCLOSURE** (including a building in a zoo, farm or other facility) containing one or more species on display to the public, plus the associated **INTERPRETATION** and, in some cases, the visitor paths and landscape surrounding the enclosure. *See also* **AFFILIATIVE EXHIBIT, BIOPARK, ENCLOSURE DESIGN, EXHIBIT STRUCTURE, GREEN EXHIBIT, HUMAN EXHIBIT, IMMERSION EXHIBIT, INTERACTIVE SIGN, MULTI-SPECIES EXHIBIT, NATURALISTIC EXHIBIT, ROTATIONAL EXHIBIT DESIGN, THEMED EXHIBIT, WALK-THROUGH EXHIBIT**

animal experience An animal encounter in a zoo or similar facility, whereby members of the public come into close contact with animals.

animal experimentation *See* **ANIMAL TESTING**. *See also* **VIVISECTION**

animal feed cake Any of a number of animal foods pressed into a solid mass, especially for agricultural use, e.g. cotton seed cake, coconut copra cake, palm kernel cake, rapeseed cake. *See also* **FIBRE NUGGETS**

Animal Finder's Guide A magazine and website for owners of exotic wild animals. Based in Indiana, USA. Contains information about animal care, legislation and animal auctions. *See also* **GAME AUCTION**

Animal Gatherings (England) Order 2006 A statutory instrument in England which regulates the use of premises for animal gatherings for the sale of agricultural animals, including licensing, time limits on gatherings, movement of animals, disinfection etc.

animal grabber Any of a number of devices for catching and restraining animals. *See also* **CATCH POLE, SNAKE HOOK**

animal handling chute *See* **CRUSH**

Animal Health An organisation in Great Britain previously known as the State Veterinary Service (and including other agencies), but now part of the **ANIMAL HEALTH AND VETERINARY LABORATORIES AGENCY (AHVLA)**.

Animal Health and Veterinary Laboratories Agency (AHVLA) An executive agency which works on behalf of the **DEPARTMENT FOR ENVIRONMENT, FOOD AND RURAL AFFAIRS (DEFRA)**, the Scottish Government and the Welsh Government. The agency was formed in April 2011, following the merger of **ANIMAL HEALTH** and the **VETERINARY LABORATORIES AGENCY**. The agency's role is to safeguard animal health and welfare, protect public health and enhance food security by conducting research, surveillance and inspections. The AHVLA also acts as the national, European and international reference laboratory for several exotic and zoonotic **NOTIFIABLE DISEASES**, and has a registration and licensing role in relation to **CITES** listed wildlife species.

Animal Health and Welfare Board for England An organisation created in 2011 which makes direct recommendations to **DEPARTMENT FOR ENVIRONMENT, FOOD AND RURAL AFFAIRS (DEFRA)** Ministers on policies affecting the health and welfare of kept animals in England including farm animals, horses and pets. It is not concerned with the welfare of zoo and circus animals.

Animal Health Trust An organisation in the UK concerned with fighting disease and injury in animals. It aims to develop new technology and knowledge for the better diagnosis, prevention and cure of animal diseases. It provides a clinical referral service for veterinary surgeons in practice and promotes postgraduate education.

animal hospital A facility which provides first aid and minor treatment for a wide range of pets and also

A

surgical treatment for more serious conditions. In England and Wales, the **ROYAL SOCIETY FOR THE PREVENTION OF CRUELTY TO ANIMALS (RSPCA)** provides a low-cost veterinary service to the public via its animal hospitals, and facilities for injured and rescued animals collected by **RSPCA INSPECTORS**. *See also* **ANIMAL WELLNESS CENTRE**

Animal Liberation A book published by Peter **SINGER** in 1975 supporting the concept of rights for animals. It examines the fate of animals in factory farms, science laboratories and other situations where they are exploited for the benefit of humans. It does not specifically discuss zoos, but was extremely important in bringing the fate of exploited animals to the attention of the public and it was instrumental in establishing **ANIMAL RIGHTS** as a legitimate subject of academic study.

Animal Liberation Front (ALF) A secret loose association of individuals (with no formal membership) who carry out illegal actions against industries which profit from animal exploitation, including the liberation of live animals from fur farms, slaughterhouses and other places, and economic sabotage in the form of property destruction.

animal loading *See* **STOCKING DENSITY**

Animal Machines A book published in 1964 by Ruth **HARRISON** which exposed the cruel practices of the **INTENSIVE FARMING** industry.

Animal Magic A popular children's television programme about animals produced by the BBC and broadcast between 1962 and 1983. The original presenter, Johnny Morris, made regular visits to Bristol Zoo, playing the part of a keeper who could talk to the animals.

animal needs index (ANI) *See* **ANIMAL WELFARE INDEX**

animal noise monitor collar A device designed by the US Air Force and used to record and study the effect of noise – particularly aircraft noise – on animals. It records location (using the **GLOBAL POSITIONING SYSTEM (GPS)**), movements, acoustic noise level and heart rate. Initially tested on free-ranging bighorn sheep (*Ovis canadensis*) in Idaho to monitor their reactions to low-level military aircraft in 1997.

animal personality Animals may be considered to possess a personality when individuals differ from one another in either single behaviours or suites of related behaviours in a way that is consistent over time. It is usually assumed that such consistent individual differences in behaviour are driven by variation in how individuals respond to information about their environment, rather than by differences in external factors such as variation in **MICROHABITAT** (Briffa and Greenaway, 2011).

animal recognition technology An animal management system that uses a video camera and computer software to distinguish between animal types. This information can then be used to open or close a gate and to count animals. Animals are funnelled towards the gate by a V-shaped fence and walk in front of a blue screen. The animal's shape is recorded by a video camera and compared to a database of shapes. Such a system is used by farmers in Australia to protect water resources and can distinguish between horses, cows, sheep, goats, pigs, kangaroos and emus. *See also* **AUTO DRAFTER, RADIO FREQUENCY IDENTIFICATION (RFID) TECHNOLOGY**

Animal Record Keeping System (ARKS) Software used for keeping animal records within an individual institution. It is a PC-based application and is multilingual. The software allows a zoo to produce a number of different reports based on its own records. ARKS4 allows individual institutions to contribute their data to the pooled **INTERNATIONAL SPECIES INFORMATION SYSTEM (ISIS)** database so that it is then available to others through the ISIS website. *See also* **MEDICAL ANIMAL RECORD KEEPING SYSTEM (MEDARKS), ZOOLOGICAL INFORMATION MANAGEMENT SYSTEM (ZIMS)**

animal records *See* **ANIMAL RECORD KEEPING SYSTEM (ARKS), PEDIGREE REGISTER, RECORD KEEPING, STUDBOOK**

animal research *See* **ANIMAL TESTING**

Animal Research Act 1995 (NSW) A law introduced in New South Wales, Australia, to protect animal welfare by ensuring that the use of animals in research is always humane, considerate, responsible and justified.

animal ride A ride given by a tame animal (e.g. horse, donkey, elephant, camel) in a zoo, on a farm or at some other facility, for entertainment.

animal rights The concept in moral philosophy that animals (or at least some animals) should have rights. *See also* **ANIMAL LIBERATION, ARISTOTLE, BASIC NEEDS TEST, BENTHAM, COBBE, DECLARATION OF RIGHTS FOR CETACEANS: WHALES AND DOLPHINS, DESCARTES, ethics, GREAT APE PROJECT, KANT, RYDER, SINGER, SPECIESISM, UNIVERSAL DECLARATION ON ANIMAL WELFARE (UDAW), UTILITARIANISM, WHO**

animal rights legal cases A number of attempts have been made to achieve rights for animals using the courts. In 2012 the **PEOPLE FOR THE ETHICAL TREATMENT OF ANIMALS (PETA) FOUNDATION** brought a lawsuit in the US District Court for the Southern District of California against *SeaWorld* on behalf of five wild-caught orcas, claiming that they were held in violation of Section One of the Thirteenth Amendment to the Constitution of the United States, which prohibits slavery and involuntary servitude. In November 2011 the Free Morgan Support Group failed in its attempt to use a Dutch court to prevent a killer whale from being transferred from the dolphinarium in Harderwijk that rescued her from shallow waters in Waddenzee to Loro Parque in Tenerife. The group argued that the transfer would breach EU wildlife trade laws and that she should be returned to the wild. In 1988 approximately 15 000 seals washed up on the beaches of the Adriatic and North Seas. A group of German environmental lawyers brought a case in the German courts naming North Sea Seals as the principal plaintiffs, with the lawyers appearing as guardians against the West German Government as they were responsible for issuing permits which resulted in the sea being polluted by heavy metals. The courts did not recognise the seals' standing (*LOCUS STANDI*) as they were not **PERSONS** and would not allow the lawyers to stand on their behalf as no specific legislation authorised this. *See also* **ASSOCIATION OF LAWYERS FOR ANIMAL WELFARE**, **DECLARATION OF RIGHTS FOR CETACEANS: WHALES AND DOLPHINS**

animal rotation exhibit *See* **ROTATIONAL EXHIBIT DESIGN**

animal sentinels Animals used to detect the presence of a toxic substance, disease or other hazard, e.g. canaries used to detect carbon monoxide and methane in coal mines. Murray Valley encephalitis (MVE) is a mosquito-borne viral disease of chickens which occurs in New South Wales, Australia. Wild birds act as a natural reservoir of the disease and it may be fatal if contracted by humans. During the mosquito breeding season the NSW government regularly tests sentinel flocks of chickens located near known bird breeding sites for MVE. *See also* **SENTRY, WORKING ANIMALS**

animal shelter A place where unwanted animals are cared for either on a temporary or permanent basis. *See also* **SANCTUARY (2)**

animal show An entertainment event which features the use of animals. Sometimes included in attractions within a zoo. In the past some zoos arranged inappropriate shows such as **CHIMPANZEES' TEA PARTIES**. Nowadays, some modern zoos use shows to inform visitors about the natural abilities of animals (e.g. climbing or flying) or to educate the public about conservation issues. At **WHIPSNADE ZOO** Asian elephants 'demolish' huts and 'trees' to illustrate human–elephant conflicts in Asia.

animal stars A number of animals have become stars in long-running television series and popular films, beginning with the *Tarzan* films in the 1930s (*see also* **CHEETAH, TARZAN OF THE APES**) and culminating in a number of popular TV series in the 1950s and 1960s (*see also* **DAKTARI, FLIPPER, LASSIE, SKIPPY THE BUSH KANGAROO**)

animal testing, animal experimentation, animal research The process of testing chemicals or procedures on animals, generally using them as models of how these chemicals or processes may affect humans. For example testing the efficacy of drugs, the toxicity of chemicals or the effectiveness of surgical procedures such as organ transplants. Some animal welfare organisations claim that all animal testing is immoral, scientifically flawed and unnecessary because other study methods are available

(e.g. the use of cell cultures). Some scientists claim that it is not possible to study complex biological systems by experimenting on cultures of cells. For example, it is not possible to determine how a drug may affect behaviour unless a whole live animal is used. *See also* **DRAIZE TEST, LD$_{50}$ TEST, THREE RS, VIVISECTION**

animal theme park A visitor attraction which consists of a zoo and an amusement park, usually operated on a commercial basis primarily for entertainment purposes, e.g. *Disney's Animal Kingdom*, Busch Gardens (Tampa).

animal therapy
1. May refer to animal-assisted activities. *See* **ANIMAL-ASSISTED THERAPY**
2. The treatment of illness or injury in an animal using complementary medicine instead of medication or surgery. *See also* **HOLISTIC VETERINARY MEDICINE**

animal trafficking The illegal international trade in animals, especially rare species, for the pet trade, use in **TRADITIONAL MEDICINE** etc. *See also* **CITES, TRADE RECORDS ANALYSIS OF FLORA AND FAUNA IN COMMERCE (TRAFFIC)**

Animal Transaction Policy A document published by the **BRITISH AND IRISH ASSOCIATION OF ZOOS AND AQUARIUMS (BIAZA)** which sets out its policy in relation to animal acquisitions, disposal of surplus animals, culling and euthanasia, use of animal dealers, and the transfer of animals between institutions.

Animal Welfare An academic journal which publishes **SCIENTIFIC PAPERS** concerned with the welfare of animals including those living in zoos. Published by the **UNIVERSITIES FEDERATION FOR ANIMAL WELFARE (UFAW)**.

animal welfare
1. The physical and psychological wellbeing of animals, including a consideration of the effects of their housing, transportation, handling, feeding, breeding etc. The welfare of an individual is defined by **BROOM** (1986) as '*its state as regards its attempts to cope with its environment.*'
2. '*Welfare is a characteristic of an animal, not something given to it, and can be measured using an array of indicators*' Broom (1991).
3. An academic discipline, taught in colleges and universities, concerned particularly with the welfare of companion animals and animals living on farms and in zoos.

Animal Welfare Act 1999 (NZ) A wide-ranging animal welfare law in New Zealand which regulates and makes provision for the care of and conduct towards animals, the use of traps, animal exports, the use of animals in research, testing and teaching, the disposal of animals, hunting, fishing, pest control and surgical procedures on animals. *See also* **ANIMAL WELFARE (ZOOS) CODE OF WELFARE 2004 (NZ)**

Animal Welfare Act 2006 A law in England and Wales which makes provisions relating to the

A

A

welfare of domestic animals under human control (including pet owners). It aims to prevent unnecessary **SUFFERING** (e.g. by banning the docking of dogs' tails (**TAIL DOCKING**) except for medical treatment, mutilation, poisoning, and animal fighting) and promote welfare (by imposing a duty on persons responsible for animals to ensure welfare, banning the sale of animals to persons aged under 16 years and by licensing certain activities). The Act provides for the inspection of premises and the seizure of animals and equipment. *See also* **HUNTING ACT 2004, WILD MAMMALS (PROTECTION) ACT 1996**

Animal Welfare Act of 1966 (USA) The only Federal law in the United States that regulates the treatment of animals in research, exhibition, transport, and by dealers. It requires minimum standards of care and treatment for certain animals which are bred for commercial sale, used in research, transported commercially or exhibited to the public. It makes provision for the licensing of dealers, marking of animals and record keeping, and bans animal fighting.

animal welfare domains *See* **FIVE FREEDOMS**

animal welfare index A method of assessing welfare which is based on a list of parameters for which the presence/absence/degree is given a numerical value or weighting. The calculated value may be compared with minimum, target or threshold totals. Examples include the **ANIMAL NEEDS INDEX (ANI)** or Tiegerechtheitsindex (TGI). *See also* **BODY CONDITION SCORE**

animal welfare needs *See* **FIVE FREEDOMS**

Animal Welfare (Zoos) Code of Welfare 2004 (NZ) A code of welfare for zoo animals in New Zealand issued under section 75 of the **ANIMAL WELFARE ACT 1999 (NZ)**.

animal wellness centre A facility which provides veterinary care for animals, often including holistic veterinary treatments. *See also* **ANIMAL HOSPITAL**

animal-assisted therapy The use of an animal to improve the health of a human, e.g. to treat depression due to loneliness, to provide mental and physical stimulation, reduce blood pressure, and improve verbal skills, social skills and attention span. May involve the use of pet animals such as cats or dogs, or contact with horses or farm animals. *See also* **DOLPHIN THERAPY**

Animalia The animal kingdom. *See also* **ANIMAL (1)**

Animals Act 1971 In England and Wales, an Act which makes provision for the civil liability for damage done by animals in general, dangerous animals, dogs and straying livestock. It allows for the killing of any dog for the protection of livestock and the detention and sale of trespassing livestock. *See also* **KEEPER (2)**

animals and religion Some animal species are protected by virtue of their status in a particular religion. For example the elephant god *Ganesh* is an important deity in Hinduism, Jainism and Buddhism

Fig. A9 Animals and religion: the elephant god *Ganesh* is an important deity in Hinduism, Jainism and Buddhism.

(Fig. A9). The Asian elephant is revered and kept at temples and used in religious ceremonies in Asia. *See also* **BUDDHISM AND ANIMALS, HINDUISM AND ANIMALS, ISLAM AND ANIMALS, JAINISM AND ANIMALS, JUDAISM AND ANIMALS, RELIGIOUS SLAUGHTER**

Animals and Society Institute An independent think tank which operates as a non-profit, independent research organisation whose aim is to advance the status of animals in public policy, and which promotes the study of animal–human relationships. The institute is based in the USA and produces educational resources, hold events and publishes academic journals (*SOCIETY AND ANIMALS JOURNAL* and the *JOURNAL OF APPLIED ANIMAL WELFARE SCIENCE (JAAWS)*).

animals in film The depiction of animals in film and television affects attitudes towards animals. Films such as *KING KONG* and *JAWS* engendered negative attitudes towards gorillas and sharks respectively. However, other films and TV series have promoted positive attitudes towards animals and the natural environment in general, e.g. *DAKTARI, FLIPPER, LASSIE, SKIPPY THE BUSH KANGAROO* and the TV series of David **ATTENBOROUGH**.

Animals (Scientific Procedures) Act 1986 A law in the UK which requires a Home Office licence for animal experiments ('regulated procedures') involving any 'protected animal': a living vertebrate other than man or a specimen of the octopus *Octopus vulgaris* once it becomes capable of independent feeding. It replaced provisions within the **CRUELTY TO ANIMALS ACT 1876** and transposed into UK law Council Directive 86/609/EEC, which makes provision for the protection of animals used for experimental or other scientific purposes.

Animals' VC *See* **DICKIN MEDAL**

anion A negatively charged ion, e.g. chloride (Cl⁻). *Compare* **CATION**

ankus, bullhook A tool used by some elephant keepers and mahouts in the training and control of elephants. Usually consists of a short wooden pole tipped with a metal spike and a metal hook. Known as a bullhook in the USA.

Annelida A phylum of animals. Most annelids are free-living segmented worms found in soil and aquatic environments, e.g. earthworms, ragworms. Some forms are parasitic (leeches). Their bodies are segmented, and constructed of muscular rings separated by thin partitions called septa. **SEGMENTATION** has allowed the development of advanced systems of co-ordinated **LOCOMOTION** with some terrestrial species using **SETAE** ('hairs') to grip the soil and marine species using a system of paddles (parapodia) attached to each segment. The **NERVOUS SYSTEM** is relatively advanced with a simple 'brain' at the head end. The **DIGESTIVE SYSTEM** consists of a tube which runs the length of the body. Annelids have an excretory system made up of individual excretory units (nephridia) in each segment.

Annex A species In relation to Council Regulation (EC) No 338/97 of 9 December 1996 on the protection of species of wild fauna and flora by regulating trade therein, all **CITES APPENDIX I** species, plus certain others (including some non-CITES species) that are considered to need a similar level of protection.

Annex B species In relation to Council Regulation (EC) No 338/97 of 9 December 1996 on the protection of species of wild fauna and flora by regulating trade therein, all **CITES APPENDIX II** species, plus certain others (including some non-CITES species) that are considered to need a similar level of protection.

Annex C species In relation to Council Regulation (EC) No 338/97 of 9 December 1996 on the protection of species of wild fauna and flora by regulating trade therein, all **CITES APPENDIX III** species, apart from those which EU countries have entered a reservation for and are included in Annex D.

Annex D species In relation to Council Regulation (EC) No 338/97 of 9 December 1996 on the protection of species of wild fauna and flora by regulating trade therein, certain non-CITES species that have been imported into the EU in high enough numbers to need monitoring.

annual plant A plant that germinates, flowers and dies within a single growing season or year. *See also* **PERENNIAL PLANT**

annual stock list *See* **STOCK LIST**

anoestrous, anestrous Relating to a female who is not experiencing an **OESTROUS CYCLE** and is not sexually receptive.

anoestrus, anestrus A period of sexual inactivity in female mammals between two periods of **OESTRUS**.

anogenital Relating to the area around the anus and genitals.

ANOVA Analysis of variance. A statistical method which compares the means of samples from several populations. It analyses the total variation shown by the data, splits this into the variation within the samples and the variation between the samples, and then compares these two components.

anovulatory The state of not ovulating. *See also* **OVULATION**

anoxic Relating to the absence of oxygen; a pathological deficiency of oxygen, e.g. anoxic water; anoxic seizure. *See also* **ANAEROBIC**

Anseriformes An order of birds: geese, swans, ducks, screamers.

antagonistic Functioning in an opposite manner or having opposite effects, e.g. the biceps and triceps are antagonistic muscles because they have opposite effects on the movement of the forelimb; **INSULIN** and **GLUCAGON** are antagonistic hormones because they have opposite effects on glucose.

Antarctic Relating to Antarctica: the continent which lies chiefly within the Antarctic Circle and is centred on the South Pole.

Antarctic Conservation Act of 1978 (USA) A US law which provides for the conservation and protection of Antarctic flora and fauna. It makes it unlawful for anyone in the US to possess, sell, offer for sale, deliver, receive, carry, transport, import into or export from the US any native mammal or bird taken in Antarctica.

Antarctic region *See* **FAUNAL REGIONS**

Antarctic Treaty 1959 An international agreement to use **ANTARCTICA** exclusively for peaceful purposes. The treaty requires parties to consider measures that should be taken regarding the preservation and conservation of living resources in Antarctica. *See also* **CONVENTION FOR THE CONSERVATION OF ANTARCTIC SEALS 1972**, **CONVENTION ON THE CONSERVATION OF ANTARCTIC MARINE LIVING RESOURCES 1980 (CCAMLR)**

antelope A member of the mammalian family Bovidae which is not a bison, buffalo, goat, sheep or cattle. It is not a recognised taxon. The term refers to many species of **EVEN-TOED UNGULATES**, most of which occur in Africa, although some species are found in Asia. They possess horns which are directed upward and backward which they do not shed, and have a graceful appearance, with long legs. *See also* **DEER**

A

antennal gland, green gland An excretory organ in the head of some crustaceans which excretes excess sulphate and magnesium and conserves calcium and potassium.

anterior At or near the head or front of the body, or front of a structure (Fig. A8). *Compare* **POSTERIOR**

anterior pituitary The anterior lobe of the pituitary gland located at the base of the brain. It consists partly of the adenohypophysis which secretes several hormones (e.g. **FOLLICLE-STIMULATING HORMONE (FSH)** and **LUTEINISING HORMONE (LH)**) in response to stimulation from other hormones secreted by the **HYPOTHALAMUS** (e.g. **GONADOTROPHIN-RELEASING HORMONE (GnRH)**). It also secretes **PROLACTIN**, thyrotrophic hormone and **ADRENOCORTICOTROPHIC HORMONE (ACTH)**.

anthelmintic An agent which kills or expels parasitic worms from the gut. Also called a vermicide if it kills the worm or a vermifuge if it expels it.

anthrax A disease caused by the bacterium *Bacillus anthracis*. It is found in mammals, especially herbivores, and is usually fatal. The bacterium may form highly resistant spores which may live in the soil for 10 years or more and still be capable of infection. Spores have been found in bone-meal, wool, hides, feeds and blood fertilisers, and they may be carried to the soil surface by earthworms. Infection may be spread by ingesting spores in food or water, through a cut or by inhalation. Three forms of anthrax are recognised. In peracute cases animals may be found dead without having shown any signs. Acute cases exhibit a raised temperature, rapid pulse, 'blood-shot' eyes, cold feet and ears, followed by prostration, unconsciousness and death. In subacute cases the animal may linger for up to 48 hours with a very high temperature and laboured respirations. Swellings may occur in the neck and lower chest. A 'carbuncle' may form at the site of infection if infected through the skin. Early administration of antibiotics may be effective.

Anthropocene An informal term for the period during which human activity has had a significant effect on the Earth's ecosystems, from the beginning of farming approximately 8000 years ago until the present. *See also* **HOMOGOCENE**

anthropoid ape A member of the primate suborder Anthropoidea (monkeys, apes and humans).

anthropology The study of humanity, including human **EVOLUTION** and our relationship with other **PRIMATES** (physical anthropology), social and cultural organisation (including relationships with non-human animals) and the development of **LANGUAGE**.

anthropomorphism The attribution of human traits to other animals. *See* **MR. ED**

anthrozoology The scientific study of the interactions and relationships between animals and people. This is a relatively new multi-disciplinary subject which encompasses **ZOOLOGY**, **PSYCHOLOGY**, sociology, and veterinary science. *See also* **ANTHROZOÖS**, **INTERNATIONAL SOCIETY FOR ANTHROZOOLOGY (ISAZ)**

Anthrozoös An academic journal which reports studies on the interactions between people and animals from a range of disciplines including **ANTHROPOLOGY**, archaeozoology, art and literature, education, **ETHOLOGY**, history, human medicine, **PSYCHOLOGY**, sociology and **VETERINARY MEDICINE**. It is the official journal of the **INTERNATIONAL SOCIETY FOR ANTHROZOOLOGY (ISAZ)**.

antiaphrodisiac A **PHEROMONE** that discourages mating, usually produced after a female has mated. In some reptiles (e.g. garter snakes (*Thamnophis*)), after copulation part of the ejaculate forms a **SPERM PLUG** in the **CLOACA (1)** that emits a substance that deters males.

antibiotics A group of organic compounds produced by spore-forming organisms (fungi and bacteria). They interfere with protein synthesis in pathogens and are used to treat a variety of bacterial infections and as antitumour agents, antiparasite agents, ruminant growth promoters, insecticides and for other purposes. Sir Alexander Fleming first observed that a strain of *Penicillium* prevented growth in the bacterium *Staphylococcus*. *See also* **BROAD-SPECTRUM ANTIBIOTIC**

antibody, immunoglobulin A protein from plasma cells (formed from a **B-LYMPHOCYTE**) that binds to a specific foreign molecule (**ANTIGEN**) as part of the body's **IMMUNE REACTION**.

anticoagulant A drug or other substance that prevents or slows down **BLOOD CLOTTING**. *See also* **WARFARIN**

anticodon The triplet sequence of **TRANSFER RNA (TRNA)** nucleotides which is capable of base-pairing with the **CODON** triplet of a **MESSENGER RNA (MRNA)** molecule by virtue of their complementary nature.

antidiuretic A hormone which reduces urine output. *Compare* **DIURETIC**

antidote A drug, or other agent, that prevents or counteracts the action of a poison.

anti-emetic A drug which prevents vomiting.

antifreeze A chemical in some animals that lowers the freezing point by changing the arrangement of water molecules in body fluids, thereby allowing them to survive in subzero temperatures. Glycerol is an important antifreeze in many insects and some frog species. Some Antarctic fishes produce antifreeze peptides and antifreeze glycopeptides.

antifungal Having the action of preventing the growth of, or destroying, fungi.

antigen A molecule, usually foreign to the body, which stimulates the production of **ANTIBODIES**.

antihistamine A drug used to give relief from allergic conditions by blocking the action of **HISTAMINE**.

Used, for example, to treat allergic skin diseases in cats and dogs.

anti-inflammatory Having the effect of reducing INFLAMMATION.

antioxidant A substance (e.g. VITAMIN C, VITAMIN E) which prevents or slows down the oxidation of stored foods etc. by molecular oxygen. It neutralises free radicals which may cause cell damage and cell death, so protects against some diseases.

anti-predator behaviour Defensive behaviour which protects against predators. *See also* ALARM CALL, CAMOUFLAGE, ESCAPE BEHAVIOUR, MIMICRY

antisepsis The containment or destruction of disease-causing organisms. *See also* ANTISEPTIC

antiseptic
  1. Any substance that kills or inhibits the growth of bacteria and other microbes. It is not toxic to the skin so may be used to clean wounds. Alcohol and iodine function as antiseptics.
  2. Relating to any such substance.

antiseptic mat A mat containing ANTISEPTIC which may be located at the entrance to a zoo enclosure, a QUARANTINE area, a farm or similar place to prevent the spread of INFECTIOUS disease on footwear.

antiserum SERUM that contains ANTIBODIES to a TOXIN or other ANTIGEN.

antivenene *See* ANTIVENIN

antivenin, antivenene An antidote or ANTISERUM to a VENOM, especially that of a snake. *See also* MILKING (2)

anti-zoo group An organisation that opposes the existence of zoos and campaigns for the closure of zoos and the transfer of their animals to SANCTUARIES. *See also* BORN FREE FOUNDATION, ZOO CHECK

antler A bony growth on the head of antelopes, deer and some other EVEN-TOED UNGULATES. Consists of dermal bone initially covered by furry skin called velvet. Shed and re-grown annually. May be used to age individuals in some species. Antlers only occur in mature males, except in reindeer (*Rangifer tarandus*). *Compare* HORN

Anura An order of amphibians: frogs and toads.

anus The opening at the end of the ALIMENTARY CANAL through which faeces are expelled.

anvil behaviour The use by an animal of a rock or other hard structure to break open a food object or to break food into smaller pieces. Birds such as thrushes (Turdidae) break snail shells open on rocks. These anvils can be identified by the presence of shell debris. Wild capuchin monkeys (*Cebus libidinosus*) use stones as tools to smash open nuts on stone anvils. The sixbar wrasse (*Thalassoma hardwicke*) has been observed in aquarium conditions using a rock as an anvil to break hard food pellets into pieces small enough to swallow. *See* TOOL USE

anxiety A strong feeling of fear or distress. Occurs as a normal response to a stressful or dangerous situation. Signs may include trembling, rapid pulse, sweating, dry mouth. *See also* STRESS

aorta
  1. Largest artery in vertebrates, which carries oxygenated blood from the heart (left ventricle) to the body tissues.
  2. Anterior part of the arthropod heart.

apathy A lack of emotion, feeling or interest; impassive, indifferent, listless. A lack of response to environmental stimuli. *Compare* RESPONSIVENESS

ape A general term for hominoid primates belonging to the families Pongidae (GREAT APES) and Hylobatidae (gibbons and siamangs, the LESSER APES).

apex predator, alpha predator, super carnivore, super predator, top carnivore, top predator A PREDATOR which, when adult, has no natural predators within its natural environment. An animal occupying the last TROPHIC LEVEL in many FOOD CHAINS or FOOD WEBS.

APHIS *See* ANIMAL AND PLANT HEALTH INSPECTION SERVICE (APHIS)

apiary, bee yard A place where the hives of honey bees are kept for their honey (Fig. A10). A collection of hives. Known as a bee yard in the USA.

apnoea, apnea The cessation of breathing for an uncertain period.

apocrine sweat gland Gland located in the armpit or groin of a mammal which produces odiferous sweat used in social communication.

Apodiformes An order of birds: swifts, hummingbirds.

aposematism The advertisement by an animal of its dangerous or unpleasant properties to others, often because it is venomous or unpalatable. This often takes the form of warning (aposematic) coloration, e.g. the black and yellow markings on the abdomen of stinging wasps and bees and the striped coloration of venomous coral snakes.

apparent dry matter digestibility (ADMD) *See* GROSS ASSIMILATION EFFICIENCY (GAE)

appeasement behaviour A BEHAVIOUR which is performed to inhibit or reduce the aggression shown by another individual of the same species, especially where escape is impossible. Generally exhibited by SUBORDINATE INDIVIDUALS in the presence of DOMINANT INDIVIDUALS and may involve the hiding or turning away of weapons such as claws and bills and the exposure of vulnerable parts of the body, such as the abdomen. In mammals appeasement behaviour often involves lowering the head and body or rolling onto the back to expose the underbelly. Sometimes appeasement involves the RITUALISATION of juvenile behaviour, e.g. food begging. *See also* AGONISTIC BEHAVIOUR

appendicitis Inflammation of the APPENDIX (1) caused by obstruction of the lumen by faeces, infection, CANCER etc.

appendicular skeleton The part of the skeleton of vertebrates which consists of the PECTORAL GIRDLE, the PELVIC GIRDLE and paired limbs or fins.

Fig. A10 Apiary. Inset: bees entering and leaving.

appendix
1. In anatomy, a small diverticulum of the CAECUM in humans, many other primates and rodents, which contains LYMPHOID TISSUE. *See also* APPENDICITIS
2. In law, additional material added at the end of a legal instrument which contains details of, for example, protected species, e.g. CITES APPENDIX I.

appetite A desire for food and incentive to eat; a combination of hunger and the perceived quality of available foods. A hungry animal has no appetite in the absence of food or indicators of its presence, e.g. other animals feeding. Appetite is an important determinant of food selection and experiments have shown that, given the opportunity, animals will select foods that meet nutritional deficiencies and avoid POISONS. Control centres for feeding and satiety are located in the HYPOTHALAMUS. *See also* CAFETERIA EXPERIMENT

appetitive behaviour Goal-seeking, EXPLORATORY BEHAVIOUR, e.g. actively searching for prey. When the GOAL is reached appetitive behaviour normally ceases. *See also* CONSUMMATORY BEHAVIOUR

approved vet, accredited vet A veterinary surgeon who has undergone an approved programme of training and is accredited by or registered with an appropriate government body or professional veterinary organisation, e.g. in the UK, one who is regis-

tered with the ROYAL COLLEGE OF VETERINARY SURGEONS (RCVS).

aquaciser *See* UNDERWATER TREADMILL

aquaculture The cultivation of aquatic animals (especially fish, molluscs and crustaceans) or plants for food.

aquaponics A food production system in which fish (and other aquatic organisms such as prawns and crayfish) are reared in tanks and the effluent from these tanks is used as fertiliser for plants grown hydroponically. *See also* FISH FARM

aquarist A person who works with and cares for fish in a captive environment, e.g. an AQUARIUM (1).

aquarium
1. Place where fish and other aquatic species are kept for research or exhibition. Derived from the terms 'aquatic' and 'VIVARIUM'. *See also* BLUE PLANET, FISH HOUSE, MARINELAND, SHEDD AQUARIUM. *Compare* OCEANARIUM
2. A single tank used for keeping fish.

Aquarium for Wildlife Conservation The name of the New York Aquarium since 1993. *See also* WILDLIFE CONSERVATION SOCIETY (WCS)

aquascape A scenic view artificially created in an aquarium or other body of water, by arranging aquatic plants, stones, rocks, drift wood, trees, roots and other features in an aesthetically pleasing and usually naturalistic way. Aquascaping is used by large aquariums and hobbyists to create naturalistic environments for fish and other aquatic organisms. A number of different styles exist, e.g. *Dutch Nature Aquarium, Jungle*, and contests are held between members of specialist clubs.

aquatred *See* UNDERWATER TREADMILL

Arab A horse breed, originating from the Arabian Peninsula, which is famous for its intelligence, speed and grace.

Arabian oryx (*Oryx leucoryx*) A rare species of large antelope native to the Middle East. During the first half of the twentieth century the Arabian oryx population declined rapidly as a result of hunting. By the 1960s only a handful of oryx remained in Oman. The last one was shot here in 1972. In 1961 the Fauna Preservation Society (now FAUNA AND FLORA INTERNATIONAL (FFI)) launched 'Operation Oryx'. They captured three wild oryx and obtained others from zoos and private collections in the Middle East. Eventually, a captive population was established at Phoenix Zoo in Arizona and Los Angeles Zoo in California. By 1964 there were 13 oryx in captivity in the USA. The Arabian oryx was first returned to the New Shaumari Reserve in Jordan in 1978. Later oryx were released in the Jiddat al-Harasis plateau in Oman. The first five animals arrived in 1980 and by 1996 there were 450 free-ranging animals. Around 50 zoos currently hold some 900 oryx in captivity.

Arachnida A class of arthropods: spiders, mites and scorpions.

arachnoid membrane The membrane that covers the brain. It contains VILLI through which CEREBRO-SPINAL FLUID (CSF) drains into the blood.

arachnophobia Fear of spiders. Some zoos (e.g. LONDON ZOO and DUDLEY ZOO) run special programmes or workshops to help people overcome their fear of spiders.

ARAZPA *See* ZOO AND AQUARIUM ASSOCIATION (ZAA)

arboreal Relating to trees, living in trees and adapted to life in trees, e.g. a spider money is an arboreal animal. *Compare* CURSORIAL

archipelago A chain or cluster of islands, e.g. GALA-PAGOS ISLANDS.

architecture *See* ZOO ARCHITECTURE

archive A collection of material, often of historical interest. Zoos often keep archives of their own publications, old photographs etc., usually in the zoo's library. Such materials may be useful to researchers.

Arctic
1. The area around the north pole.
2. Relating to this area.
Used as the theme of some zoo exhibits, e.g. the *ARCTIC RING OF LIFE* at Detroit Zoo.

Arctic Ring of Life The world's largest polar bear exhibit. A major exhibit at Detroit Zoo which contains polar bears, arctic foxes and seals. Contains an underwater acrylic tunnel from which swimming polar bears and seals can be observed, and simulated ice and tundra landscapes (Fig. B1).

Argentiniformes An order of fishes: herring smelts, barreleyes and their allies.

Aristotle (384–322 BC) A Greek philosopher and teacher who wrote widely on scientific subjects. The first person to classify animals into groups. His major works include *HISTORY OF ANIMALS, On the Parts of Animals, On the Progression of Animals* and *On the Generation of Animals.*

Arizona–Sonora Desert Museum A private, non-profit organisation dedicated to the conservation of the Sonoran Desert in Arizona. It is a world-renowned zoo, natural history museum and botanical garden whose exhibits recreate the natural landscape of the Sonoran Desert Region. It contains more than 300 animal species and 1200 kinds of plants. It was founded in 1952 by William Carr and Arthur Pack.

ark
1. A term used to refer to the role of zoos in maintaining INSURANCE POPULATIONS of species in captivity, drawing parallels with the biblical story of Noah taking pairs of animals onto his ark to escape from a flood. The Holy Bible (King James Version), Genesis Chapter 7 (v 1–3): '*And the Lord said unto Noah, Come thou and all thy house into the ark; for thee have I seen righteous before me in this generation. Of every clean beast thou shalt take to thee by sevens, the male and his female: and of beasts that are not clean by two, the male and his female. Of fowls also of the air by sevens, the male and the female; to keep seed alive upon the face of all the earth.*'
See also AMPHIBIAN ARK (AARK)
2. Animal ark; a small house for keeping animals such as poultry or pet rabbits. *See also* POULTRY ARK

ARKive *See* WILDSCREEN

ARKS number An identification number assigned to an individual animal within the ANIMAL RECORD KEEPING SYSTEM (ARKS) used by many zoos to keep track of family relationships and animal movements between institutions etc.

ARKS4 *See* ANIMAL RECORD KEEPING SYSTEM (ARKS)

arrhythmia A change from the normal rhythm of the HEARTBEAT.

art in zoos *See* ZOO ART

arteriole The smallest diameter artery which carries blood from a larger ARTERY into a CAPILLARY.

artery A blood vessel that carries blood from the heart to the body tissues. The blood is usually oxygenated but not always. Vertebrate arteries have thick elastic walls made of smooth muscle and connective tissue, with an inner lining of endothelium and a narrow LUMEN. Arteries branch to form smaller ARTERIOLES.

arthritis Inflammation of the joints associated with swelling, pain, local heat and restricted movement of the affected parts. *See also* OSTEOARTHRITIS

arthropod A member of the phylum ARTHROPODA.

Arthropoda A PHYLUM of animals. The arthropods are segmented animals possessing a hard outer skeleton (EXOSKELETON) – made of the polysaccharide chitin – with paired, jointed limbs, e.g. insects, crabs, woodlice, spiders. They are advanced animals, many of which have developed impressive swimming, running or flying abilities. In many groups the exoskeleton is toughened with mineral salts, e.g. calcium carbonate. The muscular system is highly specialised and the legs are hinged, allowing the exoskeleton to move. The exoskeleton provided this group with the necessary structural support to evolve systems for walking on land.

arthroscope A fibre-optic ENDOSCOPE used in the examination of joints.

Article 10 certificate A certificate which authorises the sale and movement of CITES ANNEX A SPECIES within the EU. *See also* ARTICLE 60 CERTIFICATE, CITES

Article 60 certificate A certificate which allows a zoo to display specimens of CITES ANNEX A SPECIES to the public and trade specimens with other Article 60 holders in the EUROPEAN UNION (EU) without the need to apply for individual ARTICLE 10 CERTIFICATES. Formerly known as an Article 30 certificate. *See also* CITES

artificial embryo twinning A relatively low-tech type of CLONING. The process is essentially the same as that which produces identical twins in nature. It is accomplished in the laboratory by manually separating a very early EMBRYO into individual cells. Each cell is then allowed to divide and develop on its own. Each embryo that results is then implanted in a SURROGATE (2) mother.

artificial insemination (AI) The introduction of semen which has been collected from a male animal into the reproductive tract of a female animal for the purpose of creating a pregnancy without the need for the animals to mate. Techniques are well established for many farm animals and have more recently been developed for the purpose of captive breeding endangered species. The method has the advantage of allowing breeders to determine which animals breed and removes the need to transport males to females. It also allows semen to be collected from wild individuals to increase the GENE POOLS of captive populations in zoos. *See also* ARTIFICIAL VAGINA (AV), BALAI DIRECTIVE 1992, CATHETER, ELECTRO-EJACULATOR, INDUCED OVULATOR, SEX-SORTED SEMEN

artificial litter *See* LITTER (1)

artificial milk *See* MILK SUBSTITUTE

artificial selection Selection for particular hereditary traits during breeding controlled by humans. Important in increasing egg production in poultry, milk yield in cattle, speed in racehorses, and in producing the characteristic features of the various breeds of domestic pets. *Compare* NATURAL SELECTION

artificial uterus A life support system for embryos maintained outside the mother's body. An artificial UTERUS consisting of a series of tanks, tubes and fluid-exchange systems equipped with monitoring equipment has been constructed at Port Stephens Fisheries Institute in New South Wales, Australia, and successfully used for the rare dwarf ornate wobbegong shark (*Orectolobus ornatus*).

artificial vagina (AV) A device which uses thermal and mechanical stimulation to induce ejaculation and collect semen from many species, especially cattle and horses, but also sheep, goats, rabbits and even cats.

artiodactyl A member of the order ARTIODACTYLA.

Artiodactyla An order of mammals. Even-toed ungulates: pigs, peccaries, hippopotamus, camels, llamas, okapi, giraffe, chevrotains, deer, musk deer, pronghorn, antelopes, cattle, bison, buffalo, goats, sheep (Fig. V2).

Arusha Conference An international conference of conservationists which considered the future of the fauna and flora of Africa, held in Arusha, Tanzania, in 1961. The speech given by the Tanzanian President Mwalimu Julius Nyerere became known as the *Arusha Manifesto* and made a commitment to conserve the wildlife of Africa.

ascites An accumulation of fluid in the PERITONEAL CAVITY. In BROILER chickens It is associated with poor VENTILATION (3), and inadequate oxygen supply (especially at high altitude). Some strains are predisposed to the condition. *See also* SLOWER-GROWING BROILER BREEDS

ascorbic acid *See* VITAMIN C

asexual reproduction Reproduction which does not involve the fertilisation of a female gamete by a male gamete, e.g. fission, budding. *See also* PARTHENOGENESIS. *Compare* SEXUAL REPRODUCTION

Asian Elephant Conservation Act of 1997 (USA) A US law designed to assist in the conservation of Asian elephants (*Elephas maximus*) by supporting and providing financial resources for the conservation programmes of range states and projects of persons with demonstrated expertise in the conservation of Asian elephants, but not captive breeding unless for release into the wild.

ASPCA *See* AMERICAN SOCIETY FOR THE PREVENTION OF CRUELTY TO ANIMALS (ASPCA)

aspect ratio (AR) The ratio of the length to the width of a bird's wing.

$$AR = \frac{\text{wing length}}{\text{wing width}}$$

A high aspect ratio indicates long wings such as those found in gliding and soaring birds, e.g. many large soaring BIRDS OF PREY such as condors and SEABIRDS such as albatrosses. A low aspect ratio indicates short, wide wings such as those seen in small PASSERINES, GAME BIRDS and small hawks. Such wings are designed for rapid take-off and manoeuvrable flight. Examples of aspect ratios:

A

pheasant = 6.8; gull = 13.8; albatross = 20. *See also* **WING LOADING (WL)**

aspergillosis A severe respiratory disease of birds caused by the fungus *Aspergillus*. Infections can occur in the ear canal, eyes, nose, sinus cavities and lungs and may even affect the brain. Aspergillosis takes the form of an acute rapidly fatal **PNEUMONIA** in young chickens and turkeys.

asphyxia Suffocation by any cause that interferes with breathing (e.g. choking, drowning, inhalation of toxic gas) causing a loss of consciousness or death due to insufficient oxygen in the body (**HYPOXIA**) and increased levels of carbon dioxide in the **BLOOD** and **TISSUES**.

Aspinall, John (1926–2000) A British zoo owner who made his fortune as a bookmaker and owner of a gambling club. He founded Howletts and Port Lympne Wild Animal Parks in Kent, UK. Aspinall became notorious because of his willingness to associate freely with dangerous animals such as gorillas and tigers in their cages and to allow his family and staff to do so. Over a 20-year period five keepers were killed: three by tigers at Howletts and two by elephants at Port Lympne Wild Animal Park. In spite of initially being treated as an outcast by the zoo community, Aspinall achieved considerable breeding success, notably with gorillas.

aspiration *See* **PULMONARY ASPIRATION**

aspiration pneumonia *See* **INHALATION PNEUMONIA**

aspirator *See* **SUCTION DEVICE**

ass *See* **DONKEY**

assay The determination of the amount of a particular constituent of a mixture, or of the potency of a drug, e.g. the amount of a particular hormone in the urine. *See also* **BIOASSAY**

assimilated energy The portion of energy ingested as food that is available to the organism for growth, reproduction and storage. *See also* **GROSS ASSIMILATION EFFICIENCY (GAE)**

assimilation Within an organism, the process of absorbing small molecules and converting them into the more complex molecules that make up that organism. *See also* **GROSS ASSIMILATION EFFICIENCY (GAE)**

assimilation efficiency *See* **GROSS ASSIMILATION EFFICIENCY (GAE)**

assist hatch, manual pipping Artificially assisting a bird or reptile to hatch from its egg by breaking the egg surface. An important aid to conservation in species which have difficulty hatching, e.g. kiwis.

assistance animal
1. An animal trained to help a human with a disability perform everyday tasks.
2. In Australia, under the Disability Discrimination Act 1992 s9(2), an assistance animal is defined as *'a dog or other animal:*
   (a) accredited under a law of a State or Territory that provides for the accreditation of animals trained to assist a person with a disability to alleviate the effect of the disability; or

(b) accredited by an animal training organisation prescribed by the regulations for the purposes of this paragraph; or
(c) trained:
   (i) to assist a person with a disability to alleviate the effect of the disability; and
   (ii) to meet standards of hygiene and behaviour that are appropriate for an animal in a public place.'

See also **ASSISTANCE DOG, GUIDE DOG, HEARING DOG, HELPER MONKEY**

assistance dog A dog trained to help a human with a disability. *See also* **GUIDE DOG, HEARING DOG**

assisted reproductive technology (ART) The collective term for a number of methods used to improve the reproductive performance of animals including **ARTIFICIAL EMBRYO TWINNING, ARTIFICIAL INSEMINATION (AI), CLONING, EMBRYO TRANSFER (ET)**, *IN-VITRO* **FERTILISATION, NUCLEAR TRANSFER**. ART is particularly important in the conservation of some very rare species. *See also* **FROZEN ZOO**

association
1. In studies of learning, the pairing of a neutral stimulus with a stimulus which normally elicits a response such that eventually the **NEUTRAL STIMULUS** (e.g. a sound) may elicit the response in the absence of the original stimulus through the process of conditioning.
2. In social behaviour, a relationship between two individuals inferred by virtue of their being observed in close proximity. *See also* **ASSOCIATION INDEX, MAINTENANCE OF PROXIMITY INDEX (MPI)**
3. A group of people or organisations that have a common interest, e.g. **ASSOCIATION OF ZOOS AND AQUARIUMS (AZA)**. *See also* **KEEPER ASSOCIATION, ZOO ORGANISATIONS**

Association for the Study of Animal Behaviour (ASAB) An organisation, founded in 1936, to promote the study of animal behaviour. The majority of its 2000 members are drawn from Britain and Europe. Publishes the journal *ANIMAL BEHAVIOUR* in collaboration with the **ANIMAL BEHAVIOR SOCIETY (ABS)**.

association index A measure of the extent to which an individual animal has spent time associating with another individual. The index is calculated for each **DYAD** in the group. In a group of three, A, B and C, the dyads are A-B, A-C and B-C. In its simplest form, the association index for A-B may be calculated as:

$$I_{AB} = \frac{2J}{N_A + N_B}$$

where J is the number of times A and B were seen together (including in a group with others), N_A is the total number of times A was seen (alone or as part of a group) and N_B is the total number of times B was seen (alone or as part of a group). If the value

A

is zero the two animals have never been recorded together; if it is one, they are always recorded together. A number of other formulae are available for calculating an association index. *See also* **INDIVIDUAL DISTANCE, MAINTENANCE OF PROXIMITY INDEX (MPI)**

Association of British and Irish Wild Animal Keepers (ABWAK) A **KEEPER ASSOCIATION** founded in 1974. Its members are interested in the keeping and conservation of wild animals and it seeks to achieve the highest possible standards in animal welfare. ABWAK is involved in the training, education and development of keepers and publishes a journal called *Ratel*.

Association of Lawyers for Animal Welfare A UK charity which brings together lawyers interested in animal protection law with the purpose of sharing experience, securing more effective animal protection laws and better enforcement of existing laws.

Association of Veterinarians for Animal Rights An association which works toward the acquisition of rights for all non-human animals by educating the public and the veterinary profession about non-human animal use.

Association of Zoos and Aquariums (AZA) A non-profit organisation, founded in the United States in 1924, dedicated to the advancement of zoos and aquariums in the areas of conservation, education, science and recreation. It was formerly the American Association of Zoos and Aquariums. It established the **SPECIES SURVIVAL PLAN® (SSP) PROGRAMS** in 1981 to manage and conserve selected threatened or endangered species. The AZA is based in Silver Spring, Maryland.

assurance population *See* **INSURANCE POPULATION**

asthma A chronic inflammatory disorder which results in obstruction of the airways when the bronchi fill with mucus and go into spasm as a result of an **ALLERGIC REACTION** causing wheezing, coughing and breathing difficulties. An asthma attack may be caused by irritants such as smoke or allergens such as animal hair, feathers, mould, pollen, dust and some foods and drugs. Asthma can be a problem for people who work with animals, food and bedding materials. It also occurs in some animals, e.g. cats, dogs, horses. *See also* **ATOPY**

Astley, Philip (1742–1814) An English ex-cavalryman who established a riding school and later the first circus in England at Astley's Royal Amphitheatre which opened in London in 1795 after the original building had burnt down.

asymmetry A form which is incapable of being bisected into two identical halves. *See also* **BILATERAL SYMMETRY, PENTAMEROUS (RADIAL) SYMMETRY, RADIAL SYMMETRY**

asymptomatic Exhibiting no **SIGNS (1)** of a disease. Strictly speaking the term symptom does not apply to animals.

asymptomatic carrier An animal that is a carrier of a disease, exhibits no **SIGNS (1)**, but may transmit it to another animal. Strictly speaking the term symptom does not apply to animals.

Ateleopodiformes An order of fishes: jellynose fishes.

Atheriniformes An order of fishes: silversides, rainbowfishes and their allies.

atlas
1. The first cervical vertebra which is located at the anterior end of the vertebral column and articulates with and supports the skull. Allows nodding movement of the head. *See also* **AXIS (2)**
2. A bound collection of maps showing, for instance, the distribution of a particular taxon of animals, e.g. an atlas of the breeding birds of Britain, an atlas of the mammals of North America. Often based on presence or absence in 10 km squares.
3. A book of photographs of anatomical or histological structures, e.g. an atlas of histology, *A Colour Atlas of Veterinary Dentistry and Oral Surgery*.

atopic Relating to **ATOPY**.

atopy A genetic predisposition toward the development of hypersensitivity to substances (**ANTIGENS**) in the environment, e.g. **ASTHMA, DERMATITIS**.

ATP *See* **ADENOSINE TRIPHOSPHATE (ATP)**

atrioventricular (AV) node A small area of specialised tissue which is located between the atria and ventricles of the heart. It receives a wave of excitation which passes from the **SINOATRIAL (SA) NODE** and conducts it to the ventricles causing them to contract after a short delay, which ensures that the ventricles have been filled with blood by the contracting atria before they contract themselves.

atrium (atria *pl.*)
1. A chamber within the **HEART**. In mammals there are two: the right atrium receives deoxygenated blood from the body and pumps it to the right ventricle which then pumps it to the lungs; the left atrium receives oxygenated blood from the lungs and pumps it to the left ventricle from where it is pumped around the body. *See also* **AURICLE (1)**
2. A general anatomical term for a chamber.
3. In architecture, a large open space inside a building.
4. Central chamber (spongocoel) of a sponge.

atrophy A reduction in the size of a structure, usually involving the destruction of cells. May be under genetic or hormonal control.

Attenborough, David (1926–) Sir David Attenborough is an English naturalist, author and broadcaster whose name is synonymous with the production of high-quality wildlife documentaries. His highly acclaimed BBC series include *The Life of Mammals, The Life of Birds, Life in the Undergrowth, The Blue Planet, Life in the Freezer, Life in Cold Blood, Planet Earth* and *Africa*. Attenborough's work has undoubtedly been extremely important in increasing awareness of conservation issues worldwide.

attendance *See* **VISITOR ATTENDANCE**

attenuated vaccine A vaccine which has been pre-pared from live microbes whose disease-producing ability has been weakened but which still retain their immunological properties.

attracting power In visitor studies, a measure of the number and kinds of visitors attracted to an **ANIMAL EXHIBIT**. Often measured as the number of visitors who stop at the exhibit expressed as a percentage of the total number who pass by. *See also* **HOLDING POWER**

ATV *See* **QUAD BIKE**

auditory Relating to hearing or the organs associated with hearing.

auditory enrichment A **SENSORY ENRICHMENT** technique that uses sound, e.g. music, to improve the environment of a captive animal. Music is played to shy, nervous animals, e.g. okapi (*Okapia johnstoni*); some cows on farms are played music to keep them calm thereby increasing milk production.

auditory ossicles *See* **EAR OSSICLES**

auditory signals *See* **ACOUSTIC SIGNALLING**

Audubon *See* **NATIONAL AUDUBON SOCIETY**

Audubon, John James (1785–1851) An American ornithologist, collector, artist and writer. He is com-memorated in the name of Audubon's oriole (*Icterus graduacauda*). *See also* **NATIONAL AUDUBON SOCIETY**

Aulopiformes An order of fishes: lizardfishes and their allies.

auricle
1. An alternative, out-dated, name for an **ATRIUM (1)** of the heart.
2. A **PINNA (1)** (external ear) or any structure of similar shape.

Australasian region *See* **FAUNAL REGIONS**

Australasian Species Management Program (ASMP) A captive breeding programme operated in Aus-tralasia by the **ZOO AND AQUARIUM ASSOCIA-TION (ZAA)**. Under the ASMP there are 14 **TAXON ADVISORY GROUPS (TAGs)** and more than 100 programmes.

Australian Animal Health Laboratory A national facility for animal health in Australia whose purpose is to protect the country against the threat of exotic and emerging animal diseases. It provides diagnostic services and surveillance capabilities.

autecology The **ECOLOGY** of individual species; the study of the relationship between a single **SPECIES** and the **ENVIRONMENT**. *Compare* **SYNECOLOGY**

auto drafter A device which automatically sorts live-stock, e.g. sheep. Animals are briefly held in a **RACE** where they are identified using **RADIO FRE-QUENCY IDENTIFICATION (RFID) TECHNOL-OGY** (e.g. by tags). An automatically operated gate system then sorts individuals into different pens based on specified criteria, e.g. sex, weight etc. This equipment may be combined with a **WALK-OVER WEIGHING SYSTEM**. *See also* **ANIMAL RECOGNI-TION TECHNOLOGY**

autoclave, steriliser A device that sterilises laboratory and veterinary equipment etc. (e.g. test tubes, **AGAR PLATES**) using steam under high pressure.

autoimmune disease An **IMMUNE REACTION** exhibited by an animal against its own molecules.

automata, animals as A view held by **DESCARTES** that animals are not conscious beings but simply self-operating machines that respond to simple instructions (like robots). This attitude to animals does not require us to care about **ANIMAL WELFARE** or how we treat them.

automated cat flap A small door which allows a par-ticular domestic cat access to a house or other place after it has been identified using **RADIO FRE-QUENCY IDENTIFICATION (RFID) TECHNOL-OGY**. *See also* **SMART DOOR**

automated feeding system A method of controlling the amount of feed individual animals receive in a livestock rearing system, often by only delivering food when a sensor linked to a computer detects the presence of a **TRANSPONDER (1)** attached to the animal. *See also* **RADIO FREQUENCY IDENTI-FICATION (RFID) TECHNOLOGY, WET FEEDING**

automatic egg turner A device that turns eggs during incubation in an incubator to ensure an even warming, simulating the action of parent birds.

automatic food dispenser A programmable device designed to provide food for an animal at predeter-mined intervals. Many designs are available for feeding a variety of pets, from cats and dogs to lizards and fishes.

automatic water drinker A wall-mounted drinking water bowl fitted with a valve which causes it to fill automatically when the water level falls after use. Used for horses, bovids and other livestock. *See also* **NOSE FILL DRINKER**

automimicry Protection conferred on the harmless (or non-toxic) members of a species by virtue of their resemblance to harmful (or toxic) members of the same species. For example, female bees possess stingers while males of the same species do not; some butterfly larvae absorb toxins from poisonous plants while other individuals of the same species feed on non-poisonous plants. *Compare* **AGGRES-SIVE MIMICRY, BATESIAN MIMICRY, MÜLLE-RIAN MIMICRY**

autonomic nervous system (ANS), involuntary nervous system The part of the vertebrate visceral nervous system (**PERIPHERAL NERVOUS SYSTEM (PNS)**) which supplies the **SMOOTH MUSCLES** and glands (the visceral motor system) and involuntarily controls organs. Divided into the **SYMPATHETIC NERVOUS SYSTEM (SNS)** and **PARASYMPA-THETIC NERVOUS SYSTEM (PNS)**.

autonomous recording unit (ARU) A sound record-ing unit designed to record at preset intervals and durations, and save the sounds to a built-in hard drive. Used to detect the presence of species, e.g. birds and bats, especially in inaccessible places such as caves and roof spaces.

autopsy See *POST-MORTEM* **(2)**

auto-reset pest traps Rat and squirrel traps which kill individuals without poison by crushing them with a powerful air-ram (operated by compressed air), remove the dead animals, and then reset themselves.

autosome Any **CHROMOSOME** that is not a **SEX CHROMOSOME**. Autosomes occur in the same number in males and females. *See also* **CHROMO-SOME NUMBER**

autotomy The voluntary self-amputation of a body part (usually a limb or **TAIL (1)**) often as a result of attack and to escape capture. Occurs in the claws of some **CRUSTACEANS** and the tails of some lizards.

autotroph, primary producer An organism capable of obtaining its energy from inorganic sources by **PHOTOSYNTHESIS** or chemosynthesis. *See also* **CHEMOAUTOTROPH** *Compare* **HETEROTROPH**

autotrophic Relating to an **AUTOTROPH**.

aversion

1. A strong dislike of something.
2. Dislike of particular foods. Many species have an **INNATE** aversion to poisonous organisms. Others are capable of aversion learning (by conditioning) as a result of exposure to the sight, smell or taste of a food which has caused illness.

aversion learning *See* **AVERSION (2)**

aversion test A test which examines whether or not an animal is prepared to experience an aversive stimulus, e.g. an electric shock, in order to achieve a particular goal (e.g. access to food).

Aves A class of vertebrates containing the birds. There are around 10000 species. They are **BIPEDAL**, **HOMEOTHERMIC**, winged animals covered in **FEATHERS**. They reproduce by laying **CLEIDOIC EGGS**. They possess a cornified bill (beak) with no teeth. Their forelimbs are developed as wings and most are capable of flight. All birds possess a reversed first toe and fewer than 26 tail **VERTEBRAE**.

avian cholera A contagious disease of birds resulting from infection by the bacterium *Pasteurella multocida*.

avian diphtheria, fowl pox A mild-to-severe, slow developing disease of birds caused by a virus belonging to the avipoxvirus group which is a subgroup of poxviruses.

avian influenza, bird flu, fowl plague This is a **NOTIFIABLE DISEASE** caused by influenza A virus. It mainly affects domesticated fowl but also occurs in ducks, geese, turkeys and most common wild birds. Affected birds may die suddenly. Signs include an elevated temperature, drooping wings and tail, fast and laboured respiration and lack of movement. The bird may tuck its head under a wing while squatting on its breast. Oedema of the head and neck is common. Vaccines are available but their use depends upon government policy. *See also* **H5N1 VIRUS**

avian tuberculosis Avian tuberculosis is usually caused by the bacterium *Mycobacterium avium*. In captivity, turkeys, pheasants, quail, cranes, and certain birds of prey are commonly affected.

aviary A place where birds are kept for display or breeding.

aviary hack The process of releasing a bird of prey by keeping it in an aviary (hack aviary) at the release site prior to release. The aviary should have a good view of the surrounding area and one side should be solid. The carer approaches the aviary from this side and provides food via a hatch or pipe. Suitable for old birds, scavengers (e.g. kites and buzzards), owls and kestrels.

aviculture

1. The keeping and breeding of birds in captivity.
2. Under s.27(1) of the **WILDLIFE AND COUNTRYSIDE ACT 1981**, '"aviculture" means the breeding and rearing of birds in captivity.'

avifauna Bird **FAUNA**, e.g. the avifauna of Morocco.

avoidance A behavioural defence against harmful circumstances which may be innate or learnt as a result of association (avoidance learning), e.g. the avoidance of places where predators occur or poisonous foods. *See also* **AVERSION (2)**

avoidance learning *See* **AVOIDANCE**

axial Located along the axis of an organism.

axial skeleton That part of the **VERTEBRATE** skeleton which consists of the **CRANIUM**, **VERTEBRAE**, **RIBS**, **STERNUM**, hyoid and associated cartilages.

axis

1. A plane of symmetry in animals: antero-posterior, dorso-ventral, medio-lateral. Important in the early development of the embryo.
2. Second vertebra, which has been modified to support the head. Allows rotation of the head. *See also* **ATLAS (1)**

axon A long thin process which extends away from the cell body of some nerve cells (**NEURONES**) towards a target structure (e.g. another neurone). The axon carries nerve impulses which are caused by a wave of depolarisation of its membrane. Myelinated axons are covered with a **MYELIN SHEATH** which increases the rate of conduction of **NERVE IMPULSES**. *See also* **ACTION POTENTIAL**

AZA *See* **ASSOCIATION OF ZOOS AND AQUARIUMS (AZA)**

AZK Australasian Zoo Keeping Association (AZK). *See also* **KEEPER ASSOCIATION**

Aztec zoo *See* **MONTEZUMA II**

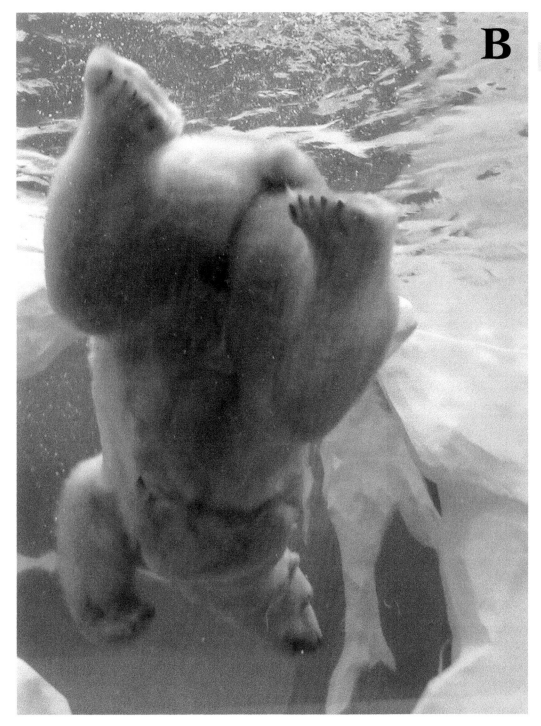

Fig. B1 B is for bear. Polar bear (*Ursus maritimus*) swimming underwater in the *Arctic Ring of Life* exhibit at Detroit Zoo.

Dictionary of Zoo Biology and Animal Management: A guide to terminology used in zoo biology, animal welfare, wildlife conservation and livestock production, First Edition. Paul A. Rees.

B

B2 A famous male tiger (*Panthera tigris*) who was the star attraction at the Bandhavgarh National Park in India. He was well-known to tourists and local people. *B2* was an important ambassador for ECO-TOURISM, demonstrating that animals can be worth more alive than dead.

bachelor group A group of males of the same species. Often kept together in a zoo because they are not currently required for breeding. Such groups may need to be formed where the social structure of the species is such that relatively few breeding males are required, e.g. bongos (*Tragelaphus eurycerus*), gorillas (*Gorilla gorilla*). Bachelor groups occur in many species in the wild, where they are able to exist independently of family groups. The relatively small amount of space available in zoo enclosures prevents the normal DISPERSAL of males that occurs in nature.

bacillus A rod-shaped, spore-forming, aerobic bacterium of the genus *Bacillus*.

backbone *See* VERTEBRAL COLUMN

baconer A pig grown specifically for its bacon.

bacteraemia, bacteremia The condition in which bacteria are present in the blood.

bacterial disease A disease caused by infection with a bacterium, e.g. from *SALMONELLA*.

bactericide, bacteriocide A substance that kills bacteria.

bacteriocide *See* BACTERICIDE

bacterium (bacteria *pl.*) A microscopically small organism which usually consists of a single cell with an outer rigid cell wall. Many taxa are important in decomposition and nutrient cycles. Others cause disease. Some species are capable of very rapid population growth.

baculum (bacula *pl.*) *See* OS PENIS

badger culling The practice of killing wild badgers (*Meles meles*) especially in the UK in the belief that they transmit bovine TUBERCULOSIS to cattle, in order to control the disease. *See also* KREBS REPORT

badger-baiting A contest between a badger (*Meles meles*) and a domestic dog that has been trained to fight. Illegal in the UK. Participants take badgers from the wild, usually locating them using dogs and digging them from their setts. *See also* BADGER CULLING, PROTECTION OF BADGERS ACT 1992

bag Animals (generally game birds) taken by hunters, e.g. a bag of three ducks shot on a hunting trip.

bag statistics Data indicating the number of different species of game birds (or other hunted animals) taken by hunters.

Bailey, Hachaliah Bailey is credited with founding the first travelling MENAGERIE in America around 1815, when he toured New England with a single elephant called *Old Bet*.

Bailey, James Anthony Partner of Phileas T. Barnum. *See also* BARNUM AND BAILEY

bait An object, usually food, used to attract an animal into a trap, to a fishing line, or to some other place. *See also* DECOY (1)

bait marking A method of determining the ranging behaviour of an animal by feeding it marked bait (e.g. bait containing a food dye) and then searching for locations where it has deposited faeces. This method was used to study badgers (*Meles meles*) in an investigation into the role of this species in the transmission of bovine TUBERCULOSIS in the UK.

bait shyness The development of an AVERSION (2) towards bait and novel food which occurs widely in birds and mammals. This makes some species very difficult to control by poisoning, e.g. rats.

bait station A lockable container used for rodent control and designed to hold RODENTICIDE securely.

Bakewell, Robert (1725–1795) A British agriculturalist during the AGRICULTURAL REVOLUTION who is credited with being the first person to systematically breed livestock, leading to many improvements in cattle, sheep and horses, and contributing to our understanding or ARTIFICIAL SELECTION.

Balai Directive 1992 An EU Directive (Council Directive 92/65/EEC) which lays down the animal health requirements governing trade in and imports into the EU of animals, semen, ova and embryos which are not subject to other EU legislation. Animals must come from registered or approved premises (e.g. a breeder registered with their local authority) and such premises must meet stringent biosecurity requirements. The Directive sets special conditions for: foxes, ferrets, mink, cats, dogs, apes, lagomorphs, wild ungulates and non-domesticated varieties of sheep, goats, camels, pig, cows and deer. It also covers captive birds and their hatching eggs and all animals susceptible to RABIES. The directive does not apply to pet animals, or domestic cattle, swine, sheep, goats, equids, poultry (including eggs), fish and fishery products, bivalves or AQUACULTURE animals.

balance The capacity of an animal to maintain a stable position with respect to gravity as a result of information received through specialised MECHANORECEPTORS. *See also* INNER EAR, STATOCYST

balance of nature The concept that ECOSYSTEMS exist as stable entities which deteriorate if disturbed and wherein the composition of the BIOLOGICAL COMMUNITY remains more or less constant with time. The British ecologist Charles Elton suggested that there is no such thing as 'the balance of nature', recognising that many ecosystems undergo significant changes with time in both the composition of communities and in the size of the populations of individual species. This is especially so during ECOLOGICAL SUCCESSION.

Bald and Golden Eagle Protection Act of 1940 (USA) A US law which prohibits anyone, without a permit, from taking (*see* **TAKE**) bald eagles (*Haliaeetus leucocephalus*) or golden eagles (*Aquíla chrysaetos*), including their parts, nests, or eggs.

baleen A comb-like structure located on the upper jaw of some whale species (baleen whales) that filters food from water or sediments.

ball and socket joint, enarthrosis, spheroidal joint A joint in the vertebrate skeleton formed from the end (head) of a **LONG BONE** (ball) and a cup-like depression in the pelvis or shoulder (socket). These joints are highly mobile and allow movements in three dimensions. *See also* **BRACHIATION**

ballarat An Australian slang term for a cat.

bamboo A fast-growing, woody, evergreen perennial grass. There are more than 70 genera and over 1450 species. Most species flower infrequently. Used as food by certain specialist feeders, e.g. the giant panda (*Ailuropoda melanoleuca*), and grown for this purpose by some zoos, e.g. San Diego. Also used to create naturalistic exhibits and to create screens to hide animals from public view.

barb
1. One of the thin processes that extends from the shaft of a feather. *See also* **BARBULE**
2. A ray-finned cyprinid fish especially of the genus *Barbus* and *Puntius*.
3. A hardy breed of horse from North Africa (Barbary coast).

Barbary lion (*Panthera leo leo*) An extinct **SUBSPECIES** of lion which once lived in North Africa. Also called the Atlas or Nubian lion. Attempts have been made to breed the subspecies back from animals in zoos which morphologically resemble the lost form.

bar-biting An abnormal behaviour in which the animal bites metal bars in enclosures and cages. Common in cattle, sheep, sows, giraffes. *See also* **CRIBBING**

barbule A hooked structure at the end of the **BARB** of a **FEATHER** which connects it to other barbs.

Bärengraben A **BEAR PIT** in Bern, Switzerland, built around 1857 and now located in Bern Zoo. Not used for bears since 2009. Listed and protected as a heritage site of national importance.

bark collar A dog collar used to train a dog to stop barking. *See also* **ELECTRONIC TRAINING COLLAR**

bark controller A hand-held device that emits ultrasonic sound commands and is used to train dogs to stop barking. *See also* **BARK COLLAR**

barn system, perchery system A system of keeping hens in large open sheds with several tiers of perches. Flocks may consist of thousands of birds. *See also* **BATTERY CHICKENS, ENRICHED COLONY CAGES**

Barnum and Bailey Showmen in the USA. From around 1875 P.T. Barnum's Great Traveling Museum, Menagerie, Caravan, and Hippodrome exhibited animals across the United States. This eventually evolved into The Ringling Bros. & Barnum and Bailey Circus ('*The Greatest Show on Earth*') which

still exists today. In 1882 Barnum and Bailey purchased the famous elephant '*JUMBO*' from **LONDON ZOO** (Fig. J3). *See also* **TRAVELLING MENAGERIE**

Barnum, Phileas T. Partner of James Anthony Bailey. *See also* **BARNUM AND BAILEY**

barren enclosure An **ANIMAL ENCLOSURE** which provides little or no stimulation for the animals it houses because it contains little, if any, **FURNITURE** or few objects.

barrier *See* **ENCLOSURE BARRIER, PUBLIC BARRIER**

Barrier Designs for Zoos A document which describes the types of **ENCLOSURE BARRIERS** that are suitable for a variety of taxa found in zoos, the appropriate dimensions for particular species (e.g. fence heights, moat widths etc.), suitable barrier materials etc. Produced by the **CENTRAL ZOO AUTHORITY (CZA)**, India.

barrier gel A preparation used to treat minor wounds which encases cuts, abrasions and scratches and encourages healing and hair regrowth. Often contains anti-bacterial and cleansing agents.

barrier nursing The use of special disinfectants, special protective clothing and specific techniques to protect vets, veterinary nurses and others from infection when dealing with an animal that is suffering from an infectious disease. This often occurs in an isolation ward to prevent transmission of the disease to other animals.

barrier standards *See* **BARRIER DESIGNS FOR ZOOS, CONTAINMENT STANDARD**

Bartlett, Abraham Dee Superintendent of London Zoo from 1859 until his death in 1897. *See* **BARTLETT SOCIETY**

Bartlett Society A society of **ZOO ENTHUSIASTS** interested in the history of zoos and animal keeping, named after Abraham Dee **BARTLETT**. It publishes a list of *First and Early Breeding Records for Wild Animals in the UK and Eire*.

basal body temperature The temperature of the body at rest. *See also* **BASAL METABOLIC RATE (BMR)**

basal ganglia A set of neural structures located deep inside the **CEREBRUM** of the brain. One of their functions appears to be to select and trigger well co-ordinated voluntary movements. The presence of **STEREOTYPIC BEHAVIOUR** appears to be correlated with disorders of the basal ganglia.

basal metabolic rate (BMR) The energy budget of an endotherm measured over a specific period of time (e.g. 24 hr) when it is motionless, fasting and not under temperature stress. This is effectively the amount of energy that the animal needs merely to stay alive when inactive. *See also* **BASAL BODY TEMPERATURE**

base
1. A substance that combines with hydrogen ions in solution.
2. The nitrogenous part of a **NUCLEIC ACID**.

basic life support *See* LIFE SUPPORT

basic needs test A test used to establish whether or not an animal should be kept in a zoo by asking if the zoo can provide for the basic physiological and psychological needs of the animal. If it can, then the basic needs test has been passed. *Compare* COMPARABLE LIFE TEST

basking The action of exposing the body to the warmth of the sun or other heat source. Amphibians and reptiles bask in the sun to increase their body temperature. *See also* SUNBATHING

bat box A container used by bats for roosting and breeding, either in the wild or in captivity. Often mounted high on a wall, on a pole or in a tree. Usually wooden and fitted with a hinged roof to allow access for cleaning, marking bats etc.

bat detector An electronic device capable of converting the inaudible sounds made by bats into frequencies which can be detected by the human ear (Fig. B2). Some are capable of recording. Bat detectors can be used to identify bat species and to distinguish particular behaviours which are accompanied by distinctive sounds (e.g. feeding).

bat house A building in a zoo that is a type of NOCTURNAL HOUSE that contains bats of one or more species, in which day and night are reversed so that visitors may see the animals in the dark when they are most active. Visitors are able to walk through free-flight bat houses and observe the bats at close quarters.

Batesian mimicry The phenomenon whereby a harmless (palatable) species of animal has evolved so that it possesses the same warning signals as a harmful (toxic) species in order to protect itself from predators, e.g. the yellow and black abdomen of harmless hornets is similar to that of wasps. Named after Henry Walter Bates, an English naturalist. *Compare* AGGRESSIVE MIMICRY, AUTOMIMICRY, MÜLLERIAN MIMICRY

Batrachoidiformes An order of fishes: toadfishes and midshipmen.

battery bear *See* BEAR FARMING

battery cage *See* BATTERY FARMING. *Compare* ENRICHED COLONY CAGES

battery chickens Chicken raised in intensive conditions indoors. *See also* BATTERY FARMING

battery farming A method of producing eggs from chickens as quickly and cheaply as possible using the minimum of space. Also known as factory farming. Birds are kept indoors in very small battery cages and may be fed antibiotics to prevent disease. *See also* LAYING HENS DIRECTIVE

Bayliss v. Coleridge (1903) A legal case tried at the Royal Courts of Justice in London in 1903 concerning the VIVISECTION of a dog in which the law was broken and which brought the suffering of animals in experiments to the attention of the general public. *See also* BROWN DOG AFFAIR

BBC Natural History Unit A production unit within the British Broadcasting Corporation which specialises in the making of natural history documentaries. Many of their programmes have been written and narrated by Sir David ATTENBOROUGH. The unit maintains extensive archives of sound recordings, film and video recordings, and still images. *See also* WILDSCREEN

beach donkey A donkey that is used to give rides to children on a beach (Fig. B3). In the UK, operators of beach donkeys must hold a licence from the local authority for each donkey working on the beach under the RIDING ESTABLISHMENTS ACT 1964 AND 1970. *See also* DONKEY SANCTUARY

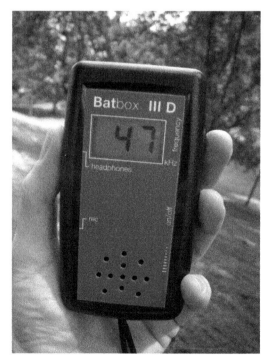

Fig. B2 Bat detector.

Fig. B3 Beach donkeys at Blackpool, UK.

beaching *See* STRANDING

beak *See* BILL

beak coping The process of trimming the overgrown beak of a bird, e.g. a BIRD OF PREY. *See also* BEAK TRIMMING

beak trimming The practice of removing the tips of the beaks of poultry to prevent injury resulting from aggression between individuals and FEATHER PECKING. *See also* BEAK COPING

bear farming The practice, in some Asian countries, of keeping bears – known as bile bears or battery bears – in order to extract BILE from their GALL BLADDERS for use in Asian TRADITIONAL MEDICINES.

bear pit An outdated enclosure in which bears are kept in a pit below the viewing level of the public. *See also* BÄRENGRABEN

bear-baiting A BLOOD SPORT in which trained dogs attack a tethered bear that has had its claws removed. Popular in England until the 19th century. Still common in rural Pakistan in spite of being illegal. *See also* BULL-BAITING, RAT-BAITING

beast of burden, draft animal, draught animal
1. An animal used for carrying loads or doing other types of heavy work such as pulling vehicles, excavating land, ploughing, logging, water-lifting etc. Includes horses, donkeys, mules, camels, cattle, buffaloes and elephants.
2. In Thailand the Beast of Burden Act 1939 defines beasts of burden as elephants, horses, cattle, buffaloes, mules or donkeys.

beast show, wild beast show An alternative name for a TRAVELLING MENAGERIE.

bee yard *See* APIARY

behaviour, behavior
1. The repertoire of responses that an animal may make to changes in its environment.
2. A specific RESPONSE made by an organism to a specific STIMULUS.
3. According to Manning (1972) *'. . . all those processes by which an animal senses the external world and the internal state of its body, and responds to changes which it perceives.'*

behaviour budget *See* ACTIVITY BUDGET

behaviour sampling The process of collecting data on an individual animal's BEHAVIOUR by breaking it up into discrete components for the purpose of study and analysis. *See also* FOCAL SAMPLING, SCAN SAMPLING

behavioural competence The ability of an animal to express appropriate behaviour in a particular situation. *Compare* BEHAVIOURAL RESTRICTION

behavioural diversity A measure of the number and variety of behaviours exhibited by an individual animal.

behavioural ecology The scientific study of the ecological and evolutionary basis of animal behaviour, and its role in adapting an organism to its environment, particularly the way in which behaviour contributes to reproduction and survival. Hypotheses in behavioural ecology assume that behaviour is optimised. *See also* CONSERVATION BEHAVIOUR, EVOLUTIONARILY STABLE STRATEGY (ESS)

behavioural economic theory, animal economics The application of economic theory to animal behaviour, e.g. elasticity of demand. It treats animals as spending time and energy on various activities in a way that parallels the expenditure of time and money by humans.

behavioural engineering A means of increasing activity in animals which requires them to work for a reward. *See* ENVIRONMENTAL ENRICHMENT, MARKOWITZ

behavioural enrichment *See* BEHAVIOURAL ENGINEERING, ENVIRONMENTAL ENRICHMENT

behavioural genetics The scientific study of the effect of the genes and inheritance on the behaviour of animals and the evolution of behaviour.

behavioural husbandry Changes in the captive environment of an animal which will affect its behaviour such as ENVIRONMENTAL ENRICHMENT or TRAINING.

behavioural plasticity The ability of an organism to vary its behaviour in response to changes in the environment.

behavioural restriction The inability of an animal to perform it full repertoire of behaviours. *See also* ETHOGRAM. *Compare* BEHAVIOURAL COMPETENCE

behavioural sampling
1. The process of collecting samples of behaviours. *See also* BEHAVIOUR SAMPLING
2. The occasional selection by an animal faced with a choice of the normally less-preferred option, as a means of keeping check on alternatives.

behaviourism, behaviorism A school of PSYCHOLOGY, founded in the United States by J. B. WATSON, which was largely concerned with the behaviour of animals in the laboratory. It avoids making inferences about the internal psychological state of animals to explain behaviour and focuses instead on observable stimuli, muscular movements and glandular secretions. It focused particularly on stimulus–response relationships. *See also* SKINNER, THORNDIKE. *Compare* ETHOLOGY

Belle Vue Zoological Gardens A zoo that existed in Manchester, England, between 1836 and 1979. The grounds also contained an amusement park, a FLEA CIRCUS, a dance hall and other facilities.

Beloniformes An order of fishes: ricefishes, flyingfishes and their allies.

Bengal A hybrid cat formed from a cross between a leopard cat (*Prionailurus bengalensis*) and a domestic cat (*Felis catus*).

Bentham, Jeremy (1748–1832) An English philosopher, lawyer and social reformer who advocated UTILITARIANISM and ANIMAL RIGHTS. In considering which animals we should protect from injury, Bentham famously considered that *'The question is not, can they reason? Nor, can they talk? But, can they suffer?'*

B

B

benthic Relating to the bottom of a lake or ocean, e.g. benthic fauna. *See also* **BENTHOS**

benthos The community of organisms that live on, in, or near the bottom of water bodies, especially the seabed; benthic organisms.

Bergmann's rule A principle described by the German biologist Christian Bergmann which states that homeothermic species within a genus (and individuals within a species) tend to be smaller in warmer climates and larger in cooler climates. This is usually most noticeable in species that inhabit a range of altitudes or latitudes. Individuals that live at high altitude or northern latitude tend to be slightly larger than those living lower down or further south. *See also* **ALLEN'S RULE, CLIMATE CHANGE**

Berne Convention Colloquial name for the **CONVENTION ON THE CONSERVATION OF EUROPEAN WILDLIFE AND NATURAL HABITATS 1979.**

Beryciformes An order of fishes: squirrelfishes and their allies.

bestiality, zoophilia Sexual intercourse between a human and a non-human animal. A crime in some jurisdictions but not others. Bestiality is an offence under the Sexual Offences Act 2003 s69 (in England and Wales). The possession of images portraying bestiality is illegal under s63(7)(d) of the Criminal Justice and Immigration Act 2008 (in England, Wales and Northern Ireland).

bestiary A medieval book containing descriptions and illustrations of beasts and monsters, some real, some imaginary. *See also* **CRYPTOZOOLOGY**

beta (β) diversity The change in species diversity between two different ecosystems, i.e. the total number of species that are unique to each of the ecosystems being compared. *See also* **ALPHA (α) DIVERSITY, BIODIVERSITY, GAMMA (γ) DIVERSITY**

beta status The second most **DOMINANT INDIVIDUAL** in a social group of animals, which is **SUBORDINATE** to the alpha individual but dominant to all others. *See also* **ALPHA STATUS**

between-observer reliability, interobserver reliability The extent to which observations or recordings made of a particular behaviour or other phenomenon by different people are identified and recorded in the same way. For example, in a behaviour study examining the **ACTIVITY BUDGET** of a group of penguins it is important that each behaviour is clearly defined in an **ETHOGRAM** and that all of the observers are using the same definitions and interpret them in the same way. This may be tested if two or more observers make recordings of the same animals at the same time and then compare their observations to check for consistency.

BIAZA *See* **BRITISH AND IRISH ASSOCIATION OF ZOOS AND AQUARIUMS (BIAZA)**

big cat One of the larger species of felids, typically those species able to roar. The definition may be confined to the genus *Panthera* (lions (Fig. L1), tigers, leopards and jaguars (Fig. J1)), or extended to include pumas (*Puma concolor*), cheetahs (*Acinonyx jubatus*), snow leopards (*Uncia uncia*) and clouded leopards (*Neofelis nebulosa*). Not a recognised **TAXON**.

Big Five A term originally coined by **WHITE HUNTERS** to refer to the most difficult and dangerous animals in Africa to hunt on foot: elephant, rhinoceros, leopard, lion, buffalo.

big game Large wild animals, especially those hunted for sport. Often used to refer to large African mammals. *See also* **BIG FIVE**

bilateral symmetry The possession of one plane (usually antero-posterior and dorso-ventral), in which an organism may be divided into two halves (left and right) which are approximately mirror images of one another. This is a property of most metazoan animals and is especially obvious in vertebrates. *Compare* **PENTAMEROUS (RADIAL) SYMMETRY, RADIAL SYMMETRY**

bile A secretion from the **LIVER** which is stored in the **GALL BLADDER** before passing to the **SMALL INTESTINE**. Contains **BILE SALTS**.

bile bear *See* **BEAR FARMING**

bile salts Conjugated compounds of bile acids which emulsify fats to aid their digestion. A major component of the bile produced by the liver.

bilharziasis *See* **SCHISTOSOMIASIS**

bill, beak, rostrum A horny protrusion of the upper and lower jaws in birds, made of **KERATIN**. Used for feeding, killing, **COURTSHIP, GROOMING,** fighting etc. *See also* **BILL DUELLING**

bill duelling Bird behaviour in which two individuals 'fence' with their bills. This may occur as a pair-bonding behaviour in mates or a means of establishing social order in non-mated pairs. Occurs in a number of taxa, notably penguins and albatrosses.

Bill of Rights for Animals *See* **TEN COMMANDMENTS**

bimodal distribution A **POPULATION** distribution which has two **MODES**, creating two peaks when represented as a graph. Occurs, for example, when the sizes of males and females are pooled together in a sexually dimorphic species in which males are larger than females. One mode represents the males and the second the females. If the sexes were graphed separately they might appear as two overlapping **NORMAL DISTRIBUTIONS**.

binary fission Asexual reproduction in which the organism divides into two approximately equal parts. Used by many microbes including **PROTOZOA** and **BACTERIA**.

binocular vision A visual system which collects information from two forward-facing eyes separated horizontally so that when the brain merges the two images it creates one three-dimensional image, thereby allowing depth perception. Restricted to vertebrates, but varies in extent between species. Important in **ARBOREAL** species (especially primates) and many predators that need to be able to judge distance and speed. In birds and fishes only part of the visual field is viewed by binocular vision.

binomial name The scientific name of an organism: the genus followed by the species name, e.g. *Panthera leo*. *See also* BINOMIAL SYSTEM OF NOMENCLATURE, LINNAEUS, TRINOMIAL NAME, VERNACULAR NAME

binomial system of nomenclature Universal system for naming species using two names (genus and species) devised by the Swedish botanist and physician Carolus LINNAEUS. This BINOMIAL NAME is often called the scientific or Latin name. Article 5.1 of the INTERNATIONAL CODE OF ZOOLOGICAL NOMENCLATURE (Principle of Binomial Nomenclature) states that: *'The scientific name of a species, . . . , is a combination of two names (a binomen), the first being the generic name and the second being the specific name. The generic name must begin with an upper-case letter and the specific name must begin with a lower-case letter.'* *See also* INTERNATIONAL COMMISSION ON ZOOLOGICAL NOMENCLATURE (ICZN), TRINOMIAL NAME

bioaccumulation The sequestering of a substance by an organism at a higher concentration than that at which it occurs in the surrounding environment. Many organisms concentrate chemicals obtained from food or water so that they reach toxic levels in the body because they are taken in faster than they are removed. These chemicals may subsequently be passed along food chains. *See also* DDT. *Compare* BIOMAGNIFICATION

bioassay
1. The determination of the biological activity or strength of a substance, e.g. a hormone or drug, by comparing its effects with those of a standard preparation on a test organism or isolated tissue.
2. A test used to determine such activity or strength. *See also* ASSAY

bioavailability
1. The proportion of a nutrient in the diet which is available for metabolic use by the body. Interaction between some substances prevents their utilisation. For example, the uptake of zinc is inhibited by the presence of excess calcium, cadmium or copper in the diet.
2. The extent to which, and rate at which, a drug that has been administered is taken up by the body and reaches the tissues and organs.

biocapacity The Earth's regenerative capacity; the area of land available to produce renewable resources and absorb carbon dioxide emissions.

biochemical oxygen demand *See* BIOLOGICAL OXYGEN DEMAND

biochemistry The scientific study of the chemicals found in biological systems and their metabolic pathways

biodiversity
1. A contraction of 'biological diversity'; the component organisms in a particular place e.g. a locality,

Table B1 Estimated number of described species of animals recognised by the IUCN in 2011 (Anon., 2011a).

Taxon	Species
Vertebrates	
Mammals	5494
Birds	10027
Reptiles	9362
Amphibians	6771
Fishes	32000
Subtotal	63654
Invertebrates	
Insects	1000000
Molluscs	85000
Crustaceans	47000
Corals	2175
Arachnids	102248
Velvet Worms	165
Horseshoe Crabs	4
Others	68658
Subtotal	1305250
Total	1368904

habitat, ecosystem or the whole of the globe (Table B1). The concept of 'biological diversity' first appeared in papers by Lovejoy (1980) and Norse and McManus (1980). The term 'biodiversity' was first credited to Walter Rosen of the National Academy of Sciences (USA). It was made popular by Edward O. WILSON. *See also* MAY
2. The UN CONVENTION ON BIOLOGICAL DIVERSITY 1992 (CBD) (Art. 2) defines biological diversity as *'the variability among living organisms from all sources including, inter alia, terrestrial, marine and other aquatic ecosystems and the ecological complexes of which they are part; this includes diversity within species, between species and of ecosystems.'*
3. Under Article 7(2) of Costa Rica's Biodiversity Law 7788 (1998), biodiversity is: *'Variability of living organisms from any source, whether they are found in terrestrial, air or marine or aquatic ecosystems or in other ecological complexes. This includes the diversity within each species, as well as between species and between the ecosystems that they form part of.'* *See also* ALPHA (α) DIVERSITY, BETA (β) DIVERSITY, GAMMA (γ) DIVERSITY

Biodiversity Convention *See* CONVENTION ON BIOLOGICAL DIVERSITY 1992 (CBD)

biodiversity hotspot A term widely used to indicate a geographical area where there is a high diversity of ENDEMIC (1) animal and plant life and a high level of threat including Madagascar, Indonesia, New Zealand, Sri Lanka, the Western Ghats (India), Central America, the Amazon, parts of Tanzania and parts of the west coast of the United States. However, some authors refer to 'hotspots' as areas of high SPECIES RICHNESS or high endemicity, without any reference to the degree of threat. Some modern multi-species zoo exhibits focus attention on the importance of hotspots, e.g. the Bronx Zoo's *Madagascar*, the *Amazonia* exhibit at the National Zoo, Washington.

biodiversity loss The loss of species, especially when due to human activity. *See also* **BIODIVERSITY**

biofloor A self-cleaning floor used in animal enclosures which functions as a biological system to prevent build-up of pathogens or parasite infestation. It consists of a top layer of peat, woodchips or similar material lying on top of a filterpad on a concrete floor with a drain below the pad.

biogenetic reserve One of a number of nature reserves in Europe which form a network created to conserve representative examples of European flora, fauna and natural areas established by the **COUNCIL OF EUROPE**.

biogeochemical cycle A nutrient cycle which involves biological and geological processes, e.g. carbon cycle, nitrogen cycle. Animal faeces and the decomposition of dead animal bodies and artefacts (e.g. antlers) play an important role in returning nutrients to the soil. *See also* **CARBON CYCLE, NITROGEN CYCLE**

biogeographical region *See* **FAUNAL REGIONS**

biographical life A life which has a 'story', i.e. a past, a present and a concept of the future. Humans have a sense of their own biographical lives. They remember their past and plan for their future. There is some evidence that some animals, e.g. chimpanzees, may have a sense of their own biographical lives because they can remember past events (e.g. previous relationships) and they appear to make plans, at least for their immediate future, e.g. when they set off on hunting trips. *See also* **COGNITION**

biohazard A biological hazard. A biological material which may be hazardous to the health of living things, particularly humans. Various levels of biohazard are recognised, including organisms that can infect humans, organisms that can infect animals, and medical waste (including waste from the treatment of animals). Materials and places where there is a risk of exposure to a biohazard should be marked by the internationally recognised symbol.

biologic environment *See* **BIOLOGICAL ENVIRONMENT**

biological clock An internal mechanism found in animals that allows them to synchronise changes in their physiology and behaviour with cyclic changes in the environment. This may allow them to measure the passage of time and often involves them monitoring the position of the sun, stars and moon, and changes in environmental temperature. *See also* **CIRCADIAN RHYTHM, CIRCANNUAL RHYTHM, INFRADIAN RHYTHM, LUNAR RHYTHM, ULTRADIAN RHYTHM, ZEITGEBER**

biological community In ecology, a group of species that occur together, e.g. sand dune community, grassland community. Some zoo exhibits attempt to simulate animal communities found in the wild. *See also* **BIOPARK, MULTI-SPECIES EXHIBIT**

biological concentration *See* **BIOMAGNIFICATION**

biological control A means of controlling pest species by using their natural predators, competitors, **PARASITES** or diseases. Often used in closed systems, e.g. greenhouses. May also be used in natural **ECOSYSTEMS** to eradicate or control an unwanted species. *See also* **MYXOMATOSIS**

biological diversity *See* **BIODIVERSITY**

biological environment, biologic environment, biotic environment All of the living things in an organism's environment, i.e. organisms of the same species and organisms of other species, including animals (non-human and human), plants, and microorganisms. *See also* **ENVIRONMENT, IMMEDIATE ENVIRONMENT** *Compare* **PHYSICAL ENVIRONMENT**

biological filter *See* **BIOLOGICAL FILTRATION**

biological filtration In relation to an aquarium or other aquatic exhibit, the process whereby a device which contains naturally occurring bacteria (in a biological filter) removes ammonia and nitrite (from fish waste) from the water. *Compare* **MECHANICAL FILTRATION**

biological magnification *See* **BIOMAGNIFICATION**

biological oxygen demand, biochemical oxygen demand (BOD) A measure of water quality, especially organic pollution, in which the number of aerobic organisms present is measured indirectly by their oxygen consumption. Measured as milligrams of oxygen consumed per litre of water sample during 5 days of incubation at 20°C. A high value indicates bacterial contamination.

biological species concept *See* **SPECIES (1)**

biology The scientific study of living things.

bioluminescence Light produced by an organism.

bioluminescent organ An **ORGAN** capable of producing light.

biomagnification, biological concentration The process by which the concentration of a **TOXIN** increases as it passes up the **FOOD CHAIN**. Low levels in water are absorbed and concentrated by **PLANKTON**. Fish eat large quantities of plankton, further concentrating the toxin. Fish-eating birds eat many fish and concentrate the toxin further and this may have damaging effects on the **PHYSIOLOGY** or reproduction of the birds. When **DDT** was widely used as an insecticide it was sprayed in water and concentrated in food chains until it ultimately reached **BIRDS OF PREY** as the **APEX PREDATORS**. It had the effect of reducing the thickness of egg shells resulting in many broken eggs and a decline in raptop populations. When DDT was banned raptor populations recovered. *Compare* **BIOACCUMULATION**

biomass The mass of biological material. May refer to the mass of a particular organism or the mass present in all or part of a particular ecosystem, or a specific area, e.g. the biomass of trees in an area of tropical forest.

biome A terrestrial ecosystem of a characteristic type: arctic tundra, northern coniferous forest, temperate forest, temperate grassland, chaparral, tropical rain forest, tropical savanna grassland and desert. Some zoo exhibits simulate the conditions in these biomes, e.g. *ARCTIC RING OF LIFE* (Detroit Zoo), *African Plains* (Dublin Zoo).

biomechanics The study of the mechanical laws concerned with the movement and structure of living organisms. The principles of biomechanics have been used to study GAIT in many species of animals.

biopark, BioPark, ecosystem exhibit, ecosystem zoo A zoo exhibit or zoo in which the animals are displayed within a representation of the ecosystem or ecosystems to which they belong, e.g. a tropical rainforest.

biophilia '*The innately emotional affiliation of human beings to other living organisms*' (Wilson, 1993). The existence of biophilia is often quoted as the reason why people feel drawn to zoos and wild places (Fig. B4). *Compare* NATURE DEFICIT DISORDER

bioprospecting
1. The removal of organisms from their natural environment for the purpose of commercial exploitation. This activity is regulated by the law in some countries. For example, in Costa Rica bioprospecting is controlled by access permits.

2. In Costa Rica the Biodiversity Law 7788 (1998) article 7(3) defines bioprospecting as '*the systematic search, classification and research for commercial purposes of new sources of chemical compounds, genes, proteins, and micro-organisms, with real or potential economic value, which are found in biodiversity.*'

biopsy The collection of a sample of biological material, usually a tissue sample from an animal, for examination for the presence of disease etc.

biosecure unit
1. An enclosure, cage or container from which pets, livestock, or other organisms, cannot escape.
2. A facility where organisms, especially rare species, are kept to protect them from infection, attack by pests etc. Chester Zoo has constructed biosecure amphibian pods to protect rare species. *See also* BIOSECURITY

biosecurity Procedures and practices intended to protect humans and animals against disease or harmful biological agents, including damage to the environment by ALIEN SPECIES. *See also* BIOSECURE UNIT, CARTAGENA PROTOCOL ON BIOSAFETY 2000

Biosecurity Act 1993 (NZ) A New Zealand law which provides a legal basis for excluding, eradicating and

Fig. B4 Biophilia. People go to zoos to reconnect with nature.

effectively managing pests and unwanted organisms in New Zealand.

biosphere That part of the Earth's surface, seas and atmosphere where living things exist.

biosphere reserve Biosphere reserves are sites established by countries and recognised under the **UNITED NATIONS EDUCATIONAL, SCIENTIFIC AND CULTURAL ORGANIZATION (UNESCO)**'s Man and the Biosphere (MAB) Programme to promote sustainable development based on local community efforts and sound science.

biotechnology The use of living organisms in the large-scale manufacture of useful products or in the development of useful processes. Focuses on enzyme technology and gene manipulation. *See also* **CLONING, CARTAGENA PROTOCOL ON BIOSAFETY 2000**

Biotechnology and Biological Sciences Research Council (BBSRC) A government-funded research council in the UK which supports the research of biological scientists and research students in universities and research institutes.

biotic environment *See* **BIOLOGICAL ENVIRONMENT**

bipedal Relating to the use of the two rear limbs for locomotion; walking upright, e.g. as in humans, ostriches. Some species are normally **QUADRUPEDAL** but may walk or run bipedally for short periods, e.g. chimpanzees, lemurs, some lizards.

biradial symmetry Radial symmetry with some paired structures on opposite sides of the body, e.g. as seen in comb jellies (Ctenophora).

biramous Having two branches, e.g. the appendages of crustaceans.

bird *See* **AVES**

bird box A nesting box located on a tree, building, pole or other structure to provide a nesting place for birds in the wild or in captivity. Bird boxes may be designed specifically for a number of different types of birds to provide for their particular requirements, e.g. barn owls, house sparrows, tits etc.

bird dog *See* **GUN DOG**

bird feeder A device for providing wild or captive birds with food (Fig. B5).

bird flu *See* **AVIAN INFLUENZA**

bird observatory A place where the passage of migratory birds is recorded. Birds are often caught in **HELIGOLAND TRAPS** and ringed so that their future movements may be traced. Often located on major **FLYWAYS**. *See also* **DECOY (2)**

bird of prey, raptor A predatory, carnivorous bird, many species of which hunt on the wing: eagles, ospreys, kites, hawks, harriers, buzzards, vultures, falcons, owls, etc. (Fig. O1).

bird of prey centre A place where **BIRDS OF PREY** are kept for exhibition, breeding and display to the public. A centre usually provides flying displays and may offer bird handling sessions to the public.

bird ringing The process of attaching a **RING** to a bird, usually a **NESTLING**, for identification pur-

Fig. B5 Bird feeder.

poses. In the UK bird ringers must be appropriately trained and licensed.

bird scarer A device for deterring birds from agricultural crops, buildings (especially roofs), zoo enclosures etc. Includes silhouettes of birds of prey fixed to windows and glass doors, mock birds of prey (Fig. B6), balloons with 'predator eyes', and rotating **SOLAR BIRD SCARERS**. *See also* **BIRD SPIKES**

bird spikes Strips of plastic or metal spikes which are fixed to roofs, lamp posts and other structures, to prevent birds from perching and roosting. *See also* **BIRD SCARER**

bird strike A collision between a bird and a vehicle, usually an aircraft in flight. Generally results in the death of the bird and damage to the aircraft. A hazard when undertaking low level **AERIAL SURVEYS** in light aircraft. Wildlife legislation generally allows for the control of birds (including otherwise protected species) for the purposes of air safety (e.g. **WILDLIFE AND COUNTRYSIDE ACT 1981**).

bird–window collision An event involving a bird flying into a window usually resulting in injury and often death. This may occur because the bird perceives the window as a gap in a wall or building or it can see a reflection of the surrounding landscape. It may be a problem in zoos with many large windows built into the exhibit designs. Wild birds may be killed trying to fly through them. The problem may be alleviated by sticking silhouettes of

Fig. B6 Bird scarer.

birds of prey on the glass. *See also* **BIRD SCARER**, **BIRD STRIKE**

birder A bird watcher. A person who searches for and observes wild birds, usually as a hobby. Many amateur bird watchers are extremely knowledgeable and expert at identifying birds visually and from their songs. Some keep detailed records of their observations and make significant contributions to the biological records of the areas where they are active. *Compare* **TWITCHER**

birding Engaging in the activities of a **BIRDER**. *Compare* **TWITCHING**

BirdLife International A global partnership of non-governmental organisations in over 100 countries and territories which works to conserve birds and their habitats. Each partner represents a unique geographical area including the **ROYAL SOCIETY FOR THE PROTECTION OF BIRDS (RSPB)** (in the UK), the Nature and Biodiversity Conservation Union (in Germany), the **NATIONAL AUDUBON SOCIETY** (in the USA), the Bombay Natural History Society (in India) and **BIRDS AUSTRALIA**.

Birds Australia Formerly the Royal Australasian Ornithologists' Union. An organisation which was founded in 1901 to promote the study and conservation of Australian native bird species and those of adjacent regions. It is Australia's oldest and largest non-governmental national **BIRDING** association and publishes the journal *Emu*.

Birds Directive *See* **WILD BIRDS DIRECTIVES**

birdsong therapy The use of recordings of birdsong as an aid to recovery in human patients.

birth control *See* **CONTRACEPTION, POPULATION CONTROL**

birth rate, natality, natality rate The rate at which individuals are added to a **POPULATION** by births, hatching of eggs etc. during a specific period of time, usually a year. Generally corrected to births per thousand in a **LIFE TABLE**.

black panther A melanic leopard (*Panthera pardus*). *See also* **MELANISM**

black-footed ferret (*Mustela nigripes*) A rare North American species of **MUSTELID** that feeds on prairie dogs (*Cynomys*). As a result of the dramatic decline in prairie dog habitat the ferret population fell until there were just 18 animals left in 1985. The **UNITED STATES FISH AND WILDLIFE SERVICE (USFWS)** has established a breeding **COLONY** of black-footed ferrets at its National Black-footed Ferret Conservation Center near Laramie, Wyoming, as part of the **RECOVERY PLAN** for this species. In addition populations have been established at Phoenix Zoo, Louisville Zoological Garden, Henry Doorly Zoo, Smithsonian National Zoological Park's Conservation Research Center, the Cheyenne Mountain Zoological Park and Toronto Zoo, Canada. By 2010 over 6500 **KITS (1)** had been born in captivity, most of which have been released into the wild.

bladder

1. A muscular bag located in the lower abdomen which collects urine produced by the kidneys (via two **URETERS**) before it is released to the outside environment through the **URETHRA**. *See also* **GALL BLADDER**

2. A bag which stores air, e.g. **SWIM BLADDER**.

blastula (blastulae *pl.*) An early stage in the development of an embryo, consisting of a mass of cells (blastomeres) generally enclosing a cavity (blastocoel).

bleeding The loss of blood from the circulatory system, as a result of injury or disease which causes rupture of the blood vessels. Internal bleeding is the loss of blood from internal organs. *See also* **BLOOD CLOTTING**

blood A fluid which is pumped around the body of an animal, through a **CIRCULATORY SYSTEM**, carrying oxygen, nutrients, waste materials and carbon dioxide. Necessary for communication between the body **CELLS** and the external environment in large animals where the distance between the external environment and the body cells is too great to rely on **DIFFUSION** for **GASEOUS EXCHANGE** and the movement of **NUTRIENTS**. The composition of blood varies between species but generally contains a pigment for carrying oxygen, cells that defend the body by producing antibodies (lymphocytes) and cells that are able to move through tissues to destroy invading organisms such as bacteria. A mechanism for **BLOOD CLOTTING** may also be present. *See also* **BLOOD GROUP, HAEMOGLOBIN**

blood cells *See* **ERYTHROCYTES, LEUCOCYTES**

blood clotting, coagulation, haemostasis, hemostasis The process by which the blood forms a solid mass or clot when exposed to the air in order to prevent blood loss and infection. In vertebrates a clot is formed from a mesh of fibrin fibres which trap blood cells. Fibrin is produced when the enzyme thrombin acts upon the protein fibrinogen in the blood **PLASMA** as a result of the rupture of blood **PLATELETS**.

blood doping The illegal practice of temporarily increasing the oxygen-carrying capacity of an animal's blood (especially a racehorse) by injecting red blood cells that have previously been withdrawn. *See also* **BLOOD DOPING AGENT**

blood doping agent A drug which increases the production of red blood cells. *See also* **BLOOD DOPING**

blood group, blood type Any of a number of types into which blood may be classified (e.g. the ABO system in humans and **GREAT APES**). A system of antigens located on red blood cells which are partly responsible for determining the compatibility between the blood of a donor and that of a recipient. Many different blood group systems exist. Incompatibility between blood groups which may occur during a **BLOOD TRANSFUSION** may cause extreme reactions in the recipient including **HAEMOLYSIS, AGGLUTINATION**, blockage of **CAPILLARIES**, tissue damage and possibly death.

blood plasma *See* **PLASMA**

blood poisoning *See* **SEPTICAEMIA**

blood pressure (BP) The pressure exerted by the circulating **BLOOD** on the **BLOOD VESSELS**. Usually refers to the arterial pressure in the **SYSTEMIC CIRCULATION**.

blood serum *See* **SERUM**

blood sport A sport that involves bloodshed or the killing of animals, e.g. **BADGER-BAITING, BEAR-BAITING, BULLFIGHT, DEER COURSING, FOX HUNTING, HARE COURSING**

blood sugar level The quantity of glucose circulating in the blood. Normally kept within narrow limits by **HORMONES**, especially **INSULIN**.

blood tests A test in which a small amount of blood is taken to determine its **BLOOD GROUP**, the presence of disease, drugs or some other substance.

blood transfusion The introduction of a quantity of blood from one individual into another to replace blood lost by injury, surgery, burns etc. Must be of a compatible **BLOOD GROUP**. *See also* **BLOOD DOPING**

blood type *See* **BLOOD GROUP**

blood typing The process by which the **BLOOD GROUPS** present in a blood sample are established.

blood vessels The collective name for tubes which carry blood around the body: **ARTERIES, VEINS, ARTERIOLES, VENULES** and **CAPILLARIES**.

bloodline All of the individuals in a genetically related group of animals over a number of generations, especially when considered in terms of some inherited characteristic or **PEDIGREE** etc. *See also* **STUDBOOK**

bloods Colloquial term for samples of blood taken from an animal for testing. *See also* **BLOOD TESTS**

bloodsicle A block of ice containing blood. Used as a **FEEDING ENRICHMENT** for some large **CARNIVORES**.

bloodstock Horses that have been especially bred for racing; pedigree horses.

blowfly Any of a number of fly species which lay their eggs in flesh or excrement on which the larvae feed. *See also* **FLY STRIKE**

blubber A subcutaneous layer of fat which acts as an insulator and an energy store in marine mammals (**CETACEANS, PINNIPEDS** and sirenians) and polar bears. Blubber thickness is an indicator of physical condition. *See also* **ABDOMINAL SKINFOLD**

Blue Cross A charity in Great Britain which provides inexpensive veterinary services, operates animal adoption centres and a pet bereavement support service, and promotes good pet care. *See also* **SOCIETY FOR COMPANION ANIMAL STUDIES**

Blue Planet A modern aquarium complex at Ellesmere Port, Cheshire. Based around a large simulated coral reef tank.

blue-green algae Cyanobacteria. Some forms may cause toxic **ALGAL BLOOMS** in freshwater. Drinking from contaminated water may cause death in a wide range of animals. *See* **ALGAL BLOOM, EUTROPHICATION**

bluetongue A viral disease spread by midges which affects all ruminants, including sheep, cattle, deer, goats and camelids. Sheep are usually most severely affected.

B-lymphocyte A type of **LEUCOCYTE** which is capable of binding to a specific **ANTIGEN** and functions as part of the **IMMUNE SYSTEM**.

BMI *See* **BODY MASS INDEX (BMI)**

boardwalk A path made of wooden boards usually slightly raised above the general ground level. Often used to cross wet or uneven ground and used to create interesting **VIEWPOINTS** in zoos. High-level boardwalks may be used to allow visitors to look over enclosure barriers.

BOD *See* **BIOLOGICAL OXYGEN DEMAND**

body condition score A numerical value which represents the physical condition of an animal's body (Table B2). It is species specific, independent of weight and should take into account age, time of year and individual differences (e.g. natural variation in size). Methods vary between species but all measure muscle and fat reserves by scoring different parts of the body. In **LIVESTOCK** it can be a useful tool in managing nutrition. Sudden changes in body condition score may indicate a health problem. *See*

Table B2 Body condition score: a simple body condition scoring system.

Score x/5	Appearance
1.0	Emaciated
1.5	Thin
2.0	Underweight
2.5	Mildly underweight
3.0	Healthy weight
3.5	Mildly overweight
4.0	Overweight
4.5	Obese
5.0	Morbidly obese

also ABDOMINAL SKINFOLD, ANIMAL WELFARE INDEX, BODY MASS INDEX (BMI), OBESITY

body mass index (BMI) An index of obesity in humans calculated by dividing the individual's weight in kilograms by the square of its height in metres. Has also been used for chimpanzees. *See also* ABDOMINAL SKINFOLD, BODY CONDITION SCORE, OBESITY

body temperature The internal temperature of an animal's body. In HOMEOTHERMS this normally remains relatively constant. A raised body temperature may indicate the presence of an INFECTION.

Bohr effect An effect which describes the reduced affinity of HAEMOGLOBIN for oxygen in the presence of carbon dioxide.

Bois de Vincennes, Paris Zoo, Vincennes Zoo One of two zoos in Paris. Originally a royal hunting park, now a public park containing a zoo. A temporary zoo was constructed in 1931 for the Colonial Exhibition. A permanent zoo was created in 1934. It is the largest public zoo in France (approximately 15 ha). *See also* JARDIN DES PLANTES

bolus (boluses *pl.*)
1. A ball of material, usually food after chewing, or faeces, e.g. a bolus of elephant dung.
2. A dose of a substance (especially a large dose) given by injection.

bond *See* PAIR BOND

bond group, kin group, kinship group A level of social organisation in some species consisting of a group of related families, e.g. in African elephants (*Loxodonta africana*) (Figs. S12 and S13).

bone The tissue which makes up the skeleton of vertebrates. It contains the protein COLLAGEN, is made up of about 70% inorganic calcium salts, and has its own blood supply. Cancellous (spongy) bone is light and forms the vertebrae, skull and the ends of LONG BONES. Compact bone is a hard, strong material found on the outside of bony structures providing strength, e.g. the shafts of the long bones.

bone marrow, marrow The spongy tissue found inside some bones (e.g. the femur), which contains immature STEM CELLS which can develop into the red blood cells, white blood cells and PLATELETS.

bone rarefaction A reduction in bone density. This occurs in OSTEOPOROSIS and also as a result of insufficient exposure to ULTRAVIOLET LIGHT in some species.

Bonn Convention Colloquial name for the CONVENTION ON THE CONSERVATION OF MIGRATORY SPECIES OF WILD ANIMALS 1979 (CMS).

bonobo (*Pan paniscus*) A pygmy chimpanzee. *See also* KANZI

bony fishes Fishes which are classified as OSTEICHTHYES.

book lung The respiratory organ of spiders in which the respiratory membranes are arranged like the pages of a book.

boomer ball A hard polyethylene ball used as an ENVIRONMENTAL ENRICHMENT device for some large mammals, e.g. primates, large felids.

boreal forest, taiga A coniferous forest BIOME that rings the northern hemisphere largely north of the 50th parallel. It is home to bears, lynx, wolves, herds of caribou and many species of songbirds.

boredom The state of being tired and uninterested as a result of being unoccupied or under-occupied. A condition normally used to describe humans rather than animals to avoid ANTHROPOMORPHISM. Increasingly also applied to animals exposed to repetitive stimulation or monotonous conditions, e.g. some captive animals held in poor conditions in circuses, some zoos and intensive farms. Considered to be an aversive state that an animal will try to avoid. Often associated with STRESS and STEREOTYPIC BEHAVIOUR.

Born Free A book and film (released in 1966) of a true story about the rearing and eventual release into the wild of an orphaned lioness cub called *Elsa* in Kenya by George and Joy Adamson. The film starred Bill Travers and Virginia McKenna. They were profoundly affected by the fate of the lioness that played *Elsa* in the film when she was sent to live in a zoo, and later founded the BORN FREE FOUNDATION and ZOO CHECK.

Born Free Foundation A charity established by the actors Bill Travers and Virginia McKenna after making the film *BORN FREE*. It takes action worldwide to stop animal suffering and protect threatened species, by rescuing individual animals and working with local communities. The foundation campaigns to change public attitudes and works to phase out zoos. *See also* ZOO CHECK

Bostock and Wombwell's Royal Menagerie A large TRAVELLING MENAGERIE which toured widely in Britain and abroad from 1805 until 1932. Created from the amalgamation of George Wombwell's Travelling Menagerie and a MENAGERIE owned by Edward Bostock. It housed a very wide range of animals including elephants, camels, lions and tigers. Eventually it was so successful that it grew to

B

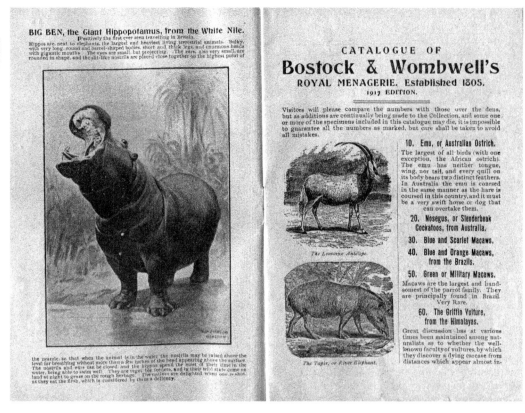

Fig. B7 Bostock and Wombwell's Royal Menagerie: pages from the 1917 catalogue. (Courtesy of Chetham's Library, Manchester.)

three menageries which toured independently and gave several royal command performances before Queen Victoria (*see* frontispiece). The sale of the menagerie to **LONDON ZOO** in 1932 probably marked the end of the large self-sufficient travelling menageries, although small menageries continued to tour with fairs (Fig. B7).

bot Botfly (Oestridae) larva. A parasite, especially of horses and cattle.

bot knife A tool for removing bot eggs.

botany The scientific study of plants.

bottle jaw An accumulation of fluid (**OEDEMA**) under the lower jaw associated primarily with an infestation of the parasitic **NEMATODE** *Haemonchus contortus* in sheep, but also with **FLUKE** (1) infestations in sheep and cattle.

bottle-feeding The provision of liquid food (often milk) to a mammal via a bottle. May be essential for the survival of animals that are rejected by the mother at birth. In some cases, a **MILK SUBSTITUTE** may be used.

botulism A severe form of food poisoning caused by ingesting toxins produced by the bacterium *Clostridium botulinum.*

BOU *See* **BRITISH ORNITHOLOGISTS' UNION (BOU)**

bout criterion interval (BCI) The minimum interval of time that separates two successive bouts of behaviour. Used to define when one bout ends and the next begins.

bout length The period of time between the beginning and end of an episode of a particular behaviour.

bovid A member of the **BOVIDAE**.

Bovidae A mammalian family consisting of cattle, buffalo, bison, antelopes, gazelles, goats, sheep and their relatives.

bovine Relating to **BOVIDS**, especially cattle.

bovine spongiform encephalopathy (BSE) BSE is caused by a **PRION** (a self-replicating protein) which causes spaces to develop in the brain tissue. The disease was first recognised in 1986. It has been called 'mad cow disease' as it affects the nervous system in cattle causing them to become hypersensitive to noise, frightened and aggressive. The head is lowered and they exhibit an abnormal gait with hindlimb swaying. At rest, muscle twitching can be seen. BSE has been transmitted to cattle because BSE-infected cow products were used to make meat and bone meal. Meat and bone meal is now banned from all animal feeds. BSE has been recorded in several antelope species in zoos. There may be a link with feline spongiform encephalopathy as some big cats in zoos developed this after eating bovine heads before BSE was identified. The disease has been transmitted to monkeys experimentally.

bovine TB *See* TUBERCULOSIS

bovine tuberculosis *See* TUBERCULOSIS

bovine viral diarrhoea (BVD) A viral infection of cattle which may cause abortion, infertility, and immunosuppression that underlies calf respiratory and enteric diseases and fatal mucosal disease.

Bowman's capsule In the kidney, the spherical structure at the end of each NEPHRON which contains the GLOMERULUS. Filtration of the blood and the formation of urine begins here.

BP *See* BLOOD PRESSURE (BP)

brachial Relating to the arm, particularly the upper arm, e.g. brachial artery.

brachial glands Glands located on the arms. They occur in primates and are used for OLFACTORY COMMUNICATION, especially for maintaining DOMINANCE HIERARCHIES and demarcating TERRITORY.

brachial heart An accessory heart found in some cephalopod MOLLUSCS which pumps deoxygenated blood from the body tissues to the gills.

brachiation A form of locomotion which involves swinging by the arms, hand over hand, from one object (e.g. a branch or cage bar) to another as in gibbons and some other ARBOREAL primates. Animals that brachiate possess long arms, long fingers and freely rotating shoulder joints.

brackish Containing a mixture of seawater and fresh water, e.g. the water found in salt marshes.

bradycardia A heart rate which is below the RESTING HEART RATE. *Compare* TACHYCARDIA

bradypnoea A lower than normal RESPIRATORY RATE. *Compare* EUPNOEA, TACHYPNOEA

brain The anterior section of the nervous system in VERTEBRATES, associated with the major sense organs, and made up of the CEREBRUM, CEREBELLUM and BRAINSTEM. In invertebrates the 'brain' is formed from the cerebral ganglia.

brainstem The part of the brain that remains after removal of the CEREBRUM and the CEREBELLUM; the vertebrate midbrain, pons and MEDULLA OBLONGATA. Location of the respiratory, vasomotor and cardiac centres.

brak A South African term for a mongrel dog.

Brambell Report The report of an investigation of farm animal welfare entitled *Report of the Technical Committee to Enquire into the Welfare of Animals kept under Intensive Livestock Husbandry Conditions* published in 1965 in the UK which led to the establishment of the FIVE FREEDOMS. The committee was chaired by F. W. Rodgers Brambell.

branchial Relating to GILLS.

branding The application of an identification mark to an animal using a heated metal instrument (branding iron) or a super-cold iron (freeze branding). *See also* FREEZE BRAND, MARK

branding cradle *See* CALF CRADLE

breathing Ventilation of the respiratory organs. In vertebrates this involves expansion of the thoracic cavity (caused by raising the rib cage and lowering the DIAPHRAGM) which results in a decrease in pressure inside the LUNGS, causing inhalation. This is followed by a reduction in the volume of the THORAX (1) which results in exhalation.

breed
1. A variety of animal, particularly LIVESTOCK or other DOMESTICATED SPECIES (e.g. cattle, sheep, cats, dogs).
2. To produce more individuals by allowing or encouraging reproduction, either for commercial or conservation purposes. *See also* CAPTIVE BREEDING PROGRAMMES

breed and cull A management strategy in zoos whereby animals are allowed to breed and the excess are then selectively culled to maintain a stable population size. In some zoos animals are allowed to breed so that they produce young as an enrichment activity for the adults and because regular births attract visitors.

breed club *See* BREED SOCIETY

breed society, breed club An organisation which works to advance the HUSBANDRY, breeding etc. of particular breeds of farm or domestic animals, maintains a register of the animals belonging to a recognised breed, and publishes standards for the physical appearance of the breed, e.g. the Hereford Cattle Society, the Lleyn Sheep Society, British Canaan Dog Society, Burmese Cat Society.

breed standard A set of guidelines which specifies the observable characteristics of a BREED of animal – for example appearance, movement and temperament – which are used to define the breed. Used by judges, exhibitors and animal breeders. The breed standard for a breed of dog would define the general appearance of the breed, its characteristics, temperament, the form of the head and skull, eyes, ears, mouth, neck, forequarters, body, hindquarters, feet, tail, GAIT and movement, the nature and colour of the coat, and the characteristic size (weight) of dogs and bitches of the breed.

Breeding and Sale of Dogs (Welfare) Act 1999 An Act in Great Britain which regulates the commercial breeding and sale of dogs; regulates the welfare of dogs kept in commercial breeding establishments; and requires the keeping of records of dogs kept at such establishments.

Breeding Centre for Endangered Arabian Wildlife (BCEAW) A centre in the United Arab Emirates which is responsible for the captive breeding of, and research on, indigenous Arabian fauna. It includes a FROZEN ZOO.

breeding cycle A cycle of change in the behaviour and reproductive physiology of an animal which may be seasonal and synchronised with environmental changes in a CIRCANNUAL RHYTHM. The cycle of seasonal breeders is generally linked to

B

seasonal changes in climate and day length to ensure that reproduction occurs when food is abundant and the environmental conditions are favourable. *See also* OESTROUS CYCLE

breeding loan An arrangement between two zoos (or other organisations) to transfer temporarily, and at no cost to the receiving zoo, an animal from one to the other for the purpose of breeding, usually as part of a CAPTIVE BREEDING PROGRAMME. *See also* GIANT PANDA LOANS

breeding records *See* PEDIGREE, STUDBOOK

breeding season The time of the year when an animal normally reproduces. Some species have no particular breeding season and may reproduce at any time of the year. *See also* MONOESTROUS, POLYOESTROUS

BREEDPLAN A genetic evaluation system for LIVESTOCK, especially CATTLE. It produces ESTIMATED BREEDING VALUES (EBVs) for a range of traits.

Breeds at Risk Register (BARR) A list of animals which belong to rare BREEDS of CATTLE, SHEEP, GOATS and PIGS kept in Great Britain which would be considered for exemption from any culling requirements in the event of an outbreak of FOOT AND MOUTH DISEASE (FMD), CLASSICAL SWINE FEVER (CSF), African swine fever or other exotic disease. Keepers of such animals must register them with the DEPARTMENT FOR ENVIRONMENT, FOOD AND RURAL AFFAIRS (DEFRA), which maintains a list of breeds at risk. Exemption is not guaranteed and would only be made if disease control was not compromised. Similar registers exist in other EUROPEAN UNION (EU) countries as EU disease control legislation allows for exemptions for rare breeds kept for conservation purposes. *See also* RARE BREEDS SURVIVAL TRUST (RBST)

breed-specific legislation In relation to dogs, laws which restrict the breeding and possession of certain breeds, e.g. many state laws in the USA, and the DANGEROUS DOGS ACT 1991 in Great Britain. Such laws are controversial because of the difficulty in accurately identifying some breeds and the lack of conclusive evidence that some dog breeds are more dangerous than others.

bridge In animal TRAINING, a device (often a sound) used to indicate the precise time that a desired BEHAVIOUR occurs prior to a reward being given. A bridge is necessary because it is often difficult to reward the animal quickly enough for it to form a direct association between the desired behaviour and the reward itself. *See also* CLICKER TRAINING

bristle
1. A type of feather which has BARBS only at the base of the shaft.
2. A short, hair-like structure.

British Alpaca Society (BAS) A UK organisation dedicated to the welfare of alpacas (*Vicugna pacos*) and the education of their owners. It provides information, and support and organises events for alpaca owners, breeders and those interested in alpacas. It maintains the PEDIGREE REGISTER for the NATIONAL HERD.

British and Irish Association of Zoos and Aquariums (BIAZA) An association for zoos and aquariums in the UK and Ireland; formerly the Federation of Zoological Gardens of Great Britain and Ireland. BIAZA's offices are in Regent's Park, London. BIAZA leads and supports its members in their conservation, education and research initiatives. It works closely with the EUROPEAN ASSOCIATION OF ZOOS AND AQUARIA (EAZA) in its conservation campaigns.

British Antarctic Survey (BAS) A unit of the NATURAL ENVIRONMENT RESEARCH COUNCIL (NERC) in the UK which conducts research on the wildlife and the physical environment of the Antarctic.

British Association for Shooting and Conservation (BASC) An organisation whose mission is to promote and protect sporting shooting and the well-being of the countryside throughout the UK and overseas. It was founded in 1908 as the Wildfowlers' Association of Great Britain and Ireland (WAGBI).

British Cattle Movement Service (BCMS) The organisation in the UK which is responsible for maintaining a database of all BOVID animals (cattle, bison and water buffalo etc.). This is achieved by issuing EAR TAG identification numbers and CATTLE PASSPORTS which remain with each animal from birth until death, thereby allowing the source of meat to be traced. The Service is part of the RURAL PAYMENTS AGENCY.

British Equine Veterinary Society An organisation based in the UK which champions high standards of equine health and welfare and the promotion of scientific excellence and education throughout the world.

British Goat Society An organisation in Britain which promotes the keeping of goats and registers pedigree stock. *See also* GOAT BREEDS

British Hen Welfare Trust A charity that re-homes commercial laying hens, educates the public about hen welfare and encourages support for the British egg industry. Formerly called the Battery Hen Welfare Trust.

British Horseracing Authority (BHA) The governing and regulatory body for horseracing in Britain. It supervises race programmes, licenses and registers participants (jockeys, trainers, horses, owners and stable staff), enforces discipline, conducts research in equine science and welfare, engages in training and education, and sets and enforces the RULES OF RACING. *See also* WHIPPING REGULATION OF

British Ornithologists' Union (BOU) One of the oldest ornithological organisations in the world (founded in 1858). It aims to promote ornithology within the scientific and birdwatching communities by encouraging the study of birds throughout the world with the object of understanding their biology and assisting their conservation. The BOU publishes the journal *Ibis*.

British Pig Association (BPA) An association founded in 1884 which is the official breed society and maintains the **HERD BOOKS** for a number of important pig breeds. The BPA represents the interests of pedigree pig breeders to the government.

British Poultry Council The trade association for producers of poultry meat from **CHICKENS**, **TURKEYS**, **DUCKS** and **GEESE** in Britain.

British Rabbit Council (BRC) The BRC promotes the breeding and exhibition of **RABBITS** in Britain; coordinates and represents the interests of rabbit breeders; and promotes education about and research into rabbit welfare. The BRC produces **BREED STANDARDS** for rabbits and operates a ringing scheme for identifying individual rabbits.

British Small Animal Veterinary Association (BSAVA) An association founded in 1957 as a professional body to serve veterinary surgeons in the UK who treat companion animals and to promote excellence in small animal practice through education and science.

British Trust for Ornithology (BTO) An independent charitable research institute which monitors wildlife populations, especially birds, in the UK. It uses professional scientists and volunteers to undertake long-term censuses, the data from which may be used by policy-makers, the government and others to inform decision-making. Some of the BTO's data sets are for periods in excess of 50 years.

British Union for the Abolition of Vivisection (BUAV) A UK-based organisation which campaigns to abolish experiments on animals. It has an international network of lawyers, campaigners, investigators, political lobbyists, scientists, researchers and supporters.

British Veterinary Association (BVA) The national representative body for the veterinary profession in Britain. Publishes the *VETERINARY RECORD*.

broad-spectrum antibiotic A drug designed to kill a wide range of different types of bacteria. Often administered to an animal when the precise cause of an infection is unclear. *See also* **ANTIBIOTICS**

broiler A **CHICKEN** raised for meat production.

bronchioles Fine branching tubes extending from the **BRONCHI** to the **ALVEOLI** in the **LUNGS**.

bronchodilator A drug which dilates the **BRONCHI-OLES** as an aid to breathing.

bronchopneumonia An inflammation of the **LUNGS** and the **BRONCHI** usually caused by the inhalation of bacteria. May be fatal.

bronchoscope A device used for examining the interior of the lungs and for taking tissue for **BIOPSY**.

bronchus (bronchi *pl.***)** One of two tubes which connect the **TRACHEA** to the **LUNGS** via the **BRONCHIOLES** and form part of the airway.

Bronx Zoo *See* **WILDLIFE CONSERVATION SOCIETY (WCS)**

brood
1. The eggs, larvae or young of an organism.
2. To guard or warm **EGGS (2)**, larvae or young.

brood parasitism The laying of eggs in the nest of another bird so that they will be incubated and the resulting chicks reared by that bird. **INTERSPECIFIC** brood parasites, e.g. cuckoo (*Cuculus canorus*), lay eggs in the nests of other species. **CONSPECIFIC** brood parasites, e.g. American coot (*Fulica americana*), lay eggs in the nest of other individuals of their own species.

brood patch A highly vascular area of bare skin on the ventral surface of some birds which is used to keep eggs warm during incubation. *See also* **INCUBATE (1)**

brooder, brooder room A building, container or other place or device used for raising young animals, especially birds. *See also* **INTENSIVE CARE BROODER**

broodmare
1. A **MARE** used for breeding.
2. According to the **AMERICAN STUD BOOK** (2008) '*A filly or mare that has been bred (mated) and is used to produce foals.*'

Broom, Donald (1942–) A zoologist who, in 1986, was appointed the first Professor of Animal Welfare in the world in the Department of Veterinary Medicine, University of Cambridge. Here he established the Centre for Animal Welfare and **ANTHROZOOLOGY** whose members work on methods for the scientific assessment of animal welfare and the management, transportation and housing of farm animals. The welfare of pets, zoo and laboratory animals have also been studied. Broom has been influential in the drafting of many animal welfare laws and standards and in the development of **ANIMAL WELFARE** as a scientific subject in many countries.

brown adipose tissue *See* **ADIPOSE TISSUE (1)**

Brown Dog Affair An incident in which the fate of a terrier at the hands of vivisectionists was exposed in a court case, *BAYLISS V. COLERIDGE* (1903). In 1906 a statue to commemorate the dog was erected in Battersea in London. It was removed shortly after but replaced by a second statue erected in Battersea Park in 1985.

brown fat *See* **ADIPOSE TISSUE (1)**

Brown, Lancelot 'Capability' (1716–1783) An 18th century English landscape architect who widely used a **HA-HA** in his garden designs as a means of providing an 'invisible' barrier between the garden itself and the surrounding countryside. Zoos now often use this device as a containment barrier.

browse
1. Branches and leaves. Provides important **NUTRIENTS** for **BROWSERS** and also acts as an **ENVIRONMENTAL ENRICHMENT**, especially if located

at the level at which it would normally occur in the wild, e.g. suspended at height for giraffes.

2. To eat branches and leaves etc. *See also* **BROWSER**

browser An animal that feeds by browsing, e.g. impala, giraffe, black rhinoceros. *See also* **BROWSE**

brucellosis Brucellosis is caused by infection with bacteria from the genus *Brucella*. Signs often take the form of an undulating fluctuation of temperature ('undulant fever'). It may also cause abortion, arthritis, infertility and a range of other signs. Brucellosis may affect a wide range of animals including goats, sheep, cattle, horses, dogs, foxes, deer, poultry, cetaceans and humans. The harbour porpoise (*Phocoena phocoena*) may carry *B. maris*; hares *B. suis*; and *B. abortus* has been found in deer, foxes, waterbuck and rodents. Care must be taken in handling and disposing of aborted foetuses, foetal membranes and discharges.

brush border A group of **MICROVILLI** on the surface of a **CELL MEMBRANE** which increases its surface area.

BSE *See* **BOVINE SPONGIFORM ENCEPHALOPATHY (BSE)**

BTO *See* **BRITISH TRUST FOR ORNITHOLOGY (BTO)**

bubblenest, foam nest A mass of bubbles made by some fishes and amphibians which is used for protecting eggs.

buccal Relating to the mouth or inside of the cheek.

buccal cavity Mouth cavity.

buccal pumping The ventilation of the **GILLS** in fishes by contracting the **BUCCAL CAVITY** and the opercular cavity.

buck The male of some species of mammals, e.g. **DEER**, **ANTELOPE**, hare.

budding Asexual reproduction in which young are produced by the formation of a small growth (bud) on or in the parent.

Buddhism and animals Buddhism teaches that humans and other animals are closely related and should be treated with equal respect. It forbids cruel acts towards animals and its followers believe that humans may be reborn in the body of non-human animals, sometimes as a result of past misdeeds.

budget An itemised summary of intended or actual expenditures on different things over a period of time. The expenditure may be financial or it may relate to time or energy. A budget may be used as a means of measuring and monitoring performance in relation to the activity to which the budget relates. *See also* **ACTIVITY BUDGET, ENERGY BUDGET**

buffalo One of a number of different species of **BOVIDS** from Africa and Asia. Often wrongly used as a name for bison (*Bison bison*) in North America.

Buffalo House The first building constructed at the **SMITHSONIAN NATIONAL ZOOLOGICAL PARK** (in 1891), to house American bison (*Bison bison*).

Buffalo Protection Act of 1877 (Canada) An Act passed in Canada to protect bison. It was largely ineffective due to a lack of enforcement.

Buffalo Protection Act of 1894 (USA) An Act passed in the USA to protect the small number of bison that remained in **YELLOWSTONE NATIONAL PARK** at that time. It was some of the first federal legislation passed to protect an important wildlife resource.

buffer A molecule or group of molecules which resists changes in the pH of a solution.

bulbourethral gland, Cowper's gland A gland within the reproductive system of males that produces a clear, **MUCOUS** fluid used as a lubricant during copulation.

bulk grazer A large grazing animal which is adapted to feeding on large quantities of long grass and which takes large mouthfuls at a time, in a more-or-less indiscriminate manner (i.e. is non-selective and cannot graze close to the ground). Bulk grazers are mainly **BOVIDS** and **CERVIDS**, but also include zebras. *Compare* **SELECTIVE GRAZER**

bull

1. The male of many species, e.g. seals, elephants, **BOVIDS**. May have a specific legal meaning.

2. In England and Wales, under s15 of the **PROTECTION OF ANIMALS ACT 1911** a 'bull' is defined as including '*any cow, bullock, heifer, calf, steer or ox*'.

bull pen An **ANIMAL ENCLOSURE** used to house bull animals, especially **ELEPHANTS**, separately from females. Particularly useful for separating bulls from the rest of the herd when they are in **MUSTH**.

bull-baiting

A blood sport which involves the use of dogs to attack a tethered bull. Banned in the UK by the **CRUELTY TO ANIMALS ACT 1835**. *See also* **BEAR-BAITING, RAT-BAITING**

bullfight

1. A traditional **BLOOD SPORT** provided as entertainment in Spain, Portugal, southern France and some Latin American countries, in which a bull bred for the purpose is taunted in a bullring by men on foot or horseback and eventually killed. Catalonia banned bullfighting in September 2011, the first region to do so in Spain.

2. In ancient Egypt, a sport involving fighting between two bulls.

bullhook *See* **ANKUS**

bulling

1. Referring to a cow in **OESTRUS** (heat).

2. A behaviour seen in cattle when one mounts another, usually when one or the other is a female in oestrus. Cows may mount bulls or other cows. Used by farmers to identify cows in oestrus; however, when one cow mounts another cow either or both could be in oestrus. The most intensive interactions are between cows which are both in oestrus.

Bulling Beacon® A device used to detect **OESTRUS** in cows, consisting of a capsule which is fixed to the hind end of a cow that releases a coloured dye in response to sustained pressure as a result of the cow being mounted.

bumble-foot A condition of the feet of birds in which an **ABSCESS** forms in the soft tissue between the toes. May be caused by the penetration of a sharp object and the resulting abscess may cause lameness. The abscess usually contains *Staphylococcus* but other microorganisms may be involved, including *Brucella abortus*. Treatment involves opening the abscess and evacuating the pus. Lack of **VITAMIN A** may make birds more susceptible to infection.

bunny-hugger A derogatory term used for someone who is concerned for animal welfare and/or wildlife conservation. *See also* **TREE-HUGGER**

buoyancy An upward force applied by a fluid in which a body (e.g. an organism) is immersed. *See also* **SWIM BLADDER**, **SWIM BLADDER DISEASE**

Bureau of Land Management (BLM) An agency of the **UNITED STATES DEPARTMENT OF THE INTERIOR (DOI)** whose mission is to sustain the health, diversity and productivity of America's public lands for the use and enjoyment of present and future generations. The Bureau protects, manages and controls wild horses and **BURROS** under the authority of the **WILD FREE-ROAMING HORSES AND BURROS ACT OF 1971 (USA)**.

burlap bag A bag made of sacking and filled with hay and food treats and then tied closed. Used as a **FEEDING ENRICHMENT** in zoos. The animal has to access the bag and sort through the hay to find the treats.

Burns Report A report by Lord Burns on the effects of hunting with dogs in the UK and on the communities whose economies depended on these activities. The report was instrumental in the passing of the **HUNTING ACT 2004** in England and Wales which banned hunting with dogs and effectively banned **FOX HUNTING** in its traditional form. *See also* **DRAG HUNT**

burro A term used in the USA for a small **DONKEY**. *See also* **WILD FREE-ROAMING HORSES AND BURROS ACT OF 1971 (USA)**

Burton, Decimus (1800–1881) An English architect and garden designer who designed the gardens and several of the original animal houses and other structures at **LONDON ZOO** including the **RAVEN'S CAGE** (1829) and the Giraffe House (1836–1837). Burton was the official architect of the Zoo from 1826 to 1841.

bushmeat Any wild animal hunted for food, medicine or traditional cultural uses; meat derived from wild animals. An important source of **PROTEIN** for some indigenous people, especially those living in tropical forests. Often taken illegally by **POACHERS**. Especially important in the decline of many **PRIMATE** species, including **GREAT APES**.

butterfly farm A place where butterflies are bred on a large scale for sale to hobbyists, zoos etc. Sometimes they are open to the public as visitor attractions. *See also* **CHRYSALARIUM**

butterfly needle A small needle used for the **INFUSION** of liquids or withdrawal of blood which has plastic butterfly wing-like extensions protruding from its sides which can be used as handles and may be taped to the patient's skin once the needle is inserted. *See also* **GIVING SET**

bycatch A capture of non-target species during fishing, e.g. dolphins. *See also* **FISH DISCARDS**

B

Fig. C1 C is for cattle. English Longhorn.

Dictionary of Zoo Biology and Animal Management: A guide to terminology used in zoo biology, animal welfare, wildlife conservation and livestock production, First Edition. Paul A. Rees.
© 2013 John Wiley & Sons, Ltd. Published 2013 by John Wiley & Sons, Ltd.

C *See* CYTOSINE (C)

caecotrophs *See* COPROPHAGY

caecum (caeca *pl.*), cecum (ceca *pl.*) A blind-ending chamber or pouch. Occurs at the beginning of the LARGE INTESTINE in vertebrates and in the midgut of insects.

caesarean section, cesarean section A surgical procedure in which a young animal is delivered through an incision in the ABDOMEN (1) of the mother.

cafeteria experiment A study in which an animal is presented with different foods or dietary constituents in separate containers in order to determine and measure the DIET (1) they select for themselves given a free choice. *See also* SELECTIVE FEEDING

cafeteria-style feeding (CSF) A system of feeding in which animals are free to choose from a range of foods. *See also* COMPLETE FEED-STYLE FEEDING

cage A container in which an animal is kept, often made of metal mesh. Frequently used for small companion animals such as hamsters, mice and rats, and for small and medium-sized aviary birds such as finches, cockatiels and parrots. Simple, barren cages made with iron bars were common in Victorian zoos but have now largely been replaced in modern zoos although they still exist in zoos in some developing countries. *See also* ANIMAL ENCLOSURE, CRUSH, RAVEN'S CAGE

cage layer fatigue A nutritional disease found in POULTRY and observed in caged birds, especially CHICKENS. Occurs in caged laying hens around the time of peak egg production and is thought to be caused by an imbalance of MINERALS (especially calcium) and ELECTROLYTES (2). It may be associated with RICKETS and OSTEOPOROSIS, causing brittle bones. Hens may be crippled and unable to stand.

calcareous Possessing a mineral structure based on CALCIUM, especially calcium carbonate, e.g. the calcareous skeleton in CORAL.

calcium An important element required by organisms. It is has a role in cell signalling, acts as a buffer and is involved in enzyme activities and many other metabolic functions. In combination with phosphorus it forms the mineral portion of bones and teeth. Calcium is needed in large quantities during growth and lactation. Bone demineralisation, tetany and death may occur if the food provided has a low calcium:phosphorus ratio, as skeletal stores are depleted.

calculus
1. An abnormal concretion of material in an animal body, normally made of mineral salts. Generally found in the KIDNEY, GALL BLADDER or urinary bladder.
2. Alternative name for TARTAR.
3. A branch of mathematics used in the study of POPULATION GROWTH.

calf and sheep staller A portable device for restraining calves or sheep consisting of a metal frame which is attached to a gate or other fixed structure and in which the animal's head is held. Removes the requirement for a second person to be present during operations such as tagging, DEHORNING, injecting, veterinary treatment etc.

calf cradle, branding cradle A metal cage-like device similar to a CRUSH which is designed to hold a calf securely during branding, ear-tagging, marking, inoculation or DEHORNING. The cradle is usually positioned at the end of a RACE and when the calf enters the cradle it is caught by pushing the side of the cradle inwards until it holds the animal firmly. The calf is initially held in the standing position and then the cradle is rotated so that the animal is held securely on its side (the 'table' position).

calf creep feeder Feeding device for calves containing a food hopper and feeding trays to which bars are fitted to exclude older animals.

California condor (*Gymnogyps californianus*) A rare condor of the west coast of North America. In 1982 only 23 birds remained in the wild. The species was saved when all the remaining wild birds were taken into captivity to establish a CAPTIVE BREEDING PROGRAMME. The RECOVERY PLAN for the condor involves captive breeding facilities at San Diego Wild Animal Park, Los Angeles Zoo, and The Peregrine Fund's World Center for Birds of Prey. The project is overseen by the UNITED STATES FISH AND WILDLIFE SERVICE (USFWS) and the California Condor Recovery Team. Around 190 birds were living in the wild in 2010 and a further 200 in captivity. *See also* PUPPET MOTHER

caliper, calliper, vernier caliper, vernier calliper A measuring device consisting of two bars that slide apart along a scale. The distance between the bars is indicated on a mechanical dial (dial caliper) or on a digital display (digital caliper). The two bars are placed at the opposite ends of the structure to be measured and the distance is read from the dial/display. Used for measuring lengths of body parts, the thickness of SUBCUTANEOUS body fat, etc.

callitrichid A member of the NEW WORLD MONKEY family Callitrichidae which contains the marmosets, tamarins and lion tamarins.

callitrichine Of or relating to a CALLITRICHID.

callosity *See* CALLUS (2)

callus
1. A mass of CONNECTIVE TISSUE and blood that forms around the exposed end of a fractured BONE while healing.
2. An area of SKIN which has become thickened as a result of repeated contact and friction, e.g. on the feet or hands. Also called a callosity. Callosities occur on the heads of right whales (*Eubalaena* spp.). Their unique patterns make callosities useful for the photo-identification of these animals. *See also* ISCHIAL CALLOSITY

calorie The quantity of heat required to increase the temperature of 1 gram of water by 1 °C. One calorie (cal) = 4.184 joules. In many countries the calorie is still widely used as a measure of the energy in food. 1 Calorie (Cal) = 1000 calories (cal).

calorific restriction The provision of a diet with a reduced CALORIFIC VALUE. Restricting calorific intake may have a number of beneficial effects, including a reduction in the incidence of disease and increased LONGEVITY.

calorific value A measure of the energy content of a substance, especially a FOOD; the amount of heat produced during the complete combustion of the unit mass of a food.

calving aid Any of a number of devices used to assist the process of calving, especially during abnormal births. Usually consists of a metal pole with a 'head' at one end which is braced against the rump of the cow. A ratchet system is used to pull on a rope attached to the calf's feet.

camel racing The racing of camels as a sport. Common in countries in the Middle East, Pakistan, Mongolia and Australia.

camel wrestling A traditional 'sport' in which camels fight using their necks in response to exposure to a female camel in heat. Common in Turkey but also occurs in the Middle East and south Asia.

camelid A member of the CAMELIDAE.

Camelidae A family of mammals containing the camels and their relatives: alpacas, llamas, vicunas.

camera trap, trail camera A camera used to record still or moving digital images of animals in the wild. Often triggered by a passive infrared sensor as the animal moves in front of the camera. May be used in animal CENSUSES where individual animals may be recognised, e.g. tigers, or to monitor the use of wildlife trails. May record time of recording and environmental temperature (Fig. C2). *See also* INFRARED RADIATION

Fig. C2 Camera trap: an image of a curious bear taken with a camera trap. Insert: camera trap.

camouflage A device, coloration or other means of concealing or disguising an animal, often by adopting the colour or texture of the natural surroundings. Often used to avoid detection by predators or remain undetected by prey. The outline of the body is often hidden by disruptive coloration or the presence of morphological projections. *See also* CRYPSIS, MIMICRY

CAMPFIRE Communal Areas Management Programme for Indigenous Resources. A scheme established in Zimbabwe under the Parks and Wildlife Act 1975 which returned the ownership and control of natural resources to some local communities. These communities establish HUNTING QUOTAS for various wildlife species. Most of the revenue from hunting comes from elephants as this species attracts the highest hunting fees. *See also* SPORT HUNTING

campylobacteriosis An infection by a bacterium of the genus *Campylobacter* which cause a range of disorders from abortion to dysentery.

Canadian Animal Genetic Resources (CAGR) Program A joint initiative of AGRICULTURE AND AGRI-FOOD CANADA (AAFC) and the University of Saskatchewan to preserve the genetic diversity of Canadian livestock and poultry breeds.

Canadian Cattle Identification Agency (CCIA) A Canadian organisation that manages a trace back system for cattle designed for the containment and eradication of animal disease (Canadian Livestock Tracking System (CLTS)). Established in 1998, CCIA has developed the only mandatory national identification programme for the cattle industry and works with the Canadian Food Inspection Agency to ensure the food safety of the Canadian cattle industry.

Canadian Livestock Tracking System (CLTS) *See* CANADIAN CATTLE IDENTIFICATION AGENCY (CCIA)

cancellous bone *See* BONE

cancer A term used to describe a number of diseases which are characterised by the presence of cells whose growth is not properly regulated and which produce CLONES (2) of DAUGHTER CELLS which invade adjacent tissues and may interfere with their functioning. Cancer cells which proliferate and remain together form benign tumours. Those that proliferate and shed cells into the blood and lymphatic system form malignant tumours. *See also* METASTASIS, OncoMouse®

candida The fungus *Candida albicans* causes a disease called candidiasis (moniliasis) in livestock and humans.

candling, lamping The use of a lamp (a candling lamp) held behind a bird's egg to determine whether or not a live EMBRYO is present.

canid A member of the CANIDAE.

Canidae A family of mammals containing dogs, wolves, foxes, coyotes and their relatives.

canine

1. Relating to **CANIDS** (dogs).

2. A type of sharp, pointed **TOOTH** at the front of a mammalian jaw, adapted to tearing flesh.

canine distemper, hardpad disease A highly contagious, frequently fatal, disease caused by a **PARAMYXOVIRUS** which targets various organ systems all at the same time. It affects **CANIDS** (dogs, foxes, wolves), **MUSTELIDS** (e.g. ferret, mink, skunk), most procyonids (e.g. raccoon, coatimundi) and some viverids (binturong).

canine separation-related behaviour (SRB) Undesirable behaviours in dogs that only occur when they are separated from their owners. They include destructive behaviour, excessive vocalisation and inappropriate toileting. SRB is a common reason why dogs are relinquished to animal shelters.

canker

1. **ULCERATION** or abscesses of the mouth, lips or tongue, eyelids or ears or the cloaca of birds, especially **INFLAMMATION** of the ears of canids and felids.

2. **INFLAMMATION** and decay of the hooves of **EQUIDS**.

canned hunting The **HUNTING** of animals that are contained within a fenced enclosure. Occurs in some African countries and in the USA where the animals are often former zoo or circus animals. *See also* **BLOOD SPORTS, INTERNET HUNTING, SPORT HUNTING**

cannibalism The act of eating an individual of the same species. Occurs in a wide range of species from insects to mammals. Cannibalism may be active (hunting, killing and eating) or passive (eating *post-mortem*). Passive cannibalism is part of normal feeding behaviour in some species. Cannibalism may be the result of sexual rivalry, overcrowding or **STRESS**. *See also* **FOETICIDE, FRATRICIDE, INFANTICIDE, PROLICIDE**

cannon netting A method of capturing birds while they are on the ground by firing a large net over them using explosively driven projectiles. The net is tethered to the ground along one edge while the parallel edge is pulled in an arc by the projectiles. Rocket-netting is essentially the same, but the net is lifted by rockets.

cannula (cannulae *pl.*) A small flexible tube used for **CANNULATION**.

cannulation The insertion of a tube or **CANNULA** into a blood vessel, duct or body cavity.

cantilever A behaviour of primates which involves springing out from a branch to catch a prey animal with the hands while holding onto a branch or other structure with the hind legs.

CAPACITY Software designed to establish target **POPULATION** sizes for managed populations in zoos.

capacity-building This is an approach to development which refers to the strengthening of skills, abilities and competencies of individuals and communities

in developing countries. Many zoos support *in-situ* programmes by assisting local communities with training, education and other employment related to conservation objectives. **CHESTER ZOO** works with communities in Assam where human–elephant conflict is common and has trained local people to track elephant movements and implement deterrent measures.

capillary A very fine blood vessel (one cell thick), located between an **ARTERIOLE** and a **VENULE**, which carries oxygen and food to the tissues, and removes waste products and carbon dioxide.

capillary bed A network of **CAPILLARIES** located between **ARTERIOLES** and **VENULES**.

capillary refill time The time taken for the surface **CAPILLARIES** in the gums to refill with blood after being squeezed until white. A slow time may indicate **SHOCK**.

caprid Any of the species of **BOVIDS** in the subfamily Caprinae: goats, sheep and their relatives.

Caprimulgiformes An order of birds: oilbird, nightjars, frogmouths, potoos.

caprine Relating to **CAPRIDS**.

captive animal

1. An animal that is held in captivity.

2. In England and Wales a 'captive animal' is defined by s.15 of the **PROTECTION OF ANIMALS ACT 1911**, as '*any animal (not being a domestic animal) of whatsoever kind or species, and whether a quadruped or not, including any bird, fish, or reptile, which is in captivity, or confinement, or which is maimed, pinioned, or subjected to any appliance or contrivance for the purpose of hindering or preventing its escape from captivity or confinement'.*

Captive Animals' Protection Society (CAPS) An organisation that campaigns on behalf of animals in **CIRCUSES (1)**, **ZOOS (1)** and the entertainments industry. Founded in 1957 and based in the UK. One of its stated aims is to end the captivity of animals in zoos and the society actively discourages the public from visiting zoos, claiming that they have no conservation or educational value. *See also* **PERFORMING ANIMAL WELFARE SOCIETY (PAWS), ZOO CHECK**

captive bolt A device for the humane stunning of **LIVESTOCK** prior to slaughter which fires a metal bolt into the brain at high velocity using either compressed air or a blank round which is struck by a firing pin. *See also* **STUNNING**

captive breeding The breeding of animals in captivity in zoos and other places, generally for conservation purposes, and in order to create and maintain an **INSURANCE POPULATION**. *See also* **CAPTIVE BREEDING PROGRAMME, CONSERVATION BREEDING**

captive breeding programme A co-ordinated breeding programme for an animal species in captivity. *See* **ARABIAN ORYX (*ORYX LEUCORYX*), BLACK-FOOTED FERRET (*MUSTELA NIGRIPES*),**

C

CALIFORNIA CONDOR (*GYMNOGYPS CALIFOR-NIANUS*), EUROPEAN ENDANGERED SPECIES PROGRAMME (EEP), EUROPEAN STUDBOOK (ESB), GIANT PANDA BREEDING CENTRES, REGIONAL ASSOCIATION OF ZOOS, SPECIES SURVIVAL PLAN® (SSP) PROGRAMS

Captive Breeding Specialist Group *See* CONSERVATION BREEDING SPECIALIST GROUP (CBSG)

capture myopathy, exertional myopathy A muscle disease caused as a result of the STRESS associated with capture in a wide range of bird and mammal taxa, especially wild UNGULATES. It is caused by the build-up of lactic acid in muscles after capture (resulting from an OXYGEN DEBT). Signs include muscle damage and stiffness, depression and shock, and it may result in death in extreme cases, sometimes weeks later.

capybara farm A farm where capybara (*Hydrochoerus hydrochaeris*) are raised for meat. Common in South America, notably Venezuela. Capybara are the largest rodents in the world and have a high growth rate. They are kept as pets in North America.

carapace
1. The dorsal shell of a turtle.
2. The dorsal plate which covers the CEPHALOTHORAX of a CRUSTACEAN.

carbohydrate An organic molecule which contains an aldehyde or ketone group, and consists of carbon, hydrogen and oxygen in the ratio 1:2:1. Monosaccharides are simple sugars such as glucose and fructose. Disaccharides consist of two monosaccharide units joined together (e.g. maltose). Polysaccharides are made up of a large number of monosaccharide units joined together, e.g. starch is made up of many glucose units.

carbon cycle The cyclic movement of carbon between the biological and physical components of the environment. Carbon from carbon dioxide in the atmosphere is incorporated into the structure of GLUCOSE as a result of PHOTOSYNTHESIS in green plants. It then passes along the FOOD CHAIN as herbivores eat plants and carnivores eat herbivores. Carbon is released back into the atmosphere in the form of CARBON DIOXIDE as organisms respire.

carbon dioxide (CO$_2$) A colourless gas present in the atmosphere and formed during cellular respiration. Used by photosynthetic plants as a source of carbon for making sugars and other biological chemicals. Used in FIRE EXTINGUISHERS.

carbon dioxide injection *See* PRESSURISED CARBON DIOXIDE KIT

carbon neutral Having no net effect on the carbon balance of the atmosphere. May refer to a building (e.g. a zoo exhibit), a process or an organisation.

carcass feeding The practice of feeding whole or part carcasses to carnivores to simulate natural feeding behaviour and diet as an ENVIRONMENTAL ENRICHMENT. *See also* MANNEQUIN

Carcharhiniformes An order of fishes: ground sharks.

carcinogen A cancer-causing agent.

carcinoma A CANCER that originates in the EPITHELIUM, especially in the breast, intestines and lungs.

cardiac Relating to the HEART.

cardiac arrest A cessation in the pumping action of the heart. May cause brain damage and death. *See also* CORONARY THROMBOSIS

cardiac cycle The series of muscular contractions which move blood through the heart, corresponding to one heart beat. *See also* DIASTOLE, SYSTOLE

cardiac muscle Heart muscle in vertebrates which is branching and specialised to propagate waves of contraction across the heart. This muscle is capable of spontaneous rhythmic contractions in the absence of nervous stimulation.

cardiac output The volume of blood pumped by the heart per unit time (l/min). The HEART RATE × STROKE VOLUME.

cardiomyopathy Disease of the heart muscle.

cardiovascular Relating to the heart and blood vessels.

care for life In relation to animals, especially those living in zoos, a commitment to care for the animals until the natural end of their lives. *See also* CHIMP HAVEN, ELEPHANT SANCTUARY

caregiver An alternative term for a keeper or other person who cares for animals.

caries The progressive decay and decomposition of TEETH and BONE.

carina, keel, sternal keel The long, thin process that extends ventrally from the STERNUM of a bird, providing a large surface area for the attachment of the FLIGHT MUSCLES responsible for the down stroke of the wings during flight.

carnassial tooth A specialised cheek tooth found in CARNIVORES used for crushing bone etc.

Carnivora An order of mammals which includes dogs, bears, raccoons, weasels, civets, otters, mongooses, hyenas, cats.

carnivore
1. An animal that eats meat or other animals.
2. A member of the order CARNIVORA.

carotene One of a group of CAROTENOIDS.

carotenoid A hydrocarbon related to VITAMIN A which often occurs as a red, orange or yellow pigment.

carotid body A receptor in the sinus of the carotid artery that senses the level of carbon dioxide in the blood.

carrier
1. In genetics, an individual who is carrying a condition or characteristic but does not exhibit it, and may pass on the recessive gene which causes it to his or her offspring. *See also* ASYMPTOMATIC CARRIER
2. In relation to animal transportation, a person or organisation which undertakes such transportation.

carrion The bodies of dead animals.

carrying capacity In ecology, the maximum number of individuals of a particular species that can be sustained indefinitely by a particular habitat, under specific environmental conditions. This may vary from time to time due to changes in food or water supply, the number of available nesting or denning sites, the number of available territories etc.

Carson, Rachel (1907–1964) An American biologist who campaigned against the pollution of ecosystems with chemicals, especially in the USA. She brought the plight of wildlife and, in particular, the effect of **PESTICIDES** on **FOOD CHAINS**, to the attention of the public in her book *SILENT SPRING*. Carson is widely credited with helping to found the environmental movement. *See also* **LEOPOLD**

Cartagena Protocol on Biosafety 2000 A **PROTOCOL** to the **CONVENTION ON BIOLOGICAL DIVERSITY 1992 (CBD)** which aims to ensure the safe handling, transport and use of **LIVING MODIFIED ORGANISMS (LMOs)** resulting from modern **BIOTECHNOLOGY** that may have adverse effects on biological diversity and human health.

cartilage A flexible **CONNECTIVE TISSUE** found in the vertebrate **SKELETON**. *See also* **CHONDROCYTE**

cartilaginous Consisting of **CARTILAGE**.

cartilaginous fishes Fishes in which the skeleton consists of cartilage and never develops into bone, e.g. sharks and rays (**CHONDRICHTHYES**).

case law The reported decisions of certain courts and other judicial bodies. Some case law sets precedents which must be followed in other similar cases. *See also* **ANIMAL RIGHTS LEGAL CASES**, *BAYLISS V. COLERIDGE* **(1903)**, **CLIMATE CHANGE**, *LEGAL EAGLE*, **LOCUS STANDI**, **NORTHERN SPOTTED OWL** (*STRIX OCCIDENTALIS CAURINA*), *WYOMING FARM BUREAU FEDERATION V. BABBITT* **(1997)**

casque An air-filled cavity enclosed by minimal **CANCELLOUS BONE** which occurs on the dorsal **MAXILLARY** beak of all but one of the 54 extant species of hornbills (Bucerotidae). The casque is a common location of self-induced injury, **CONSPECIFIC** trauma, environmental damage and disease.

Casson, Sir Hugh (1910–1999) A British architect who designed the elephant and rhinoceros pavilion at **LONDON ZOO** in 1965. This is protected as a grade II* **LISTED BUILDING** but is no longer used for elephants or rhinos. It is an architecturally important building but unsuitable for elephants. The zoo moved its elephants to **WHIPSNADE ZOO** in 2001.

caste A distinct group of individuals within a **EUSOCIAL** species (e.g. bees, termites, ants), which is different in appearance and behaviour from other groups within the species, and which perform a specific role in the **COLONY**, e.g. worker ants, soldier ants, drones.

castration The surgical removal of the **TESTES**. May be performed as a **CONTRACEPTIVE** measure or to remove diseased tissue. Sometimes used to make male livestock (e.g. sheep) and some companion

Table C1 Major cat breeds.

Abyssinian	Ocicat
Asian	Oriental Longhair
Balinese	Oriental Shorthair
Bengal	Persian
Birman	Ragamuffin
British Shorthair	Ragdoll
Burmese	Russian Blue
Cornish Rex	Selkirk Rex
Devon Rex	Siamese
Egyptian Mau	Siberian
Exotic Shorthair	Singapura
Korat	Snowshoe
LaPerm	Somali
Maine Coon	Sphynx
Manx	Tonkinese
Norwegian Forest	Turkish Van

animals (e.g. dogs and rabbits) less aggressive and more manageable. The term is sometimes also used to refer to the removal of the **OVARIES** but this is more usually called an **OVARIECTOMY**.

cat
1. The domestic cat (*Felis catus*). *See also* **BALLARAT**, **CAT BREEDS**, **FERAL CAT**
2. Any member of the **FELIDAE**.
See also **BIG CAT**

cat breeds A wide range of cat breeds of different coat colours and hair lengths have been produced for showing. Major cat breeds are listed in Table C1.

CAT scanner *See* **CT SCANNER**

catabolism The breakdown of organic compounds during **METABOLISM**. *Compare* **ANABOLISM**

catadromous Relating to a species that migrates from freshwater to seawater to spawn, e.g. eels, salmon. *Compare* **ANADROMOUS**, **DIADROMOUS**

Catalogue of Life *See* **SPECIES 2000 AND ITIS CATALOGUE OF LIFE**

cataract An abnormal opaque area of the lens of the eye that affects vision.

catarrh Inflammation of the **MUCOUS** lining of the nose and throat.

catarrhine A member of the primate infraorder Catarrhini, which includes **OLD WORLD MONKEYS**, **GREAT APES**, gibbons and humans. Individuals possess narrow, downward pointing nostrils. *Compare* **PLATYRRHINE**

catastrophic moult A moult which involves the loss of feathers all at once (e.g. as in African penguins (*Spheniscus demersus*)) or the **SLOUGHING** off of the skin and hair at once (e.g. as in northern elephant seals (*Mirounga angustirostris*)).

catatonia A neurological condition characterised by immobility. *See* **IMMOBILON**

catch pole A piece of equipment designed for capturing and restraining large animals consisting of a long pole with a noose at the end.

C

Category 1, 2 and 3 species Animal species listed in Appendix 12 of the *SECRETARY OF STATE'S STANDARDS OF MODERN ZOO PRACTICE* (**SSSMZP**) in the UK on the basis of their risk to the public. Category 1 species are species which pose the highest risk and include big cats, wolves and ostriches. The Health and Safety Executive advises that keepers should only have physical contact with category 1 species in exceptional circumstances, e.g. when they are anaesthetised.

cathemeral Active periodically and intermittently throughout the day and night, e.g. characteristic of some lemur species. *Compare* **CREPUSCULAR, DIURNAL, MATUTINAL, NOCTURNAL, VESPERTINE**

catheter A thin flexible tube inserted into a narrow opening or body cavity usually to drain a fluid (especially urine) but sometimes to introduce something. May be used in **ARTIFICIAL INSEMINATION (AI)** to inject semen into the vagina where this extends deep inside the body, e.g. in elephants and rhinos. In rhinos the distance the semen must travel is around 1.5 m.

cation A positively charged ion, e.g. sodium (Na^+). *Compare* **ANION**

Cats Protection The leading cat welfare charity in the UK, with a network of over 250 volunteer-run branches, 29 adoption centres and a homing centre. Founded in 1927 as the Cats' Protection League.

Cats' Protection League *See* **CATS PROTECTION**

cattery A place where cats are bred or cared for in their owner's absence.

cattle
1. Large heavily built grazing animals including both wild and domestic species (and many breeds), the latter being kept and farmed for their meat, milk and hides.
2. Legal definitions of cattle vary. In the UK, under the Dogs Act 1906 (s.7), *'the expression "cattle" includes horses, mules, asses, sheep, goats, and swine.'*

cattle breed A wide range of cattle breeds have been produced of different sizes and builds, and adapted to different environmental conditions (Table C2; Figs. C1 and C3).

cattle grid A metal frame surrounding a series of parallel metal bars over which vehicles may drive but large animals (such as cattle and sheep) will not pass. Often used at the entrance to farms, between adjacent fields, and in **SAFARI PARKS** where vehicles drive between **ANIMAL ENCLOSURES**.

Cattle Health Certification Standards (CheCS) The self-regulatory body for cattle health schemes in the UK and Ireland, established by the cattle industry for the control and eradication of non-statutory diseases by the maintenance of a set of standards.

cattle passport A document issued by the **BRITISH CATTLE MOVEMENT SERVICE (BCMS)** which relates to a single animal and records its mother, place of birth and all movements throughout its lifetime.

Table C2 Major pedigree breeds of beef and dairy cattle.

Pedigree beef breeds

Aberdeen Angus	Dexter	Red Angus
Africander	Drakensberger	Red Poll
Aubrac	Droughtmaster	Retinta
Barzona	English Longhorn	Romagnola
Bazadaise	Galloway	Salers
Beef Shorthorn	Gelbvieh	Sanganer
Beefalo	Gloucester	Santa Cruz
Beefmaster	Hays Converter	Santa Gertrudis
Belgian Blue	Hereford	Senepol
Belmont Red	Highland	Shetland
Belted Galloway	Hybridmaster	Simbrah
Blonde d'Aquitaine	Limousin	Simmental
Bonsmara	Lincoln Red	South Devon
Boran	Lowline	Speckle Park
Braford	Luing	Square Meaters
Brahman	Maine-Anjou	Sussex
Brahmousin	(Rouge des Prés)	Tarentaise
Brangus	Marchigiana	Texas Longhorn
British White	Miniature Hereford	Tuli
Buelingo	Mirandesa	Wagyu
Canchim	Mongolian	Watusi
Caracu	Murray Grey	Welsh Black
Charolais	Nelore	Whitebred
Chianina	Nguni	Shorthorn
Composite	Parthenais	Zebu
Corriente	Piemontese	
Devon	Pinzgauer	

Pedigree dairy breeds

Ayrshire	Girolando	Meuse Rhine Issel
Brown Swiss	Guernsey	Milking Devon
Busa	Holstein	Montbéliarde
Canadienne	Illawarra	Normande
Dairy Shorthorn	Irish Moiled	Norwegian Red
Dutch Belted	Jersey	Randall
Estonian Red	Kerry	Sahiwal
Friesian	Lineback	

cattle prod, stock prod A hand-held device – usually a metal or fibreglass rod – used to strike or poke cattle and other livestock to make them move. Electric cattle prods apply a high-voltage, low-current electric shock from the tip. Some animal welfare groups consider the use of electric prods to be physically and mentally harmful.

cattle station A large farm in Australia where cattle are reared. The Australian equivalent of an American ranch. Many are in remote locations. *See also* **TRAVELLING STOCK ROUTE (TSR)**

cattle tracing system A system for tracing the origin, movements and destinations of cattle, to protect human food chains and promote **BIOSECURITY**. *See also* **BRITISH CATTLE MOVEMENT SERVICE (BCMS), CATTLE PASSPORT, CANADIAN CATTLE IDENTIFICATION AGENCY (CCIA), NATIONAL ANIMAL IDENTIFICATION AND**

C

Fig. C3 Cattle breeds: White Park (top left); Simmental (top right); Holstein Friesian (middle left); Red and White (middle right); Hereford (bottom left); British Blond (bottom right).

C

Fig. C4 CCTV: closed-circuit television cameras monitoring European beavers (*Castor fiber*) at Martin Mere, UK.

TRACING (NAIT), NATIONAL ANIMAL IDENTI-FICATION SYSTEM (NAIS)

caudal Associated with or located near the tail (Fig. A8).

Caudata An order of amphibians: salamanders, newts, sirens, hellbenders, mudpuppies.

cauterisation The destruction of living tissue using a directly applied heated instrument, electric current, a laser beam or a caustic chemical. Used to prevent infection of wounds etc.

cavy *See* GUINEA PIG

CBSG *See* CONSERVATION BREEDING SPECIALIST GROUP (CBSG)

CCTV Closed-circuit television. Used to monitor animals when sick, pregnant or inaccessible. May be used to allow zoo visitors to see animals while OFF-SHOW, in their dens, nests, etc. and to monitor visitor behaviour and movements for security reasons (Fig. C4). May be connected to a recording device to record and analyse behaviour. Also used to monitor moving doors when transferring animals between enclosures. CCTV has been used on board fishing vessels to monitor FISH DISCARDS.

CDC *See* CENTERS FOR DISEASE CONTROL AND PREVENTION (CDC)

cecotropes *See* COPROPHAGY

cecum *See* CAECUM

cell The basic functional and structural unit from which all living things are made. It consists of CYTOPLASM which is surrounded by a CELL MEMBRANE and includes a variety of ORGANELLES such as a NUCLEUS, NUCLEOLUS, MITOCHONDRIA, ENDOPLASMIC RETICULUM (ER), GOLGI BODY and RIBOSOMES. Plant cells possess a cell wall outside the cell membrane which is made predominantly of CELLULOSE. Somatic cells divide by MITOSIS (to produce more body cells) and MEIOSIS (to produce GAMETES). *See also* EUKARYOTE, PROKARYOTE

cell cycle The cycle of cellular activity during MITOSIS that commences with interphase in a new daughter cell and proceeds through prophase, metaphase,

anaphase and telophase, resulting in the division of this cell into two.

cell division The process by which an existing CELL (PARENT CELL) produces new cells (DAUGHTER CELLS). This occurs during growth (MITOSIS), where two identical DIPLOID cells are formed, and in the production of sex cells (MEIOSIS), where HAPLOID gametes are formed. *See also* CELL CYCLE

cell fractionation A variety of methods used to separate cell ORGANELLES using homogenisation (the splitting open of the CELLS) followed by DIFFERENTIAL CENTRIFUGATION.

cell membrane, plasma membrane, plasmalemma The outer limiting boundary of cells, which consists of a semi-permeable structure composed of lipids and proteins. Water passes freely through the membrane but the passage of most other molecules is controlled. *See also* ACTIVE TRANSPORT, DIFFUSION, OSMOSIS

cellular respiration *See* RESPIRATION (1)

cellulase An ENZYME that is capable of digesting CELLULOSE.

cellulose A CARBOHYDRATE found widely in plants, composed of chains of glucose molecules.

census A systematic count of animals, either in the field as part of an ecological study, or in a zoo or other animal collection during the process of STOCKTAKING. In the field it may be a TOTAL COUNT, or based on samples taken using CAMERA TRAPS, TRANSECTS etc. *See also* MARK–RECAPTURE TECHNIQUE

Center for Conservation Medicine A centre within Tufts Cummings School of Veterinary Medicine at Tufts University, which was established in 1997, and pioneered the concept of CONSERVATION MEDICINE.

Center for Zoo Animal Welfare A resource centre established by the Detroit Zoological Society in the USA for captive ANIMAL WELFARE knowledge, research and best practices. The Center recognises that zoos have unintended effects on the animals in their care. It conducts and facilitates research, trains professionals in captive animal welfare and has established two annual welfare awards. The Center's Advisory Committee is composed of zoo and aquarium professionals, scientists, sociologists and animal advocacy leaders.

Centers for Disease Control and Prevention (CDC) A US federal agency that works to protect human health and safety, located in Atlanta, Georgia. Its work focuses on the prevention and control of disease, especially infectious disease, including ZOONOSES such as H1N1 VIRUS.

central nervous system (CNS)
1. The brain and spinal cord of vertebrates.
2. The brain of invertebrates. *See also* CEPHALIC GANGLION, PERIPHERAL NERVOUS SYSTEM (PNS).

Central Science Laboratory *See* FOOD AND ENVIRONMENT RESEARCH AGENCY (FERA)

Central Zoo Authority (CZA) An autonomous statutory body that licenses and regulates zoos in India.

centrifuge A laboratory device for spinning materials in small tubes at high speed in order to separate out different fractions, e.g. to separate out blood cells from plasma. Larger particles move to the bottom of the tube faster than smaller particles.

Cephalaspidomorphi A class of fishes: lampreys and their allies.

cephalic ganglion An enlargement of the central nervous system in the head, which is called the brain in many taxa.

cephalopod A member of the CEPHALOPODA.

Cephalopoda A class of marine MOLLUSCS which possess tentacles: cuttlefish, nautiloids, squids, octopuses.

cephalothorax The fused head and thorax found in spiders and many CRUSTACEANS.

Ceratodontiformes An order of fishes: Australian lungfishes.

cercaria (cercariae *pl.***)** The larva of flukes (e.g. *Schistosoma*) that swims from the molluscan intermediate host to either the final (primary) host or another intermediate host.

cercopithecine
1. A member of the OLD WORLD MONKEY subfamily Cercopithecinae which includes baboons, macaques, guenons and langurs.
2. Relating to such a monkey.

cercus (cerci *pl.***)** Either of a pair of antenna-like structures located at the posterior end of some insects and other arthropods. Their function may be sensory, defensive or reproductive.

cerebellum (cerebella *pl.***)** The part of the vertebrate brain that is responsible for co-ordinating learned movements.

cerebral Relating to the brain.

cerebral hemisphere Either of two halves of the CEREBRUM in vertebrates.

cerebrospinal Relating to the brain and spinal cord.

cerebrospinal fluid (CSF) The fluid in the VENTRICLES of the vertebrate BRAIN and the central canal of the SPINAL CORD. It serves as a shock absorber for the CENTRAL NERVOUS SYSTEM (CNS) (1), and also delivers nutrients filtered from the blood and removes wastes.

cerebrum The forebrain of vertebrates; responsible for voluntary movements and perception.

cervical
1. Relating to the neck region, e.g. cervical VERTEBRAE.
2. Relating to the CERVIX, e.g. cervical smear, cervical cancer.

cervid A member of the mammalian family Cervidae.

cervid chronic wasting disease *See* CHRONIC WASTING DISEASE (CWD)

Cervidae A mammalian family which includes deer, reindeer and their relatives.

cervix The neck of the UTERUS where it meets the VAGINA in mammals.

cesarean section *See* CAESAREAN SECTION

Cestoda The tapeworms. A class of parasitic worms (PLATYHELMINTHES) without a gut, the adults of which are endoparasites of vertebrates. Most possess a scolex ('head') with hooks or suckers for attachment.

cestode A member of the CESTODA.

Cetacea An order of aquatic mammals: whales, dolphins, porpoises.

cetacean
1. A member of the CETACEA.
2. Relating to such an animal.

cetacean rights *See* DECLARATION OF RIGHTS FOR CETACEANS: WHALES AND DOLPHINS

cetacean stranding *See* STRANDING

chaeta (chaetae *pl.***)** A chitinous hair-like bristle characteristically found on the body surface of oligochaete and polychaete worms.

Chagas disease *See* TRYPANOSOMIASIS

chain reaction *See* REACTION CHAIN

chain responses *See* REACTION CHAIN

chain-chewing A STEREOTYPIC BEHAVIOUR which involves repetitive chewing on a chain. Chains may be installed experimentally as a focus for oral activities attached to an automatic DATA LOGGER.

chaining *See* PICKETING

chain-linked fence A type of fencing used widely for zoo enclosures and constructed from lengths of wire woven together to form diamond-shaped holes. The wire thickness and hole size varies depending upon the species being contained. Such fences often incorporate a RETURN at the top.

chancre A hard ulcer or sore on the skin which occurs in some infectious diseases, and especially at the site of a tsetse fly bite. *See also* TRYPANOSOMIASIS

Characiformes An order of fishes: characins and their allies.

Charadriiformes An order of birds associated with water: jacanas, snipes, plovers, dotterels, avocets, coursers, pratincoles, stilts, curlews, sheathbills, gulls, terns, skimmers, skuas, auks.

charge In animal behaviour, a rapid, threatening advance made by an animal towards another animal or human, which is perceived as a THREAT. Often used as a threat display and stops short of physical contact, e.g. in ELEPHANTS, CHIMPANZEES, gorillas.

charismatic megafauna Large well-known animal species (particularly LARGE MAMMALS) which attract a disproportionate share of the public's attention. The concept is widely used by conservationists to gain support for wildlife CONSERVATION efforts.

cheating A strategy in which some individuals in a social group gain an advantage by not conforming to social norms of behaviour. For example, they may not reciprocate after others have shared food with them. *See also* RECIPROCAL ALTRUISM. *Compare* SHARING

checklist A list of species constructed by scientists or enthusiasts (e.g. bird-watchers) who study a particular taxon or geographical area, e.g. a checklist of the birds of Turkey, a checklist of the mammals of the SERENGETI National Park.

cheek pouch *See* POUCH (2)

Cheetah A chimpanzee who starred in the *Tarzan* films with Johnny Weissmuller and Maureen O'Sullivan from 1932 to 1934. He died in 2011 in Suncoast Primate Sanctuary, Palm Harbor, Florida, aged 80. *See also* TARZAN OF THE APES

chela The last joint of an ARTHROPOD limb which forms grasping pincers.

chelicera (chelicerae *pl.*) The paired mouthparts found in spiders and other chelicerates (e.g. mites, scorpion, horseshoe crabs).

chelipeds The first pair of walking legs, bearing pincers, in certain species of crabs, crayfish and other decapod CRUSTACEANS.

chelonian, chelonid
1. A member of the reptilian order Chelonia (turtles, terrapins and tortoises).
2. Of, or relating to, members of the Chelonia.
3. A general term used for shelled reptiles.

chelonid *See* CHELONIAN

chemical digestion The breakdown of large food molecules into smaller molecules by ENZYMES to aid ABSORPTION (2) through the gut.

chemoautotroph A bacterium or other organism capable of deriving energy from chemical reactions and synthesising organic molecules by oxidising inorganic molecules (chemosynthesis), e.g. hydrogen sulphide, ferrous iron. They are the primary producers in hostile environments, e.g. deep sea vents.

chemoreception The ability to detect the presence and concentration of chemicals in the environment.

chemoreceptor A SENSORY RECEPTOR which detects chemicals in the environment.

chemosynthesis *See* CHEMOAUTOTROPH

chemotherapy An anti-cancer treatment that uses CYTOTOXIC drugs.

Chester Zoo A major zoo and conservation charity in England operated by the North of England Zoological Society which engages in a wide range of *EX-SITU* CONSERVATION projects. It was founded by George Mottershead and opened in 1931. The design of the enclosures was greatly influenced by the work of HAGENBECK. *See also* CAPACITY-BUILDING

chick crumb A type of commercially produced food pellet for birds.

chicken The chicken (*Gallus gallus*) is a domesticated bird, kept for its meat and eggs, which is descended from the jungle fowl of southeast Asia. *See also* POULTRY

Table C3 Major chicken breeds.

Hard Feather	*True Bantam*
Indian Game	Belgian Bantam
Modern Game	Booted Bantam
Old English Game	Dutch Bantam
Rumpless Game	Japanese Bantam
	Pekin
Soft Feather Heavy	Rosecomb
Australorp	Selbright
Barnevelder	
Brahma	*Rare Breeds*
Cochin	Andalusian
Croad Langshan	Asil
Dorking	Campine
Faverolles	Houdan
Frizzle	Ixworth
Marans	Jersey Giant
Orpington	Lakenvelder
Plymouth Rock	Malay
Rhode Island Red	Marsh Daisy
Sussex	New Hampshire Red
Wyandotte	Norfolk Grey
	North Holland Blue
Soft Feather Light	Orloff
Ancona	Shamo
Appenzeller	Sicilian Buttercup
Hamburgh	Spanish
Leghorn	Sultan
Minorca	Sumatra
Poland	Transylvanian Naked Neck
Scots Dumpy	Vorwerk
Scots Grey	Yokohama
Silkie	
Welsummer	

chicken breeds There are hundreds of different chicken breeds (Fig. P9). Some major breeds are listed in Table C3.

chicken coop A general term for a house in which chickens are kept. Usually constructed of wood and wire mesh and provided with access doors for removing eggs from nest boxes. May also include a RUN made of framed wire mesh. *See also* POULTRY ARK

chicken harvester *See* MECHANICAL CHICKEN HARVESTER

children's zoo, pets' corner, petting zoo A collection of domestic and other harmless animals which children (and other visitors) may touch and feed. May be located within a larger zoo.

Chimaeriformes An order of fishes: chimaeras.

chimera, chimaera An organism which is made up of two or more distinct GENOMES, which has been created by experimental manipulation early in its development. In January 2012 the Oregon National Primate Research Center in the USA reported the creation of the world's first primate chimeras using RHESUS MONKEY (*MACACA MULATTA*) cells. *See also* TRANSGENIC ORGANISM

Chimp Haven The National Chimpanzee Sanctuary of the USA, located in Louisiana. It is an independent, non-profit organisation which provides lifetime care for chimpanzees who are unwanted pets, or have retired from medical research or the entertainment industry.

chimpanzee Either of two extant species of GREAT APES: the common chimpanzee (*Pan troglodytes*) (Fig. Q1) and the bonobo (*P. paniscus*). Controversially used in medical research, space exploration and exhibited in zoos. *See* CHIMPANZEE HEALTH IMPROVEMENT MAINTENANCE AND PROTECTION ACT OF 2000 (USA), CHIMPANZEES' TEA PARTY, CHIMPONAUT

Chimpanzee Health Improvement Maintenance and Protection Act of 2000 (USA) A law in the USA (the CHIMP Act) which provides a system of sanctuaries for surplus chimpanzees previously used in biomedical research. The Act lays down requirements for sanctuaries, defines criteria for chimpanzees that may be transferred to a sanctuary and the exceptions under which a chimpanzee may be brought out of 'retirement'.

chimpanzees' tea party An ANIMAL SHOW in a zoo consisting of a group of chimpanzees eating food and drinking from cups sitting on chairs around a table, sometimes dressed in human clothes. Common in the first half of the 20th century but now considered inappropriate in a modern zoo.

chimponaut A chimpanzee used by the US Air Force in the NASA space programme and sent into space to determine the safety of space flight for humans. Many of these chimps have now been moved to SANCTUARIES in the USA.

Chinese medicine *See* ACUPUNCTURE, TRADITIONAL CHINESE MEDICINE (TCM)

Chipperfield, Jimmy (1912–1990) The owner of Chipperfield's Circus and a member of a great British CIRCUS (1) dynasty. In its heyday in the 1950s the circus employed 250 people and owned the largest tent in the world (seating 9000 people). Chipperfield is credited with inventing the SAFARI PARK. His organisation supplied large numbers of wild-caught animals to safari parks in England in the 1960s. *See also* LIONS OF LONGLEAT

Chiroptera An order of mammals: bats.

chi-squared test (χ^2 test) A simple statistical test which compares observed measurements with values predicted on the basis of probability theory and allows the detection of significant differences between the two.

chitin A polysaccharide that is an important component of the CUTICLE of arthropods. It also occurs in the integuments of many other invertebrate taxa.

chloramphenicol An antibiotic.

chloride A compound of chlorine (e.g. sodium chloride, magnesium chloride) which is a mineral nutrient that helps to control acid–base balance and catalyses certain ENZYMES. Deficiencies and toxicities are rare.

choice In animal behaviour studies, the power or liberty to choose between alternatives. The availability of choice may be an important factor in the wellbeing of some captive animals. *See also* PREFERENCE, WELLNESS

choice chamber A piece of apparatus consisting of transparent plastic tray and lid, divided into quarters, which is used for simple behaviour experiments with invertebrates. Different environmental conditions may be created in each quarter of the chamber for experiments on KINESES.

chokepoint A bottle neck; a narrow passage, point of congestion or obstruction. May be used to describe a narrow place in a pathway which restricts VISITOR CIRCULATION around a zoo, or a point in the migratory route of a species where movement is restricted. The Straits of Gibraltar are a chokepoint on the African–Eurasian FLYWAY used by birds migrating between Africa and Europe.

Chondrichthyes A class of fishes containing the cartilaginous fishes, e.g. sharks and rays.

chondrocyte A cell found in CARTILAGE which secretes COLLAGEN and proteoglycans (complexes of CARBOHYDRATE and PROTEIN that combine with collagen to make a fibrous network).

Chordata A PHYLUM of animals. Most of the chordates are VERTEBRATES: MAMMALS, BIRDS, REPTILES, AMPHIBIANS and FISHES. All chordates share a number of common features in their embryonic development and at some point in their life cycle they have: a dorsal hollow nerve cord; gill slits used to circulate water during feeding and respiration; and a stiffening hollow rod along the dorsal surface called a notochord.

chordate
1. A member of the CHORDATA.
2. Of or relating to a member of the CHORDATA.

chorion A membrane found in vertebrates through which respiratory gases are exchanged in the eggs of reptiles and birds. In placental mammals it develops into part of the PLACENTA.

chromatin The material from which chromosomes are made; DNA and PROTEIN.

chromatography A process for separating the components of a mixture.

chromatophore A cell containing pigment which may, under neural or hormonal control, be covered or uncovered, or dispersed or concentrated to change the colour of the integument. Widely occurring in amphibians, fishes, reptiles, crustaceans and cephalopods.

chromosomal aberration *See* CHROMOSOME MUTATION

chromosome A structure consisting of DNA and PROTEINS, elongated or rounded in form, which contains the genetic information required by the cell for its own development and functioning and which carries this information to **DAUGHTER CELLS** when it divides by **MITOSIS** and to **GAMETES** when it divides by **MEIOSIS**.

chromosome mutation, chromosomal aberration A change in the number or structure of **CHROMOSOMES**. Often occurs in the crossing-over stage of **MEIOSIS**. Sections of chromosomes may be duplicated, deleted, inverted (broken off and reinserted so that the sequence of genes runs DCBA instead of ABCD) or translocated from one chromosome to another (or within the same chromosome). *See also* **GENE MUTATION**

chromosome number The total number of chromosomes in the **GENOTYPE** of an organism, including the **AUTOSOMES** and the **SEX CHROMOSOMES**. This varies between species and is halved in the gametes. *See also* **DIPLOID, HAPLOID**

chromosome pair The two **CHROMOSOMES** that make up a pair in a **DIPLOID** cell.

chronic condition A disease or disorder, of long duration or frequent occurrence (sometimes **ASYMPTOMATIC**); often becoming more serious over a long period of time. It may or may not be severe. *Compare* **ACUTE CONDITION**

chronic wasting disease (CWD), cervid chronic wasting disease An important **PRION** disease (a transmissible spongiform encephalopathy (TSE)) of captive and free-ranging **CERVIDS**.

chrysalarium A place in a **BUTTERFLY FARM**, or similar facility, where butterfly or moth **CHRYSALISES** are kept under controlled conditions of temperature and humidity until the adults emerge.

chrysalis The **PUPA** of a butterfly or moth. *See also* **CHRYSALARIUM**

chute *See* **CRUSH**. *See also* **TRANSFER CHUTE**

chyle The milky-appearing fluid containing lymph and digested fats which is taken up by the lacteals from the intestine and passes up the thoracic duct to empty into the venous system.

chyme The watery, partly digested food and acid (from the stomach wall) which passes from the stomach to the small intestine.

chymosin *See* **RENNIN**

chymotrypsin A pancreatic enzyme that is secreted into the small intestine and digests proteins.

chytrid fungus *See* **CHYTRIDIOMYCOSIS**

chytridiomycosis, chytrid fungus A disease caused by a fungus, *Batrachochytrium dendrobatidis*, which has been linked to declines in amphibian populations around the world since 1999. This organism, chytrid fungus, was first discovered in a museum specimen of the frog *Xenopus laevis* from 1938 collected in South Africa. It is thought to have spread around the world when international trade in this species began in the 1930s. Chytridiomycosis is often fatal. It spreads through water and through contact between amphibians. Mortalities have been reported from many zoos including facilities in

Japan, Australia, USA and Europe (including the UK). *See also* **AMPHIBIAN ARK (AArk)**

Ciconiiformes An order of birds: herons, storks, hammerhead, ibises, spoonbills, New World vultures.

ciliate A ciliated **PROTOZOAN**; any protozoan in the phylum Ciliophora.

cilium (cilia *pl.*) A short, hair-like structure which occurs on the surface of some cells, usually in large numbers. Movement of the cilia can provide a means of **LOCOMOTION** in very small organisms (e.g. ciliated protozoa such as *Paramecium*) or move material over the surface of the cell where it is part of a tissue in a large animal (e.g. ciliated epithelia in the respiratory tract). *Compare* **FLAGELLUM**

circadian rhythm A biological rhythm that has a period of approximately 24 hours. *See also* **BIOLOGICAL CLOCK, PINEAL GLAND**. *Compare* **CIRCANNUAL RHYTHM, INFRADIAN RHYTHM, LUNAR RHYTHM, ULTRADIAN RHYTHM**

circannual rhythm A biological rhythm that has a period of approximately one year, e.g. an annual **MIGRATION (1)** or **HIBERNATION**. *See also* **BIOLOGICAL CLOCK**. *Compare* **CIRCADIAN RHYTHM, INFRADIAN RHYTHM, LUNAR RHYTHM, ULTRADIAN RHYTHM**

circling A type of **STEREOTYPIC BEHAVIOUR** which involves an animal walking repetitively in a circle, sometimes stepping in its own footprints. *See also* **STEREOTYPED ROUTE-TRACING**

circulation

1. Movement or passage through a series of vessels, especially blood in the **CIRCULATORY SYSTEM**.

2. *See* **VISITOR CIRCULATION**

circulatory system The **HEART, BLOOD** and **BLOOD VESSELS** which transport gases, food waste materials, hormones and other chemicals around the body. *See also* **CLOSED CIRCULATORY SYSTEM, OPEN CIRCULATORY SYSTEM**

circus

1. An entertainment show consisting of a range of human and animal acts, e.g. trapeze, clowns, acrobats, horse riding acts, acts involving trained lions, tigers, bears and other species. Often travels from place to place in specialised vehicles and holds performances in a tent, but may use a permanent building. The first circus in England was established by Philip **ASTLEY**. Animal acts are now rare in British circuses largely as a result of public objections and the activities of animal welfare organisations (e.g. the **CAPTIVE ANIMALS' PROTECTION SOCIETY (CAPS), ROYAL SOCIETY FOR THE PREVENTION OF CRUELTY TO ANIMALS (RSPCA)**). The animals were often exhibited in their cages as a **TRAVELLING MENAGERIE**. *See also* **CHIPPERFIELD, RADFORD REPORT, STANDARDS FOR EXHIBITING CIRCUS ANIMALS IN NSW, TRAVELING EXOTIC ANIMAL PROTECTION ACT (USA), WELFARE OF WILD ANIMALS IN TRAVELLING CIRCUSES (ENGLAND) REGULA-**

TIONS 2012. In law, the definition depends upon the legislation concerned.

2. In England and Wales the **ZOO LICENSING ACT** 1981 (s.21(1)), '. . . *"circus" means a place where animals are kept or introduced wholly or mainly for the purpose of performing tricks or manoeuvres at that place'*. Under the **DANGEROUS WILD ANIMALS ACT** 1976 (s.7(4)) the definition is essentially the same.

3. In Nova Scotia, Canada, a circus is defined under the Standards Exhibiting Circus Animals in Nova Scotia (1999) as '*Any mobile establishment in which animals held and exhibited therein are made to perform behaviours at the behest of human handler/trainers for the entertainment and/or education of members of the public.*'

4. A large circular or oval stadium constructed in Roman times in which contests between gladiators and animals, and other spectacles were held. They were constructed in many places apart from Rome (including Britain) and very large numbers of animals were killed for the entertainment of the audience. Other animals were trained to perform tricks. The best-known circus is the **CIRCUS MAXIMUS**. *See also* **COLOSSEUM**

Circus Maximus A large Roman stadium surrounding a sand-covered arena which could hold 200 000 spectators. Used for chariot races, displays of equestrian and acrobatic skills and animal spectacles. Construction began in 329BC. *See also* **CIRCUS**, **COLOSSEUM**

circus parade A procession of **CIRCUS** performers and animals that used to take place through a town or city to publicise the arrival of the circus. Such parades often included elephants (Fig. C5).

CITES Convention on International Trade in Endangered Species of Wild Fauna and Flora 1973 (CITES). A convention which restricts international movements of protected species including movements involving zoo animals. CITES prohibits international commercial trade in the rarest species and requires licences for the movement of some other rare species. The convention regulates trade in whole animals and plants, living or dead, and recognisable parts and derivatives. The protected species are listed in three appendices: I, II and III (Table C4). Trade in endangered species is regulated by the requirement for import and export licences. The strictest restrictions apply to Appendix I species (Art. III). Parties to CITES meet biennially and may agree to add or remove species from the appendices or move them for one appendix to another as their status improves or deteriorates. *See also* **CITES APPENDIX I, CITES APPENDIX II, CITES APPENDIX III**

CITES Appendix I Includes all species threatened with extinction which are or may be affected by trade (Table C4).

CITES Appendix II Includes all species which may become threatened with extinction if trade is not strictly regulated (and other species which must be subject to strict regulation in order to achieve this objective) (Table C4).

CITES Appendix III Includes other species which any party strictly protects within its own jurisdiction and which requires the co-operation of other parties in the control of trade (Table C4).

citric acid cycle *See* **KREBS CYCLE**

clade In **CLADISTICS**, the taxonomic group represented by a branch in a **CLADOGRAM** along with all of the branches that descend from it.

cladistics, phylogenetic systematics An objective system of **CLASSIFICATION** which examines the evolutionary relations between taxa by determining whether the characters they possess are **HOMOLOGOUS (1)** characters or **ANALOGOUS** characters.

Fig. C5 Circus parade: the circus comes to an American town, c1900. (Source: Library of Congress Prints and Photographs Division, Washington DC.)

Table C4 CITES: Examples of taxa listed on CITES Appendices I, II and III (May 2012).

Appendix I	
Red panda	*Ailurus fulgens*
Tiger	*Panthera tigris*
Common otter	*Lutra lutra*
Madagascar red owl	*Tyco soumagnei*
Resplendent quetzal	*Pharomachrus mocinno*
Appendix II	
Hippopotamus	*Hippopotamus amphibius*
Striped civet	*Fossa fossana*
Fennec fox	*Vulpes zerda*
Southern elephant seal	*Mirounga leonine*
African penguin	*Spheniscus demersus*
Appendix III	
Golden jackal	*Canis aureus* (India)
Northern naked-tailed armadillo	*Cabassous centralis* (Costa Rica)
Alligator snapping turtle	*Macrochelys temminckii* (USA)
Satyr tragopan	*Tragopan satyra* (Nepal)
Galapagos sea cucumber	*Isostichopus fuscus* (Ecuador)

cladogram A branching diagram that illustrates the **PHYLOGENETIC** relationships of taxa, in which two groups branch from a single point. It begins with a single node, from which two or more lines branch, one or more of the lines lead to other nodes, from which two or more other lines may branch, and so on.

clan Social groups are called clans when cultural linkages also reflect common ancestry and/or are shared by individuals that live together. Clans occur in hyenas (*Crocuta crocuta*), and killer whales (*Orcinus orca*) form vocal clans based on similarities in their vocalisations. In African elephants (*Loxodonta africana*) clans are families and **BOND GROUPS** which use the same dry season home range (Fig. S12).

clasper A modified fin found in male sharks and some other cartilaginous fishes which is used in transferring sperm to the female.

class In **TAXONOMY**, a group of related **ORDERS**; subdivision of a **PHYLUM**, e.g. class **MAMMALIA** (Table C5).

classical conditioning, Pavlovian conditioning, type I conditioning A form of **LEARNING** in which a normal **RESPONSE** to a **STIMULUS** (the unconditional stimulus) becomes associated (or paired) with a new stimulus (the conditional stimulus). In his famous experiment, the Russian physiologist Ivan Petrovich **PAVLOV** conditioned a dog to associate the sound of a bell with the presence of food, thereby causing it to salivate when it heard the sound. *See also* **OPERANT CONDITIONING**

classical fitness *See* **DIRECT FITNESS**

classical swine fever (CSF), hog cholera A highly contagious, often fatal, viral disease of domesticated pigs, wild boars (*Sus scrofa*) and some other members of the Suidae. It is a **NOTIFIABLE DISEASE**. Signs include coughing, skin discoloration, abortion, weakness of the hind quarters and constipation followed by diarrhoea.

Table C5 A classification of the okapi (*Okapia johnstoni*) (partly based on Nowak, 1999).

Kingdom	Animalia
Phylum	Chordata
Subphylum	Vertebrata
Class	Mammalia
Subclass	Theria
Infraclass	Eutheria
Order	Artiodactyla
Suborder	Ruminantia
Infraorder	Pecora
Superfamily	Giraffoidea
Family	Giraffidae
Genus	*Okapia*
Species	*johnstoni*

classification

1. Any method which systematically organises the diversity of living and extinct organisms into groups based on a particular set of rules.

2. A statement of the particular taxonomic groups to which an organism belongs, e.g. the phylum, class, order, family etc. (Table C5).
See also **BINOMIAL SYSTEM OF NOMENCLATURE, CLADISTICS, Linnaeus, SYSTEMATICS, TAXONOMY**

clear-felling The wholesale clearance of forests by commercial logging companies. This activity has resulted in **HABITAT FRAGMENTATION** and the encroachment of humans into tropical forests resulting in greater access to species used as **BUSHMEAT**.

cleidoic egg The self-contained egg of an **AMNIOTE** **(2)** (egg-laying mammal, bird or reptile) or insect.

Clever Hans A horse owned by Wilhelm von Osten in Germany who appeared to be able to count, spell, make arithmetical calculations and communicate other things by stamping his hoof the correct number of times. In 1907 Oskar Pfungst, a psychologist, demonstrated that *Clever Hans* was responding to subtle cues from his owner and could only respond correctly if von Osten knew the answer to the question.

clicker A small hand-held metal device that makes a clicking sound. The sound is produced when a small, thin metal sheet is depressed by the thumb and then released. Used in **CLICKER TRAINING**.

clicker training A means of **TRAINING** an animal by causing it to form an association between the sound of a **CLICKER**, the performance of a specific behaviour and the receipt of a reward. The clicker sound acts as a **BRIDGE** that indicates the specific behaviour that is being rewarded.

climate change Climate change may be defined as: '*a statistically significant variation in either the mean state of the climate or in its variability, persisting for an extended period (typically decades or longer). Climate change may be due to natural processes or external forcing, or to persistent anthropogenic changes in the composition of the atmosphere or in land-use*' (Anon., 2001). Global warming is affecting some bird migrations. Researchers in Finland have found that some waterfowl delayed migrations by up to a month compared with 30 years ago. Studies of birds in North America have suggested that climate change is causing many species to become lighter and grow smaller wings in response to temperature increase (*see* **BERGMANN'S RULE**). In 2007 the **UNITED STATES FISH AND WILDLIFE SERVICE (USFWS)** removed the grizzly bears (*Ursus arctos*) in **YELLOWSTONE NATIONAL PARK** from the endangered species list. The US 9th Circuit Court of Appeals subsequently reversed this decision citing climate change as its reason. Climate change has accelerated a beetle infestation that destroys the bears' white-bark pine food source. The only other species protected by law in the US (under the **ENDANGERED SPECIES ACT OF 1973 (USA)**) as a result of climate change is the polar bear (*U. maritimus*) due to the retreat of Arctic ice.

climax community The relatively stable group of organisms which exists in equilibrium with existing environmental conditions present at the final stage of an **ECOLOGICAL SUCCESSION**.

clinical Concerned with the observation or treatment of disease.

clinical shock *See* **SHOCK**

clinical signs *See* **SIGN (1)**

cloaca
1. The vent. The terminal part of the gut in most vertebrates (except most mammals but including monotremes), where the alimentary canal, urinary and reproductive systems open into a single aperture.
2. The terminal section of the gut in some invertebrates, e.g. in nematodes and sea cucumbers.

cloacal kiss The coming together of the **CLOACAS (1)** of male and female individuals during mating, especially in birds and amphibians, during which sperm is transferred from the male to the female.

clone
1. An animal which is genetically identical to another, usually produced artificially; identical twins are clones.
2. A group of genetically identical organisms or **CELLS**.

cloning The process of producing a **CLONE**. Cloning is the creation of an exact genetic copy of another organism. This occurs naturally whenever identical twins are produced. Cloning has the potential to assist with captive breeding programmes, especially when the total number of individuals of a species that remains is very small. Cloning can be achieved in two different ways: **ARTIFICIAL EMBRYO TWINNING** or **NUCLEAR TRANSFER**.

close season, closed season The period during the year when hunting of a particular species of animal (e.g. **DEER, WILDFOWL**) is not allowed. Close seasons were established in England and Wales by the Game Act 1831. The timing and duration of the close season varies between species and jurisdictions. For example, the close season for pheasants (*Phasianus colchicus*) in England and Wales is 1 February–1 October, but for woodcock (*Scolopax rusticola*) it is 1 February–30 September. However, in Scotland the close season for woodcock is 1 February–31 August. The purpose of close seasons is generally to protect animals during the **BREEDING SEASON**. *See also* **GAME BIRD**

closed circulatory system A **CIRCULATORY SYSTEM** in which the blood is contained within a heart and a network of blood vessels. Such systems occur in all vertebrates. *Compare* **OPEN CIRCULATORY SYSTEM**

closed season *See* **CLOSE SEASON**

clotting *See* **BLOOD CLOTTING**

clotting factor Any of a group of **PROTEINS** in the **BLOOD** which is essential to the **CLOTTING** process.

cloud forest A tropical forest ecosystem characterised by an almost constant cloud cover, even during the dry season, which usually occurs near coastal mountain peaks.

cloven-hoofed species A species of mammal whose hoof is divided into two distinct toes. Found in members of the **ARTIODACTYLA**.

Clupeiformes An order of fishes: sardines, herrings, anchovies and their allies.

cluster analysis A mathematical method which allocates items into groups (clusters) such that all members of a group are more similar to each other (in some respect or other) than they are to members of other groups. These clusters may be further combined into larger clusters which are similar in a hierarchical fashion, until all of the larger clusters form a single group. When illustrated as a graph these relationships form a branching structure called a dendrogram.

clutch A group of eggs produced by a bird, reptile or amphibian at the same time.

clutch size The number of eggs (of a bird or reptile) in a **CLUTCH**.

CMS *See* **CONVENTION ON THE CONSERVATION OF MIGRATORY SPECIES OF WILD ANIMALS 1979 (CMS)**

Cnidaria A phylum of simple aquatic organisms, including coral-forming species and jellyfish. Cnidarians possess body organs. Their tentacles are covered in stinging cells called nematocysts used for catching food. They have a mouth and a blind-ending gastrovascular cavity. Most cnidarians exhibit an alternation of generations in which they alternate in form between a polyp – a fixed (immobile) stage – and a medusa. The polyp has tentacles and is attached to the substratum and the medusa is free-swimming and has a body that looks like the typical jellyfish. Corals are made up of a vast number of tiny polyps.

CNS *See* **CENTRAL NERVOUS SYSTEM (CNS)**

coadaptation The parallel evolution of two species which allows them to coexist, e.g. predators and their prey.

coagulation *See* **BLOOD CLOTTING**

coalition An association of individuals of the same species who act together for their mutual benefit, e.g. coalitions of (usually related) male lions (*Panther leo*) may co-operate to take possession of a pride by driving away the resident males.

coarse fishing The catching of freshwater fish other than salmon or trout. Coarse fishing is generally subjected to a **CLOSE SEASON**.

cobalt A mineral nutrient. A component of vitamin B$_{12}$ and therefore required for its synthesis. **RUMINANTS** have a relatively high requirement due to the inefficient production of B_{12} in the RUMEN, and poor absorption of the vitamin in the SMALL INTESTINE.

Cobbe, Frances Power (1822–1904) A writer who campaigned against vivisection and founded the SOCIETY FOR THE PROTECTION OF ANIMALS LIABLE TO VIVISECTION in 1875. She is generally considered to be the founder of the ANIMAL RIGHTS movement.

coccidiosis A parasitic disease of livestock and poultry, caused by a coccidian PROTOZOAN, which affects the intestines.

cochlea A tubular auditory organ found within the skull of mammals, birds and some reptiles. *See also* INNER EAR

cockfight A blood sport consisting of a contest between two cockerels in a ring called a cockpit. Now illegal in many countries including the UK, most of Europe, the USA and Australia.

Cockfighting Act 1952 An Act to make it unlawful in Great Britain to have possession of any instrument or appliance designed or adapted for use in connection with the fighting of a domestic fowl. Cockfighting was banned in Great Britain by the CRUELTY TO ANIMALS ACT 1835. *See also* COCKFIGHT

cocoon A hollow structure containing a developing or resting stage of an organism, e.g. butterflies, earthworms, spiders.

code of practice A document produced for guidance. This is not law, however, some codes of practice may be based on legal requirements. *See also* POLICY (2)

codominance
1. In genetics, the phenomenon whereby HETEROZYGOTES exhibit fully the phenotypic effects of both ALLELES at a gene LOCUS, rather than one allele being DOMINANT and the other RECESSIVE. *Compare* DOMINANCE (1)
2. In animal behaviour, a relationship in which two or more individual animals occupy the same position in a DOMINANCE HIERARCHY.

codon A group of three adjacent bases (triplet) on MESSENGER RNA (mRNA) or DNA that specifies the location of a particular AMINO ACID or indicates the beginning or end of the synthesis of a PROTEIN. *See also* ANTICODON, MICROSATELLITE MARKER

Coe, Jon Jon Coe is an influential zoo exhibit designer. He is a landscape architect and formerly held an academic post at the University of Pennsylvania. Coe worked for a number of major companies involved in zoo design, including JONES & JONES ARCHITECTS AND LANDSCAPE ARCHITECTS LTD of Seattle, before founding his own company, Jon Coe Design Pty Ltd., in Victoria, Australia. He has worked on over 150 design projects for over 60 zoos, aquariums, museums and similar organisations. These have included the *Gorillas of Cameroon* exhibit at Zoo Atlanta, the Asian Forest and Elephant Exhibit at Taronga Zoo, and the African Plains

Table C6 Coefficient of relatedness (r).

Relationship	r
Non-relatives	0
Full siblings	0.5
Parent – offspring	0.5
Half-siblings	0.25
Uncle/ aunt – nephew/ niece	0.25
Grandparent – grandchild	0.25
Cousin – cousin	0.125

Master Plan at Metro Toronto Zoo. Coe has published a wide range of academic papers on zoo exhibit design.

coefficient of (genetic) relatedness A measure of the extent to which two animals are genetically related. The probability that two individuals have inherited the same ALLELE from a common ancestor (Table C6). *See also* ALTRUISM, INCLUSIVE FITNESS

Coelacanthiformes An order of fishes: coelacanths.

coelomate Possessing a fluid-filled body cavity (coelom). Formed within the mesoderm of some animals. In mammals, the coelom forms the peritoneal, pleural and pericardial cavities. *Compare* ACOELOMATE

coexistence In ecology, the phenomenon whereby organisms, which might potentially compete for resources, live together in the same environment. This is the result of evolution which has modified their ecological requirements so that they do not completely overlap. *Compare* COMPETITION, COMPETITIVE EXCLUSION PRINCIPLE

cognition The mental process of knowing. This includes aspects such as awareness, PERCEPTION, INTELLIGENCE, PROBLEM SOLVING and judgement. These mental process cannot be observed directly but are presumed to occur within some animals and involve the manipulation of specific knowledge. Any kind of mental abstraction of which an animal appears capable, e.g. possessing a mental picture of the geography of an area, appearing to understand what another individual is thinking. Whether or not animals can think is a controversial question. *See also* DECEIT, EMPATHY, MORALITY

cognitive enrichment ENVIRONMENTAL ENRICHMENT where an appropriate cognitive challenge (which requires reasoning, problem solving ability etc.) results in measurable beneficial changes to an animal's wellbeing.

cognitive research Studies of COGNITION in animals.

cohort A group of animals born at the same time, e.g. in the same year or the same BREEDING SEASON.

cold desert *See* DESERT

coldblood A HORSE TYPE consisting of heavy draught horses. *See also* SHIRE HORSE

Coleoptera A large order of insects: beetles.

colic Sever spasmodic pain in the abdomen. Sometimes caused by an obstruction in the intestine.

Coliiformes An order of birds: mousebirds.

colitis INFLAMMATION of the LARGE INTESTINE. Signs include diarrhoea and abdominal pain.

collagen A strong, fibrous, inelastic PROTEIN which is a major constituent of BONE and CONNECTIVE TISSUE.

collapse

1. In relation to blood vessels, airways or lungs, to become flattened. *See also* PROLAPSE

2. To fall or drop into a state of unconsciousness.

3. To drop into a state of exhaustion or helplessness.

collection In a zoo context, an animal collection (and possibly a plant collection), i.e. a zoo, aquarium etc. Considered by some zoo professionals to be an outdated term. Some prefer to use the term LIVING RESIDENT POPULATIONS. *See also* INSTITUTIONAL COLLECTION PLAN (ICP), REGIONAL COLLECTION PLAN (RCP)

collection plan *See* INSTITUTIONAL COLLECTION PLAN (ICP), GLOBAL COLLECTION PLAN, REGIONAL ANIMAL SPECIES COLLECTION PLAN (REGASP), REGIONAL COLLECTION PLAN (RCP)

College of African Wildlife Management (CAWM) A college founded in 1963 at Mweka, near Moshi, Tanzania, to train African wildlife managers, as a result of the ARUSHA CONFERENCE.

colobomas Eye lesions. *See also* MULTIPLE OCULAR COLOBOMA (MOC)

colon Part of the LARGE INTESTINE.

colony A collective term for a group of individuals of certain species, e.g. gulls, seals, ants, termites, bees etc. *See also* SUPER-ORGANISM

coloration The colour patterns on the surface of an animal's body. May be important in THERMOREGULATION because dark-coloured surfaces absorb more radiant heat than those that are light coloured. May also be important in ADVERTISEMENT, CAMOUFLAGE, MIMICRY.

Colosseum The largest of the Roman amphitheatres where large numbers of elephants, lions, tigers, bears, rhinos and hippos were killed for entertainment (Fig. C6). The arena comprised a wooden floor

Fig. C6 Colosseum: the hypogeum in the centre of the picture was a network of tunnels and cages where animals were kept, and was originally covered by a wooden floor. (Courtesy of Sandra Hickman.)

covering an underground network of tunnels and cages (hypogeum) where animals and gladiators were kept prior to contests. *See also* **CIRCUS MAXIMUS**

colostrum A yellowish milky fluid produced by the mammary glands of mammals immediately before and after giving birth before true milk is secreted. It is rich in nutrients and maternal antibodies and is important in providing immunity to the neonate.

colostrum supplement A food supplement for young **LIVESTOCK** (e.g. lambs and calves) which contains **NUTRIENTS** and **ANTIBODIES**.

colt
1. An uncastrated male horse, pony, donkey or mule which is under 2 years of age.
2. The **AMERICAN STUD BOOK** (2008) defines a colt as '*an entire male horse 4 years old or younger*'.

Columbiformes An order of birds: sandgrouse, pigeons (Fig. V1), doves.

coma A prolonged state of deep unconsciousness from which the individual cannot be awakened, possibly the result of a head injury, **STROKE**, brain damage etc.

comb claw *See* **TOILET CLAW**

Combined Flock Book The register of eight sheep breeds (Boreray, Castlemilk Moorit, Manx Loaghtan, Norfolk Horn, North Ronaldsay, Portland, Soay and Whitefaced Woodland) for which the **RARE BREEDS SURVIVAL TRUST (RBST)** acts as the breed society.

comfort behaviour Behaviour concerned with body care, e.g. bathing, **DUST BATHING**, **GROOMING**, **PREENING**, **SUN BATHING** etc.

Common Agricultural Policy (CAP) A system of subsidies used to support agriculture in the Member States of the EU, with the aim of providing farmers with a reasonable standard of living, producing good quality food at fair prices, stabilising markets, securing food supplies and protecting the rural heritage.

Common Fisheries Policy (CFP) The system within the EU by which fisheries are managed which brings together a range of measures designed to achieve a thriving and sustainable European fishing industry. This includes the use of **FISHING QUOTAS** and preventing the further expansion of the European fishing fleet.

Commonwealth Scientific and Industrial Research Organisation (CSIRO) The Australian government body which undertakes scientific research. It was founded in 1926 and consists of a large number of Divisions concerned with a wide range of scientific disciplines, some of which conduct research on livestock, sustainable agriculture, marine systems, entomology and ecosystems.

communal nesting This occurs when two or more females (especially mammals) and their dependent young occupy a common nest or burrow. It does not

Fig. C7 Companion animal.

necessarily involve communal care of the young. Occurs widely in rodents, e.g. degus (*Octodon degus*). *See also* **CO-OPERATIVE BREEDING**

communicable disease A disease which is infectious or contagious and can be transmitted from one organism to another through physical contact, droplet infection, etc.

communication The transmission of information from one animal to another, designed to influence the current or future behaviour of the recipient. It may be visual, auditory, olfactory, gustatory or electrical. *See also* **DISPLAY**, **INTERSPECIES COMMUNICATION**, **INTERSPECIES SEMANTIC COMMUNICATION**, **LANGUAGE**, **NON-VERBAL LANGUAGE**, **OLFACTORY COMMUNICATION**

community *See* **BIOLOGICAL COMMUNITY**

compact bone *See* **BONE**

companion animal
1. An alternative term to 'pet'. Any of a range of animals kept by humans as a companion including dogs, cats, rabbits, parrots etc. (Fig. C7). The legal definition of a companion animal varies between jurisdictions.
2. Under New York State's Agriculture and Markets Law §350 (5) '*"Companion animal" or "pet" means any dog or cat, and shall also mean any other domesticated animal normally maintained in or near the household of the owner or person who cares for such other domesticated animal. "Pet" or "companion animal" shall not include a "farm animal" as defined in this section.*'
3. In Australia, under s.5 of the **COMPANION ANIMALS ACT 1998 (NSW)** a companion animal means '*(a) a dog, (b) a cat, (c) any other animal that is prescribed by the regulations as a companion animal.*'

Companion Animal Welfare Council (CAWC) An independent organisation established in the UK in 1999 and funded through a charitable trust (the Welfare Fund for Companion Animals), which is dedicated to improving our understanding of **COMPANION ANIMAL** welfare and of the role of companion animals in society. It assesses existing welfare legislation and makes recommendations regarding changes to the law.

Companion Animals Act 1998 (NSW) An Act in New South Wales, Australia, which provides for the identification and registration of COMPANION ANIMALS (3) and establishes the duties and responsibilities of their owners.

comparable life test A test used to establish whether or not an animal should be kept in a zoo by asking if the zoo can provide a life at least as good as the life the animal could expect in the wild. If it can, the zoo passes the comparable life test in relation to that animal. *Compare* BASIC NEEDS TEST

comparative psychology A term used by psychologists for the study of animal BEHAVIOUR.

compass sense The ability to move in a particular compass direction without reference to landmarks, using the position of the sun, stars, magnetic fields or some other environmental cue. *See also* HOMING, MAGNETIC ORIENTATION, NAVIGATION

Compassion in World Farming (CIWF) A UK-based charity which campaigns against cruel practices in farming including the live transport of livestock. *See also* TRANSPORT OF LIVE ANIMALS

competition The process that occurs when two or more animals use a scarce resource, e.g. food, space, nesting places, etc. Competition may be between individuals of the same species (INTRASPECIFIC COMPETITION) or between individuals of different species (INTERSPECIFIC COMPETITION). *See also* COMPETITIVE EXCLUSION PRINCIPLE, NICHE SEPARATION. *Compare* COEXISTENCE

competitive exclusion principle In ecology, the theory that species which are complete competitors cannot coexist because they require exactly the same resources from the environment. This should result in one species surviving and the other becoming extinct. Evolution has produced species that have adapted so that they can coexist. *See also* INTERSPECIFIC COMPETITION, INTRASPECIFIC COMPETITION, NICHE SEPARATION. *Compare* COEXISTENCE

complementarity In animal breeding, the additive genetic effect of selecting breeds which possess traits that complement each other when CROSS-BREEDING.

complementary medicine Alternative therapies that may be used to treat animals instead of using medication or surgery. May include massage, muscle stimulation, chiropractic therapy, homeopathy, herbal remedies, healing therapies, nutritional therapy, flower essences and acupuncture. Many of these practices have no scientific foundation.

complete feed-style feeding The provision of processed food which offers no choice to the animal but provides all of the required nutrients. *Compare* CAFETERIA-STYLE FEEDING (CSF)

composite breed In relation to LIVESTOCK, a new BREED composed of other existing breeds. Once created, the composite breed can be mated like a conventional (straight) breed.

compound microscope A standard microscope which is used for examining specimens using transmitted light (i.e. from beneath the specimen) through one or two eyepieces (binocular) and objective lenses of magnifications generally between ×5 and ×100.

computed tomography scan *See* CT SCAN, TOMOGRAPHY

computer-assisted remote hunting *See* INTERNET HUNTING

concentrates Concentrated food products from which most of the water or other liquid content has been removed, e.g. PROTEIN CONCENTRATES. Generally produced in the form of pellets or a powder. Widely fed to pets, farm animals, farmed fish and species kept in zoos as a FOOD SUPPLEMENT or an alternative to fresh food.

conception The fertilisation of an EGG (1) by a SPERMATOZOON; the beginning of PREGNANCY (1).

condition score *See* BODY CONDITION SCORE

conditioning *See* ASSOCIATION (1), CLASSICAL CONDITIONING, OPERANT CONDITIONING, PAVLOV

cone A type of PHOTORECEPTOR cell found in the RETINA that is capable of detecting colour. *See also* ROD (1)

Conference of the Parties A periodic meeting of the parties to an INTERNATIONAL CONVENTION, at which the treaty may be amended, e.g. the parties to **CITES** meet biennially to make amendments to the species listed in the appendices.

conflict A state in which an animal is motivated to perform more than one activity simultaneously. This may result in DISPLACEMENT ACTIVITY.

confounding variable In an experiment or study, a variable that causes a nuisance effect which makes it impossible to distinguish a potential effect of interest.

congeneric Belonging to the same GENUS. *See also* CONSPECIFIC

congenital Present at or before birth.

congenital disease A disease which is present at or before birth.

congestive heart failure *See* HEART FAILURE

coniferous plant An evergreen tree or shrub with needle-like leaves which produces its pollen and seeds in cones. *Compare* DECIDUOUS PLANT

conjunctiva The membrane which covers the front of the eye and the inside of the eyelids.

conjunctivitis Inflammation of the CONJUNCTIVA which may be caused by dust, sand, pollen, seeds and other atmospheric contaminants, but also flies, ticks and worms and bacterial infections.

Connect A journal published by the ASSOCIATION OF ZOOS AND AQUARIUMS (AZA).

connective tissue Tissue which has a support, packing and defensive function in vertebrates, largely made up of COLLAGEN and elastin fibres.

C

consciousness A mental state recognised in humans in which individuals are aware of their own existence, their environment, sensations and thoughts, and which may (or may not) exist in some other species. *See also* **COGNITION, RED SPOT TEST**

conservation
1. The protection of wildlife from irreversible harm (Hambler, 2004). The term includes rational use (*see* **CONVENTION ON THE CONSERVATION OF ANTARCTIC MARINE LIVING RESOURCES 1980 (CCAMLR)**).
2. According to the **WORLD CONSERVATION STRATEGY (1980)** conservation is: '*The management of human use of the biosphere so that it may yield the greatest sustainable benefit to the present generation while maintaining its potential to meet the needs and aspirations of future generations.*'

conservation behaviour A branch of **CONSERVATION BIOLOGY** which examines the mechanisms, development, function and **PHYLOGENY** of behavioural variation in an attempt to devise tools to conserve wildlife. *See also* **BEHAVIOURAL ECOLOGY**

conservation biology The scientific study of those aspects of the biology of species that affect their survival (e.g. reproduction, disease, human activity etc.), the management of **ECOSYSTEMS** and **HABITATS**, and the assessment and protection of **BIODIVERSITY**.

conservation breeding A term that has replaced '**CAPTIVE BREEDING**' which refers to the breeding of animals in captivity for conservation purposes. *See also* **EUROPEAN ENDANGERED SPECIES PROGRAMME (EEP), SPECIES SURVIVAL PLAN® (SSP) PROGRAMS**

Conservation Breeding Specialist Group (CBSG) A **SPECIALIST GROUP** of the **INTERNATIONAL UNION FOR THE CONSERVATION OF NATURE AND NATURAL RESOURCES (IUCN)** whose mission is to save threatened species by increasing the effectiveness of conservation efforts worldwide. It does this by developing and disseminating innovative and interdisciplinary science-based tools and methodologies; providing culturally sensitive and respectful facilitation that results in conservation action plans; promoting global partnerships and collaborations; and fostering contributions of the **CONSERVATION BREEDING** community to species conservation. It was formerly known as the Captive Breeding Specialist Group.

conservation corridor, wildlife corridor A strip of land which connects two or more isolated areas within a fragmented habitat, thereby allowing the **DISPERSAL** of animals and plants and enabling **GENE FLOW** between them, e.g. a forest corridor connecting two isolated areas of forest. Important in the conservation of wild animal species, especially when population sizes are small. Corridors may be artificially created as a conservation measure or an area of land may be protected so that it may act as a corridor.

conservation grazing The use of grazing **LIVESTOCK** to manage woodlands, grasslands and other habitats by virtue of their ability to practise **SELECTIVE GRAZING** on the available plant species. Some conservation organisations employ a conservation grazing manager to manage their animals, e.g. the Woodland Trust. Traditional breeds are often used such as English longhorn cattle (Fig. C1). *See also* **GRAZING FACILITATION, HEFTING, PLEISTOCENE PARK**

Conservation International A large international conservation non-governmental organisation founded in 1987 and based in the USA.

conservation medicine, veterinary conservation medicine An emerging discipline which studies the health relationships that occur at the interface of animals, humans and ecosystems, encompassing aspects of human health, animal health, environmental health, wildlife medicine and **CONSERVATION BIOLOGY**.

Conservation (Natural Habitats, &c.) Regulations 1994 A **STATUTORY INSTRUMENT** which transposed the **HABITATS DIRECTIVE** into national law in England, Wales and Scotland.

Conservation of Habitats and Species Regulations 2010 A **STATUTORY INSTRUMENT** which consolidates all of the amendments made to the **CONSERVATION (NATURAL HABITATS, &C.) REGULATIONS 1994** in respect of England and Wales.

conservation officer In the context of a zoo, a person who raises conservation awareness, raises funds for conservation projects and participates in *IN-SITU* **CONSERVATION** programmes including **CAPACITY-BUILDING**.

conservation sensitive Referring to a taxon which is declining in the wild due to habitat loss, **POACHING**, small **POPULATION** size or for some other reason and would benefit from increased conservation effort. *See also* **RED LIST**

consorts A male and a female of a species who form a relationship during the breeding season or, in mammals, during her **OESTRUS** whereby they move and feed together and engage in frequent sexual activity and mutual **GROOMING**. *See also* **PAIR BOND**

conspecific Belonging to the same species. *See also* **CONGENERIC**. *Compare* **HETEROSPECIFIC**

consummatory behaviour Behaviour which brings a period of searching or **APPETITIVE BEHAVIOUR** to an end, e.g. locating food will bring to an end **FORAGING BEHAVIOUR**. *See also* **GOAL**

contact call A **VOCALISATION** made by an animal (e.g. a monkey) in order to remain in contact with others in its group. Particularly useful in complex habitats such as forests.

containment The means by which an animal in a zoo is held safely and separated from visitors and staff by barriers.

***Containment Facilities for Zoo Animals*, Standard 154.03.04 (NZ)** A standard for the containment of zoo animals in New Zealand which has been approved in accordance with the **HAZARDOUS SUBSTANCES AND NEW ORGANISMS ACT 1996 (NZ)**.

containment standard Specification for the design of an **ANIMAL ENCLOSURE** with respect to the nature, size and strength of the barrier (fencing etc.). Generally applies to zoos. May be laid down in detail for various taxa in some jurisdictions, e.g. New Zealand. *See also* **CONTAINMENT FACILITIES FOR ZOO ANIMALS STANDARD 154. 03. 04 (NZ)**.

containment, primary Primary containment is the main barrier to escape from an **ANIMAL ENCLO-SURE**, e.g. a **WET MOAT** around the perimeter of an enclosure for chimpanzees. *Compare* **CONTAINMENT SECONDARY**

containment, secondary Secondary containment is a second barrier to escape which is operative if the primary barrier is breached, e.g. a 'hot wire' used as a secondary barrier on the visitor side of a **WET MOAT**. *Compare* **CONTAINMENT PRIMARY**

continental drift The theory that the continents originated from a single land mass which split up and moved apart across the surface of the Earth as a result of geological process. This 'drift' is largely responsible for the global distribution of plant and animal species and for the similarities and differences found between taxa from different parts of the world. The unique faunas of many islands (e.g. Madagascar, Australia, the Galapagos Islands) are the result of their isolation from other land masses for long periods of time. *See also* **FAUNAL REGIONS**, **WALLACE'S LINE**

continuous variable A variable which may have any value, between certain limits, e.g. length, height, mass, volume. *Compare* **DISCRETE VARIABLE**

contraception A mechanism for preventing pregnancy. *See also* **CASTRATION**, **HORMONE IMPLANT**, **IMMUNO-CONTRACEPTION**, **NEUTER**, **STERILISE (2)**

contraceptive
1. Having the effect of preventing **PREGNANCY (1)**.
2. A drug or device having such an effect.

contrafreeloading Working for something, e.g. food, that could be obtained for little or no effort. Individuals of some species will expend considerable time, effort and energy to obtain small amounts of food scattered around an enclosure when other food is freely available at a feeding station.

control In scientific studies, the subjects which are not exposed to the experimental condition, e.g. to determine the effect of a **FEEDING ENRICHMENT** on the behaviour of a group of monkeys they would be studied before the enrichment was provided (the control condition) and when the enrichment was present (experimental condition). Alternatively two identical groups could be used one exposed to the enrichment and the other not (the control). *See also* **SCIENTIFIC METHOD**

controlled drug In the UK, a **DRUG** which is listed in the Schedules to the Misuse of Drugs Regulations 2001. Controlled Drugs are classified into five Schedules. Schedule 1 Controlled Drugs have the highest level of restriction. A veterinary surgeon has the authority to supply Schedule 2, 3, 4 and 5 Controlled Drugs. The Royal Pharmaceutical Society of Great Britain maintains a live database that indicates the legal classification of medicines.

contusion A bruise.

convention In international law, an agreement between two or more states to cooperate in dealings with each other in a particular area of activity, e.g. wildlife law, trade, environmental protection. Also known as a treaty.

Convention for the Conservation of Antarctic Seals 1972 An agreement made, in response to obligations created by the **ANTARCTIC TREATY**, to take measures to protect several species of Antarctic seals: southern elephant seal (*Mirounga leonine*), leopard seal (*Hydrurga leptonyx*), Weddell seal (*Leptonychotes weddelli*), crabeater seal (*Lobodon carcinophagus*), Ross seal (*Ommatophoca rossi*) and Southern fur seals (*Arctocephalus* spp.). These include the control of hunting, creation of seal reserves, imposition of close seasons etc. *See also* **CONVENTION ON THE CONSERVATION OF ANTARCTIC MARINE LIVING RESOURCES 1980 (CCAMLR)**

Convention for the Protection of the World Cultural and Natural Heritage 1972 A **UNITED NATIONS (UN)** (UNESCO) convention whose aim is to identify, protect, conserve, present and transmit to future generations the world's cultural and natural heritage. It protects sites of global importance because of their great cultural, historical or natural interest. They are designated as World Heritage Sites. Some are important because of their natural history interest, including the Great Barrier Reef (Australia), Wood Buffalo National Park (Canada), Sichuan **GIANT PANDA SANCTUARIES** (China), Virunga National Park, Democratic Republic of Congo, **NGORONGORO CONSERVATION AREA (NCA)** (Tanzania) and the **SERENGETI** National Park (Tanzania).

Convention on Biological Diversity 1992 (CBD) An international agreement whose objectives are '*the conservation of biological diversity, the sustainable use of its components and the fair and equitable sharing of the benefits arising out of the utilization of genetic resources, including by appropriate access to genetic resources and by the appropriate transfer of relevant technologies, taking into account all rights over those resources and to technologies, and by appropriate funding*' (Art. 1). The United Nations Conference on Environment and

Development (UNCED) was held in June 1992, in Rio de Janeiro, Brazil, and has come to be known as the 'Earth Summit'. At this summit the Convention on Biological Diversity 1992 was signed by 155 states and the EU, along with a programme of action for governments which is called 'Agenda 21'. By May 2013 all states had become parties to the convention except the USA, Andorra, the Holy See (Vatican) and South Sudan. *See also* **DARWIN INITIATIVE**

Convention on International Trade in Endangered Species of Wild Fauna and Flora 1973 (CITES). *See* **CITES**

Convention on the Conservation of Antarctic Marine Living Resources 1980 (CCAMLR) An international agreement made, in response to obligations created by the **ANTARCTIC TREATY 1959**, to take measures for the conservation (including rational use) of all of the marine living resources of the Antarctic (including birds).

Convention on the Conservation of European Wildlife and Natural Habitats 1979 An international agreement which aims to ensure the conservation of wild animal and plant species and their natural habitats. Protected species are listed in three appendices: I – Strictly protected flora species; II – Strictly protected fauna species; and III – Protected fauna species (migratory species). Appendix IV lists prohibited means and methods of killing and capture, and other forms of exploitation. Also known as the Berne Convention.

Convention on the Conservation of Migratory Species of Wild Animals 1979 (CMS) A convention which aims to protect migratory species. It provides strict protection for listed species in danger of extinction and aims to persuade range states to conclude agreements for the conservation of other species that have an unfavourable conservation status. Also known as the Bonn Convention.

Convention on the Law of the Sea 1982 (UNCLOS) An international agreement which defines the rights and responsibilities of nations in their use of the world's oceans. It establishes guidelines for businesses, the environment and the management of marine natural resources. It establishes jurisdictional limits on the ocean area that countries may claim, which include a 12-mile territorial sea limit and a 200-mile exclusive economic zone limit. *See also* **FISH STOCKS AGREEMENT 1995**

Convention on Wetlands of International Importance Especially as Waterfowl Habitat 1971 (Ramsar Convention) A convention whose aim is to protect **WETLANDS**, their flora and fauna, and to promote their wise use. Article 2.1 of the Convention requires the Contracting Parties to designate suitable wetlands for inclusion in a 'List of Wetlands of International Importance' which is maintained by the **INTERNATIONAL UNION FOR THE CONSERVATION OF NATURE AND NATURAL RESOURCES (IUCN)**. Ramsar sites include the Danube Delta

(Romania), Lake Naivasha (Kenya), Okavango Delta System (Botswana), Morecambe Bay (UK), Kakadu National Park (Australia) and the Everglades National Park (USA).

convergent evolution The development of similar characteristics by genetically unrelated taxa due to adaptation to similar environments, e.g. wings by insects and bats. *See also* **ANALOGOUS (1)**. *Compare* **DIVERGENT EVOLUTION**

Conway, William G. (1929–) Dr. William Conway is a zoologist, ornithologist and conservationist who played an important part in promoting the development of captive breeding programmes for endangered species. He was formerly Director of the New York Zoological Society and was responsible for modernising many of the exhibits at the Bronx Zoo. He later became president of the **WILDLIFE CONSERVATION SOCIETY (WCS)**. Conway led the development of the accreditation programme for the **ASSOCIATION OF ZOOS AND AQUARIUMS (AZA)** and has written extensively on zoos. He is currently a Senior Conservationist with the Wildlife Conservation Society.

co-operative behaviour Behaviour in which individuals of a species work together in a co-ordinated manner towards a common goal for their mutual benefit, e.g. hunting, **PARENTAL CARE**, **ANTI-PREDATOR BEHAVIOUR**. *See also* **ALTRUISM**, **KIN SELECTION**

co-operative breeding A social system in which some adults breed while other non-breeders assist in the care of young. *See also* **COMMUNAL NESTING**

co-ordination, motor co-ordination Skilful or balanced movement of the body.

copepods A large subclass of small crustaceans some of which occur in vast quantities in **PLANKTON**. Live copepods are available commercially as food for fishes, corals and other invertebrates.

coping *See* **BEAK COPING**

coping hypothesis A hypothesis that contends that the performance of **STEREOTYPIC BEHAVIOUR** is a mechanism which helps an animal cope with a captive environment. It is suggested that the physiological response produced by the 'coping' behaviour is less dangerous to the individual than the normal physiological response to the **STRESSOR**. This may be because the animal's perception of the stressor is modified to reduce the severity of its effects. *See also* **SELF-NARCOTISATION**

copper A mineral nutrient. Important in connective tissue and melanin synthesis. A component of enzymes that mobilise stored iron. Deficiencies are rare but may result in hepatic iron accumulation. Deficiency may occur where **RUMINANTS** are fed on molybdenum-rich soil or if there is excessive dietary zinc.

coprophagia *See* **COPROPHAGY**

coprophagy, coprophagia Coprophagy is the ingestion of faeces. It is natural for some species to do this. Some young mammals, e.g. pigs, dogs, non-

Fig. C8 Coprophagy: an adult male western lowland gorilla (*Gorilla gorilla gorilla*) eating his own faeces in a zoo.

human primates and elephants, eat the faeces of their parents when they are very young. This helps to populate their guts with the bacteria necessary for digestion. Some species (especially rabbits, hares and some rodents) produce soft faecal pellets of partially digested food (caecotrophs, cecotropes, or night faeces) in the night or early morning which they ingest to extract more nutrients and especially to increase their intake of vitamin K and B vitamins. They also produce hard pellets. Others animals, for example great apes, may develop coprophagy as a self-stimulatory response to living in captivity (Fig. C8).

coprophilia Playing with faeces: abnormal behaviour observed in some captive species. *See also* **COPROPHAGY**

copulation Coitus; mating; sexual intercourse. The process by which a male introduces sperm into the body of the female during mating.

copulation plug *See* **SPERM PLUG**

copying *See* **IMITATION, SOCIAL FACILITATION**

Coraciiformes An order of birds: kingfishers, bee-eaters, rollers, hoopoe, todies, motmots, hornbills.

corals A group of cnidarians whose polyps produce a calcareous skeleton which forms coral reefs. Kept by some hobbyists and important in the construction of some **AQUASCAPES**.

Corbett, Edward James (Jim) (1875–1955) A British hunter and conservationist who was a colonel in the British Indian Army and was frequently called upon to kill man-eating lions and leopards to protect local villagers. He wrote a number of books including the *Man Eaters of Kumaon* (1944). He played a key role in the establishment of India's oldest **NATIONAL PARK (NP)**, which was renamed Jim Corbett National Park in his memory.

core temperature The temperature of deep structures within the body, as opposed to the peripheral temperature of the skin.

cornea In vertebrates, the layer of transparent material in front of the eye, composed of layers of **COLLAGEN**, which refracts light onto the lens.

coronary
1. Relating to the heart.
2. Denoting vessels, nerves etc. that encircle an organ or other body part.

coronary thrombosis, heart attack A blockage, by a blood clot, of one of the two coronary arteries which supplies blood to the heart.

corpus luteum, yellow body A temporary endocrine gland which forms from the ruptured **GRAAFIAN FOLLICLE** and produces progesterone which, in mammals, maintains the uterine **ENDOMETRIUM** during pregnancy, but deteriorates if **FERTILISATION** does not occur. *See also* **OESTROUS CYCLE**

corral
1. A pen for keeping **LIVESTOCK** on a farm, ranch or zoo.
2. To put or keep livestock in a corral, or to gather them together in a confined space.

correlation In **STATISTICS**, a mathematical method that generates a single value (correlation coefficient) to represent the relationship between two variables. This value ranges from +1 (a perfect positive correlation) to −1 (a perfect negative correlation). A value of 0 indicates no correlation. The higher the absolute value of the correlation (i.e. disregarding the sign) the stronger the relationship between the variables. A positive correlation between variables A and B means that as A increases, so too does B. A negative correlation between these variables means that as A increases B decreases and vice versa. A correlation coefficient that is zero, or close to zero, suggests that there is no relationship between changes in A and B. A high correlation does not necessarily imply that changes in one variable causes changes in the other. They may both be correlated to a third variable that has not been considered. This type of analysis could be used to examine questions such as 'Is there a relationship between the amount of time an animal spends feeding and the amount of time it spends exhibiting **STEREOTYPIC BEHAVIOUR**?'

correlation coefficient *See* **CORRELATION**

corridor *See* **CONSERVATION CORRIDOR**

corticosteroids A group of **HORMONES** (including **CORTISOL**) produced by the **ADRENAL CORTEX** which are involved in a number of physiological systems including the **IMMUNE REACTION** and the suppression of **INFLAMMATION**.

cortisol A stress-response hormone that reflects the activity of the **HPA AXIS**. It increases **BLOOD PRESSURE (BP)** and **BLOOD SUGAR LEVELS** and has an immunosuppressive function. **STRESS** in mammals can be measured using non-invasive methods by analysing the levels of cortisol found in

C

serum, saliva or urine. As stress increases cortisol levels increase. Faecal cortisol metabolites have recently been identified as an index of stress.

corvid A member of the bird family Corvidae; crows, rooks and their relatives.

COSHH Regulations The Control of Substances Hazardous to Health (COSHH) Regulations 2002 (as amended). The law in Great Britain which regulates the use of hazardous substances at work, to prevent ill health, and places a duty on employers to protect their employees from such substances. It includes controls over the use and handling of a wide range of chemicals, including DRUGS, PESTICIDES, feed additives, cleaning agents, paints, glues, oil and materials that produce dusts, fumes, and even the handling of living materials such PATHOGENS and plants which may cause an ALLERGIC REACTION. The regulations require RISK ASSESSMENTS to be undertaken to examine how workers may be exposed to a hazardous substance and how exposure may be reduced, for example, by using protective clothing.

Council of Europe An organisation with 47 member countries representing almost the entire continent of Europe which seeks to develop common and democratic principles based on the European Convention on Human Rights and other reference texts on the protection of individuals. It is responsible for a number of international animal welfare laws and laws protecting wildlife, and has published guidelines on the reintroduction of animals to the wild, e.g. EUROPEAN CONVENTION FOR THE PROTECTION OF ANIMALS DURING INTERNATIONAL TRANSPORT 1971, EUROPEAN CONVENTION FOR THE PROTECTION OF ANIMALS KEPT FOR FARMING PURPOSES 1976. *Compare* EUROPEAN UNION (EU)

count
1. Animal count: an estimate of the number of animals in a particular location, e.g. a bird count.
2. The ability to determine the number of objects. Some birds appear to be able to count the number of eggs they have laid, e.g. American coots (*Fulica americana*).

counter-utilitarianism A philosophical position which gives greater consideration to the rights of the individual than to those of the group. This is the position adopted by animal welfare organisations that care about the suffering of individuals rather than the survival of species. *Compare* UTILITARIANISM

Countryside Council for Wales The statutory nature conservation agency for Wales.

county show *See* AGRICULTURAL SHOW

coursing The use of dogs to pursue and catch animals, especially GAME. Sometimes practised as a competitive sport in which dogs compete against each other to catch the prey. *See also* DEER COURSING, HARE COURSING, HUNTING ACT 2004

courtship Behaviour which leads to or initiates mating between animals. It may have a number of purposes including mate attraction, mate selection, mate assessment, and the synchronisation of reproductive behaviour. Synchronisation in sticklebacks (*Gasterosteus* spp.) occurs as a result of a REACTION CHAIN. Courtship may be necessary to reduce aggression, reinforce the PAIR BOND, and prevent wasted matings resulting in HYBRIDISATION. In birds the COURTSHIP DISPLAY of a male may induce a female to lay infertile eggs if they are kept in separate cages. Female whooping cranes (*Grus americana*) will not ovulate until the male has performed a 'dance'. Courtship behaviour in some species may include behaviours which have resulted from RITUALISATION, e.g. FOOD BEGGING. *See also* COURTSHIP FEEDING

courtship display A type of behaviour displayed by individuals of one sex (usually males) to attract the other during COURTSHIP (Fig. C9). May occur at specific locations called LEKS.

courtship feeding During courtship, the presentation of food by one partner to another, e.g. FOOD BEGGING is performed by some gulls.

Cousteau, Jacques-Yves (1910–1997) A French explorer and naturalist who specialised in studying and filming marine life and was a pioneer of marine conservation. Co-inventor of the Aqua-Lung.

cover
1. To mate with.
2. Places for animals or people to hide, e.g. thick vegetation.

cow lifter A device used on farms to lift sick or fallen cattle to their feet. May consist of a strong belt (or belts) passed under the animal and then fixed to a conventional hydraulic lifting device attached to a tractor, e.g. a front end loader or hay forks.

Cowper's gland *See* BULBOURETHRAL GLAND

coywolf, eastern coyote A hybrid produced when a western coyote (*Canis latrans*) is crossed with a grey wolf (*C. lupus*) or a red wolf (*C. lupus rufus*).

CPH number County parish holding number. A number assigned by the RURAL PAYMENTS AGENCY (an executive agency of the DEPARTMENT FOR ENVIRONMENT, FOOD AND RURAL AFFAIRS (DEFRA)) to a place which keeps cattle, pigs, sheep or goats for the purpose of identifying such places in case of a disease outbreak.

Crandall, Lee Sanders (1887–1969) Crandall was the Head of the Department of Birds at the New York Zoological Society until his retirement in 1952. He published widely and is best known as the author of *The Management of Wild Mammals in Captivity* (1964). He also wrote *A Zoo Man's Notebook* (1966) with William Bridges and *Pets. Their History and Care* (1917).

cranial Relating to the cranium (Fig. A8).

cranial fermenter *See* RUMINANT

Fig. C9 Courtship display of a male Indian peafowl (*Pavo cristatus*).

cranium The dome-shaped part of the skull which consists of several fused bones and protects the brain.

crate training Training an animal to enter a transportation crate prior to transportation, to reduce the stress associated with capture. Usually achieved by giving the animal free access to the open crate and encouraging it to feed inside.

creance A long light cord on which a bird of prey is secured during training.

creativity The ability to be creative, original or imaginative. Some cetaceans appear to exhibit creative behaviour. In experiments, rough-toothed dolphins (*Steno bredanensis*) were able to identify when they were being rewarded for originality in their behaviour. They were only rewarded when they exhibited behaviours which were not part of their normal repertoire.

crepuscular A term used to describe an animal that is primarily active at twilight, i.e. dawn and dusk. *Compare* CATHEMERAL, DIURNAL, MATUTINAL, NOCTURNAL, VESPERTINE

cribber, crib biter An animal who engages in CRIBBING.

cribbing, crib-biting A STEREOTYPIC BEHAVIOUR of EQUIDS, generally related to poor management. The horse repeatedly bites the MANGER or other hard horizontal surface, and pulls back while drawing in air (WIND SUCKING). *See also* BAR-BITING

cribbing collar *See* CRIBBING STRAP

cribbing strap, cribbing collar A device which is held between the teeth of a horse to prevent CRIBBING.

Crick, Francis (1916–2004) Co-discoverer, with James WATSON, of the structure of DNA while working at the University of Cambridge.

criminal offence An act or omission punishable under criminal law, e.g. in the UK, possessing the eggs of any wild bird is an offence under the WILDLIFE AND COUNTRYSIDE ACT 1981.

critical habitat *See* ENDANGERED SPECIES ACT OF 1973 (USA)

Critically Endangered (CR) A RED LIST category used by the INTERNATIONAL UNION FOR THE CONSERVATION OF NATURE AND NATURAL RESOURCES (IUCN): Critically endangered (CR) – Best available evidence indicates that the taxon is facing an extremely high risk of extinction in the wild.

Crocodilia An order of reptiles: crocodilians (alligators, caimans, crocodiles, gharials).

crocodilian
1. A member of the CROCODILIA.
2. Relating to a member of the CROCODILIA.

crop A distensible extension of the **OESOPHAGUS** in birds (especially grain-eating species) and some insects, which is used for storing food. The crop of some bird species contains grit or small stones swallowed to aid the mechanical breakdown of food. *See also* **CROP FEEDING**

crop feeding The introduction of food into the **CROP** of a sick or hand-reared bird using a **CROP NEEDLE** or a **CROP TUBE**. *See also* **FORCE FEEDING**

crop needle A blunt needle used for feeding a bird using a syringe. *See also* **CROP TUBE, CROP FEEDING**

crop tube A flexible (e.g. silicone) tube used for tube feeding a bird using a syringe. *See also* **CROP NEEDLE**

crop-raiding The act of entering a cultivated area to eat the edible parts of crop plants, often destroying entire crops in the process. Occurs extensively where animals such as monkeys and elephants live near human settlements resulting in significant **HUMAN–ANIMAL CONFLICT**.

cross, cross-breed
1. An individual produced by **CROSS-BREEDING**. *See also* **MONGREL**
2. The act of **CROSS-BREEDING**.

cross-breed *See* **CROSS**

cross-breeding The mating of individuals from two different breeds in order to produce offspring which combine the best characteristics of both. *See also* **COMPLEMENTARITY, HETEROSIS, TERMINAL SIRE**

cross-fostering A technique used to increase the **FECUNDITY** of a rare species whereby the young offspring of an individual of a rare species are fostered by adults of another species which is usually closely related. For example, ring-necked parakeets (*Psittacula krameri*) are used to foster echo parakeets (*P. eques*) and Barbary doves (*Streptopelia risoria*) are used for pink pigeons (*Columba mayeri*).

crossing over A mutual exchange of genetic material between **HOMOLOGOUS (2)** chromatids during **MEIOSIS** which results in the recombination of **GENES**. *See also* **GENE LINKAGE**

cross-sectional study A study that examines all of the individuals in a population (or a sample) at a particular point in time, e.g. the incidence of a particular disease in a species. *Compare* **LONGITUDINAL STUDY**

crude protein, total protein The total protein content in a food (e.g. an animal feed, milk). This is calculated from the determined nitrogen content multiplied by a factor derived from the mean percentage of nitrogen in food proteins. On average proteins contain 16% nitrogen so the nitrogen content of a sample is multiplied by 6.25 to calculate the crude protein content.

cruelty
1. Behaviour which deliberately causes unnecessary **PAIN**, distress or **SUFFERING** and is performed with pleasure or callous indifference. The legal definition of cruelty (in relation to animals) varies between jurisdictions.
2. Under New York State's Agriculture and Markets Law §350 (2)' *"Torture" or "cruelty" includes every act, omission, or neglect, whereby unjustifiable physical pain, suffering or death is caused or permitted;'*
3. In Great Britain, the Wild Mammals (Protection) Act 1996 (s.1) protects wild mammals from cruel acts: *'If, save as permitted by this Act, any person mutilates, kicks, beats, nails or otherwise impales, stabs, burns, stones, crushes, drowns, drags or asphyxiates any wild mammal with intent to inflict unnecessary suffering he shall be guilty of an offence.'*
4. In New South Wales, Australia, section 4(2) of the **PREVENTION OF CRUELTY TO ANIMALS ACT 1979 (NSW)** defines acts of cruelty as including *'any act or omission as a consequence of which the animal is unreasonably, unnecessarily or unjustifiably:*
   a. beaten, kicked, killed, wounded, pinioned, mutilated, maimed, abused, tormented, tortured, terrified or infuriated,
   b. over-loaded, over-worked, over-driven, over-ridden or over-used,
   c. exposed to excessive heat or excessive cold, or
   d. inflicted with pain.'
See also **ANIMAL CRUELTY, ANIMAL CRUELTY LAWS, NEGLECT**

Cruelty to Animals Act 1835 A UK Act which amended **MARTIN'S ACT** and aimed to protect farm and some other domestic animals from cruel treatment, and which banned **COCKFIGHTING, BULL-BAITING, BADGER-BAITING** and **BEAR-BAITING**.

Cruelty to Animals Act 1876 A UK Act which limited animal experimentation and established a licensing system for animal experiments. Later replaced by the **ANIMALS (SCIENTIFIC PROCEDURES) ACT 1986**. *See also* **ANIMAL TESTING, MARTIN**

Crufts An annual international championship dog show organised and hosted by the **KENNEL CLUB** in the UK. Named after its founder, Charles Cruft.

crush cage *See* **CRUSH**

crush, animal handling chute, chute, crush cage, restraint chute, squeeze cage, squeeze chute, standing stock A device used for restraining a large animal (e.g. **HOOFSTOCK**), which may include a fixed wall assembly and an opposing movable wall assembly that is pushed towards the fixed wall to confine the animal (Fig. C10). In some designs, when the animal is fixed in the device it may be rotated so that the handler or vet may gain access to its feet, legs and other parts of its body. Very useful for handling large dangerous animals such as elephants. *See also* **FEED CHUTE, RACE, TRANSFER CHUTE**

Crustacea A class of **ARTHROPODS** containing crabs, lobsters, barnacles, woodlice and their relatives.

crustacean A member of the **ARTHROPOD** class **CRUSTACEA**.

Fig. C10 Crush.

cryobiology The study of the effects of freezing and low temperatures on biological systems, including the cryopreservation of biological specimens, including **SPERMATOZOA** and **OVA**. This technology is becoming increasingly important in conservation as scientists endeavour to build collections of preserved **DNA** and other materials from animals (and plants) in danger of extinction.

cryopreservation *See* **CRYOBIOLOGY**

crypsis A type of **CAMOUFLAGE** in which the animal resembles part of the environment, e.g. some insects resemble leaves.

cryptid An animal species whose existence has been suggested but not scientifically proven, e.g. yeti, Loch Ness monster. *See also* **CRYPTOZOOLOGY**

cryptorchid A developmental abnormality in which one or both of the **TESTES** do not descend into the scrotum.

cryptozoology The study of 'hidden' or 'undiscovered' animals (cryptids) such as the yeti, Loch Ness monster and sea serpents, whose existence has not yet been proven but which appear in myth and legends. Cryptozoologists study the evidence for the existence of these animals such as footprints, photographs, eye-witness reports, hair and faecal samples. Many species that are now well known were only discovered by western scientists relatively recently, e.g. the okapi (*Okapia johnstoni*) in 1901, Komodo dragon (*Varanus komodoensis*) in 1912.

CT scanner, CAT scanner Computed **TOMOGRAPHY** (computer-assisted tomography or computed axial tomography) scanner. A machine that allows the examination of the inside of the body in three dimensions by producing **X-RAY** images of cross-sectional 'slices' of the brain or other soft tissues. *See also* **MAGNETIC RESONANCE IMAGING (MRI) SCANNER**

cube home A small plastic cube used as a shelter for small animals, e.g. small primates. Can be suspended from branches etc.

Cuculiformes An order of birds: turacos, louries, cuckoos.

cud Regurgitated food being chewed for the second time in a **RUMINANT**, having been returned back to the **BUCCAL CAVITY** from the **RUMEN**.

cudding Chewing the **CUD**; **MASTICATION** of regurgitated food in **RUMINANTS**.

cull
1. To kill a proportion of an animal population or an entire population, because it is surplus to requirements for breeding in a zoo, for commercial reasons, or because it carries or may become infected with disease, is a pest or is considered too numerous in the wild. *See also* **MARTIN THE WOLF**
2. An operation in which animals are culled, e.g. a seal cull, an elephant cull.

cull cows Cows that are removed from a herd because they are inferior in genetic quality or no longer productive.

cultural behaviour Behaviour that is transmitted from one generation to another through non-genetic means by **SOCIAL LEARNING**. This results in the creation of simple forms of traditions which may vary between populations of the same species, e.g. regional **INTRASPECIFIC** differences in bird song, knowledge of **MIGRATION (1)** routes, differences in **TOOL USE** between populations. *See also* **CLAN, MEME, SOCIAL FACILITATION**

culture A set of behaviours which is population specific and has been acquired by **SOCIAL LEARNING**. For example, chimpanzees (*Pan troglodytes*) in Kibale Forest in Uganda use sticks to extract honey from fallen logs whereas those in the nearby Budongo Forest use 'sponges' made of chewed leaves. *See also* **CLAN, MEME**

curator A person responsible for part or all of a zoo or other collection, e.g. curator of mammals.

curiosity A form of **EXPLORATORY BEHAVIOUR** which is voluntary and may be motivated by a desire to learn. This may be related to **FEAR** as many animals will approach objects and other animals of which they should be frightened (e.g. predators).

curry comb A tool used for grooming horses, especially for removing mud and loose hair (Fig. C11).

cursorial Adapted to running, e.g. African hunting dogs (*Lycaon pictus*). *Compare* **ARBOREAL**

cutaneous Relating to the **SKIN**.

C

Fig. C11 Curry combs: a plastic curry comb for brushing horses (left) and a metal spiral curry comb, for cleaning combs only (right).

cuticle A protective layer of **CHITIN**, proteins, lipids and other materials secreted onto the epidermis of arthropods which provides water-proofing and resistance to **DESICCATION**.

Cuvier, Georges (1769–1832) Cuvier was a French zoologist and naturalist. He was largely responsible for founding vertebrate palaeontology and comparative anatomy as a scientific disciplines and demonstrated that past life forms had become extinct. In 1795 Cuvier moved to Paris and shortly thereafter he was appointed as professor of comparative anatomy, at the Musée National d'Histoire Naturelle (National Museum of Natural History). In 1803 he became warden of the **MENAGERIE** of the **JARDIN DES PLANTES**, which, at that time, included a monkey and bird house, bear pits and a rotunda for herbivores including elephant and giraffe. Cuvier's most famous academic work was a systematic survey of zoology, *Le Règne Animal* (*The Animal Kingdom*), published in 1817.

CWD *See* **CHRONIC WASTING DISEASE (CWD)**

cyber hunting *See* **INTERNET HUNTING**

cycling Exhibiting an **OESTROUS CYCLE**.

Cypriniformes An order of fishes: carps, minnows and their allies.

Cyprinodontiformes An order of fishes: killifishes and their allies.

cyst
1. An epithelium-lined sac containing liquid or semi-solid material.
2. The structure resulting from the encapsulation of a **BACTERIUM**, **PROTOZOAN** or **PARASITE** (e.g. a hydatid cyst). *See also* **HYDATID DISEASE**

cytology The scientific study of **CELLS**; cell biology.

cytoplasm The constituents of a **CELL** which are enclosed within the **CELL MEMBRANE**, excluding the **NUCLEUS**.

cytosine (C) A **NUCLEOTIDE** base which pairs with G (guanine) in **DNA**.

cytotoxic Capable of killing cells. Cytotoxic drugs are used to treat **CANCER**.

Fig. D1 D is for donkey. A donkey used to give rides to children.

Dictionary of Zoo Biology and Animal Management: A guide to terminology used in zoo biology, animal welfare, wildlife conservation and livestock production, First Edition. Paul A. Rees.
© 2013 John Wiley & Sons, Ltd. Published 2013 by John Wiley & Sons, Ltd.

daily routine The pattern of activity which an individual animal exhibits during a single day and which is repeated each day. This is affected by environmental changes and tends to follow a CIRCADIAN RHYTHM and is influenced by its BIOLOGICAL CLOCK. *See also* ACTIVITY BUDGET

Daktari An American children's television series made by Ivan Tors Films Inc. in the 1960s which starred Marshall Thompson as Dr. Marsh Tracy, a vet running an animal study centre in East Africa with his daughter Paula (Cheryl Miller) and staff, *Clarence* the cross-eyed lion and *Judy* the chimpanzee. 'Daktari' is Swahili for doctor.

dam

1. Mother, with respect to animals in general.
2. The AMERICAN STUD BOOK 2008 defines a dam as '*A female horse that has produced, or is producing, foals*'.

DanBred International (DBI) A company which produces and exports Danish pig breeds (purebred Danish Landrace, Danish Large White, Danish Duroc and cross breeds) and pig semen to over 40 countries around the world. *See also* PIG IMPROVEMENT COMPANY (PIC)

dander Small fragments of organic material derived from the skin, hair or feathers of animals, SEBUM and bacteria which may cause an ALLERGIC REACTION in sensitive persons.

dangerous animal categorisation A system used in the UK which divides animal species found in zoos into three categories based on their risk to the public should they come in contact with them. The list of species may be found in Appendix 12 of the *SECRETARY OF STATE'S STANDARDS OF MODERN ZOO PRACTICE (SSSMZP)*. *See also* CATEGORY 1, 2 AND 3 SPECIES

dangerous animals *See* CATEGORY 1, 2 AND 3 SPECIES, DANGEROUS ANIMAL CATEGORISATION, DANGEROUS DOG, DANGEROUS DOGS ACT 1991, DANGEROUS SPECIES, DANGEROUS WILD ANIMAL, DANGEROUS WILD ANIMALS ACT 1976

dangerous dog

1. A dog that is considered a danger to humans. Legal definitions vary between jurisdictions.
2. In Great Britain it an offence under the DANGEROUS DOGS ACT 1991 to be in charge of a dog when it is dangerously out of control in a public place.
3. In New South Wales, Australia, under s.33 of the COMPANION ANIMALS ACT 1998 (NSW) '*a dog is "dangerous" if it:*
   (a) has without provocation, attacked or killed a person or animal (other than vermin), or
   (b) has, without provocation, repeatedly threatened to attack or repeatedly chased a person or animal (other than vermin), or
   (c) has displayed unreasonable aggression towards a person or animal (other than vermin), or
   (d) is kept or used for the purposes of hunting.*

[* *except if only used to locate flush, point or retrieve birds or vermin; vermin includes small pest animals only (e.g. rodents)*].
See also STATUS DOG, STATUS DOG UNIT, WEAPON DOG

Dangerous Dogs Act 1991 A law in Great Britain which prohibits the possession of dogs belonging to types bred for fighting and other dangerous dogs, and requires all dogs to be kept under proper control. The Act bans the keeping, breeding and transfer of ownership of the following breeds: PIT BULL TERRIER (American), Japanese Tosa, Dogo Argentino, Fila Braziliero and any other fighting breeds which may be designated by the Secretary of State. Dangerous dogs are classified by 'type' (which is determined by a court based on the characteristics of the animal) not the breed. Individual animals may be exempted and registered on the Index of Exempted Dogs if a court determines that it is not a danger to the public. The Act also makes it an offence to be in charge of a dog when it is dangerously out of control in a public place. The Act was passed in response to a number of serious incidents of injury and death caused by dogs.

dangerous species Under s.6(2) of the ANIMALS ACT 1971 '*a dangerous species is a species –*
(a) which is not commonly domesticated in the British Islands; and
(b) whose fully grown animals normally have such characteristics that they are likely, unless restrained, to cause severe damage or that any damage they may cause is likely to be severe.'
See also DANGEROUS ANIMAL CATEGORISATION, DANGEROUS DOG, DANGEROUS WILD ANIMAL

dangerous wild animal In Great Britain, under the Dangerous Wild Animals Act 1976 (s.7(4)), a '*"dangerous wild animal" means any animal of any kind for the time being specified in . . . the Schedule to* [the] *Act.*' (For examples *see* DANGEROUS WILD ANIMALS ACT 1976).

Dangerous Wild Animals Act 1976 A law in Great Britain which regulates the keeping of certain kinds of DANGEROUS WILD ANIMALS by members of the public. It requires a licence to be issued by the local authority before species listed in the SCHEDULE to the Act may be kept. These include some kangaroos, all apes, many monkeys and lemurs, large canids such as wolves and jackals, elephants, big cats, giraffes and rhinoceroses, ostriches and cassowaries, alligators and crocodiles, many venomous snakes and also some scorpions and spiders. The Act gives the local authority powers to inspect premises and to seize and dispose of animals if kept in contravention of the Act. The Act does not apply to zoos, circuses, pet shops or licensed laboratories.

Darwin, Charles Robert (1809–1882) An English naturalist (Fig. D2) who travelled around the world in *HMS Beagle* collecting specimens of animals,

Fig. D2 Darwin.

plants and fossils before publishing, in 1859, a detailed account of the process of **EVOLUTION** by **NATURAL SELECTION** in a book entitled *On the Origin of Species by Means of Natural Selection, or the Preservation of Favoured Races in the Struggle for Life*. The title of the sixth edition was abbreviated to *The ORIGIN OF SPECIES*. Darwin's work forms the basis of our modern-day understanding of evolutionary processes and the formation of new species. He was a great supporter of **LONDON ZOO**. *See also* **LAMARCK, WALLACE**

Darwin Initiative A funding scheme administered by the **DEPARTMENT FOR THE ENVIRONMENT, FOOD AND RURAL AFFAIRS (DEFRA)** as part of the UK's obligation to assist developing countries with biodiversity conservation in order to meet its objectives under the **CONVENTION ON BIOLOGICAL DIVERSITY 1992 (CBD)**, **CITES** and the **CONVENTION ON THE CONSERVATION OF MIGRATORY SPECIES OF WILD ANIMALS 1979 (CMS)**. Some major UK zoos have been able to attract funding from this scheme to support *IN-SITU* **CONSERVATION** projects.

Darwinian fitness The capacity of an individual animal to produce offspring. This changes with age and is affected by the individual's ability to survive to reproductive age, mating success, the fecundity of the mated pair and the survival of offspring to reproductive age. *See also* **NATURAL SELECTION**

Darwin's finches A group of bird species from the **GALAPAGOS ISLANDS** described by **DARWIN** and used to demonstrate evolution because of the way in which the form of their bills was adapted to the type of food: insect-eaters had long narrow bills, seed-eaters had stronger bills etc.

Dasmann, Raymond F. (1919–2002) An American ecologist and conservationist who experimented with **GAME RANCHING** as a means of food production in Africa. He published *African Game Ranching* in 1964.

Dasyuromorphia An order of mammals: Australasian carnivorous mice, Tasmanian devil, marsupial 'mice' and 'cats', numbat, thylacine.

Data Deficient (DD) A **RED LIST** category used by the **INTERNATIONAL UNION FOR THE CONSERVATION OF NATURE AND NATURAL RESOURCES (IUCN)**: Data Deficient (DD) – Inadequate data available to make a direct or indirect assessment of the risk of extinction of the taxon, based on its distribution and/or its abundance. A taxon in this category may be well studied and its biology well known. DD is not a category of threat. More information is needed. Future research may indicate that a **THREATENED** category is appropriate.

data logger An electronic device for recording data at intervals, e.g. temperature, animal movements etc. The data is downloaded to a computer for analysis.

database A collection of records kept on paper or on a computer. May be used to collect, store and analyse data (e.g. family history information, medical records) on zoo animals, e.g. **ANIMAL RECORD KEEPING SYSTEM (ARKS)**, **EGGS**, **INTERNATIONAL SPECIES INFORMATION SYSTEM (ISIS)**, **MEDICAL ANIMAL RECORD KEEPING SYSTEM (MEDARKS)**, **ZOOLOGICAL INFORMATION MANAGEMENT SYSTEM (ZIMS)**, or farm animals, e.g. **AFIFarm, BREEDPLAN, PIG GRADING INFORMATION SYSTEM (PIGIS)**

daughter cells Cells produced by the division of a single cell.

Dawkins' organ A device for recording animal behaviour by pressing keys on a keyboard connected to a computer, invented by Richard **DAWKINS**.

Dawkins, Richard (1941–) Evolutionary biologist and ethologist. Inventor of the **DAWKINS' ORGAN**. Formerly Professor for the Public Understanding of Science at the University of Oxford. He popularised the concept of the selfish gene in his book *The Selfish Gene* (1976) and coined the term '**MEME**'.

dawn chorus The combined songs of birds heard at dawn.

DDT Dichlorodiphenyltrichloroethane. A synthetic organochlorine contact pesticide used widely to control malaria and typhus during the Second World War. Indiscriminate spraying, particularly in the USA, caused DDT to enter many food chains. **BIOACCUMULATION**, especially in fatty tissues, and **BIOMAGNIFICATION** along the food chain resulted

D

in eggshell thinning in birds of prey (and other top carnivores) and their subsequent decline (e.g. bald eagles, peregrine falcons, California condors). The environmental effects were brought to the attention of the American public by Rachel **CARSON** in her book *SILENT SPRING*, and DDT use was banned in the US in 1972 except for some public health use, although it continued to export it.

de Waal, Frans (1948–) A Dutch primatologist at the **YERKES NATIONAL PRIMATE RESEARCH CENTER**. He has studied primate **COGNITION**, **EMPATHY**, and **MORALITY** in primates, especially chimpanzees and bonobos and is the author of many books including *Chimpanzee Politics* (1982), *Peacemaking among Primates* (1989), and *Good Natured: The Origin of Right and Wrong in Humans and Other Animals* (1996).

death feigning A behaviour in which an animal remains motionless as if dead, usually after capture by a predator. This may allow escape because many predators will not eat dead prey, or may not eat captured prey immediately. *See also* **HYPNOSIS**, **TONIC IMMOBILITY**

deceit An evolutionary strategy adopted by some animals in which they are able to increase their fitness at the expense of another by sending misleading signals, e.g. by **MIMICRY** (posing as a more dangerous or toxic species) or distracting a predator from young by feigning injury. Some primates appear to use deceit to avoid sharing food or to copulate surreptitiously with females to which they would normally have no access. Some scientists believe that such behaviour involves **COGNITION**. These primate behaviours have been extensively studied in captivity. *See also* **CHEATING**

deceptive swelling A sexual swelling (**PERINEAL TUMESCENCE**) which appears to be shown in females of some primate species, when they are experiencing **LACTATIONAL AMENORRHOEA**, to reduce the probability that their infants will be killed when new males take over a group. They appear to be sexually available but do not become pregnant. This appears to be a female strategy to reduce **INFANTICIDE**, e.g. in hamadryas baboons (*Papio h. hamadryas*).

decibel (dB) A measure of sound intensity, sound pressure or loudness which uses a logarithmic scale (base 10).

deciduous plant A plant which sheds its leaves annually, usually in autumn (fall), as a protection against winter conditions. *Compare* **CONIFEROUS PLANT**

Decisions of the European Court of Justice Decisions concerning **EUROPEAN LAW** which assist in its interpretation and application. *See also* **EUROPEAN COURT OF JUSTICE (ECJ)**

Declaration of Rights for Cetaceans: Whales and Dolphins A declaration concluded by experts during a conference entitled *Cetacean Rights: Conference on Fostering Moral and Legal Change* at the Helsinki Collegium for Advanced Studies, University of Helsinki, Finland in May 2010. The declaration asserts that all **CETACEANS** as persons have the right to life, liberty and wellbeing. It opposes the keeping of cetaceans in captivity and their use as a resource.

declawing The practice of removing the claws from dangerous animals (e.g. lion and tiger cubs) so that they can be handled by members of the public, e.g. at **ROADSIDE ZOOS**.

decomposer A bacterium, fungus or invertebrate involved in the decomposition of organic matter.

decoy
1. A live animal or model of an animal used to attract other wild animals (particularly **WILDFOWL**) to a particular place by hunters or for capture.
2. A tunnel-like structure made of netting into which wild birds fly and are later captured. *See also* **HELIGOLAND TRAP**

decoyman A person who attracts birds into a **DECOY (2)** for the purpose of capturing them.

deep litter housing system A system used for raising hens, pigs and other livestock which is based on the repeated spreading of straw or sawdust onto the floor of the living area as it becomes soiled, building up a deep layer of **LITTER (1)**.

deer A member of the mammalian family Cervidae, which includes moose, reindeer, chital, elk and red deer. Most species grow new antlers and shed them every year. Some are commercially important (e.g. red deer (*Cervus elephus*), Fig. D3). *See also* **DEER COURSING**, **DEER FARMING**, **DEER FENCE**, **DEER PARK**, **DEER STALKING**

deer coursing The use of dogs to pursue and kill deer as a country 'sport'. Banned in England and Wales by the **HUNTING ACT 2004**.

deer farming The commercial production of deer (usually red deer (*Cervus elephus*) in the UK) primarily for slaughter for their meat (venison)(Fig. D3).

deer fence A tall fence erected to exclude deer from agricultural land, forests, roads and other places where they may do harm or be injured.

Deer Initiative An organisation which seeks to ensure the delivery of a sustainable well-managed wild deer population in England and Wales.

deer park An area of open space, usually walled and/ or fenced, generally consisting of grassland and woodland, where deer (especially red deer (*Cervus elephus*) and fallow deer (*Dama dama*)) roam freely, and which is open to the public. Often the grounds are associated with a stately home (e.g. Tatton Park, Cheshire) and they may once have been **ROYAL HUNTING GROUNDS**. The deer are managed to prevent over-population and damage to vegetation, especially trees. In the UK deer parks are not required to have a **ZOO LICENCE**.

deer stalking The pursuit of deer (especially red deer (*Cervus elephus*) in Scotland) by stealth for sport, at the conclusion of which the deer is shot,

Fig. D3 Deer: a herd of red deer (*Cervus elephus*).

typically with a high-powered rifle with a telescopic sight.

de-extinction The idea of bringing an **EXTINCT** species back to life by cloning preserved cells or by reconstructing **GENOMES** from ancient **DNA**. De-extinction has never been achieved but some scientists believe that genetic engineering techniques may make it possible in the future. The concept formed the basis of the film *JURASSIC PARK*.

defecation The release of faeces from the gut to the outside environment. *See also* **LATRINE BEHAVIOUR, MIDDEN**

Defenders of Wildlife A non-profit membership organisation in the United States which was founded in 1947. It is dedicated to the protection of all native animals and plants in their natural communities and champions the **ENDANGERED SPECIES ACT OF 1973 (USA)**. It works to protect and restore America's native wildlife, safeguard habitats, resolve conflicts, work across international borders and educate and mobilise the public. It has been particularly active in supporting grey wolf reintroductions. *See also WYOMING FARM BUREAU FEDERATION V. BABBITT* **(1997)**.

defensive circle A protective formation used by some large herbivorous mammals (e.g. elephants, bison, (*Bison bison*)) when protecting themselves from pred-

ators. Adult animals form a circle and face outwards with young animals enclosed in the centre.

defibrillator An electrical device for applying an electric current to an animal's thorax through electrodes in order to stimulate a heart that has either stopped beating or is beating abnormally.

deficiency disease A disease caused by the absence of an essential nutrient such as a **VITAMIN** or an **ESSENTIAL AMINO ACID**.

deficit financing A term used to describe the phenomenon whereby the popular species kept by zoos generate income and the surplus is used to help maintain the large groups of less popular species. This subsidisation process has been referred to as 'deficit financing' (Ironmonger, 1992).

deforestation The clearance of forest, usually for commercial reasons. *See also* **CLEAR-FELLING**. *Compare* **AFFORESTATION**

DEFRA (defra) *See* **DEPARTMENT FOR ENVIRONMENT, FOOD AND RURAL AFFAIRS (DEFRA)**

degenerative joint disease *See* **OSTEOARTHRITIS**

dehorning The act of removing the **HORN** of an animal, e.g. a rhino. May be done for illegal reasons (poaching) or to prevent animals from being killed for their horns. May also be done in captivity for safety reasons. *See also* **DISBUDDING, POLLED LIVESTOCK**

dehydrated Suffering from excessive water loss resulting in insufficient water in the body. *See also* DESICCATION

deimatic display A DISPLAY designed to scare away a predator, e.g. the eye spots located on the wings of some butterflies and moths which appear as a face when the wings are suddenly opened.

delayed implantation, embryonic diapause A condition in which the embryo remains in a state of dormancy prior to IMPLANTATION. This can extend GESTATION by up to a year. There are two types of embryonic DIAPAUSE. Facultative diapause is usually associated with metabolic stress. If copulation takes place when a female is still suckling an existing offspring, implantation may be delayed until the individual is weaned. This type occurs in some marsupials, rodents and insectivores. Obligate diapause allows the mother to delay birth until the environmental conditions are optimal. This type occurs in some PINNIPEDS, MUSTELIDS, URSIDS and armadillos.

deleterious gene A gene which is damaging to the individual because it causes a disease or in some other way impairs survival or reduces FITNESS. *See also* FIXATION

delivery Birth, especially in mammals.

deme A spatially discrete group of interbreeding organisms; an interbreeding local population of a species whose members share a GENE POOL.

demersal Living underwater, at or near the bottom of the sea or a lake. Usually applied to fish.

demographically extinct A population which is destined to become extinct by virtue of having insufficient reproductive capacity because the individuals have low fecundity, offspring survival is too low, too many of the population are in their post-reproductive phase, or for some combination of these or other reasons. This may become a problem in zoo populations as they age. *See also* FUNCTIONALLY EXTINCT, MINIMUM VIABLE POPULATION (MVP)

demography
1. The study of the age structure, growth, natality, MORTALITY and other related aspects of a POPULATION.
2. The age structure characteristics of a population.

den The place used by some animals for shelter and/ or breeding, e.g. polar bears (*Ursus maritimus*).

density A characteristic of a POPULATION of organisms: the number of organisms per unit area, e.g. 20 gazelles per km^2; 6 snails per m^2. *See also* STOCKING DENSITY

dental Relating to TEETH.

dental comb *See* TOOTH COMB

dental formula A coded representation of the number and arrangement of teeth in one side of the jaw of an animal, especially a mammal. The formula takes the form: number of teeth in upper jaw/ number of teeth in lower jaw, for example,

I 3/3   C 1/1   P 3/3   M 2/2 = 36 (European otter, (*Lutra lutra*)) where I = INCISORS, C = CANINES, P = PREMOLARS, M = MOLARS.

dental tooth comb *See* TOOTH COMB

dentary One of several tooth-bearing bones of the lower jaw in vertebrates. Mammals possess a single dentary on each side.

dentine A bone-like substance which is the main constituent of teeth. It is located under the ENAMEL and consists mainly of calcium phosphate in a fibrous matrix. IVORY is dentine.

dentist A person who is qualified in, and licensed to practice, DENTISTRY. Human dentists sometimes treat animal patients, especially exotic species kept in zoos. *See also* KERTESZ

dentistry The study and treatment of diseases and malformations of the oral cavity, especially the teeth, gums and associated structures. *See also* DENTIST

dentition The number, type and arrangement of TEETH. *See also* DENTAL FORMULA, HETERODONT DENTITION, HOMODONT DENTITION

deoxygenated Not carrying oxygen, e.g. deoxygenated BLOOD.

Department for Environment, Food and Rural Affairs (DEFRA/defra) The government department responsible for farming, fisheries, food production, rural communities, the ENVIRONMENT, BIODIVERSITY and zoos in the UK. *See also* FOOD AND ENVIRONMENT RESEARCH AGENCY (FERA), NATURAL ENVIRONMENT RESEARCH COUNCIL (NERC)

dependent variable A variable whose value depends upon the value of another variable, the INDEPENDENT VARIABLE (Fig. D4). For example, at any particular latitude, temperature (the dependent

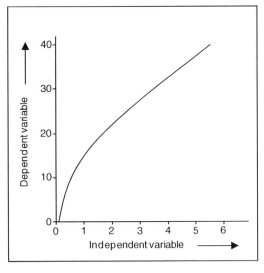

Fig. D4 Dependent and independent variables.

Fig. **D5** Depressed vertical fence barrier between two adjacent enclosures.

variable) decreases with height above sea level (the independent variable).

depressed fence *See* **DEPRESSED VERTICAL FENCE BARRIER**

depressed vertical fence barrier A vertical fence around an animal enclosure which is so placed that visitors can see over it into the enclosure with an unimpeded view usually because the viewing position is relatively higher than the fence. It may be located in a ditch between adjacent enclosures as an 'invisible' barrier (Fig. D5).

depression A psychological state in an animal which may be the result of a chemical imbalance. It may be exhibited by changes in behaviour such as aggression, **ANXIETY**, destructive behaviour, excessive sleeping, lethargy, excessive or lack of grooming, appetite loss and pacing.

Der Zoologische Garten A German journal which publishes articles on zoos and animals living in zoos. It was founded in 1859.

dermatitis **INFLAMMATION** of the skin (a rash) which may have a variety of causes and may be one of a number of different types.

dermatology The study and treatment of skin and skin diseases.

Dermoptera An order of mammals: flying lemurs.

Descartes, René (1596–1650) A French philosopher and scientist who considered animals to be unthinking **AUTOMATA** without consciousness.

desert Usually a hot desert, e.g. Sahara. A place with too little rainfall to support most plants. May be divided into extremely arid lands (which have at least 12 consecutive months without rainfall); arid lands (which have less than 250 mm of annual rainfall); and semiarid lands (which have a mean annual rainfall of between 250 and 500 mm). The term 'desert' is usually used for arid and extremely arid lands. Semiarid grasslands are referred to as **STEPPES**. Cold deserts (polar deserts) are deserts where the main form of **PRECIPITATION** is snow,

Fig. **D6** Designer dog: a labradoodle (a Labrador retriever × poodle hybrid).

e.g. Antarctica. *See also* **ARIZONA–SONORA DESERT MUSEUM**

desertification The process by which land is turned into **DESERT** as a result of over-grazing, loss of topsoil, etc.

desexed Rendered permanently incapable of reproduction (s.5, **COMPANION ANIMALS ACT 1998 (NSW)**).

desiccation The process of water loss that results in extreme dryness, e.g. marine organisms on a shore may undergo desiccation if they become stranded on the beach when the tide goes out and they are exposed to heat from the sun. *See also* **DEHYDRATED**

designer cat A cat which has been specially bred to satisfy a need; a specific type of cat which has become popular because of a fashion or as a fashion item, e.g. **BENGAL**, **TOYGER**

designer dog The result of a cross between existing purebreds, e.g. Labrador retriever × poodle = labradoodle (Fig. D6); cocker spaniel × poodle = cock-a-poo; pug × beagle = puggle. Poodles have been widely used in the breeding of designer dogs because they do not shed their hair and hybrids produced using them are therefore less likely to cause an **ALLERGIC REACTION** in people.

D

destructive sampling Taking and/or measuring material collected in samples in such a manner as to kill or damage them so that they may not be returned to the environment or used in any other way which requires them to be in their original condition. *See also* SAMPLING

detector beam activated surveillance system A security system which directs CCTV cameras to a location where a detector beam has been broken and activates an ALARM to alert staff to the presence of an intruder. Such a system is used in London Zoo's *Gorilla Kingdom* exhibit to alert staff to any members of the public who reach too far over the WET MOAT at the perimeter of the exhibit.

detergent A water-soluble cleansing agent that combines with impurities and dirt to make them more soluble. If used to clean oiled birds detergents must be used with care as they may interfere with the waterproofing and insulation properties of feathers.

determinate growth Growth which has a specific end-point beyond which no further growth occurs. HIGHER VERTEBRATES tend to grow rapidly as juveniles but stop growing when they reach adult size. *Compare* INDETERMINATE GROWTH

deterministic In relation to an algorithm, MODEL or process, one where the outcome is fixed, predictable and entirely determined by its initial state and input values. *Compare* STOCHASTIC

detoxification The process of making a substance non-toxic. Gut enzymes in some species detoxify potentially poisonous plants. One of the functions of the liver is to detoxify the blood.

detritivore An organism that feeds on DETRITUS.

detritus Dead plants and animals and material derived from them and their decomposition, e.g. leaf litter and other material on a forest floor.

development The process by which organisms become transformed from simple unorganised structures to complex, organised, integrated structures. *See also* MATURATION, ONTOGENY

dewlap A flat of loose skin which hangs down from the throat of certain cattle, dogs, rabbits and other animals.

deworming *See* WORMING

diabetes mellitus Sugar diabetes. A condition in which BLOOD SUGAR LEVELS are raised due to a deficiency of the hormone INSULIN.

diadromous Capable of living in, and migrating between, freshwater and saltwater. *Compare* ANADROMOUS, CATADROMOUS

diagnosis In veterinary medicine, the determination of the nature or cause of a disease or condition.

dialect In animal communication, a local variation in vocalisation, e.g. in bird song or whale song. Generally learned by IMPRINTING.

diapause A period of suspended DEVELOPMENT, especially in insects and some other invertebrates, but also applied to mammals. *See also* DELAYED IMPLANTATION

diaphragm A membrane made of muscle and tendon that separates the abdomen from the thoracic cavity (thorax) in mammals and whose downward movement assists in the inflation of the lungs during BREATHING.

diarrhoea, diarrhea, scour The frequent passing of loose, watery faeces. Often a sign of illness or disease. If severe it may cause dehydration and eventual death.

diastema A gap in the rows of teeth in some mammals, typically herbivores, separating the cheek teeth from the incisors and canines. The lower canines in APES typically fit into gaps in the teeth in the upper jaws known as 'monkey gaps'.

diastole The expansion of the heart chambers (atria and ventricles) when they relax and fill with blood from the veins. *See also* CARDIAC CYCLE. *Compare* SYSTOLE

DICE *See* DURRELL INSTITUTE OF CONSERVATION AND ECOLOGY (DICE)

dichotomous key An identification key which uses a series of questions to which there can only be two possible answers, e.g. Does the insect have one pair of wings or two? Does the bird have a red breast (yes or no)?

Dickin, Maria (1870–1951) Founder of the PEOPLE'S DISPENSARY FOR SICK ANIMALS (PDSA). While working as a social worker in the East End of London she encountered many sick and ill-treated animals. She established a 'dispensary' in Whitechapel and a fleet of horse-drawn mobile clinics to provide free treatment for animals owned by the poor. She published *The Cry of the Animal* around 1940. *See also* DICKEN MEDAL

Dickin Medal A medal awarded by the PEOPLE'S DISPENSARY FOR SICK ANIMALS (PDSA) to animals who have exhibited exceptional bravery in wartime. Named for the founder of the PDSA, Maria DICKIN. A total of 63 medals had been awarded by 2011 (54 of which were awarded between 1943 and 1949) to 32 pigeons, 18 dogs, three horses and one cat. On 24 February 2010 a Labrador called *Treo* (ROYAL ARMY VETERINARY CORPS (RAVC)) was awarded the medal for saving the lives of men of the Royal Irish Regiment by locating roadside improvised explosive devices while deployed as forward protection in Helmand Province, Afghanistan.

dicotyledonous plant A flowering plant which has two cotyledons (seed leaves), and true leaves that are broad with branching veins, with the floral parts divided into sets of five, sometimes four and rarely three. *Compare* MONOCOTYLEDONOUS PLANT

Didelphimorphia Order of mammals: American opossums.

diestrus *See* DIOESTRUS

diet
1. The combination of foods taken in by an animal for its NUTRITION (1).

2. A restricted quantity and/or quality of foods formulated to achieve a particular end, e.g. weight loss. *See also* CALORIFIC RESTRICTION, OBESITY

dietary drift The gradual change of an animal's diet over time by keepers so that it moves away from the original prescribed diet.

dietary supplement *See* FOOD SUPPLEMENT

differential centrifugation The process of separating out the component parts of cells (ORGANELLES) on the basis of their mass using a centrifuge. Large, relatively heavy structures, e.g. MITOCHONDRIA, will pass to the bottom of a centrifugation tube at relatively slow speeds whereas smaller structures, e.g. RIBOSOMES, require higher speeds. *See also* CELL FRACTIONATION

differential resource utilisation *See* RESOURCE PARTITIONING

diffusion The movement of molecules from an area of high concentration to an area of low concentration by random movements.

digestibility The extent to which a food material is capable of being broken down (digested) into its component parts for absorption through the gut. *See* GROSS ASSIMILATION EFFICIENCY (GAE)

digestion The process by which large molecules in food are broken down into smaller molecules which may be absorbed through the gut wall into the circulatory system (or LYMPHATIC SYSTEM).

digestive system The organ system responsible for breaking food down into particles which are small enough to absorb into the body. Made up of the alimentary canal and associated organs (e.g. LIVER and PANCREAS) that produce digestive enzymes.

Digit A male GORILLA studied by Dian FOSSEY in Rwanda. He was killed and decapitated by poachers in 1977. His death was depicted in the 1988 film *GORILLAS IN THE MIST* and helped to bring the plight of great apes to the attention of the public.

digital stethoscope *See* STETHOSCOPE

digital X-ray A RADIOGRAPH taken with a machine which produces a digital image.

digitigrade Relating to contact between the toes and the ground, rather than the sole of the foot, especially while walking. *See also* BIPEDAL, KNUCKLE WALKING, PLANTIGRADE, UNGULIGRADE

dihybrid cross A genetic cross which considers the inheritance of two genes for different characters. When both parents are HETEROZYGOUS for each of the two genes (e.g. AaBb × AaBb), the phenotypes produced in the F_1 generation occur in the ratio $9:3:3:1$. *Compare* MONOHYBRID CROSS

dimming thermostat A switch design to control the heat released from a heater or heat lamp by increasing or decreasing the electrical current it receives rather than simply switching it on or off. *See also* THERMOSTAT

dimorphic species *See* SEXUAL DIMORPHISM

dioestrus, diestrus A short period of quiescence between the end of one OESTROUS CYCLE and the beginning of the next in species that have more than one cycle per BREEDING SEASON.

dip A bath through which animals pass in order to submerge them briefly in a chemical solution of insecticide and fungicide, e.g. a sheep dip. *See also* RACE

diploid In relation to a CELL, possessing both CHROMOSOMES of each pair which make up the GENOME.

Diprotodontia An order of mammals: koala, wombats, possums, wallabies, kangaroos.

Diptera A large order of insects: true flies.

direct fitness, classical fitness A measure of the number of genes passed on from a particular individual directly to the next generation by producing and supporting its own offspring. *Compare* INCLUSIVE FITNESS, INDIRECT FITNESS

directional selection A type of NATURAL SELECTION in which the distribution of a character is pushed in a particular direction as a result of the elimination of PHENOTYPES at one extreme of the distribution of the character. For example, small individuals may be eliminated by very cold weather, resulting in an increase in mean size (Fig. S4). *Compare* DISRUPTIVE SELECTION, STABILISING SELECTION

directive *See* EUROPEAN DIRECTIVE

disaccharide *See* CARBOHYDRATE

disbudding The removal of horn buds in young mammals such as goats, cattle and sheep. Horns are removed for safety and economic reasons. Disbudding is usually achieved by CAUTERISATION of the growth ring in the horn using a hot disbudding iron. Animal welfare organisations consider this practice to be painful and cruel. *See also* DEHORNING, POLLED LIVESTOCK

disc *See* INTERVERTEBRAL DISC

discontinuous variable *See* DISCRETE VARIABLE

discovery centre An alternative name for an education centre in a zoo or similar institution.

discrete variable, discontinuous variable A variable which may only take certain values, e.g. the number of animals in a litter. *Compare* CONTINUOUS VARIABLE

discrimination In animal behaviour, the ability to distinguish between different stimuli and make different responses to them.

disease A disorder or illness, often caused by infection.

Disease Control Orders Legal restrictions in the UK placed on the movements of livestock to prevent the spread of disease.

disinfectant A substance which kills microorganisms and is used on work surfaces, floors, containers etc. to prevent the spread of infection.

Disinfectant Era A period in zoo history in the 1920s and 1930s when cages and enclosures were designed for ease of cleaning, typified by the covering of the walls and floors of indoor animal accommodation

D

with ceramic tiles, rather than to meet the needs of the animals. Some zoos still have enclosures of this type to this day, particularly the indoor accommodation for monkeys and apes. *See also* **MODERNIST MOVEMENT**

disinfectant mat A mat soaked in disinfectant which is designed to disinfect the footwear of anyone who walks across it. Used at the entrance to quarantine areas, nurseries etc.

dispersal The movement of adult organisms or their propagules from their place of birth (origin) or natal group. In bird species, female dispersal is prevalent whereas in mammal species male dispersal is more common. This process prevents **INBREEDING** between close relatives. It is important that zoos simulate this process as part of their **CAPTIVE BREEDING PROGRAMMES** to maintain the genetic health of captive populations. *See also* **PHILOPATRY (1)**

dispersion The spatial distribution of objects or organisms. Dispersion may be regular, random or clumped (clustered).

displacement activity, displacement behaviour A behaviour which appears to be irrelevant in the context in which it occurs and is the result of conflicting drives (such as the desire to approach something and fear of it) or being prevented from accomplishing something, e.g. if disturbed during courtship a bull elephant may stop and dig in the soil. It sometimes involves **RITUALISATION** allowing displacement behaviour to be incorporated into a **COURTSHIP DISPLAY**, e.g. ritualised **PREENING** in ducks. It includes **REDIRECTED BEHAVIOUR** and **SELF-DIRECTED BEHAVIOUR**.

display A behaviour (often ritualised) which an animal uses to convey specific information, usually to conspecifics. *See also* **COURTSHIP DISPLAY**

Display, Education and Research Populations An **ASSOCIATION OF ZOOS AND AQUARIUMS (AZA)** population management level for zoo animal populations which do not require genetic or demographic management and are useful for research, conservation education and display.

disruptive selection Selection which produces two divergent phenotypic extremes within a population as a result of selection against intermediate **PHENOTYPES**. It increases the **VARIANCE** of a trait and is a possible mechanism for **SYMPATRIC** speciation (Fig. S4). *See also* **NATURAL SELECTION**, *Compare* **DIRECTIONAL SELECTION, STABILISING SELECTION**

dissecting microscope, stereo microscope A binocular (stereoscopic) microscope of relatively low power (usually ×10 – ×40) used to aid the **DISSECTION (1)** of small organisms or structures. May be fitted with two lamps: one above the stand to provide reflected light for illuminating the specimen from above, and a second for providing transmitted light for illuminating the specimen from below. May be trinocular, with a third 'eyepiece' or port for a camera system.

dissection
1. The cutting apart and exposing of tissues and organs for anatomical study or experimentation. *See also* **VIVISECTION**
2. The result of the process of dissection, e.g. a dissection of a mammalian heart.

dissection kit A set of instruments used for **DISSECTION (1)** including scalpels, forceps, scissors, mounted needles etc.

distal Located away from some point of reference, e.g. the foot is distal to the pelvis (Fig. A8). *Compare* **PROXIMAL**

distemper *See* **CANINE DISTEMPER, STRANGLES**

distress A motivational state in an animal which is the result of **STRESS** that it is unable to deal with. This may be caused by the physical environment (e.g. very high or very low temperature) or the biological environment (e.g. social interactions with **CONSPECIFICS** or **KEEPERS (1)**).

distribution
1. Of a species, the geographical area it inhabits.
2. Of a variable, the pattern of occurrence of values, e.g. **NORMAL DISTRIBUTION**.

diuretic A drug that increases urine production. *Compare* **ANTIDIURETIC**

diurnal
1. Daily; during the day.
2. Of animals, active during the day, e.g. a diurnal species (as opposed to a **NOCTURNAL** species). *Compare* **CATHEMERAL, CREPUSCULAR, MATUTINAL, VESPERTINE**

divergent evolution The splitting of a line of evolution into two or more distinct groups with the passage of time, eventually producing different taxa. *See also* **DISRUPTIVE SELECTION, HOMOLOGOUS (1)**. *Compare* **CONVERGENT EVOLUTION**

diversity index Any of a number of mathematical methods of assessing biodiversity that takes into account the number of individuals and the number of species present in a habitat at a particular point in time, e.g. the Shannon Wiener index. *See also* **ALPHA (α) DIVERSITY, BETA (β) DIVERSITY, GAMMA (γ) DIVERSITY, LIVING PLANET INDEX**

diving reflex A physiological mechanism found in many aquatic mammals (e.g. seals, dolphins and otters) that submerge themselves in water, whereby on submersion the **HEART RATE** is reduced (**BRADYCARDIA**) and peripheral vasoconstriction diverts blood from the body surface to the internal organs (blood shift). A similar phenomenon also occurs in diving birds such as penguins.

division of labour
1. The phenomenon whereby different types of **CELLS, TISSUES** and **ORGANS** in the body perform different functions, e.g. respiration, excretion, reproduction etc.
2. The phenomenon in some animal societies whereby different types of individuals perform different roles, e.g. in ants, workers, soldiers, repro-

ductive individuals etc. *See also* **EUSOCIALITY,**
SUPER-ORGANISM

DMI *See* **DRY MATTER INTAKE (DMI)**

DNA Deoxyribonucleic acid; the genetic material
which contains the code for the development of
an organism. It is constructed from two chains of
NUCLEOTIDES arranged as a double helix.

DNA fingerprinting, DNA profiling, genetic finger-
printing A molecular technique used to identify
GENOMES by comparison to a known standard or
to compare DNA from different sources. This technology may be used to sex individuals from species
that are **MONOMORPHIC,** distinguish between subspecies, examine the relatedness of individuals, and
identify animal parts and products from rare species.
It may be used to identify an individual animal and
establish its parentage. In the UK, the **KENNEL**
CLUB offers a DNA Profiling Service and a DNA
Parentage Analysis Service, for any breed or cross-
breed of dog. *See also* **MICROSATELLITE MARKER,**
POLYMERASE CHAIN REACTION (PCR), USFWS
FORENSICS LABORATORY

DNA microsatellite marker *See* **MICROSATELLITE**
MARKER

DNA profiling *See* **DNA FINGERPRINTING**

DNS Did not survive. Used as an abbreviation referring to young animals that survived for a relatively
short time, especially in the stocktaking records of
zoos. *See also* **STOCK RECORDS**

docent A volunteer worker (especially in the USA)
who leads guided tours and provides other assistance
to visitors to zoos, museums and similar facilities.

docking *See* **TAIL DOCKING**

dodo (*Raphus cucullatus*) A species of flightless bird
ENDEMIC (1) to Mauritius which became extinct in
1681 as a result of human activity (Fig. D7). It has
become a symbol of extinction and is the emblem
of *DURRELL*.

dog
1. A domestic dog (*Canis familiaris*). *See also* **DOG**
BREEDS
2. A male domestic dog, canid, otter or weasel.

dog agility training The training of a dog to run
through tunnels and along narrow boards, jump
fences and negotiate other obstacles (Fig. D8), either
for entertainment and as part of general **OBEDI-**
ENCE TRAINING, or as part of its training as a
WORKING DOG employed by the police, armed
services or other agency.

dog breeds Dogs have been bred for a variety of purposes including as companion animals, for hunting,
guarding livestock, herding livestock, pulling loads,
protection and as fashion accessories (Fig. D9). In
the UK the **KENNEL CLUB** recognises 210 breeds.
The **AMERICAN KENNEL CLUB** fully recognises
only 174 breeds (Table D1).

dog grasper A device for catching stray and dangerous dogs. *See* **CATCH POLE**

D

Fig. D7 Dodo: model of a dodo (*Raphus cucullatus*).

Fig. D8 Dog agility training.

Fig. D9 Dog breeds: Great Dane (left); Old English sheepdog (top right); St. Bernard (middle right); Chinese crested (bottom right).

dog warden A person employed to capture stray and feral dogs and return them to their owners, rehome or dispose of them, and issue warnings to owners in relation to nuisances caused by dogs.

dogfight, dogfighting An organised, usually illegal, contest between two dogs that have been bred and trained to fight, watched by persons who gamble on the outcome. Still common in some developing countries, e.g. Afghanistan and Pakistan. *See also* PIT BULL TERRIER

dognapping The illegal taking of pet dogs from their owners. Dogs are often sold to new owners or used for dogfighting, and have been sold to laboratories for ANIMAL TESTING experiments or medical research, although tougher legislation in many countries now makes this difficult. Sometimes dogs are held for ransom or simply returned to owners who offer a reward. Dogs used in greyhound racing and WORKING DOGS have been particularly targeted. Dognapping is a particular problem in the UK and the USA. In the USA the ANIMAL WELFARE ACT OF 1966 (USA) offers some protection to dogs by restricting the handling and sale of animals for research. Pet and valuable dogs may be protected by MICROCHIP implants or other means of identification (e.g. TATTOOS or DNA sampling) so that they may be identified if found.

Dogs (Protection of Livestock) Act 1953 *See* WORRYING LIVESTOCK

Table D1 Major dog breeds.

Breed	Origin	Breed	Origin
Companion dogs		Sloughi	Morocco
Australian shepherd (miniature)	USA	Swedish elkhound	Sweden
Australian silky terrier	Australia	Whippet	UK
Bichon frisé	Tenerife	*Gundogs*	
Bolognese	Italy	American cocker spaniel	USA
Cavalier King Charles spaniel	UK	Brittany	France
Chihuahua	Mexico	English cocker spaniel	UK
Coton de tuléar	Madagascar	English pointer	UK
French bulldog	France	English setter	UK
German spitz	Germany	English springer spaniel	UK
Havanese	Cuba	French spaniel	France
King Charles spaniel	UK	German pointers	Germany
Lhasa apso	Tibet	Golden retriever	UK
Löwchen	France	Gordon setter	UK
Maltese	Malta	Hungarian vizsla	Hungary
Papillon	France	Hungarian wirehaired vizsla	Hungary
Pekingese	China	Irish setter	Ireland
Pomeranian	Germany	Labrador retriever	UK
Poodle (miniature and toy)	France	Large Munsterlander	Germany
Pug	China	Small Munsterlander	Germany
Shih tzu	Tibet/China	Weimaraner	Germany
Terriers		*Herding dogs*	
Airedale terrier	UK	Bearded collie	UK
American pit bull terrier	USA	Belgian shepherd dog	Belgium
American Staffordshire terrier	USA	Border collie	UK
Belgian griffon	Belgium	German shepherd dog	Germany
Border terrier	UK	Old English sheepdog	UK
Boston terrier	USA	Puli	Hungary
Bull terrier	UK	Rough collie	UK
Cairn terrier	UK	Welsh corgis	UK
German pinscher	Germany		
Irish terrier	Ireland	*Working dogs*	
Jack Russell terrier	UK	Akita	Japan
Kerry blue terrier	Ireland	Alaskan malamute	USA
Manchester terrier	UK	American Eskimo (standard)	USA
Parson Russell terrier	UK	Australian kelpie	Australia
Scottish terrier	UK	Bernese mountain dog	Switzerland
Skye terrier	UK	Boxer	Germany
Smooth fox terrier	UK	British bulldog	UK
Staffordshire bull terrier	UK	Cane corso Italiano	Italy
West Highland white terrier	UK	Chow chow	China
Yorkshire terrier	UK	Dalmatian	Croatia or India
		Doberman	Germany
Hounds		Giant schnauzer	Germany
Afghan hound	Afghanistan	Great Dane	Germany
American foxhound	USA	Labradoodle	Australia
Basenji	Zaire	Leonberger	Germany
Basset hound	UK	Neapolitan mastiff	Italy
Beagle	UK	Newfoundland	Canada
Black-and-tan coonhound	USA	Poodle (standard)	Germany
Bloodhound	Belgium	Pyrenean mastiff	Spain
Borzoi	Russia	Pyrenean mountain dog	France
Dachshunds	Germany	Rhodesian ridgeback	South Africa
Deerhound	UK	Rottweiler	Germany
Drever	Sweden	St. Bernard	Switzerland
English coonhound	USA	Samoyed	Northern Russia/
Finnish hound	Finland		Siberia
Greyhound	UK	Shar pei	China
Irish wolfhound	Ireland	Shiba inu	Japan
Saluki	Iran	Siberian husky	Siberia

D

Dogs Trust The largest dog welfare organisation in the UK. Founded as the National Canine Defence League in 1891. It runs 17 rehoming centres and subsidised neutering campaigns.

***Dolly* the sheep** A sheep at the **ROSLIN INSTITUTE** who was the first animal to be cloned from an adult **SOMATIC CELL,** by **NUCLEAR TRANSFER** (somatic cell nuclear transfer (SCNT)). She was born in 1996. A year earlier two other sheep, *Megan* and *Morag,* were cloned at the Institute from embryonic (as opposed to adult) cells.

dolphin therapy An **ANIMAL-ASSISTED THERAPY** that involves direct physical contact with dolphins for people suffering from a variety of conditions including cerebral palsy, autism, Down's syndrome, depression, emotional stress, chronic fatigue syndrome, neurasthenia and various phobias. Also called dolphin-assisted therapy.

dolphinarium A facility that keeps and exhibits dolphin species.

domestic animal

1. An animal that is kept by, or living, with humans. May be defined by statute in some animal welfare legislation.

2. In England and Wales, s15 of the Protection of Animals Act 1911 defines a 'domestic animal' as *'any horse, ass, mule, bull, sheep, pig, goat, dog, cat, or fowl, or any other animal of whatsoever kind or species, and whether a quadruped or not which is tame or which has been or is being sufficiently tamed to serve some purpose for the use of man'.*

3. In s1 of the National Parks of Canada Domestic Animals Regulations 1998, *'"domestic animal" means an animal of a species of vertebrates that has been domesticated by humans so as to live and breed in a tame condition and depend on humankind for survival.'*

See also **ANIMAL, CAPTIVE ANIMAL, DANGEROUS WILD ANIMAL, WILD ANIMAL**

domestic law *See* **NATIONAL LAW**

domesticated species

1. A species that is tame and has been brought under human control.

2. The UN Convention on Biological Diversity 1992 (Art. 2) defines *'domesticated or cultivated species'* as *'species in which the evolutionary process has been influenced by humans to meet their needs'.*

See also **DOMESTIC ANIMAL, DOMESTICATION**

domestication The process whereby some species of animals have been modified by **SELECTIVE BREEDING** for some human purpose. This modification may be anatomical, physiological or behavioural altering characteristics such as size, productivity, **TAMENESS**, fertility and **LONGEVITY**. It may have been undertaken in order to make the species more useful for food production, work, exhibition, as a **COMPANION ANIMAL**, or for some other human purpose (Table D2).

dominance

1. In genetics, the phenomenon whereby one genetically controlled character (the dominant char-

Table D2 Domestication: estimated dates at which selected species were first domesticated. (Adapted from Hirst (no date))

Taxon	Date	Location
Alpaca	1500 BC	Peru
Bactrian camel	3000 BC	Southern Russia
Banteng	3000 BC	Thailand
Cat	8500 BC	Fertile Crescent
Cattle	7000 BC	Eastern Sahara
Chicken	6000 BC	Asia
Dog	14–30000 BC?	undetermined
Donkey	4000 BC	Northeast Africa
Dromedary camel	3000 BC	Saudi Arabia
Duck	2500 BC	Western Asia
Goat	8000 BC	Western Asia
Goose	1500 BC	Germany
Guinea pig	5000 BC	Andes Mountains
Honey bee	3000 BC	Egypt
Horse	3600 BC	Kazakhstan
Llama	3500 BC	Peru
Pig	7000 BC	Western Asia
Reindeer	1000 BC	Siberia
Sheep	8500 BC	Western Asia
Silkworm	3500 BC	China
Turkey	100 BC–AD 100	Mexico
Water buffalo	2500 BC	Pakistan
Yak	2500 BC	Tibet

acter) is exhibited in the **PHENOTYPE** of an individual at the expense of an alternative character (the recessive character) when the individual is **HETEROZYGOUS** for the relevant **GENE**, i.e. is carrying one **DOMINANT ALLELE** and one **RECESSIVE ALLELE** for the same gene. *Compare* **CODOMINANCE (1)**

2. In animal behaviour, the phenomenon whereby some individuals in a social group have a higher status than others, usually achieved by aggression or the threat of aggression.

dominance hierarchy A social system in which some individuals have a higher status (**RANK**) than others, giving them preferential access to mates, food and other resources. **SUBORDINATE INDIVIDUALS** often exhibit **APPEASEMENT BEHAVIOUR** (which may be ritualised) when they encounter dominant animals. Dominance hierarchies were first demonstrated in domestic fowl (*Gallus gallus domesticus*). Once individuals have established their place in a hierarchy, fighting is rare. There are different types of dominance hierarchy, e.g. it may be linear (e.g. A → B → C → D) or there may be a single dominant male (A → B, C, D). In primate societies the social structure is often maintained by the **GROOMING** of dominant individuals by subordinates. *See also* **ALPHA STATUS, BETA STATUS, PECKING ORDER**

dominant allele The allele whose character is expressed when only one copy is present in the GENOME and paired with a RECESSIVE ALLELE.

dominant individual An individual in an hierarchically organised SOCIAL GROUP whose status is higher than that of another, SUBORDINATE INDIVIDUAL. *See also* ALPHA STATUS, DOMINANCE HIERARCHY

donkey, ass, moke A domesticated equid (*Equus africanus asinus*) descended from the African wild ass (*E. africanus*). Widely used throughout the world as a pack and draught animal, especially in developing countries (Figs. B3 and D1). *See also* BEACH DONKEY, DONKEY SANCTUARY, MULE

Donkey Sanctuary, The An international animal welfare charity based in the UK which works to protect DONKEYS and MULES. It operates a number of farms and provides welfare advice and support to operators of BEACH DONKEYS.

dopamine A NEUROTRANSMITTER found in the vertebrate brain. It plays a role in motivation, reward and responses to surprising events.

doping The illegal application of any substance which might enhance an animal's performance, especially a racehorse. *See also* BLOOD DOPING

dormancy A general term for a resting condition in which an animal is alive but relatively inactive metabolically. Often induced by the onset of unfavourable conditions to ensure survival. *See also* AESTIVATION, HIBERNATION, WINTER LETHARGY

dorsal Relating to or near the back of an animal, e.g. the dorsal fin of a shark. The opposite of VENTRAL (Fig. A8).

double banding The fitting of two identification bands to the legs of a bird. One may be a metal band with an identification number, the other a colour-coded band used for identification at a distance. Some organisations add a REWARD BAND to the legs of wild birds. The WILDFOWL AND WETLANDS TRUST (WWT) uses one band to indicate sex (depending on which leg it is on) and the other bears an identification number.

double door system, double door entry system An entry system used in some ANIMAL ENCLOSURES to allow the entry and exit of keepers or visitors while still containing the animals (Fig. D10). A person entering the enclosure passes through the outside door into a small holding area, closes this door, and then opens a second (inside) door. The two doors are never open at the same time. In some systems the locks on the doors are electronically linked so that it is not physically possible to open both doors together, and the movement of visitors is controlled by a 'traffic light' system. Double door systems for visitors are commonly used at the entrances and exits of WALK-THROUGH EXHIBITS that contain free-flying birds, bats, butterflies or free-ranging animals such as small primates. They are also used for keeper access to the enclosures of

Fig. D10 Double door entry system to a lion (*Panthera leo*) enclosure.

dangerous animals such as large primates and carnivores. *See also* CONTAINMENT

double-clutching The practice of removing eggs from a nest in order to induce a hen bird to lay a second clutch. Used to increase breeding rate in CONSERVATION BREEDING PROGRAMMES. The first clutch may be hatched in an INCUBATOR (2) or incubated by a SURROGATE (1).

down Very soft, fine FEATHERS or hair.

draft animal *See* BEAST OF BURDEN

drag hunt A sporting activity that involves horse riders following a scent trail as a substitute for hunting an animal such as a fox. Practised as an alternative to fox hunting as a result of the passing of the HUNTING ACT 2004 in England and Wales. *See also* BURNS REPORT

Draize test A test used by companies which produce cosmetics, pharmaceuticals etc. to determine the sensitivity of the eye or skin to their products using animals.

draught animal *See* BEAST OF BURDEN

dreaming The experience of thoughts, emotions and images during SLEEP. In humans this is accompanied by RAPID EYE MOVEMENTS (REM) and a characteristic pattern of brain activity. Some mammals also experience REM during sleep and appear to be dreaming.

drenching *See* WORMING

dressage A competitive equestrian sport in which the horse makes a prescribed series of movements when ridden within a standard arena, demonstrating the horse's natural athletic ability and willingness to perform with minimal direction from the rider. *See also* **LIPIZZANER**

drinker A container or device for providing drinking water to animals. It may be a simple **DRINKING BOTTLE** or a self-filling trough attached to a mains supply of water. *See also* **AUTOMATIC WATER DRINKER**, **NIPPLE DRINKING SYSTEM**, **NOSE FILL DRINKER**

drinking bottle An inverted plastic bottle for providing water for small animals. Water passes down a metal tube which is sealed with a small metal ball at its end, and water is released when the animal moves the ball by licking it. *See also* **DRINKER**

drip
1. A device for supplying a liquid (e.g. a drug, plasma) slowly and often continuously into a vein. *See also* **CANNULATION**, **CATHETER**, **INFUSION**
2. The act or process of giving a liquid in this manner.
3. The liquid given in this manner.

drive An outdated term used in **MOTIVATION** theory which may be defined as the psychological force or internal state of tension that motivates an animal to perform a particular behaviour.

drive-through enclosure A **ZOO** or **SAFARI PARK** enclosure through which visitors may drive in their own vehicles (or vehicles which are provided), and which usually contains **LARGE MAMMALS** such as antelope, deer, rhinoceros, lions, and large birds such as **RATITES**, often in **MULTI-SPECIES EXHIBITS** groups.

droppings Faeces.

drove A moving herd of animals, especially cattle or sheep. *See also* **DROVING ROUTE**

drover A person who drives livestock on foot, horseback or using a vehicle, to move them to new pastures, markets etc.

droving route, droving road A traditional route over which **LIVESTOCK** are (or were) moved on foot, generally from farms to market. *See also* **TRAVELLING STOCK ROUTE** (TSR)

drug
1. A substance that has a physiological effect on the body when ingested, injected or otherwise applied to the body and which is used in the diagnosis, treatment or prevention of disease.
2. To administer a drug.

drugs, regulation of *See* **VETERINARY MEDICINES DIRECTORATE**, **VETERINARY MEDICINES REGULATIONS 2011**

dry matter (DM)
1. A measure of the mass of something when completely dried.
2. All of the content of food except for the water.

dry matter intake (DMI) Feed intake minus its water content.

dry moat An outdated type of barrier frequently used for **PACHYDERMS**. Usually a concrete channel with steep sides which is wide enough to prevent the animal from stepping across it and deep enough to prevent it from climbing out. Dangerous because animals may fall in and be injured or killed. *Compare* **WET MOAT**

duck Mostly aquatic species of birds which are members of the avian family Anatidae which also includes the geese and swans.

duck canopy A shelter provided for ducks, often designed to float on water.

duck stamp A stamp purchased by hunters in the USA in order legally to hunt ducks. These stamps are required by law under the **MIGRATORY BIRD HUNTING AND CONSERVATION STAMP ACT OF 1934 (USA)**. The money raised is used to purchase duck refuges.

Dudley Zoo A zoo in the West Midlands (UK) in the grounds of Dudley Castle. Dudley Zoo contains 12 iconic **TECTON GROUP** buildings (now **LISTED BUILDINGS**), the largest collection in the world. They include an Elephant House, Sea Lion Pool, Bear Ravine, Birdhouse, Polar Bear Triple Complex and the Zoo Entrance (Fig. L4). When it was opened in 1937 Dudley Zoo was described as the 'most modern zoo in Europe'.

Duke Lemur Center The world's largest **PROSIMIAN** sanctuary dedicated to conservation and research on rare and endangered species. Located at Duke University, Durham, North Carolina and established in 1966. Funded by Duke University and the National Science Foundation.

dull emitter heat lamp bulb A ceramic bulb which emits heat but no light and may be used to heat a **BROODER** or small exhibit, e.g. a vivarium. As no light is produced the lamp does not interfere with the natural diurnal changes in light levels. *See also* **INFRARED LAMP**

dummy egg An artificial egg used to remind hens where to lay their eggs. Useful for young birds and ex-battery hens.

duodenum Part of the small intestine.

Durrell Gerald **DURRELL** opened the Jersey Zoological Park in 1959 (now known as *Durrell*). The zoo was dedicated to conservation from the outset and specialises in keeping, breeding and reintroducing endangered species. *Durrell* has been extremely influential in persuading zoos to refocus their efforts on endangered species and to become involved in field conservation. The Durrell Wildlife Conservation Trust now works to protect critically endangered animal and plant species in 16 countries.

Durrell, Gerald (1925–1995) Gerald Durrell was a successful animal trader, author and television presenter who established his own zoo on Jersey in the Channel Islands. He worked for a short time as a keeper at **WHIPSNADE ZOO** and then began making animal collecting trips to West Africa. Some of his best-known books are *My Family and Other Animals* (1956) and *A Zoo in My Luggage* (1960).

Durrell Institute of Conservation and Ecology (DICE) An organisation run by the University of Kent in association with *DURRELL* which provides education and training in conservation skills, especially for people from developing countries.

Durrell Wildlife Conservation Trust *See DURRELL*

dust bathing Covering the body in soil (or similar material) by throwing it over the body or rolling in it. The purpose may be to protect the skin from the sun or to remove parasites. Also called dusting.

dusting
1. *See* **DUST BATHING**
2. In animal nutrition, the application of a coating of a powdered nutrient supplement to the outside of a food item (e.g. a cricket) before feeding it to another animal. Necessary because such food items are often a suitable prey food for other species (e.g. amphibians) but of low nutrient quality. *See also* **GUT LOADING**

duty of care In law, a requirement to take reasonable care to avoid any act or omission that it could reasonably be foreseen might injure another person. *See also* **NEGLIGENCE, OCCUPIERS' LIABILITY**

dwell time In **VISITOR STUDIES**, the amount of time a visitor spends engaging with an exhibit, e.g. watching animals, reading signage etc. Average dwell times at zoo exhibits are generally low (often only a few minutes), especially if animals are not visible or if they are visible but inactive. *See also* **VISITOR BEHAVIOUR**

dyad A pair of individuals (regardless of sex), especially when recorded together in a study of social organisation. *See also* **ASSOCIATION INDEX, CONSORTS, SOCIOGRAM**

dyed fish, painted fish Fish that have been injected with or dipped in dye to make them more colourful and attractive to hobbyists. May make fish more prone to infections such as the viral disease lymphocystis.

dynamic life table *See* **LIFE TABLE**

dysecdysis The abnormal shedding of the skin, e.g. in reptiles. *See also* **ECDYSIS**

dysplasia The abnormal growth of **ORGANS, TISSUES** or **CELLS**. *See also* **HIP DYSPLASIA**

dyspnoea, dyspnea Difficult or laboured **BREATHING**; shortness of breath.

dystocia Abnormal, slow or difficult birth, often caused by a disorder of the uterus or ineffective contractions of the uterus in mammals, or difficulty in laying eggs. Occurs in live-bearing and **OVOVIVI-PAROUS** reptiles, birds and fishes. *See also* **EGG BINDING**

dzo/dzomo A hybrid formed by mating a yak with a domestic cow. A male hybrid is called a dzo, a female is a dzomo.

D

Fig. E1 E is for elephant. Bull African elephant (*Loxodonta africana*), Tarangire National Park, Tanzania.

Dictionary of Zoo Biology and Animal Management: A guide to terminology used in zoo biology, animal welfare, wildlife conservation and livestock production, First Edition. Paul A. Rees.
© 2013 John Wiley & Sons, Ltd. Published 2013 by John Wiley & Sons, Ltd.

E. coli *Escherichia coli*; a bacterium present in the gut of most mammals which is normally harmless, but some strains can cause serious infections in animals and humans, e.g. O157:H7 lives in the gut of healthy cattle but can cause serious illness in children and the elderly.

ear notch A piece of the ear (**PINNA (1)**) of an animal that has been removed as an identification **MARK**. The location of several notches on one or both ears can be used as a code to create a numerical system; e.g. individual notches in specific positions on the left ear may indicate the numbers 10, 20, 40 and 70, those on the right ear 1, 2, 4 and 7. The identification number 36 would consist of four notches in the positions indicating 10, 20, 2 and 4 (which total 36).

ear ossicles, auditory ossicles The bones of the middle ear. In mammals three tiny bones, the malleus, incus and stapes, transmit sound from the **TYMPANUM** to the **COCHLEA**. In non-mammalian vertebrates only the stapes, or its homologue, exists.

ear tag An identification **MARK** (usually made of plastic or metal) fixed to an animal's ear, generally bearing a unique number (Fig. M2). *See also* **RECORD KEEPING**

Earth jurisprudence The examination of the philosophy and value systems that underpin most legal and governance systems and ensuring that they support the integrity and health of the Earth. *See also* **WILD LAW**

Earth Summit, Rio The United Nations Conference on Environment and Development (UNCED). An international meeting of governments and **NGOs** held in Rio de Janeiro, Brazil, in 1992 at which the UN **CONVENTION ON BIOLOGICAL DIVERSITY 1992 (CBD)** was adopted.

Earthwatch An international environmental charity, founded in 1971, which engages ordinary people in scientific field research and education to promote the understanding and action necessary for a sustainable environment.

eastern coyote *See* **COYWOLF**

EAZA *See* **EUROPEAN ASSOCIATION OF ZOOS AND AQUARIA (EAZA)**

EAZA Nutrition Group A group of specialists established to improve communication, education and research concerning zoo animal nutrition within zoos in Europe. It produces a number of nutrition books that contain the scientific contributions to the European Zoo Nutrition Conferences and other publications. Formerly known as the European Nutrition Group.

ecdysis The shedding of the **CUTICLE** (**MOULTING**) in **ARTHROPODS** during growth.

ECG *See* **ELECTROCARDIOGRAM (ECG), ELECTROCARDIOGRAPH (ECG)**

echinoderm A member of the phylum **ECHINODERMATA**.

Echinodermata A phylum of animals. The echinoderms are marine animals most of which have a radially symmetrical structure, e.g. starfish, brittlestars, sea cucumbers, sea urchins. They possess a sophisticated water-vascular system which consists of an array of canals and tube feet which terminate in tiny suckers. This system provides both a means of **LOCOMOTION** and a means of catching prey. Echinoderms have an internal skeleton of calcareous plates which form a rigid skeletal box. Spiny or wart-like projections extend outward for protection.

echolocation A method of orientation whereby an animal produces a high-pitched sound and locates nearby objects by the echoes produced. It has been demonstrated in bats, dolphins and other marine mammals, and some birds.

ECJ *See* **EUROPEAN COURT OF JUSTICE (ECJ)**

eclipse plumage The **PLUMAGE** of male ducks, adopted in the summer, when they moult and become temporarily flightless.

ECOLEX A web-based comprehensive source of national and international environmental and natural resources law and policy operated by the **FOOD AND AGRICULTURE ORGANISATION (FAO)**, **INTERNATIONAL UNION FOR THE CONSERVATION OF NATURE AND NATURAL RESOURCES (IUCN)** and **UNITED NATIONS ENVIRONMENT PROGRAMME (UNEP)**.

ecological niche The role an organism plays in an ecosystem and the set of environmental conditions it is able to tolerate. *See also* **FUNDAMENTAL NICHE, REALISED NICHE**

ecological succession The ecological process whereby a **BIOLOGICAL COMMUNITY** undergoes orderly and more or less predictable change in species composition when a new habitat is colonised (primary succession) or a habitat is disturbed, e.g. by deforestation (secondary succession). Succession culminates in a **CLIMAX COMMUNITY**.

ecology The scientific study of the relationships between **ORGANISMS** and each other and organisms and their **ENVIRONMENT**.

eco-roof *See* **GREEN ROOF**

ecosystem
   1. A **BIOLOGICAL COMMUNITY** and its physical environment.
   2. The UN **CONVENTION ON BIOLOGICAL DIVERSITY 1992 (CBD)** (Art. 2) defines an ecosystem as '*a dynamic complex of plant, animal and microorganism communities and their non-living environment interacting as a functional unit.*'
   3. In the USA, Executive Order 13112 of February 3, 1999 defines an ecosystem as '. . . *the complex of a community of organisms and its environment.*'

ecosystem exhibit *See* **BIOPARK**

ecosystem services The functions of an ecosystem which provide a benefit to humans, e.g. the filtration of water by reedbeds, protection of coastlines by marshlands, sequestration of carbon dioxide by forests. These services have an economic value

E

which should be considered when ecosystems are damaged or destroyed.

ecosystem zoo *See* BIOPARK

ecotone A narrow transitional zone between different ecological communities, e.g. a woodland and a savanna, or the edge of a forest where trees have been felled. Where zoos are divided into different 'BIOMES' the areas where they meet should be designed as ecotones where possible to enhance visitors' experience.

ecotourism Tourism based on the tourist's general interest in the ecology and wildlife of an area, usually to undisturbed areas of natural beauty or high biodiversity, and involving minimal ecological impact and damage.

ecotype A locally adapted form of a species which has a wide distribution.

ectoderm The outer layer of cells in the embryo of a triploblastic animal, which gives rise to the epidermis, most of the nervous system and the extreme anterior and posterior parts of the gut. *Compare* ENDODERM

ectoparasite, exoparasite A parasite that lives on the outside of its HOST (e.g. on its skin or hair). *Compare* ENDOPARASITE

ectopic
1. Located away from the normal position in the body, e.g. an ectopic kidney is located somewhere other than its normal position, possibly as a result of a congenital abnormality.
2. Ectopic pregnancy; a pregnancy in which the embryo implants in the FALLOPIAN TUBES instead of the wall of the uterus.
3. In cardiology, a HEARTBEAT that has originated somewhere other than the SINOATRIAL (SA) NODE.

ectotherm *See* POIKILOTHERM

ectothermic Relating to POIKILOTHERMS (ectotherms). *Compare* ENDOTHERMIC (1)

Eden Project A visitor attraction in Cornwall, UK, consisting of a series of large transparent geodesic domes which house artificial ecosystems including birds and reptiles.

Edinburgh family pen system A system of pig production designed by Alex Stolba in which piglets and fatteners grow up in species-specific family groups that correspond to the normal social organisation of domestic pigs. *See also* HURNIK–MORRIS PIG HOUSING SYSTEM

education department The administrative unit of a zoo or other facility which is responsible for all its educational activities including classroom sessions for schools, design of INTERPRETATION, presentations and keeper talks, production of educational materials and web content. Zoo education officers are often qualified teachers. Zoos in the EUROPEAN UNION (EU) are required to have an educational function under the ZOOS DIRECTIVE.

education officer A person who provides educational services within an EDUCATION DEPARTMENT in a zoo or other facility.

education outreach animal An individual animal who is used primarily for education purposes and who may be used for contact sessions with members of the public and possibly taken out of the zoo on educational visits to schools etc. *See also* AMBASSADOR SPECIES, DISPLAY EDUCATION AND RESEARCH POPULATIONS, PROGRAM ANIMAL

EEG *See* ELECTROENCEPHALOGRAM (EEG), ELECTROENCEPHALOGRAPH (EEG)

EEP *See* EUROPEAN ENDANGERED SPECIES PROGRAMME (EEP)

effective population size The effective size of a POPULATION is a measure of how well the population maintains GENETIC DIVERSITY from one generation to the next. It is usually lower than actual population sizes because, for example, the SEX RATIO may be unequal and some individual animals have more offspring during their lifetime than others. Also small populations lose genes as a result of chance events. For example, the last two individuals possessing the gene for blue eyes might be killed in a storm, thereby removing this gene from the population. This process is known as GENETIC DRIFT. The effective population size is the size of an ideal population that would lose genetic variation by genetic drift at the same rate. In other words, a population of 300 individuals may have an effective population size of, say, 250 because there are too few females in the population. So this population of 300 actually loses genetic variation at the same rate as an ideal population of 250. The effective population size is essentially a measure of the number of individuals that are effectively contributing genes to the next generation. Effective population size (N_e) may be calculated as:

$$N_e = \frac{4(N_f \times N_m)}{N_f + N_m}$$

where N_m = number of breeding males, N_f = number of breeding females. Strictly speaking this formula only applies to stable, randomly mating populations with non-overlapping generations.

effector organ An organ that responds to a stimulus, e.g. a muscle or a gland.

efferent Conduction away from. For example, an efferent blood vessel carries blood away from the heart. An efferent nerve conducts impulses away from the CENTRAL NERVOUS SYSTEM (CNS) towards the muscles. *Compare* AFFERENT

egestion The passing of waste out of the gut (as faeces) which has never been part of the constituents of the cells of the body. *Compare* EXCRETION

egg
1. A female gamete; an ovum or egg cell.
2. A reproductive cell or developing embryo inside a protective structure (sometimes a shell) in birds, reptiles, amphibians, fishes and many invertebrates.

egg binding, egg impaction The inability of an egg to pass through the reproductive system at the normal rate, often due to an obstruction. May result in **RUPTURE** or **PROLAPSE** of the reproductive tract or lead to infection. May occur in birds, reptiles and fishes. *See also* **DYSTOCIA**

egg collector A person (usually male) who collects the eggs of wild birds as a hobby, usually illegally. *See also* **EGGING**

egg impaction *See* **EGG BINDING**

egg tooth A small hard projection on the beak of embryonic birds and the upper jaw of embryonic reptiles which is used to break through the egg surface during hatching.

egg-breaking behaviour
1. The destruction of eggs by a parent bird. This is a problem in some captive birds and has been recorded in whooping cranes (*Grus americana*) and domestic chickens. *See also* **EGG-EATING BEHAVIOUR**
2. The use of stones by birds to break open the eggs of other species before consuming the contents (e.g. Egyptian vultures (*Neophron percnopterus*)).

egg-eating behaviour The consumption of eggs by birds. Birds may eat their own eggs or the eggs of conspecifics or other species. A common problem in chickens.

egging The act of taking eggs from a bird's nest. In the past the eggs of wild birds were an important source of food for humans, especially the eggs of colonial seabirds which were easy to collect because of the high density of nests. *See also* **EGG COLLECTOR**

EGGS Software which supports record-keeping and egg clutch management for a single institution and augments collection records kept in **ARKS4** and **SINGLE POPULATION ANALYSIS AND RECORD KEEPING SYSTEM (SPARKS)**.

EIA *See* **ENVIRONMENTAL IMPACT ASSESSMENT (EIA), ENVIRONMENTAL INVESTIGATION AGENCY (EIA)**

EID technology *See* **RADIO FREQUENCY IDENTIFICATION (RFID) TECHNOLOGY**

ejaculate
1. To discharge semen.
2. A quantity of semen released in a single ejaculation.

ejaculation The discharge of semen from the male reproductive system.

EKG *See* **ELECTROCARDIOGRAM (ECG), ELECTROCARDIOGRAPH (ECG)**

Elasmobranchii A subclass of cartilaginous fishes: sharks, rays and skates.

electric fence, hot wire A wire fence carrying a low voltage electrical current. Used as a barrier around the perimeter of animal enclosures and to prevent animals from damaging trees and other vegetation and structures. *See also* **ELECTROCUTION**

electric fishing A method of capturing fishes alive by stunning them with an electric current delivered by electrodes submerged in water, usually a river or stream. This technique is widely used by agencies responsible for protecting and managing freshwater fish stocks, e.g. the **ENVIRONMENT AGENCY** in England and Wales.

electric stunner A device which stuns an animal using an electric current prior to slaughter. Widely used for poultry.

electrocardiogram (ECG), EKG The graph or image produced by an **ELECTROCARDIOGRAPH (ECG)** which shows the pattern of contraction of the **HEART**.

electrocardiograph (ECG), EKG Apparatus which records the electrical variations of the **HEART** as it is beating as an electronic image or tracing (**ELECTROCARDIOGRAM (ECG)**). May be used to detect faulty heart valves, **ARRHYTHMIAS** etc.

electrocution The process of killing an animal by electric shock. This may occur accidentally or intentionally as a means of execution. Wild elephants have been killed in India by coming into contact with low-hanging overhead power lines and by illegal high-tension electric fences constructed by farmers to protect their crops. In February 2011 two racehorses, *Fenix Two* and *Marching Song*, died of cardiac arrest after being electrocuted as a result of standing on an underground electricity cable in a paddock at Newbury Racecourse, Berkshire, UK. In 1903 Thomas Edison executed a circus elephant named *Topsy*, after she had killed three men, by passing 6600 volts through her body. *See also* **ELECTRIC FENCE, ELECTRONIC TRAINING COLLAR**

electro-ejaculation *See* **ELECTRO-EJACULATOR**

electro-ejaculator An electrical device designed to stimulate an animal to ejaculate in a process called electroejaculation. This process is used to collect **SEMEN** for **ARTIFICIAL INSEMINATION (AI)**.

electroencephalogram (EEG) A graphical record of the electrical activity of the brain produced by an **ELECTROENCEPHALOGRAPH (EEG)**.

electroencephalograph (EEG) A device that measures changes in brain activity which is used to study brain function and disease.

electrolytes
1. A solution of chemical salts.
2. Minerals in the body that have an electric charge, e.g. calcium, sodium, potassium, bicarbonate. Important in maintaining **HOMEOSTASIS** and the proper functioning of the heart, nervous system etc.

electromagnetic senses Senses which are able to detect electromagnetism. Migratory birds have been shown to use magnetic fields to guide them on their **MIGRATIONS (1)**. Some fishes use an electric sense to communicate and to detect prey. Others can produce an electric shock to stun their prey. *See also* **ACOUSTIC-LATERALIS SYSTEM, COMPASS SENSE, ELECTROCUTION**

electron micrograph *See* **ELECTRON MICROSCOPE**

electron microscope A device which uses electrons focused by magnets – instead of light focused by lenses – to produce images of very small structures. The image produced is called an electron micrograph. May be used to produce images of sub-cellular structures such as cell **ORGANELLES**. *See also* **SCANNING ELECTRON MICROSCOPE**

electron transport system, electron transport chain, respiratory chain A system of electron carriers which passes electrons or protons, derived from metabolites in **GLYCOLYSIS** or the **KREBS CYCLE**, from one to another in a sequence, resulting in the production of **ADENOSINE TRIPHOSPHATE (ATP)**. Such systems are found in **MITOCHONDRIA**.

electronic identification technology *See* **RADIO FREQUENCY IDENTIFICATION (RFID) TECHNOLOGY**

electronic training collar, shock collar A dog collar which incorporates a device capable of delivering an electric shock or other aversive stimulus such as a water spray, vibration or **ULTRASONIC** sound in order to modify behaviour. The device may be activated by a radio signal sent from a hand-held transmitter operated by a handler or trainer, or it may be activated when the dog crosses a perimeter wire. It may be used for training (training collar), to control barking (**BARK COLLAR**), or as part of a pet containment system, e.g. within a garden. The use of shock collars on dogs and cats is illegal in Wales under the Animal Welfare (Electronic Collars) (Wales) Regulations 2010.

electronic treat dispenser A device which automatically releases small items of food as treats for animals. Usually programmable so that the user can determine when the treats are released.

electrosurgery A surgical method which uses a high-frequency electric current to cut tissue, induce coagulation and destroy tissue (e.g. tumours) with minimum blood loss. *See also* **ELECTROSURGICAL UNIT (ESU)**

electrosurgical unit (ESU) A device which supplies and monitors the electric current used in **ELECTROSURGERY**.

elephant One of two extant species of **PACHYDERM**: the African elephant (*Loxodonta africana*) (Fig. E1) and the Asian elephant (*Elephas maximus*). Some authorities recognise a third species, the forest elephant (*L. cyclotis*), but others consider this form a subspecies of the African elephant (*L. a. cyclotis*). They have a complex social organisation (Figs. S12 and S13) and are threatened by **IVORY** poaching in the wild. Elephants have been popular animals at zoos since Victorian times, the most famous being *JUMBO*. The keeping of elephants in zoos is becoming increasingly controversial especially since a number of publications have highlighted the poor conditions in some zoos, low calf survival and reduced longevity compared with the wild (e.g. *LIVE HARD, DIE YOUNG*). Historically birth rates in zoos have been low. However **ARTIFICIAL INSEMINATION (AI)** techniques have now been developed for both species and used successfully in a number of zoos. Zoo elephants have killed a large number of keepers in zoos and mahouts in Asia, and there is currently a move in favour of handling them using **PROTECTED CONTACT** techniques in zoos. *See* **AFRICAN ELEPHANT CONSERVATION ACT OF 1989 (USA)**, **ASIAN ELEPHANT CONSERVATION ACT OF 1997 (USA)**, *ELEPHANT-FREE ZOOS*, **ELTRINGHAM**

elephant barn An alternative (American) name for an elephant house.

Elephant Bill A book published in 1950 recounting the true story of the exploits of Lt-Col. J. H. Williams, the most famous of the British elephant officers working with the XIVth Army in Burma during World War II.

elephant camp A place where elephants are kept by a government forest department or commercial company for the purpose of moving and handling felled trees as part of a forestry operation. Found in India, Thailand and other countries in Asia.

Elephant Managers Association (EMA) The EMA is an international non-profit organisation of professional elephant handlers, administrators, veterinary surgeons, researchers and elephant enthusiasts. It is dedicated to the welfare of elephants through improved conservation, husbandry, research, education and communication.

elephant polo A 'sport' played in Thailand, Nepal, India and Sri Lanka in which Asian elephants (*Elephas maximus*) with human riders are used to play polo in competitions.

elephant sanctuary A facility where 'rescued' elephants from circuses and zoos are allowed to live in large enclosures and usually in naturalistic social groups. Breeding is not normally allowed. The Elephant Sanctuary® in Tennessee is America's 'natural habitat refuge' for African and Asian elephants. It occupies 1093 ha (2700 acres) of land in Hohenwald. Between 1995 and 2012 the sanctuary took in 24 elephants that had retired from zoos and circuses. Riddle's Elephant and Wildlife Sanctuary is a smaller facility of around 134 ha (330 acres) in Arkansas which held 12 elephants in April 2012. It hosts an annual International School for Elephant Management. Elephant sanctuaries also exist in the range states of the species, e.g. Thailand and South Africa.

Elephant-free Zoos A campaign organised by the **PEOPLE FOR THE ETHICAL TREATMENT OF ANIMALS (PETA) FOUNDATION** whose aim is to remove all elephants from zoos by phasing out elephant exhibits and abandoning **CAPTIVE BREEDING PROGRAMMES**, and to provide elephants currently in captivity with a more humane existence. PETA believes that zoos cannot provide for the basic needs of elephants and that zoo elephants make no contribution to elephant conservation or the education of the public. Some zoo professionals agree with this view. Many zoos have stopped keeping elephants, particularly in the USA and the

UK, e.g. San Francisco Zoo, Detroit Zoo, **LINCOLN PARK ZOO, LONDON ZOO, DUDLEY ZOO,** Bristol Zoo, Longleat Safari Park. In some cases the zoo has transferred its animals to a larger facility; others have been sent to an **ELEPHANT SANCTUARY**. In 2009 in India, the **CENTRAL ZOO AUTHORITY (CZA)** announced that 140 elephants living in 26 zoos would be transferred to wildlife parks and sanctuaries. *See also* **BASIC NEEDS TEST**

elevator A small hand-held metal instrument used in the extraction of teeth.

Elizabethan collar A collar shaped like a truncated cone which is used to prevent an animal (especially a cat or dog) from licking or biting its body.

Elopiformes An order of fishes: tarpons and their allies.

Elsa A lioness who was the subject of the film *BORN FREE*. *See also* **ADAMSON**

Eltringham, S. Keith (1929–2006) A Canadian zoologist, pilot, wildlife manager and conservationist who pioneered the use of **AERIAL SURVEYS** to **CENSUS** wildfowl (for the **WILDFOWL AND WETLANDS TRUST (WWT)**) and large mammal populations in Africa. He drew the plight of the African elephant to the attention of the public during the 1970s when poaching was rife.

emaciation In relation to the condition of an animal, being thin and feeble.

embolism (emboli *pl.*) An obstruction in a blood vessel caused by a blood clot or the presence of a foreign object (embolus), preventing the flow of blood. It may be formed in one part of the body and moved to another part by the blood flow. Formed from blood, fat, gas or **TUMOUR** tissue. May be fatal if present in a major artery, e.g. in the brain (cerebral embolism) or lungs (pulmonary embolism).

embryo A structure produced by mitotic cell divisions of an egg usually after fertilisation by a sperm while still within the mother's body. *See also* **ZYGOTE**

embryo transfer (ET) This involves the insertion of a viable embryo into a recipient female. The embryo may have been taken from a pregnant donor or it may be the result of *IN-VITRO* FERTILISATION. If the embryo is inserted into the recipient female at the correct stage of her **OESTROUS CYCLE** it may produce a **PREGNANCY**. This technique allows the repeated harvesting of embryos from the same female donor by curtailing pregnancy within the first few days. Hormone therapy must be used to induce the female to repeatedly produce new embryos and the timing of insemination is also critical so **ARTIFICIAL INSEMINATION (AI)** techniques are used. ET allows the selection of both the female genes and the male genes and is a valuable tool in conservation breeding programmes. It may involve the transfer of an embryo into a female recipient of the same species or a different (**SURROGATE (2)**) species (**INTERSPECIES EMBRYO TRANSFER**).

embryonic diapause *See* **DELAYED IMPLANTATION**

emergency radio codes Numerical, colour or other verbal codes use by zoos internally to communicate the existence of an emergency. For example Houston Zoo (Texas) uses code 99 for any emergency involving an escaped dangerous animal, persons in animal enclosures, or any violent act clearly hazardous to animals or visitors; code 88 for an emergency involving an animal who has escaped from their primary containment, but poses no hazard to human life; code red for a fire; and code blue for a medical problem.

emergency response team *See* **ZOO EMERGENCY RESPONSE TEAM**

emetic A substance which can induce vomiting. *Compare* **ANTI-EMETIC**

emigration The movement of individuals out of a **POPULATION**. *Compare* **IMMIGRATION**

emotion In humans, emotion is an aroused state of the body in which certain characteristic inner feelings are experienced, along with particular physiological signs, e.g. increased **HEART RATE**. Some species, especially highly social species, appear to experience emotions and are capable of communicating them to others through their behaviour, including **FACIAL EXPRESSIONS** in some cases. *See also* **EMPATHY, GRIEF, PAIN, SPINDLE CELLS**

empathetic behaviour A behavioural response to the distress of others, e.g. when their chicks are distressed, adult hens exhibit increased heart rate, increased alertness, decreased preening, decreased **EYE TEMPERATURE** and increased **VOCALISATIONS** directed to the chicks. This behaviour clearly has welfare implications when exhibited by animals kept in laboratories or farms, especially if they are kept in conditions where individuals are routinely exposed to others in distress, perhaps due to overcrowding or injury. *See also* **EMPATHY**

empathy The capacity to be affected by and share the emotional state of another animal. This occurs in chimpanzees, elephants and even chickens. *See also* **EMOTION, EMPATHETIC BEHAVIOUR**

empirical study A study based on **EXPERIMENT**, observation or experience rather than theory.

empirical zoo concept A zoo with a scientific foundation, where policies and programmes are based on research findings gathered from studies of animals in the field, the laboratory and the zoo. This **EVIDENCE-BASED MANAGEMENT** approach to the welfare and wellbeing of animals living in zoos has been propounded by Dr. Terry **MAPLE**.

empty In relation to female animals, especially **LIVESTOCK**, not pregnant. *See also* **PSEUDOPREGNANCY** *Compare* **IN CALF, PREGNANCY**

enamel The hard material covering the crown of the tooth, consisting mainly of calcium phosphate. *See also* **DENTINE**

enarthrosis *See* **BALL AND SOCKET JOINT**

E

encephalitis INFLAMMATION of the brain usually caused by an INFECTION.

enclosure *See* ANIMAL ENCLOSURE, CAGE

enclosure barrier A physical structure which is used to contain animals within an enclosure, e.g. a FENCE, a HA-HA. *See also* CATTLE GRID, CONTAINMENT PRIMARY, CONTAINMENT SECONDARY, DEPRESSED VERTICAL FENCE BARRIER, DOUBLE DOOR SYSTEM, DRY MOAT, ELECTRIC FENCE, FEED BARRIER, PSYCHOLOGICAL BARRIER, PUBLIC BARRIER, WET MOAT

enclosure design The plan of an enclosure including its dimensions, shape, components etc. Needs to consider the physical and behavioural needs of the animals to be kept, e.g. their ENVIRONMENTAL ENRICHMENT needs, a suitable SUBSTRATUM, adequate size, suitable FURNITURE etc. *See also* BIOPARK, IMMERSION EXHIBIT, MULTISPECIES EXHIBIT, NATURALISTIC EXHIBIT, ROTATIONAL EXHIBIT DESIGN, THEMED EXHIBIT, ZOOLEX

enclosure size The dimensions of an enclosure. Minimum size requirements for zoo enclosures are generally provided in the HUSBANDRY GUIDELINES for particular species. Enclosure utilisation may be measured by calculating the SPREAD OF PARTICIPATION INDEX (SPI).

encroachment In relation to the natural environment, the loss of wildlife habitat (e.g. tropical forests) due to the expansion of human activities, e.g. growth of settlements, agriculture etc.

endangered
1. In relation to organisms, one that is in danger of becoming extinct.
2. A RED LIST category used by the INTERNATIONAL UNION FOR THE CONSERVATION OF NATURE AND NATURAL RESOURCES (IUCN): Endangered (EN) – Best available evidence indicates that the taxon is facing a very high risk of extinction in the wild.

Endangered Species Act of 1973 (USA) A law in the USA which provides for the conservation of species that are endangered or threatened throughout all or a significant part of their range and also the conservation of ecosystems of which they are a part. The UNITED STATES FISH AND WILDLIFE SERVICE (USFWS) shares responsibility for implementing the ESA with the NATIONAL MARINE FISHERIES SERVICE (NMFS). They are required to create endangered species RECOVERY PLANS to recover endangered species, including the designation of critical habitat. *See also* CLIMATE CHANGE, EXPERIMENTAL POPULATION, NORTHERN SPOTTED OWL (*STRIX OCCIDENTALIS CAURINA*), *WYOMING FARM BUREAU FEDERATION V. BABBITT (1997)*

Endangered Species Preservation Act of 1966 (USA) Law in the USA which was the forerunner of the ENDANGERED SPECIES ACT OF 1973 (USA).

Endangered Species Recovery Plans *See* ENDANGERED SPECIES ACT OF 1973 (USA), RECOVERY PLAN

ENDCAP A network of European organisations dedicated to the protection and conservation of wildlife in the wild, and opposed to the unnecessary exploitation of wild animals in captivity including in zoos, dolphinaria and circuses.

endemic
1. In relation to a species, one that naturally occurs in a particular restricted area, e.g. an island or continent.
2. An endemic species (or other taxon).
3. In relation to disease, one that regularly occurs in a particular region or within a particular group of people or animals.

endocrine gland A ductless gland that produces HORMONES and releases them directly into the blood-stream. *Compare* EXOCRINE GLAND

endocrine system A body system which consists of ductless glands that secrete HORMONES into the CIRCULATORY SYSTEM which regulate the activities of the body including the OESTROUS CYCLE, BLOOD SUGAR LEVEL, BEHAVIOUR etc. *Compare* EXOCRINE GLAND

endocrinology The scientific study of endocrine glands, hormones and their actions.

endoderm, entoderm The innermost layer of cells of the embryo which develops into the majority of the digestive system and associated glands (e.g. LIVER, PANCREAS) in the adult. Also forms the YOLK SAC and ALLANTOIS in mammals and birds.

endogenous stimulus A stimulus arising from inside the body. *Compare* EXOGENOUS STIMULUS

endometrium The MUCOUS MEMBRANE lining of the UTERUS in mammals.

endoparasite A parasite which lives inside the body of its host, e.g. in the gut, lungs or blood. *Compare* ECTOPARASITE

endoplasmic reticulum (ER) A cytoplasmic membrane which occurs in two forms inside the cell. Rough ER (which contains RIBOSOMES) and is involved in protein synthesis, and smooth ER (which has no ribosomes) and is involved in protein and lipid transport in the cell and detoxification of some drugs.

endorphins Peptides that function as NEUROTRANSMITTERS which are produced by the PITUITARY GLAND and HYPOTHALAMUS in vertebrates and have similar pain-relieving effects to morphine. They are released in response to PAIN, exercise, excitement and sexual activity.

endoscope A general term for an optical device used to make internal examinations of the body. *See also* ARTHROSCOPE, BRONCHOSCOPE, GASTROSCOPE

endoscopy A physical examination using an ENDOSCOPE.

endotherm *See* HOMEOTHERM

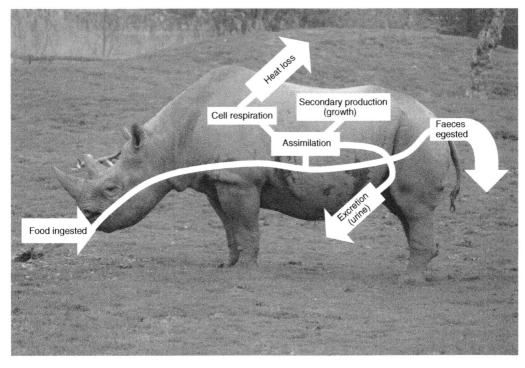

Fig. E2 Energy budget: the fate of the energy entering the body of an animal.

E

endothermic
1. Relating to **HOMEOTHERMS**. *Compare* **ECTO-THERMIC**
2. Relating to a chemical reaction which absorbs heat. *Compare* **EXOTHERMIC**

endotoxic shock *See* **SEPTIC SHOCK**

endotracheal tube A tube (catheter) passed down the trachea through the mouth or nose in order to aid breathing by maintaining an open airway, or administer a drug to the lungs, to remove mucus or prevent aspiration of stomach contents. *See also* **INTUBATION**

energy budget The partitioning of, or balance sheet for, the energy obtained and then used for various functions in an organism's body (Fig. E2), in an ecosystem or for the entire Earth.

energy flow The movement of energy between components of a **FOOD CHAIN** or **FOOD WEB**. *See also* **ENERGY BUDGET**

enriched colony cages, enriched colony systems A production system for keeping chickens which provides an enriched (rather than barren) environment. *See also* **BARN SYSTEM, BATTERY CHICKENS, LAYING HENS DIRECTIVE**

enrichment *See* **ENVIRONMENTAL ENRICHMENT**

enrichment tyre, feeder tyre A tyre used as enrichment in animal enclosures. Plastic feeder tyres are commercially available but many keepers make their own from old car tyres. They are useful for bears, primates, elephants and other animals. Animals obtain small pieces of food from small holes in the sides of the tyre.

enteric bacteria *See* **ENTEROBACTERIA**

enteritis **INFLAMMATION** of the small intestine.

enterobacteria, enteric bacteria Gram-negative rod-shaped bacteria of the family Enterobacteriaceae many of which live harmlessly in the gut of humans and other animals. Includes some pathogens, e.g. *E. COLI, SALMONELLA, Shigella*.

entire An adjective used to describe an animal that has not been **NEUTERED** and is therefore theoretically capable of breeding.

entoderm *See* **ENDODERM**

entomology The scientific study of insects.

environment An organism's environment is made up of biological components (animals, plants, microbes etc.) and physical components (air, water, rock etc.). All the things that can affect an organism's chances of survival and reproduction. *See also* **BIOLOGICAL ENVIRONMENT, ECOSYSTEM, IMMEDIATE ENVIRONMENT, INTERNAL ENVIRONMENT, PHYSICAL ENVIRONMENT**

Environment Agency A public body responsible to the Secretary of State for the Environment, Food and Rural Affairs in England (*see* **DEPARTMENT FOR ENVIRONMENT, FOOD AND RURAL**

AFFAIRS (DEFRA)) and the National Assembly for Wales. It is concerned with the improvement and protection of the environment, including pollution control, fisheries, water management (including flood and coastal management) and some aspects of wildlife conservation (especially fisheries). It was created in 1996 by an amalgamation of the National Rivers Authority (NRA), Her Majesty's Inspectorate of Pollution (HMIP) and the waste regulation authorities.

environmental education The process of changing the attitude and knowledge of the public with respect to the protection of the environment and the conservation of resources, including wildlife. Education is a mandatory function of zoos in the European Union since the passing of the **ZOOS DIRECTIVE**.

environmental enrichment, behavioural enrichment An animal husbandry principle that seeks to enhance the quality of captive animal care by identifying and providing the environmental stimuli necessary for optimal psychological and physiological wellbeing (Shepherdson, 1998) (Fig. E3). It may take the form of **NUTRITIONAL ENRICHMENT**, **OCCUPATIONAL ENRICHMENT**, **SENSORY ENRICHMENT**, **SOCIAL ENRICHMENT**, **STRUCTURAL ENRICHMENT**. *See also* **AUDITORY ENRICHMENT**, **NATURALISTIC ENRICHMENT**, **SPIDER**

Environmental Impact Assessment (EIA) An identification and assessment of the possible positive and negative effects that a project may have on the environment. EIA is required by law for large construction projects in many countries.

Environmental Investigation Agency (EIA) A **NON-GOVERNMENTAL ORGANISATION (NGO)** that campaigns to protect the natural world from environmental crime and abuse. It campaigns against whaling and the illegal trade in wildlife, to protect forests and to prevent climate change.

Environmental Protection Agency (EPA) The US Environmental Protection Agency is a federal body whose mission is to protect human heath and the environment. It was created in 1970.

environmental variation The variation in a **POPULATION** which is attributable to the environment, e.g. differences in weight between individuals that are caused by differences in nutrition. This variation is not affected by selection as it is not influenced by **GENES**.

enzootic hepatitis *See* **RIFT VALLEY FEVER**

enzyme A **PROTEIN** which catalyses a chemical reaction in a **CELL** or elsewhere in the body, e.g. in the gut. Some enzymes detoxify chemicals that are otherwise poisonous to animals. *See also* **DETOXIFICATION**

epidemiology The scientific study of the factors that determine the frequency and distribution of disease in a population. Such studies are important in deter-mining the source of a disease outbreak, and devising methods for disease control. *See also* **INDEX CASE**, **PREVALENCE**, **WELFARE EPIDEMIOLOGY**

epidermis The outer layer of surface tissue, consisting of several layers in vertebrates but often only one cell thick in invertebrates but covered by a **CUTICLE**.

epiglottis A small flap of cartilaginous tissue which pushes against the wall of the pharynx in mammals during swallowing to prevent food from being taken into the trachea.

epilepsy A group of neurological and metabolic disorders characterised by recurrent **SEIZURES** (fits), foaming at the mouth, loss of muscle control and unusual behaviour. Often of genetic origin in dogs.

epileptic fit *See* **EPILEPSY**

epinephrine *See* **ADRENALINE**

epiphyseal plate *See* **GROWTH PLATE**

epiphysis

1. The end of a **LONG BONE** which forms part of the joint. *See also* **GROWTH PLATE**

2. *See* **PINEAL GLAND**

epithelial tissue Tissue derived from the **ECTODERM** or **ENDODERM** that forms the body covering (skin) or the lining of body cavities (e.g. the intestine).

epithelium (epithelia *pl.*) An **EPITHELIAL TISSUE**.

Equality Act 2010 A law in England and Wales which, among other things, makes it unlawful to discriminate against a disabled person in the provision of goods and services and in other ways. This includes access to visitor attractions such as zoos.

equatorial Of or near the equator.

equestrian Relating to equids, specifically, domesticated horses.

equid A member of the family Equidae (horses, zebras, asses and their relatives).

equine distemper *See* **STRANGLES**

equine treadmill A **TREADMILL** designed for horses.

equine-assisted therapy The use of equids, especially horses, as a treatment or therapy for a range of human ailments including alcoholism and depression. *See also* **ANIMAL-ASSISTED THERAPY**

equitation The art and practice of horsemanship.

ER *See* **ENDOPLASMIC RETICULUM (ER)**

eruption

1. In ecology, a sudden rapid increase in the numbers of an organism, e.g. locusts.

2. In dental anatomy, the emergence of a new tooth through the gum.

3. In dermatology, the appearance of a rash or other skin blemish.

4. In epidemiology, a widespread outbreak of a disease.

erythrocyte, red blood cell The most numerous type of cell found in vertebrate blood. In adult mammals there is no nucleus. Erythrocytes contain the pigment **HAEMOGLOBIN** which carries most of the oxygen in the blood. The surface antigens of these cells specify the **BLOOD GROUP**.

Fig. E3 Environmental enrichment. Top left to bottom right: A tiger (*Panthera tigris*) retrieving food from the top of a pole; a western lowland gorilla (*Gorilla gorilla gorilla*) drinking from a bottle; a giraffe (*Giraffa camelopardalis*) browsing at high level; a black-tailed prairie dog (*Cynomys ludovicianus*) with a plastic tunnel; an orangutan (*Pongo pygmaeus*) using a piece of cloth as a sun shade; an Asian elephant (*Elephas maximus*) playing with a tractor tyre; a suspended metal barrel containing food treats for elephants; a commercially produced enrichment ball in a tiger enclosure.

ESA *See* ENDANGERED SPECIES ACT OF 1973 (USA)

ESB *See* EUROPEAN STUDBOOK (ESB)

escape behaviour A type of defence behaviour which may occur when a predator is detected or when a predator attacks. It may involve remaining motionless (e.g. in hares) or running to a prepared retreat such as a burrow. Some species make evasive manoeuvres when chased, e.g. sudden changes of direction. Zoo enclosures should be designed so that animals may escape to a refuge area if disturbed or frightened. *See also* FLIGHT DISTANCE, PORPOISING, PRONKING

escape, escapee
1. An organism that has escaped from a captive environment and established itself in the wild. Sometimes termed an escapee, but some authorities consider this incorrect usage.
2. An individual that has escaped from a zoo or other animal collection.

escapee *See* ESCAPE

Esociformes An order of fishes which contains pikes, pickerels and their allies.

esophagus *See* OESOPHAGUS

ESS *See* EVOLUTIONARILY STABLE STRATEGY (ESS)

Essay on the Principle of Population *An Essay on the Principle of Population as it Affects the Future Improvement of Society* was published in 1798 by Thomas **MALTHUS**. He suggested that the human population was growing geometrically (exponentially) while the growth in food supply was arithmetic (linear). The inevitable consequence of this was that much of the human population would be maintained in a state of poverty because of an oversupply of labour and a shortage of food. He maintained that ultimately population size was controlled by famine, war or disease. Both Charles **DARWIN** and Alfred Russel **WALLACE** quoted Mathus' work as being influential in the development of the theory of NATURAL SELECTION.

essential amino acid An AMINO ACID that is essential in the diet of a particular species because it is incapable of synthesising it. Without such amino acids an animal may exhibit a DEFICIENCY DISEASE. Some species have a specific requirement for essential amino acids: lysine, methionine, tryptophan, leucine, isoleucine, phenylalanine, threonine, histidine, valine and arginine, e.g. lysine is important for growth and milk production (LACTATION) in mammals.

estimated breeding value (EBV) A value calculated by **BREEDPLAN** which estimates an animal's breeding value expressed as the difference between its genetics and the genetic base to which it is compared, expressed in the units of the trait being examined. For example, a value of +10 kg for a 400-day weight means the animal is genetically superior by 10 kg at 400 days compared with the genetic base of the relevant population. The calculation takes into account the **HERITABILITY** of the trait. The heritability of 400-days weight is approximately 30%. So, if an animal weighed 50 kg above the average of its contemporaries at 400 days (and no other information was available on the performance of its relatives) its EBV would be calculated as +50 kg × 30% = +15 kg.

estivation *See* AESTIVATION

estrogens *See* OESTROGENS

estrous *See* OESTROUS

estrous cycle *See* OESTROUS CYCLE

estrus *See* OESTRUS

estrus synchrony *See* OESTRUS SYNCHRONY

ESU *See* EVOLUTIONARILY SIGNIFICANT UNIT (ESU)

et al. A Latin abbreviation for 'and other things' or 'and other people'. Commonly used in academic texts when a number of workers have co-operated in scientific research, e.g. Jones *et al.* (2012).

ethical review The purpose of the ethical review process is to determine and address any moral issues that need to be considered before an activity is undertaken, e.g. will a proposed study have any adverse effects on the subject animals? Will the identity of persons who complete a questionnaire be protected? Can the euthanasia of a sick animal be justified? Should zoo visitors be allowed to feed penguins? A zoo should have an Ethical Review Committee.

ethics The branch of philosophy which is concerned with **MORALITY**. *See also* ANIMAL RIGHTS, OXFORD CENTRE FOR ANIMAL ETHICS

Ethiopian region *See* FAUNAL REGIONS

ethogram A list and description of all of the behaviours a species may exhibit; the behavioural repertoire of a species (Table E1). Used in behaviour studies. May only consider certain types of behaviour, e.g. only social behaviours, depending on the purpose of the study. *See also* ACTIVITY BUDGET, ETHOSEARCH

ethology The scientific study of the behaviour of animals in their natural environment. *Compare* BEHAVIOURAL ECOLOGY, COMPARATIVE PSYCHOLOGY, PSYCHOLOGY

EthoSearch An online database of ETHOGRAMS developed by scientists at **LINCOLN PARK ZOO**.

EthoTrak An easy-to-use, menu-driven, flexible, digital system for collecting basic behavioural data on a **PERSONAL DIGITAL ASSISTANT (PDA)** within zoological institutions, developed by the Chicago Zoological Society, Brookfield Zoo. It has been specifically designed to pool data collected by different institutions in order to improve sample sizes.

ethylene tetrafluoroethylene (ETFE) A fluorocarbon-based polymer, used for exhibit roofs and other structures, which allows the transmission of ULTRAVIOLET LIGHT, thereby facilitating VITAMIN D synthesis in many animals.

Table E1 An ethogram used for studying elephant behaviour in captivity (based on Rees, 2009).

Behaviour	Description
Aggression	Hitting/pushing as a result of an antagonistic encounter (but not as part of play)
Bathing	Standing/laying in pool/squirting water from pool over body with trunk
Digging	Digging in soil using the foot (but not as part of dusting behaviour)
Drinking	Collecting water in the trunk and squirting it into the mouth
Dusting	Collecting soil and throwing it over the body/rubbing it into the skin (while standing still or walking), including digging in soil for this purpose
Feeder ball	Feeding or attempting to feed at a metal feeder ball containing small quantities of food
Feeding	Collecting solid food with the trunk and placing it in the mouth while standing or walking (does not include suckling or activity at the feeder ball)
Locomotion	Walking (except while feeding, dusting or stereotyping)
Lying down	Lying down on the ground (on its side or prone)
Playing	Chasing another elephant/mock fighting with another elephant (but not as a result of an antagonistic encounter or as part of courtship)
Rolling	Rolling in soil or mud (but not as part of playing with another elephant)
Sex	Courting or being courted/mounting another elephant or being mounted by another elephant of either sex
Standing	Standing motionless (but not while stereotyping, feeding or dusting)
Stereotyping	Repetitive behaviour with no obvious purpose: weaving, head bobbing, pacing backwards and forwards or in an arc, walking in circles
Suckling	Calf suckling from mother or another female. Measured separately from feeding

etiology *See* AETIOLOGY

etorphine (hydrochloride) *See* IMMOBILON

EU *See* EUROPEAN UNION (EU)

EU law *See* EUROPEAN LAW

eucaryote *See* EUKARYOTE

eucaryotic *See* EUKARYOTIC

eukaryote, eucaryote An organism in which the genetic material is contained within a NUCLEUS (or nuclei) that is separated from the cytoplasm of the cell by a nuclear membrane. All large complex organisms (animals, plants and fungi) are eukaryotes. *See also* PROKARYOTE

eukaryotic, eucaryotic Relating to EUKARYOTES.

euphagia *See* FOOD SELECTION

eupnoea A normal RESPIRATORY RATE. *Compare* BRADYPNOEA, TACHYPNOEA

European Association of Zoo and Wildlife Veterinarians (EAZWV) A non-profit organisation which is dedicated to advancing veterinary knowledge and skill in the field of zoo and wild animals, improving zoo animal husbandry and the management of wild animal populations.

European Association of Zoos and Aquaria (EAZA) A zoo organisation that has over 300 member institutions in 35 countries. It was founded in 1992 and its mission is to facilitate cooperation within the European zoo and aquarium community towards the goals of education, research and conservation. EAZA publishes the *JOURNAL OF ZOO AND AQUARIUM RESEARCH*.

European Coalition for Farm Animals *See* EUROPEAN NETWORK FOR FARM ANIMAL PROTECTION (ENFAP)

European Commission An institution of the EUROPEAN UNION (EU) which functions as its executive body, proposes new legislation and implements existing laws.

European Convention for the Protection of Animals during International Transport 1971 A Council of Europe convention which establishes general conditions for the international transport of animals and special conditions for their transport by road, air, sea and rail, in order to prevent suffering.

European Convention for the Protection of Animals kept for Farming Purposes 1976 A Council of Europe convention whose purpose is to require the adoption of common provisions for the protection of animals kept for farming purposes, particularly in modern intensive stock-farming systems. It requires that animals be housed and provided with food, water and care in a manner which is appropriate to their physiological and ethological needs in accordance with established experience and scientific knowledge.

European Convention for the Protection of Vertebrate Animals used for Experimental and other Scientific Purposes 1986 A Council of Europe convention, the purpose of which is to help the parties to harmonise the introduction of national measures to guarantee that animals are properly and humanely treated and that, where procedures which may possibly cause pain, suffering, distress or lasting harm to an animal are unavoidable, they are kept to a minimum.

European Court of Justice (ECJ) The institution of the EUROPEAN UNION (EU) which interprets EU law to ensure that it is applied in the same way in all Member States and settles legal disputes between EU governments and EU institutions.

European Directive A EUROPEAN UNION (EU) law which requires an individual Member State of the Union to change its NATIONAL LAW in order to comply with its provisions, e.g. the HABITATS DIRECTIVE, WILD BIRDS DIRECTIVES, ZOOS DIRECTIVE.

European Elephant Keeper and Manager Association (EEKMA) An association whose aims are: to create a platform for elephant keepers and represent their interests; to improve the husbandry and breeding of elephants; to train and educate elephant keepers; and to promote research on elephants. The Association publishes a newsletter called *Elephant Journal*.

European Endangered Species Programme (EEP) A system of intensive population management of zoo animals which began in Europe in 1985 co-ordinated by the EUROPEAN ASSOCIATION OF ZOOS AND AQUARIA (EAZA). Each EEP has a Species Co-ordinator who is responsible for collecting information. A species committee makes recommendations about which individuals should be used for breeding and the exchange of individual animals between zoos. The role of TAXON ADVISORY GROUPS (TAGs) within EAZA is to develop REGIONAL COLLECTION PLANS (RCPs) for the species that are kept in European collections. *See also* SPECIES SURVIVAL PLAN® (SSP) PROGRAMS

European law, EU law The law of the EUROPEAN UNION (EU) which consists primarily of EURO-PEAN DIRECTIVES, EUROPEAN REGULATIONS and CASE LAW decided by the EUROPEAN COURT OF JUSTICE (ECJ). Also called EC (European Community) law. *Compare* INTERNATIONAL LAW, NATIONAL LAW

European Network for Farm Animal Protection (ENFAP) An organisation formed from groups in 25 EU countries to address farm animal welfare issues across Europe. Formerly the European Coalition for Farm Animals (founded in 1993).

European Nutrition Group *See* EAZA NUTRITION GROUP

European Regulation A type of EUROPEAN LAW which becomes effective in a Member State as soon as it comes into force without each Member State having to change its NATIONAL LAW.

European Studbook (ESB) A type of STUDBOOK kept by the EUROPEAN ASSOCIATION OF ZOOS AND AQUARIA (EAZA) within which species populations are managed by the studbook keeper to a lesser extent than those which form part of EURO-PEAN ENDANGERED SPECIES PROGRAMMES (EEPs). May eventually be upgraded to an EEP.

European Studbook Foundation (ESF) A non-profit organisation which contributes to the conservation of reptiles and amphibians in captivity, especially

Table E2 The Member States of the European Union (July 2013).

Austria	Germany	Poland
Belgium	Greece	Portugal
Bulgaria	Hungary	Romania
Croatia	Ireland	Slovakia
Cyprus	Italy	Slovenia
Czech Republic	Latvia	Spain
Denmark	Lithuania	Sweden
Estonia	Luxembourg	United Kingdom
Finland	Malta	
France	Netherlands	

endangered species, by building and maintaining genetically viable captive populations.

European Union (EU) A political and economic union of 28 European states (Table E2). Formerly known as the European Community and, before that, the European Economic Community. It issues laws and makes judicial decisions with which the Member States must comply, primarily in the form of EURO-PEAN DIRECTIVES, EUROPEAN REGULATIONS and DECISIONS OF THE EUROPEAN COURT OF JUSTICE. *See also* HABITATS DIRECTIVE, WILD BIRDS DIRECTIVES, ZOOS DIRECTIVE. *Compare* COUNCIL OF EUROPE

European Union of Aquarium Curators An organisation formed in 1972 to promote professional improvement between specialists working in public aquariums.

eusocial Exhibiting or relating to EUSOCIALITY.

eusociality A social system in which there is a DIVISION OF LABOUR (2) amongst the individuals, e.g. workers, a reproductive QUEEN (2) etc., e.g. bees, ants, naked mole rats.

euthanasia The painless killing of an animal. This may be because it is terminally ill, suffering from a very serious injury from which it will not recover, or because it is surplus to requirements. Zoo organisations produce policies in relation to the use of euthanasia, e.g. BIAZA's ANIMAL TRANSACTION POLICY.

Eutheria An infraclass of mammals that includes all of the PLACENTAL MAMMALS.

eutrophication The process by which an aquatic ecosystem becomes enriched by nutrients causing an overgrowth of plants and deoxygenation of water.

event recorder An electronic or mechanical device for recording a behaviour, movement or other occurrence for later analysis. *See also* DAWKINS' ORGAN, TRAIL MONITOR

even-toed ungulate A member of the mammalian order ARTIODACTYLA. *Compare* ODD-TOED UNGULATE

evidence-based management In relation to the care and exhibition of animals, the use of husbandry systems, enclosure designs, diets, enrichment etc. that scientific studies have shown to be effective, rather than reliance upon anecdote and trial-and-error methods. *See also* EMPIRICAL ZOO CONCEPT

eviscerate
1. Remove the bowels.
2. Remove the contents of an organ, e.g. the contents of the stomach.
3. Remove an organ from a body, e.g. an eye. *See also* AUTOTOMY
4. Protrude through a surgical incision or wound.

evolution A cumulative change in the genetic composition of a POPULATION of organisms over time (from generation to generation) which eventually leads to the development of new forms, especially species as a result of SPECIATION, and generally leads to greater complexity. *See also* EXTINCT (1), SELECTION

evolutionarily significant unit (ESU) A group of organisms that is considered distinct for conservation purposes: the minimum unit of conservation management. The term could refer to a species, subspecies, geographical race or population. An ESU should be substantially reproductively isolated from other conspecific populations and it should represent an important component of the evolutionary history of the species. Modern analyses of ESUs rely upon information from molecular genetics, specifically MITOCHONDRIAL DNA (mtDNA).

evolutionarily stable strategy (ESS) An inherited strategy (usually behavioural) which, if practised by most members of a population, cannot be supplanted during evolution by a different strategy.

evolutionary biology The scientific study of the evolution of organisms especially in relation to changes in their molecular makeup, genetics, ecology, behaviour and taxonomy.

excretion The removal from the body of the waste products of metabolism. *Compare* EGESTION

excretory system A system of organs which remove nitrogenous waste from the body. In vertebrates this is achieved by the kidneys whose filtrate (urine) passes, in AMNIOTES (2), via the ureters to the bladder.

exertional myopathy *See* CAPTURE MYOPATHY

Exeter Exchange A building in London which, among other things, contained a large permanent MENAGERIE from 1773 until it was demolished in 1829.

exhibit
1. To put on display (to the public). In relation to a zoo, it is illegal to display species for which the appropriate CITES documentation does not exist (*See also* ARTICLE 60 CERTIFICATE).
2. *See* ANIMAL ENCLOSURE

exhibit design *See* ENCLOSURE DESIGN

exhibit structure The form of an exhibit, especially in a zoo. An exhibit may consist of a foreground, background, FURNITURE etc.

Exhibited Animals Protection Act 1986 (NSW) A law in New South Wales, Australia, which regulates the exhibition of animals in zoos, marine parks, circuses and other places. It makes provision for the licensing and inspection of premises exhibiting animals.

exocrine gland A gland of epithelial origin which secretes substances directly or through a duct onto an epithelial surface (not into the bloodstream). *Compare* ENDOCRINE GLAND

exogenous stimulus A stimulus that arises outside the body, in the environment. *Compare* ENDOGENOUS STIMULUS

exoparasite *See* ECTOPARASITE

exoskeleton The hard skeleton located in the skin or covering the outside of the body. In arthropods it is the CUTICLE which is secreted by the EPIDERMIS. In many vertebrates it consists of bony plates beneath the epidermis, e.g. in tortoises and armadillos. *See also* SCUTES

exothermic Relating to a chemical reaction which releases heat. *Compare* ENDOTHERMIC (2)

exotic, exotic animal
1. A species that is not native in the particular country where it is located, e.g. a flamingo living wild would be an exotic in the UK. *See also* ACCIDENTAL, ALIEN SPECIES, ESCAPE
2. Animals that are wild rather than domesticated. The definition depends upon the context. Lizards are exotic if kept as pets in the UK but not in the tropics.

exotic pet *See* EXOTIC (2)

experiment
1. A trial carried out to test a theory or discover something.
2. The process of conducting such a trial.

experimental condition *See* CONTROL

experimental farm A research facility for conducting scientific studies with the aim of improving agricultural methods, crop production and animal husbandry, often operated by a government department or a university, e.g. the Central Experimental Farm (Ottawa, Canada), Cranfield University Experimental Farm (UK).

experimental population A population defined under the ENDANGERED SPECIES ACT OF 1973 (USA) in the USA. The ESA allows for the designation of populations of reintroduced animals of endangered or threatened species as non-essential 'experimental populations', which may be treated differently from other populations of the same species (e.g. killed if they attack livestock), provided they are geographically separate. *See also* WYOMING FARM BUREAU FEDERATION V. BABBITT (1997)

explainer *See* PRESENTER

exploratory behaviour A form of APPETITIVE BEHAVIOUR which may or may not be directed at a particular situation or resource. It is exhibited when animals search for food, shelter etc. and the exploratory behaviour ceases when the resource is found or the situation is arrived at. Some species appear to explore for its own sake. *See also* CONSUMMATORY BEHAVIOUR, CURIOSITY

E

exponential growth The unrestricted growth of a POPULATION. This may occur when a species is introduced into a new area and has no predators (Fig. P8). *See also* POPULATION GROWTH

export licence In relation to animals, a licence required by law (e.g. **CITES**) to export a SPECIMEN (2) of a species from one country to another. *Compare* IMPORT LICENCE

ex-situ Moved from its original place. *Compare* IN-SITU

ex-situ **conservation**
1. Conservation which takes place in captivity.
2. The UN CONVENTION ON BIOLOGICAL DIVERSITY 1992 (CBD) (Art. 2) defines *ex-situ* conservation as *'the conservation of components of biological diversity outside their natural habitats.'* *Compare* IN-SITU CONSERVATION

extant In relation to a species, one that still lives on Earth, i.e. the opposite of EXTINCT.

extensive agriculture, extensive farming The rearing of farm animals (especially sheep and cattle) in large open fields and in other conditions at relatively low densities, with low inputs of fertiliser, labour and capital. *Compare* INTENSIVE FARMING

external fixator A device – usually a metal frame – which holds pins or wires that pass through a bone to hold a fracture in correct alignment while it heals. It may be static or dynamic. Dynamic fixators are used to manage GROWTH PLATE injuries. *Compare* INTERNAL FIXATOR

external pipping In hatching chicks, the beginning of the process of breaking through the egg shell at which time a very small outward dent appears in the egg before it cracks. *See also* ASSIST HATCH, INTERNAL PIPPING

extinct
1. In relation to organisms, no longer in existence. Can relate to the whole planet, or a particular country or locality. Sometimes refers to wild specimens only, i.e. may still exist in captivity. *See also* DEMOGRAPHICALLY EXTINCT, EXTINCT IN THE WILD (EW)
2. A RED LIST category used by the INTERNATIONAL UNION FOR THE CONSERVATION OF NATURE AND NATURAL RESOURCES (IUCN): Extinct (EX) – Exhaustive surveys throughout the historic range of the taxon have failed to record a single individual. No reasonable doubt that the last individual has died.

Extinct in the Wild (EW) A RED LIST category used by the INTERNATIONAL UNION FOR THE CONSERVATION OF NATURE AND NATURAL RESOURCES (IUCN): Extinct in the Wild (EW) – Known only to survive in cultivation, captivity or as naturalised population(s). Exhaustive surveys throughout the historic range of the taxon have failed to record a single individual.

extinction
1. In evolution, the process of becoming EXTINCT.
2. A period in geological time when the extinction of species was widespread.
See also DEMOGRAPHICALLY EXTINCT, EXTINCT IN THE WILD (EW), EXTINCTION OF LEARNED BEHAVIOUR, MASS EXTINCTION, SIXTH EXTINCTION. *Compare* DE-EXTINCTION

extinction of learned behaviour The process which causes learned behaviour to disappear when it is no longer useful. LEARNING occurs when a behaviour is reinforced. If the REINFORCEMENT is removed the animal no longer makes the connection between the behaviour and a particular consequence. A captive animal may be trained to perform a particular behaviour by giving it a reward. If the reward is removed the animal will eventually stop performing the behaviour. After extinction has occurred, relearning the behaviour will occur at a faster rate than the original learning. *See also* TRAINING

extinction rate The number of taxa lost to extinction per unit time, normally expressed as species per year.

exudativore An animal which feeds on plant exudates, e.g. sap, resin or tree gum.

eye shine, night shine The reflection of light from the RETINA of an animal at night, often seen in torchlight and when using infrared video equipment. *See also* TAPETUM LUCIDUM

eye temperature Eye temperature may be measured using infrared thermography which measures the heat emitted by blood carried in the superficial capillaries around the eye. It has been used as a measure of welfare in some species, e.g. cattle, chickens, humans, monkeys, wapiti (*Cervus canadensis*). It may be caused by the redirection of blood from the capillary bed as a result of vasoconstriction mediated by the SYMPATHETIC NERVOUS SYSTEM (SNS) when the animal experiences pain or distress. *See also* THERMAL IMAGING CAMERA

Fig. F1 F is for ferret (*Mustela putorius furo*).

Dictionary of Zoo Biology and Animal Management: A guide to terminology used in zoo biology, animal welfare, wildlife conservation and livestock production, First Edition. Paul A. Rees.

F

F₁ generation *See* FIRST FILIAL GENERATION

F₂ generation *See* SECOND FILIAL GENERATION

facial expression A visual DISPLAY made using the face. PRIMATES have a complex arrangement of facial muscles and so can communicate with facial expressions, e.g. chimpanzees make a facial expression known as the 'compressed lips face', indicating aggression, the 'play face' during play, and the 'full open grin' when frightened or very excited. Care should be taken not to interpret primate facial expressions as if they were human. The normal facial expression of howler monkeys (*Alouatta* spp.) looks like an expression of sadness to humans. *See also* ANTHROPOMORPHISM

factory farming A farm organised on industrial lines; INTENSIVE FARMING. *See also* BATTERY FARMING

fady A belief, superstition or custom in Malagasy culture which prohibits a particular behaviour, e.g. in Madagascar some people believe that lemurs are reincarnations of their ancestors and this protects the animals from persecution. *See also* TOTEM

faecal cortisol CORTISOL found in faeces whose presence is used as a measure of STRESS.

Falconiformes An order of birds: raptors (osprey, eagles, kites, hawks, Old World vultures, falcons, secretary bird, buzzards, caracaras).

falconry The practice of keeping, breeding and flying BIRDS OF PREY, especially for hunting.

fallen stock LIVESTOCK that has died from natural causes. May include animals that are destroyed as a result of injury or do not enter the food chain for some reason. In the UK, defined as an animal that has either died of natural causes or disease on a farm, or been killed on a farm for a reason other than human consumption. In the EU there are strict regulations on the disposal of fallen stock in order to control the spread of disease. *See also* NATIONAL FALLEN STOCK COMPANY (NFSCo)

Fallopian tubes Paired ducts in female mammals which have funnel-shaped openings that open behind the ovary and carry ova from the ovary to the uterus by ciliary and muscular action. Fertilisation often occurs here.

false pregnancy *See* PSEUDOPREGNANCY

family

1. In taxonomy, a group of related GENERA; a subdivision of an ORDER. Family names end in '*idae*', e.g. Felidae, Canidae, Ursidae (Table C5).

2. A group of related animals, usually consisting of a mother and her offspring.

FAO *See* FOOD AND AGRICULTURE ORGANISATION (FAO)

farm animal

1. A type of animal normally kept on a farm for the production of food or other products of use to humans, e.g. HORSES, CATTLE, SHEEP, PIGS, GOATS, CHICKENS, TURKEYS. Other species are

also farmed. *See also* DEER FARMING, FISH FARM, FUR FARM, OSTRICH FARM, OYSTER FARM, PACA FARMING

2. Under New York State's Agriculture and Markets Law Art. 6, 350§(4) '*"Farm animal", means any ungulate, poultry, species of cattle, sheep, swine, goats, llamas, horses or fur-bearing animals. . . which are raised for commercial or subsistence purposes. Fur-bearing animals shall not include dogs or cats.*'

Farm Animal Welfare Committee An expert committee which provides independent, authoritative and impartial advice on farm animal welfare issues to the DEPARTMENT FOR ENVIRONMENT, FOOD AND RURAL AFFAIRS (DEFRA) and the devolved administrations in Scotland and Wales. It replaced the FARM ANIMAL WELFARE COUNCIL on 1 April 2011.

Farm Animal Welfare Council An advisory non-departmental public body which advised the British Government on farm animal welfare issues until 31 March 2011. It was responsible for the BRAMBELL REPORT which proposed the 'FIVE FREEDOMS'. The Council was reconstituted as an expert committee on 1 April 2011; the FARM ANIMAL WELFARE COMMITTEE.

Farm Bureau *See* AMERICAN FARM BUREAU

farm park, farm zoo

1. A farm which is open to the public and may or may not keep exotic animals as well as farm breeds.

2. A farm which specialises in keeping rare breeds of farm animals. *See* RARE BREEDS SURVIVAL TRUST (RBST)

Farming and Wildlife Advisory Group (FWAG) A charity founded by a group of farmers in 1969 which provided environmental advice to the farming community in Great Britain for over 40 years until its demise in 2011. Some regional groups continue to operate.

farming of native species Many wild animal species are farmed within their natural range e.g. crocodiles, deer, pacas. *See also* ALLIGATOR FARM, DEER FARMING, GAME RANCHING, PACA FARMING

Farmland Bird Index An aggregated index of population trend estimates of a selected group of breeding bird species dependent on agricultural land for nesting or feeding, i.e. kestrel (Fig. P7), grey partridge, lapwing, stock dove, woodpigeon, turtle dove, skylark, yellow wagtail, whitethroat, jackdaw, starling, greenfinch, goldfinch, linnet, yellowhammer, reed bunting, corn bunting, rook and tree sparrow. The European Union has adopted the Farmland Bird Index as a 'long-list structural indicator' for Europe.

farrier A person who shoes horses and cares for their hooves.

Farrier Registration Council *See* FARRIERS (REGISTRATION) ACT 1975

Farriers (Registration) Act 1975 An Act in Great Britain to prevent and avoid suffering by and cruelty to horses arising from the shoeing of horses by unskilled persons; to promote the proper shoeing of horses; to promote the training of farriers and shoeing smiths; and to provide for the establishment of a Farrier Registration Council.

farrowing Giving birth to piglets.

fascioliasis *See* FLUKE DISEASE

fasciolosis *See* FLUKE DISEASE

fat *See* LIPID. *See also* ADIPOSE TISSUE

fat cattle, live cattle Cattle which are FINISHED and ready for slaughter. Now sometimes called LIVE CATTLE because of the negative connotations of the term 'fat'. *Compare* FALLEN STOCK, STORE CATTLE

fauna
1. The animals found in a particular area or during a particular geological period, e.g. the fauna of Madagascar; the fauna of the Carboniferous.
2. A book containing descriptions and identifying features of animals from a particular locality or period. *See also* FIELD GUIDE, FLORA (2)

Fauna and Flora International (FFI) An international wildlife conservation organisation. Formerly the Fauna and Flora Preservation Society. Largely responsible for saving the ARABIAN ORYX (*ORYX LEUCORYX*). Publishes *ORYX – The International Journal of Conservation.*

faunal regions, biogeographical region, zoogeographical regions Areas of the world which are defined by their distinctive FAUNA and are separated from each other by major geomorphological features such as mountain ranges, deserts or oceans. The number of regions recognised varies between authorities but the following regions are widely accepted: Palaearctic (Europe, North Africa, most of Arabia, and Asia north of the Himalayas); Ethiopian or Afrotropical (Africa south of the Sahara, including Madagascar and the south-west corner of the Arabian peninsula); Oriental or Indomalayan (India and Asia south of the Himalayas, and the Australasian archipelago except New Guinea and Sulawesi); Australasian or Australian (Australia, New Zealand and associated islands); Nearctic (North America as far south as Mexico, and Greenland); Neotropical (South America and Central America to central Mexico); Antarctic (Antarctica). The Palaearctic and Nearctic regions are sometimes combined to form the Holoarctic. The land bridge that formerly linked Alaska to Siberia has enabled considerable movement of species resulting in a great similarity in their fauna. Some zoos organise their collections into faunal regions or house species from the same region together. *See also* WALLACE'S LINE

fear A motivational state aroused by specific stimuli which give rise to defence or ESCAPE BEHAVIOUR. Accompanied by behaviour such as ALARM CALLS and changes in FACIAL EXPRESSION (e.g. in primates), physiological changes such as increased RESPIRATION RATE and HEART RATE, PILOERECTION (in mammals), and the release of ADRENALINE into the blood. Prolonged fear may give rise to STRESS, especially in captive animals.

feather Any of the flat epidermal appendages which form a bird's PLUMAGE. Vaned feathers consist of a partly hollow horny shaft from which extends a vane of interlocking barbs. Down feathers occur underneath these and are of a much finer structure. Feathers are a distinguishing characteristic of birds. Freshly plucked feathers are a useful source of **DNA** for analysis. *See also* PRIMARY FEATHER, SECONDARY FEATHER

feather pecking An abnormal behaviour of some laying hens (especially in intensive conditions) which peck others, sometimes removing their feathers. This can result in poor plumage, patches of feather loss, skin damage and sometimes death. BEAK TRIMMING has been traditionally used to alleviate this problem.

feather picking A self-mutilating behaviour in birds, especially PSITTACINES, often caused by underlying inflammatory skin disease. *See also* SELF-MUTILATION

fecundity Reproductive fertility; the potential capacity of an organism for reproduction. *See also* BIRTH RATE, INTRINSIC RATE OF NATURAL INCREASE

Federation Cynologique Internationale (FCI) The World Canine Organisation. It is based in Belgium and has 86 members and partners (one per country) each of which issues pedigrees and trains judges. The FCI recognises 341 dog breeds. Each breed is the 'property' of a specific country. Owner countries write the BREED STANDARDS. Each member country holds international shows, working/hunting trials and races/coursing. Results of contests and lists of judges are held by the FCI.

Federation of Zoological Gardens of Great Britain and Ireland The former name of the BRITISH AND IRISH ASSOCIATION OF ZOOS AND AQUARIUMS (BIAZA).

feed barrier A fence which allows animals access to their food but prevents it from being soiled and trampled. Widely used in cattle sheds (Fig. F2). *See also* CALF CREEP FEEDER

feed cake *See* ANIMAL FEED CAKE

feed chute A passage through which food may be provided by keepers (or sometimes members of the public) to a captive animal without the need to enter its enclosure. Especially used for dangerous animals such as bears and big cats.

feed conversion efficiency A measure of the efficiency with which an animal converts food into its own body mass.

feeder tube *See* FORCE FEEDING

feeder tyre *See* ENRICHMENT TYRE

feeding *See* FEEDING ENRICHMENT, FORAGING BEHAVIOUR, FORAGING STRATEGY

feeding devices Figs. B5 and E3. *See* AUTOMATIC FEEDING SYSTEM, AUTOMATIC FOOD DISPENSER, BIRD FEEDER, BURLAP BAG, CALF

F

Fig. F2 Feed barrier: a diagonal feed barrier.

CREEP FEEDER, ENRICHMENT TYRE, ENVIRONMENTAL ENRICHMENT, FEED CHUTE, FORCE FEEDING

feeding disorder Any abnormal feeding behaviour, e.g. refusal to feed, PICA (1).

feeding enrichment The provision of food in such a manner that it acts as an ENVIRONMENTAL ENRICHMENT, e.g. scatter feeding of small pieces of food so that animals have to spend time foraging for it, suspending branches from an elevated location, hiding food in a PUZZLE FEEDER (Fig. E3). *See also* SWAY BRANCH, SWAY FEEDING POLE

felid An animal belonging to the mammalian family FELIDAE (Fig. J1). *See also* BIG CAT

Felidae A family of mammals which consists of the cats, e.g. lions, tigers, pumas, leopards, cheetahs. *See also* BIG CAT

feline infectious enteritis *See* FELINE PANLEUCOPENIA

feline panleucopenia, feline infectious enteritis A highly contagious viral disease of all felids which is caused by a PARVOVIRUS and is sometimes fatal. It can also affect related families: mustelids (e.g. ferret, mink), procyonids (e.g. raccoon, coatimundi) but not canids. The disease was controversially used

to control introduced FERAL CATS on Marion Island that were threatening to cause the extinction of the population of burrowing petrels.

Feliway A synthetic cat facial PHEROMONE available as a spray intended for use with domestic CATS to prevent territorial spraying and aggression. Sometimes used as an enrichment for captive FELIDS.

female
1. Relating to an organism capable of producing eggs or bearing offspring.
2. A female animal.
See also SEX DETERMINATION. *Compare* MALE

female philopatry *See* PHILOPATRY (1)

femur (femora pl.) The upper bone of the hindlimb of TETRAPODS.

fen A low, marshy or flooded area of land. Typically frequented by wildfowl and wading birds.

fence A barrier made of wood, metal or other material designed to contain animals, separate different groups or species, protect vegetation, or prevent visitors from reaching animals. Fence design depends on its function. *See also* DEER FENCE

ferae naturae A legal term for animals of a wild nature. Legal title to individuals regarded as *ferae naturae* may be claimed by taking possession or

Fig. F3 Feral cat (*Felis catus*). Marrakesh, Morocco.

taming, or while they are on one's estate. *See also* **ANIMAL, CAPTIVE ANIMAL, DOMESTICATED ANIMAL, TAMENESS, WILD ANIMAL**

feral Relating to an animal from a domesticated species which has reverted to living wild, e.g. **FERAL CATS** are domestic cats which are living and breeding in a wild state. Feral cats do considerable damage to native fauna on some islands, and feral goats damage vegetation.

feral cat A domestic cat (*Felis catus*) who has reverted to living wild (Fig. F3). Cats have been taken to many remote islands around the world including **WHALING** and **SEALING** stations and other human settlements. This resulted in the establishment of feral populations when the settlers left, often causing significant destruction to the indigenous fauna, especially ground-nesting birds. In some areas concerted efforts have been made to eradicate feral cats (including the use of poisons, dogs and disease), especially where rare species have been threatened. In Scotland feral cats interbreed with the wildcat (*Felis sylvestris*) and are threatening its survival. *See also* **FELINE PANLEUCOPENIA, INTERBREEDING, TNR PROGRAMME**

fermentation The anaerobic process by which complex organic molecules are broken down by enzymes into simpler compounds to produce energy. It results in the production of two ATP molecules for each molecule of glucose, and produces less energy than the aerobic process of cellular respiration. Mammals rely upon symbiotic microbes to provide the cellu-

lase enzymes required to break down cellulose as their guts are unable to secrete them. These enzymes ferment the cellulose and other carbohydrates, producing volatile fatty acids which are then absorbed by the gut and used as an energy source. Species of mammals that rely on microbial fermentation for their nutrition (e.g. **RUMINANTS**) must supply the microbes in their gut with sufficient amounts of fibre or gastric disorders may develop. *See also* **LACTIC ACID FERMENTATION**

ferret
1. A domesticated mustelid (*Mustela putorius furo*) related to the polecat which is widely kept as a pet and sometimes used for hunting rodents, rabbits and other species (ferreting) (Fig. F1). Their ownership is illegal or strictly controlled in some countries because of the threat to wildlife (e.g. New Zealand).
2. Any of a number of species of mustelids, including the **BLACK-FOOTED FERRET (*MUSTELA NIGRIPES*)**.

fertilisation The combining of the **HAPLOID** genetic material of the female and male **GAMETES** to produce a **ZYGOTE** which is **DIPLOID** and will develop into a new organism.

feticide *See* **FOETICIDE**

fetus *See* **FOETUS**

fever A condition in which the body temperature is elevated above normal for the species and which is often a sign of **INFECTION**.

FFI *See* **FAUNA AND FLORA INTERNATIONAL (FFI)**

fiber *See* **FIBRE**

fibre, fiber, roughage Fibre consists largely of cellulose, lignin and hemicellulose obtained from plants. It plays an important role in the process of digestion and is also a source of energy. Cellulose and hemicellulose can be digested by microbial **FERMENTATION** in herbivorous mammals but lignin is almost impossible to digest. **MONOGASTRIC** species – those with simple guts – do not use fibre as a major source of energy. However, primates may digest substantial amounts of fibre.

fibre nuggets A partial hay alternative which provides an additional source of **FIBRE** which is low in energy, sugar and starch, especially for equids.

fibula (fibulae *pl.*) A **LONG BONE** which, together with the **TIBIA**, forms the lower hindlimb in tetrapod vertebrates. It is the smaller, and posterior, of the two bones.

field guide A book used to identify organisms in the field. Includes vernacular and scientific names, descriptions of identifying features, identification photographs and/or drawings, distribution maps, descriptions of **VOCALISATIONS**, information about **BREEDING SEASON**, food habits, ecological requirements and behaviour, e.g. *Collins Field Guide to the Larger Mammals of Africa*. *See also* **DICHOTOMOUS KEY, FAUNA (2), FLORA (2), IDENTIFICATION GUIDE**

F

field marks Distinguishing features of a species which assist its identification in the field.

field of view In relation to vision, the area that can be viewed. A wide field of view is important to large grazers as they need to remain vigilant in case of a predator attack. The positioning of the eyes on the side of the head allows such animals to see both forwards and backwards. Binocular vision requires overlapping fields of view which provides a three-dimensional image. This requires forward-facing eyes and is important in arboreal species such as monkeys, lemurs and apes, allowing them to judge distance accurately.

field propagation and release A new captive breeding and **REINTRODUCTION** technique for species that do not breed well in captivity. Captive adults are bred in large field enclosures. They are allowed to raise their young, which are then released to the wild. Useful for species that need natural habitat for breeding or rely on parent-learned behaviours. Used for the eastern loggerhead shrike (*Lanius ludovicianus migrans*) in Canada.

field shelter A small raised cover or building for poultry used to provide shelter from the elements and aerial predators. May also be a larger building provided on farmland as shelter for pigs, sheep, cattle, horses, donkey and other livestock. *See also* **POULTRY ARK**

fighting dog A dog breed that has been specially bred for fighting, e.g. **PIT BULL TERRIER**. *See also* **BREED-SPECIFIC LEGISLATION, DANGEROUS DOGS ACT 1991, DOGFIGHT, STATUS DOG**

filial
1. Relating to a son or daughter.
2. In genetics, pertaining to the sequence of generations following the parental generation. Each generation is denoted F_x, where x denotes the number of generations from the parental generation. *See also* **FIRST FILIAL GENERATION, SECOND FILIAL GENERATION**

filly A young female horse which is under 4 years of age and therefore too young to be called a **MARE**.

film and animals *See* **ANIMALS AND FILM**

filter feeder An animal that feeds by filtering small particles of food or very small organisms (**PLANKTON**) from water, e.g. many fishes, baleen whales and flamingos.

filtration *See* **BIOLOGICAL FILTRATION, LIFE SUPPORT SYSTEM, MECHANICAL FILTRATION**

fin An appendage possessed by fishes and some other aquatic organisms, which is used for locomotion, steering and balance. Fins often occur in pairs and are supported by structures made of cartilage, bone or a horny material. *See also* **SHARK FINNING**

Fin, Fur and Feather Folk An organisation founded in 1889 by Mrs. Edward Phillips and others in Croyden, England, which campaigned against the plumage trade and, together with the **PLUMAGE LEAGUE**, helped to found the **ROYAL SOCIETY FOR THE PROTECTION OF BIRDS (RSPB)**.

finished Of livestock, ready for slaughter.

finishing phase In agricultural animal production, the period of development of an animal (e.g. a cow), that occurs after the **GROWING PHASE,** when it experiences a short period of maximum weight gain which allows well-grown animals to maximise meat yield and optimise fat cover. This culminates in the animal reaching its **TARGET WEIGHT**.

finishing unit In livestock production, a facility which houses and stores animals and prepares them for slaughter, e.g. a pig finishing unit. *Compare* **GROWER UNIT**

fire extinguisher A portable device containing water, foam or carbon dioxide under pressure used for putting out fires. Some zoos use CO_2 fire extinguishers to control aggressive animals in an emergency, e.g. to separate large animals when fighting.

first filial generation, F_1 generation The hybrid offspring of a cross of true-breeding parents.

fish (fish, fishes *pl.*)
1. A member of the Chondrichthyes (cartilaginous fishes), Osteichthyes (bony fishes) or Agnatha (hagfish and lampreys). Some authorities consider only the Osteichthyes to be true fish. The plural 'fish' is used to apply to several individuals of the same species, while the plural 'fishes' is used to apply to several species of fish. Often the legal definition of 'fish' does not conform to the zoological definition.
2. The Sea Fisheries Regulation Act 1966 (in relation to England and Wales) defines 'sea fish' as fish of any kind found in the sea, and includes shellfish (defined as crustaceans and molluscs) but excludes salmon and migratory trout.
3. In the **FISH STOCKS AGREEMENT 1995** the definition of fish '*includes molluscs and crustaceans except those belonging to sedentary species as defined in article 77 of the CONVENTION ON THE LAW OF THE SEA 1982 (UNCLOS) (i.e. organisms which, at the harvestable stage, either are immobile on or under the seabed or are unable to move except in constant physical contact with the seabed or the subsoil. Art 77(4))'.*
4. In Singapore, the Animals and Birds (Live Fish) Rules 2011 (S27/2011) issued under the Animals and Birds Act (Chapter 7) defines 'live fish' as '*any varieties of marine, brackish water or fresh water fishes, crustacea, aquatic mollusca, turtles, marine sponges, trepang and any other form of aquatic life, including the young and eggs thereof, imported or exported whilst living and not intended for human consumption.*'
5. In English case law (*Caygill v. Thwaite* (1885)) the court held that a statute that prohibited the taking of fish applied equally to crayfish (which are crustaceans).
6. To attempt to catch fish. *See also* **FISHING**

fish discards Fish which are unintentionally caught during commercial fishing operations and are returned to the sea – usually dead – because they are from a species or of a size that the fishermen cannot take legally, or their taking would exceed the quota for the species. A practice that is a particular problem because of strict regulation in the EURO-PEAN UNION (EU). *See also* BYCATCH, FISHING QUOTA, OVERFISHING

fish farm A facility where fish are raised commercially in tanks or ponds, usually for human consumption. *See also* AQUAPONICS

fish foot spa, fish pedicure A 'therapy' whereby clients place their feet in tanks of warm freshwater containing toothless *Garra rufa* fish (also known as doctor or nibble fish). Concern has been expressed that these fish may spread diseases. Foot pedicures are common in Asia but have been banned in some states in the USA.

Fish House The 'Fish House' at LONDON ZOO was the first aquarium – 'aquatic vivarium' – in the world. It was established in 1853 after two men approached the zoo for advice on keeping tropical fish in tanks. The Fish House contained over 300 types of fishes and marine invertebrates.

fish ladder A structure which allows migratory fishes (e.g. salmon) to pass obstructions across rivers (such as locks and dams) generally by leaping up a series of shallow steps.

fish pedicure *See* FISH FOOT SPA

Fish Stocks Agreement 1995 Agreement for the Implementation of the Provisions of the United Nations CONVENTION ON THE LAW OF THE SEA 1982 (UNCLOS) of 10 December 1982 Relating to the Conservation and Management of Straddling Fish Stocks and Highly Migratory Fish Stocks.

Fisher, James (1912–1970) A British zoologist and broadcaster who worked as assistant curator of LONDON ZOO, and was a leading member of the ROYAL SOCIETY FOR THE PROTECTION OF BIRDS (RSPB) and the INTERNATIONAL UNION FOR THE CONSERVATION OF NATURE AND NATURAL RESOURCES (IUCN). He also worked for the Ministry of Agriculture, was a member of the National Parks Commission and vice-chairman of the Countryside Commission. He was a prolific author and in 1969 he published *The Red Book – Wildlife in Danger*.

fishing The act of catching fish, usually for food by humans, but also by other animals (e.g. seabirds, otters, bears). Commercial fishing is regulated by law. *See also* COARSE FISHING, COMMON FISH-ERIES POLICY (CFP), ELECTRIC FISHING, FISH STOCKS AREEMENT 1995, FISHING QUOTA, LONGLINE FISHING, OVERFISHING

fishing quota The share of the total catch from an area of the sea that a particular country is entitled to take under the law. *See also* COMMON FISHER-IES POLICY (CFP), FISH DISCARDS, V-NOTCHING

fishless cycling The process of achieving MATURA-TION of a biological filter in an aquarium by adding synthetic wastes to the water to encourage bacterial growth rather than by adding fishes. *See also* BIOLOGICAL FILTRATION

fission–fusion social group A pattern of social group-ing in animals whereby individuals form subgroups whose members belong to a larger unit group of stable composition. Movement occurs between sub-groups and unit groups so that changes in group size and composition occur frequently.

fistula An unnatural connecting channel between a body cavity and the outside of the body or between two body cavities. May be congenital or caused by injection or injury, e.g. cows sometimes damage their teats, creating a fistula through which milk escapes. Sometimes created artificially by a vet. PAVLOV studied digestion in animals by creating a fistula in the stomach so he could draw off the contents.

fitness In EVOLUTIONARY BIOLOGY, the capacity of an ORGANISM to pass its GENES to the next generation (generally by leaving offspring). In this context the term does not refer to physical fitness but to genetic fitness. The sum of DIRECT FITNESS and INDIRECT FITNESS.

Five Freedoms, animal welfare domains The freedoms deemed by the BRAMBELL REPORT to be important for the welfare of captive animals: freedom from hunger and thirst; freedom from dis-comfort; freedom from PAIN, injury or disease; freedom to express normal behaviour; freedom from FEAR and DISTRESS.

Five Kingdom Classification A modern system of classification which divides all organisms into one of five KINGDOMS: Bacteria, PROTOCTISTA, Anima-lia, Fungi and Plantae.

fixation In genetics, in relation to a gene which has the alleles A and a, fixation has occurred when all of the alleles in a population are A, i.e. the 'a' allele has been lost. If a population consists only of individuals which are AA all individuals it is fixed for A and a has been lost. If a population consists only of aa individu-als it is fixed for a and A has been lost. This may occur as a result of GENETIC DRIFT. If a DELETERIOUS GENE becomes fixed in a population, this may cause impaired survivorship and reduced FITNESS. *See also* HARDY–WEINBERG EQUILIBRIUM

fixator A device for supporting fractures during healing. *See* EXTERNAL FIXATOR, INTERNAL FIXATOR

fixed action pattern (FAP) An activity which has a relatively fixed pattern of coordination which appears to be stereotyped and is innate. Performed in response to a SIGN STIMULUS or releaser (a sign from one individual to another). Aggressive behav-iour in the male three-spined stickleback (*Gasteros-teus aculeatus*) is triggered by the red colour of an opponent's belly.

flagellum (flagella *pl.*) A long, hair-like structure whose movement acts as a means of LOCOMOTION in some small organisms (e.g. in flagellates). In other animals movement of the flagellum may pass mate-rial (e.g. food particles) over the surface of the cell, e.g. in feeding in sponges. *Compare* CILIUM

F

flagging *See* TAIL-FLAGGING
flagship species A charismatic species which is popular with the public and serves as a symbol and focus for raising awareness about CONSERVATION issues and stimulates action, e.g. giant pandas, African elephants, black rhinos. Focusing *IN-SITU* CONSERVATION efforts on these species may benefit whole ecosystems.
flea circus A circus and zoo sideshow in which fleas perform tricks in a small arena. Some flea circuses do not use fleas and are simple mechanical and electrical tricks.
flea collar A collar, impregnated with INSECTICIDE, commonly worn by domestic cats to deter fleas.
fledge
  1. Of a young bird, to reach the stage of development where it is capable of flight.
  2. Of a young bird, to reach this age.
  3. To raise a young bird to this age.
fledgling A young bird that has left the nest.
flehmen A behaviour exhibited by males of many mammalian species (especially ungulates and felids) when they test scent marks and the urine, faeces or genitals of females for odours. Flehmen is characterised by the male raising his head, turning back the lips, wrinkling the nose and the temporary cessation of breathing (Fig. F4). This facilitates the transfer of PHEROMONES and other scents to the VOMERO-NASAL ORGAN. *See also* COURTSHIP

Fig. F4 Flehmen in a Bactrian camel (*Camelus bactrianus*).

flight distance The distance at which an animal will take flight when approached by a human, predator etc. May vary seasonally, e.g. herring gulls (*Larus argentatus*) have a relatively long flight distance outside the BREEDING SEASON when they do not hold territories but this becomes very short when they have nests, eggs and young. Flight distances should be considered in the design of enclosures so that animals are not continually stressed by the presence of visitors or predators in nearby enclosures. Flight distances will generally decrease as animals become habituated to the presence of visitors or other species. *See also* FLIGHT ZONE, HABITUATION, INDIVIDUAL DISTANCE, TAMENESS
flight muscles In birds, the sets of muscles responsible for creating the upstroke and the downstroke of the wings during flight. Those responsible for the downstroke are attached to the CARINA (keel).
flight restraint A method of preventing an animal (usually a bird) capable of flight from escaping from an open enclosure by restricting its ability to fly or preventing flight completely. This may be achieved by surgical means (permanent change to or removal of a tendon, PATAGIAL MEMBRANE or wing bones) or non-surgical means (clipping or trimming feathers, removal of barbs on primary feathers). Removal of the distal bone of the wing is called pinioning (Fig. F5). *See also* HOBBLE, TETHER
flight zone The area around an animal within which it will display alarm and attempt to escape if approached. Important to consider when herding animals within an enclosure to avoid panic. *See also* FLIGHT DISTANCE, INDIVIDUAL DISTANCE, TAMENESS
Flipper A 1960s American television series made by Ivan Tors Films Inc. about the adventures of a bottlenose dolphin who befriends the Chief Warden of a fictional Marine Preserve in Florida and his two sons. Based on a 1963 film of the same name.
flipper A wide flat limb which has evolved for swimming, as in a dolphin, whale, seal, penguin.
flock
  1. A collective term for some animal groups, e.g. sheep, goats, birds.
  2. The action of grouping together as a flock.
flora
  1. The plants found in a particular area or during a particular geological period, e.g. the flora of Kenya; the flora of Jersey; the flora of the Cretaceous.
  2. A book containing descriptions and identifying features of plants from a particular locality or period, e.g. *A Flora of New Zealand*. *See also* FAUNA (2), FIELD GUIDE
flu *See* INFLUENZA. *See also* AVIAN INFLUENZA, H1N1 VIRUS, H5N1 VIRUS
fluid therapy Treatment intended to restore normal body fluid balance and volume, given intravenously or orally. *See also* GIVING SET, INFUSION

Fig. F5 Flight restraint: a pinioned Chilean flamingo (*Phoenicopterus chilensis*).

fluid warmer A device for pre-warming blood and other fluids before they are administered intravenously to prevent **HYPOTHERMIA**.

fluke

1. Any of a number of different taxa of parasitic flatworms. *See also* **FLUKE DISEASE**

2. One lobe of the tail of a whale or dolphin

fluke disease, fascioliasis, fasciolosis An infestation with *Fasciola hepatica*, the common liver fluke of sheep. It is found in most herbivorous animals including pigs, goats, cattle, horses, rabbits, hares, beavers, kangaroos, elephants and humans. It is generally found in the bile ducts of the liver, but also occurs in other organs, causing anaemia and hepatitis. The **INTERMEDIATE HOSTS** are various species of freshwater snails (e.g. *Limnaea* spp.). Cercariae may be ingested with water or when encysted on grass. Control may involve the use of **ANTHELMINTICS** in infected animals and land drainage to control snails. **MOLLUSCICIDES** can be used to kill snails.

fly mask *See* **FLY VEIL**

fly rug A mesh rug designed to be worn by a horse to reduce fly nuisance.

fly strike, myiasis The phenomenon whereby flies are attracted to faeces-soiled areas of wool or fur where they lay their eggs. These subsequently develop into maggots that eat into the flesh. This is a significant welfare issue in farm animals, especially sheep, and also in some pets, such as rabbits. Fly strike may be fatal and is the cause of considerable economic loss in the livestock industry. The incidence of fly strike may be reduced by procedures such as **MULESING** or **TAIL DOCKING**. However, more welfare-friendly ways to reduce incidence are by management to reduce diarrhoea, (e.g. worm control), and trimming wool around the hindquarters (dagging). Rabbits should be checked daily for faeces soiling.

fly veil, fly mask A mesh hood used to keep flies off a horse's face and head.

flyway A flight path used by birds during **MIGRATION (1)**, e.g. the African-Eurasian flyway, the East Asia-Pacific flyway.

foal

1. A young horse or other **EQUID**.

2. According to the **AMERICAN STUD BOOK** (2008) a foal is '*a young horse of either sex in its first year of life*'.

foaling In an **EQUID**, the process of giving birth to a **FOAL**.

foam fractionator *See* **PROTEIN SKIMMER**

foam nest *See* BUBBLE NEST

focal animal In studies of animal behaviour, the individual animal whose behaviours are being recorded during a particular period of time. *See also* FOCAL SAMPLING

focal sample
1. An animal that is the subject of FOCAL SAMPLING
2. The data collected in relation to a focal animal.

focal sampling A method of studying animal behaviour by sampling the activities of a single individual (or DYAD, group, litter or other unit) for a specified period of time.

foeticide, feticide An act that causes the death of a foetus. This may occur by ABORTION due to harassment of the mother or forced COPULATION by a CONSPECIFIC male. Occurs in captive zebra. *See also* CANNIBALISM, FRATRICIDE, INFANTICIDE, PROLICIDE

foetus, fetus The embryo of a VIVIPAROUS mammal when it is in the later stages of development and resembles the adult form.

foie gras A food delicacy. The term literally means fatty liver. It is the liver of a duck or goose enlarged to many times its normal size by FORCE FEEDING.

foliage Leaves.

follicle
1. A small cavity in an organ or tissue, e.g. the bulbous structure at the base of a hair.
2. A structure in the ovary which produces an egg. *See also* GRAAFIAN FOLLICLE

follicle-stimulating hormone (FSH) A hormone produced by the adenohypophysis which stimulates the maturation of GRAAFIAN FOLLICLES in the ovary in females and promotes the formation of spermatozoa in the testes in males.

folliculogenesis
1. The stimulation of the development of the FOLLICLE (2) in the ovary by hormones or drugs.
2. Follicle development in the ovary, usually under the influence of FOLLICLE-STIMULATING HORMONE (FSH).

Food and Agriculture Organisation (FAO) A UNITED NATIONS (UN) organisation whose aim is to raise levels of human nutrition, improve agricultural productivity, improve the lives of rural populations and contribute to the growth of the world economy.

Food and Drug Administration *See* UNITED STATES FOOD AND DRUG ADMINISTRATION (USFDA)

Food and Environment Research Agency (FERA) Formerly the Central Science Laboratory. FERA is an executive agency of the DEPARTMENT FOR ENVIRONMENT, FOOD AND RURAL AFFAIRS (DEFRA) whose purpose is to support and develop a sustainable food chain, a healthy natural environment, and to protect the global community from biological and chemical risks. It tests foods, studies and manages wildlife and diseases and inspects plant and bee health.

food begging Behaviour which stimulates another member of the same species to donate food, usually a parent or mate. The young of some species induce adults to regurgitate food, e.g. in herring gulls (*Larus argentatus*), hunting dogs (*Lycaon pictus*) and hyenas (*Crocuta crocuta*). Forms part of COURTSHIP in some species as a result of RITUALISATION. *See also* SIGN STIMULUS

food chain A sequential linear representation of some of the organisms which feed upon each other in an ecosystem resulting in a simplistic representation of the flow of energy from the sun to a green plant, herbivore, carnivore and then top carnivore. Species at the top of food chains tend to be generalists in relation to feeding, while those at the bottom tend to be specialists. *See also* FOOD WEB

food passage time *See* GUT PASSAGE TIME

food selection, euphagia, nutritional wisdom The ability to choose useful foods (and possibly select a balanced diet) and avoid noxious foods, either due to innate preferences or as a result of LEARNING. *See also* AVERSION (2), CAFETERIA EXPERIMENT

food spoilage The deterioration in the quality of food as a result of the action of bacteria and fungi etc. May result in contamination with MYCOTOXINS.

food supplement, dietary supplement A material added to the DIET (1) to replace those required constituents that are absent or present in insufficient quantities, e.g. COLOSTRUM SUPPLEMENT, CONCENTRATES, DUSTING (2), ESSENTIAL AMINO ACIDS, GUT LOADING, MINERAL, PREBIOTIC, PROBIOTIC, VITAMIN

food types *See* CARBOHYDRATE, FIBRE, LIPID, MINERAL, PROTEIN, VITAMIN

food web The complex of feeding relationships found in an ecosystem, made up of many interconnected FOOD CHAINS.

foot
1. The terminal part of the leg of a terrestrial vertebrate, beyond the ankle.
2. A muscular organ in a mollusc (e.g. a snail) used for locomotion.

foot and mouth disease (FMD) FMD is caused by an aphthovirus. It is highly contagious and can affect all CLOVEN-HOOFED SPECIES. Infected animals have small fluid-filled blisters (vesicles) in the mouth and on the feet. Females may also have vesicles on the skin of the udder or teat. The virus is present in the vesicles and the fluid which is released when they burst. When there are lesions in the mouth the virus spreads via saliva. It is also spread in faeces, urine and from lesions in the feet, and it is excreted in milk. FMD may be spread by wind, watercourses, people, vehicles and migratory birds. The disease is transmissible to humans but infection is usually mild. Vaccination of livestock is practised in some countries but many control the disease by slaughtering. Outbreaks of FMD in Britain in recent years have resulted in the temporary closure of **DEER PARKS** and zoos.

foot care The maintenance of healthy feet. Especially important in some species, e.g. elephants and ungulates. *See also* **LAMINITIS**

footfall, visitor footfall Footfall is strictly the sound of footsteps. In visitor studies, the number of visitors, i.e. increased visitor footfall means increased visitor numbers. *See also* **VISITOR ATTENDANCE**

footprint-impacted species Species whose populations are declining because of unsustainable hunting, fishing or logging, e.g. many dolphin species.

forage
1. A crop grown for consumption by livestock, e.g. grass or hay.
2. Food taken by grazers or **BROWSERS**.
3. The activity of searching for food.

foraging behaviour Behaviour concerned with the capture, collection and consumption of food. *See also* **FORAGING STRATEGY**

foraging strategy A type of evolutionary strategy which takes into account the costs and benefits of foraging. The act of foraging must not cost an animal more energy than is supplied by the food obtained. Some lizard species have evolved a sit-and-wait strategy, whereby they wait for their prey to come to them; other species are active hunters and search for their prey.

foramen magnum The opening at the base of the skull of a **VERTEBRATE** where it articulates with the **VERTEBRAL COLUMN** and through which the **SPINAL CORD** passes to connect with the **BRAIN**. Its orientation is important in determining whether or not an animal is capable of walking upright while facing forwards.

force feeding, gavage feeding, gavaging, tube feeding The process whereby food is directly inserted into an animal's stomach via a tube (feeder tube). This method of feeding may be used when an animal is sick and refuses to eat, but is also used in the production of **FOIE GRAS**. *See also* **CROP FEEDING**

forceps delivery A delivery of a young mammal where forceps are inserted through the vagina and used to grasp the head and pull it through the birth canal.

forensics *See* **WILDLIFE FORENSICS**

forensics laboratory *See* **USFWS FORENSICS LABORATORY**

formic acid An irritant volatile acid released in ants as a defence mechanism.

formicarium, formicary
1. The place where a colony of ants lives; an anthill or nest.
2. A **VIVARIUM** containing ants.

Fossey, Dian (1932–1985) An American zoologist who studied mountain gorillas in the Virunga Mountains, Rwanda, for 18 years and worked to prevent poaching. She became one of the world's foremost primatologists, founded the Karisoke Research Center and briefly lectured at Cornell University. Fossey was murdered in 1985 in Rwanda. Fossey and her work were the subject of the film *Gorillas in the Mist* (1988) based on her book of the same title.

fossil The remains of a once living organism from a previous geological age, or evidence of its existence, e.g. skeletons, trails, footprints, faeces, impressions, borings, casts.

fossorial A fossorial animal is one which lives underground (e.g. **AARDVARK**, badger, naked mole rat). A fossorial structure is one which is adapted to digging and burrowing (e.g. large forelimbs). Some species are fossorial to aid temperature regulation and live underground to avoid excessive cold or heat.

foster parent An animal or human who takes the place of the original parent of an individual. *See also* **CROSS-FOSTERING, PUPPET MOTHER**

founder *See* **FOUNDER POPULATION**

founder population The original individuals from which a population has descended, e.g. the founders of a captive breeding programme.

four questions *See* **TINBERGEN'S FOUR QUESTIONS**

four whys? *See* **TINBERGEN'S FOUR QUESTIONS**

four-by-four (4×4) A four-wheel drive vehicle. A motor vehicle in which all four wheels are driven by the engine and which may be equipped with low ratio gears, locking differentials and other devices which allow it to be driven off the road e.g. *Land Rover, Toyota Land Cruiser,* **UNIMOG**. Often used in fieldwork, **SAFARI PARKS** and large **ANIMAL ENCLOSURES** of pastureland.

fowl plague *See* **AVIAN INFLUENZA**

fowl pox *See* **AVIAN DIPHTHERIA**

fowl typhoid A disease of birds, especially chickens, caused by the bacterium *Salmonella gallinarum*.

fox hunting The pursuit and killing of a fox (*Vulpes vulpes*) by **FOXHOUNDS** controlled by huntsmen, usually on horseback, but sometimes on foot. Very controversial in the UK. Banned in Scotland since 2002 by the **PROTECTION OF WILD MAMMALS (SCOTLAND) ACT 2002** and in England and Wales since 2005 as a result of the **HUNTING ACT 2004**, following the **BURNS REPORT**, and replaced by **DRAG HUNTS** and hunting with **BIRDS OF PREY**. The law still allows flushing from cover with dogs. Fox hunting is practised in many parts of the world including Germany, Canada, Australia and New Zealand. Foxes were introduced into New Zealand specifically to provide quarry for hunts. In the United States there are many hunts. In many cases they pursue coyotes (*Canis latrans*) which are rarely caught. Fox hunting is not a controversial activity in the USA.

foxhound A dog bred for hunting foxes (Fig. F6). *See also* **FOX HUNTING**

Fig. F6 Foxhounds.

fracture A break in a bone. Fractures are classified into a wide range of different types including simple (a straightforward break), compound (where the bone pierces the skin), depressed (where the bone is forced below the level of the surrounding surface, as in a skull injury) and GREEN STICK FRACTURES. *See also* FIXATOR, SPLINT, STABILISATION (1)

Frankenstein food A derogatory term applied to food produced using GENETIC ENGINEERING techniques, such as cloned farm animals, to highlight concerns about its safety.

Fraser Darling effect The phenomenon whereby some species require the presence of a large number of CONSPECIFICS in order to breed successfully, e.g. many seabird species, flamingos. Named after the zoologist Fraser Darling.

fratricide, siblicide The killing of one's siblings. *See also* CANNIBALISM, FOETICIDE, INFANTICIDE, PROLICIDE

free contact A method of handling animals in which keepers interact freely with them without barriers. This is hazardous when handling potentially dangerous taxa such as great apes, elephants, rhinoceroses, etc. (Fig. F7). *Compare* PROTECTED CONTACT

Free Willy A Warner Bros. film released in 1993 about a boy who orchestrates the escape of a killer whale named *Willy* from an aquarium in an amusement park. *Willy* was played by *Keiko*, a male orca captured in 1979. As a result of publicity generated by the film a foundation was established to rehabilitate *Keiko* and return him to the wild. This caused considerable controversy because some researchers believed that he had been in captivity too long for a return to the wild to be successful. He died after STRANDING on the Norwegian Coast in 2003.

Freedom Food Freedom Food Ltd. is a farm animal welfare assurance and associated food labelling scheme operated by the ROYAL SOCIETY FOR THE PREVENTION OF CRUELTY TO ANIMALS (RSPCA) in the UK in conjunction with farmers, hauliers, abattoirs, manufacturers and retailers.

free-flight aviary A zoo exhibit containing free-flying birds through which visitors are able to walk.

free-flight bat house *See* BAT HOUSE

free-living, free-ranging In relation to an animal, an alternative term to 'wild' which is preferred by some ANIMAL ADVOCATES and animal welfare organisations.

free-range A method of extensive farming in which livestock (especially chickens) are allowed to live outside in open enclosures and feed relatively naturally.

Fig. F7 Free contact and protected contact: a keeper handling African elephants (*Loxodonta africana*) using free contact (left); a bull Asian elephant (*Elephas maximus*) being trained using protected contact (right).

free-ranging *See* FREE-LIVING

freeze brand An identifying mark made by using a metal brand which has been cooled with dry ice or liquid nitrogen. *See also* BRANDING

freshwater Relating to water in ponds, rivers and lakes as opposed to SALTWATER.

Friends of the Earth An environmental group in the UK which campaigns for environmental justice, the conservation of energy and natural resources, greener farming, action against climate change and against pollution and environmental damage.

friendship Many social animals (e.g. chimpanzees, elephants) choose to associate with, and have particular attachments to, certain individuals of the same species rather than others. Some scientists have described these relationships as friendships. *See also* ASSOCIATION INDEX

Frisch, Karl Ritter von (1886–1982) An Austrian ethologist who shared the Nobel Prize for Physiology or Medicine in 1973 with Konrad LORENZ and Niko TINBERGEN which they received for their contributions to ETHOLOGY. He specialised in insect behaviour and was able to demonstrate that bees navigate using the sun and communicate the location of food sources to other bees by performing 'dances'.

Frozen Ark A FROZEN ZOO established by a consortium of zoos, museums and universities including the ZOOLOGICAL SOCIETY OF LONDON (ZSL), the ZOOLOGICAL SOCIETY OF SAN DIEGO (ZSSD), the NATURAL HISTORY MUSEUM (LONDON), the LABORATORY FOR THE CONSERVATION OF ENDANGERED SPECIES (LACONES), CHESTER ZOO and the University of Nottingham.

Frozen Zoo
1. A storage facility where the genetic material of organisms is kept at low temperatures for a long period as a conservation measure, e.g. the BREEDING CENTRE FOR ENDANGERED ARABIAN WILDLIFE (BCEAW). *See also* FROZEN ARK
2. Part of the Conservation and Research for Endangered Species facility at the Zoological Society of San Diego. Its mission is '*To help preserve the legacy of life on Earth for future generations by establishing and maintaining genetic resources in support of worldwide efforts in research and conservation.*' The Frozen Zoo® consists of: DNA, viable cell cultures, semen, embryos, oocytes and ova, blood and tissue specimens.

frugivore An animal that feeds on fruit.

frustration A motivational state which arises when an animal is unable to obtain a particular GOAL, for example food that it can see. This is not the same as deprivation, where there is no expectation of food. Frustration arises when an animal's behaviour does not result in the consequences that it has learned to expect by experience.

fry A young fish in which the YOLK SAC is fully absorbed.

full genome sequencing *See* GENOME SEQUENCING

functional cause, ultimate cause In relation to behaviour, an attempt to explain how and why it evolved. *Compare* PROXIMATE CAUSE

functional reference A form of COMMUNICATION in which the SIGNAL varies depending upon the object to which it refers, e.g. different ALARM CALLS depending upon the type of predator present.

functionally extinct A population or species which is doomed to become EXTINCT because it contains too few individuals to reproduce, they are too widely spread geographically to meet for breeding, they are all of the same sex, the individuals are all diseased, or for some other reason. *See also* DEMOGRAPHICALLY EXTINCT, MINIMUM VIABLE POPULATION (MVP)

fundamental niche The range of conditions and resources that an organism free of interference from other species could utilise. *Compare* REALISED NICHE

fungus (fungi *pl.*) A group of EUKARYOTES which make up the Kingdom Fungi. They lack chlorophyll and vascular tissue. Some are single cells (e.g. yeasts) while others form a body mass of filamentous hyphae with specialised fruiting bodies (e.g.

mushrooms). Some species are involved in decomposition and food spoilage, while others cause disease, e.g. **ASPERGILLOSIS, CHYTRIDIOMYCOSIS**. *See also* **CANDIDA, MYCOTOXIN**

funnel trap *See* **HELIGOLAND TRAP**

fur ball A collection of hair which forms in the stomach of some mammals (e.g. felids) as a result of **GROOMING**. It may be vomited up when it becomes large.

fur farm A place where furbearing animals are raised in cages for the fur industry.

furniture Objects such as climbing frames, resting platforms and other structures located within an enclosure for use by the animals, especially in a zoo. *See also* **ENVIRONMENTAL ENRICHMENT**

FWAG *See* **FARMING AND WILDLIFE ADVISORY GROUP (FWAG)**

F

Fig. G1 G is for goat. A British Toggenburg nanny goat.

Dictionary of Zoo Biology and Animal Management: A guide to terminology used in zoo biology, animal welfare, wildlife conservation and livestock production, First Edition. Paul A. Rees.
© 2013 John Wiley & Sons, Ltd. Published 2013 by John Wiley & Sons, Ltd.

G *See* GUANINE (G)

Gadiformes An order of fishes: cods, hakes and their allies.

Gaia theory, Gaia hypothesis The theory, developed by James LOVELOCK, that the Earth is a self-regulating system that is capable of keeping the climate and chemical composition of the oceans and atmosphere suitable for living things.

gait The pattern of movement of the limbs during locomotion. Gait may be variously described as walking, running, trotting, crawling, or jumping. An abnormal gait may be indicative of the presence of a skeletal abnormality. Changes in gait may be indicative of the presence of disease. *See also* ALTERNATING TRIPOD GAIT, BRACHIATION, SALTATION (3)

Galapagos Islands An archipelago of volcanic islands off the west coast of South America, where Charles DARWIN discovered many endemic species, including DARWIN'S FINCHES, and gathered information which helped him to formulate his theory of evolution. *See also* LONESOME GEORGE

gall bladder A small organ on the underside of the liver in vertebrates, that is used to store BILE.

gallery forest A forest which grows along the banks of a watercourse (stream or river) in an area otherwise devoid of trees.

Galliformes An order of birds: megapodes, chachalacas, guans, curassows, turkeys, grouse.

Gallup, Gordon *See* RED SPOT TEST

game

1. Animals (especially birds and mammals) that have been traditionally hunted. The term has no taxonomic meaning but is defined by law in many countries. *See also* GAME BIRD

2. In England, the Game Act 1831, which is still in force, defines game as hares, pheasants, partridges, grouse, heath or moor game, black game and bustards.

3. In Swaziland, the Game Act 1953 defined game as consisting of royal game (e.g. elephant, giraffe, buffalo, black rhinoceros, cranes, ibis, owls), large game (kudu, blue wildebeest, plains zebra) and small game (e.g. impala, bushbuck, reedbuck). Subsequently the Game (amendment) Act 1991 reclassified game species into specially protected game, royal game and common game.

game auction

1. An auction of live game species raised on farms, especially in African countries. Game may be purchased by ex-farmers who are restocking their land with wildlife to establish a game-viewing/ECOTOURISM business.

2. An auction of game carcasses.

3. An auction of big game hunts by a government.

game bird Defined under the WILDLIFE AND COUNTRYSIDE ACT 1981 as any pheasant, partridge, grouse (moor game), black (heath) game or ptarmigan.

game cropping *See* GAME RANCHING

game fence A fence constructed to a standard which is substantially higher than that of a STOCK FENCE which effectively controls the movement of animals into and out of a defined area. Widely used in Africa to prevent wild ungulates from grazing on farmland and to prevent wildlife from leaving fenced protected areas.

game ranching, game cropping The commercial use of wild GAME animals for food and other products by the management and cropping of natural populations, especially LARGE MAMMALS, notably antelopes. Endemic large herbivores make more efficient use of the natural vegetation and are therefore more productive than domestic livestock, i.e. they produce more meat per unit area of land. Early studies of game ranching in Africa were made by the American ecologist Raymond F. DASMANN who wrote the book *African Game Ranching* (1964). *See also* CAMPFIRE

game reserve

1. An area set aside to protect wild animals and used for GAME VIEWING.

2. An area set aside for hunting purposes.

game viewing A type of ECOTOURISM which involves watching and photographing wildlife, usually from a vehicle or boat, but sometimes on foot.

gamekeeper A person who manages an estate for the benefit of certain animal species (usually birds) of economic value, e.g. grouse. This may involve the control of certain pest and predatory species.

gamete A sex cell: SPERMATOZOON or EGG (1) (OVUM). A HAPLOID cell which carries one of each pair of CHROMOSOMES that comprise the GENOME of the ORGANISM.

gamma (γ) diversity A measure of the overall biological diversity for the different ecosystems within a geographical region. *See also* ALPHA (α) DIVERSITY, BETA (β) DIVERSITY, BIODIVERSITY

gangrene An area of dead tissue infected with bacteria. May occur following burns, scalds, wounds, frostbite etc.

gapes *See* GAPEWORM

gapeworm, gapes, syngamiasis A parasitic nematode (*Syngamus trachea*) which infects the trachea and bronchi of certain bird species causing a disease in which the bird gasps for breath due to blockage of its AIRWAY.

Garden of Intelligence, Ling Yu A zoo built by the Chinese Emperor Wen-Wing in the 12th century BC.

gas bladder *See* SWIM BLADDER

gaseous exchange The exchange of gases, especially oxygen and carbon dioxide, as a result of DIFFUSION, between the body cells and the external environment, either over the surface of the body (skin) or internally where gases are transported to the tissues in blood by a CIRCULATORY SYSTEM.

gassed down A colloquial term referring to an animal that has been anaesthetised using gas.

Gasterosteiformes An order of fishes: sticklebacks, pipefishes and their allies.

gastrointestinal Relating to the **DIGESTIVE SYSTEM**. *See also* **GASTROINTESTINAL TRACT**

gastrointestinal tract, GI tract Strictly the stomach and intestines but sometimes includes the whole of the **DIGESTIVE SYSTEM**.

Gastropoda A class of molluscs: snails, slugs and their relatives.

gastroscope A type of **ENDOSCOPE** used for examining the stomach and intestine.

Gaussian distribution *See* **NORMAL DISTRIBUTION**

gavage feeding *See* **FORCE FEEDING**

gavaging *See* **FORCE FEEDING**

Gaviiformes An order of birds: divers.

GB Poultry Register A register in Great Britain of **POULTRY** (**1** and **2**) keepers who keep 50 or more birds. The species which must be registered include chickens (including bantams), turkeys, ducks, geese, guinea fowl, quail, partridges, pheasants, pigeons (reared for meat), ostriches, emus, rheas and cassowaries.

geld Remove the testes; castrate.

gelding A castrated **EQUID**.

gene A section of **DNA** (or **RNA** in some **VIRUSES**) which codes for a single function or a group of related functions; the fundamental unit of **HEREDITY**. *See also* **ALLELE, GENETIC CODE**

gene bank, gene library A collection of genetic material in the form of cloned **DNA** fragments stored for future use in breeding programmes, genetic manipulation etc. *See also* **FROZEN ARK, FROZEN ZOO**

gene expression The process by which the information from a gene is used to produce a gene product, e.g. a protein or enzyme. The **GENETIC CODE** is stored in the **DNA** and interpreted by the process of gene expression to give rise to the **PHENOTYPE**.

gene flow The transfer of genes between populations of the same species by **INTERBREEDING** and **MIGRATION (2)**.

gene frequency, allele frequency The frequency of a particular **ALLELE** in a population; the ratio of a particular allele to all the other alleles of the same gene in the population. Where only two alleles exist the frequency of the **DOMINANT ALLELE** is p and the frequency of the **RECESSIVE ALLELE** is q, where $p + q = 1$. *See also* **HARDY–WEINBERG EQUILIBRIUM**

gene library *See* **GENE BANK**

gene linkage The occurrence of two **GENES** on the same **CHROMOSOME**. It they are located close together they are likely to remain together when gametes are formed during **MEIOSIS** and pass together into the same gamete. If they are far apart, **CROSSING OVER** may occur, and the alleles for the two genes may be exchanged between the two chromatids and end up in different gametes.

gene manipulation *See* **GENETIC ENGINEERING**

gene mutation A change in the **GENETIC CODE** in **DNA** which may alter the product of a gene, prevent the gene from functioning correctly (or completely) or have no effect, depending upon the nature of the alteration. *See also* **CHROMOSOME MUTATION**

gene pool All of the genes present in a particular defined group of organisms of the same species, e.g. a wild population, a zoo population, a herd of cattle. *See also* **FOUNDER POPULATION**

general adaptation syndrome A complex series of changes that occurs in vertebrates when they are exposed to **STRESS**. This has three stages: (1) The flight or fight response: an acute activation of the sympathetic nervous system and the adrenal medulla resulting in the secretion of catecholamines. (2) The resistance phase: activation of the neuroendocrine system occurs: the **HPA AXIS** (hypothalamic–pituitary–adrenal). **ADRENOCORTICOTROPHIC HORMONE (ACTH)** is secreted by the **PITUITARY GLAND**. This stimulates the release of a number of glucocorticoid hormones from the adrenal cortex, especially **CORTISOL**. These stimulate the conversion of amino acids into glucose to provide energy. Other pituitary hormones may be released that inhibit growth and suppress reproduction. (3) The final stage: if adaptation to the stressor does not occur or it is not removed, gastric ulceration may occur and there may be a lowering of immunological function.

generation time

1. The time interval between consecutive generations of organisms; the time between the birth of an individual and the birth of its offspring; the average age at which a female gives birth to her first offspring.

2. The time it takes a cell to complete one full growth cycle; the time interval between a cell being produced by **MITOSIS** and when it divides into two daughter cells.

GENES Software which is used for the genetic analysis and management of pedigrees. May be used as a stand-alone program but is most easily used as an accessory program to the **SINGLE POPULATION ANALYSIS AND RECORD KEEPING SYSTEM (SPARKS)** software for the management of **STUDBOOKS**.

genetic bottleneck A reduction in the genetic variability of a population caused by its having evolved from a small number of founders. As a consequence the population is susceptible to **INBREEDING DEPRESSION**. In some wild species this has caused unusually high levels of abnormalities and low fertility. *See also* **FOUNDER POPULATION**

genetic code The sequence of nucleotide bases in a molecule of **DNA** that determines the sequence of **AMINO ACIDS** in the structure of each **PROTEIN** in a cell.

genetic diversity Variation in the **GENES** carried by individuals within a **POPULATION**, or some other

G

subdivision of a **SPECIES**, which acts as the raw material for **SELECTION** and therefore **EVOLUTION**.

genetic drift An alteration in **GENE FREQUENCIES** in a small **POPULATION** as a result of chance events, i.e. not as the result of selection, **IMMIGRATION** or mutation. For example, if one or two individuals carry a particular **ALLELE** and they are by chance killed in a storm this allele will disappear from the population. Genetic drift may result in the **FIXATION** of alleles. *See also* **EFFECTIVE POPULATION SIZE**, **HARDY–WEINBERG EQUILIBRIUM**

genetic engineering, gene manipulation The creation of hybrid DNA molecules using DNA from different sources, including different species. DNA molecules may be split using **RESTRICTION ENZYMES** and then recombined to produce novel combinations of genes (recombinant DNA). This technique is an important tool in genetics and is used to produce **TRANSGENIC ORGANISMS**.

genetic evaluation system A computer-based system designed to assist in breeding management. *See* **BREEDPLAN**

genetic fingerprinting *See* **DNA FINGERPRINTING**

genetic load The pool of deleterious genes in a population; the mean number of lethal mutations per individual in a population. This increases with **INBREEDING**.

genetic marker *See* **MOLECULAR MARKER**

genetic material The UN **CONVENTION ON BIOLOGICAL DIVERSITY 1992 (CBD)** (Art. 2) defines genetic material as '*any material of plant, animal, microbial or other origin containing functional units of HEREDITY.*' *See also* **DNA**, **RNA**

genetic mixing, genetic pollution, genetic swamping The uncontrolled flow of genes into wild populations of organisms from domestic, feral, invasive, non-native, genetically engineered and other species. This is generally considered undesirable and threatens the existence of some species due to **HYBRIDISATION**.

genetic pollution *See* **GENETIC MIXING**

genetic resources The UN **CONVENTION ON BIOLOGICAL DIVERSITY 1992 (CBD)** (Art. 2) defines genetic resources as '*GENETIC MATERIAL of actual or potential value.*' *See also* **NAGOYA PROTOCOL 2010**

genetic swamping *See* **GENETIC MIXING**

genetic variation The variation that occurs within a **POPULATION** which is attributable to differences in the **GENES** possessed by the individuals within it. This variation is the raw material for **EVOLUTION**.

genetically modified (GM) *See* **GENETICALLY MODIFIED ORGANISM (GMO)**

genetically modified organism (GMO) An organism whose genetic characteristics have been altered by **GENETIC ENGINEERING**. This may involve the insertion of a gene from another organism or a synthetic gene. *See also* **ONCOMOUSE®**, **SPIDER-GOAT**, **TRANSGENIC ORGANISM**

genetics The scientific study of inheritance.

genome The genetic makeup of an organism; all of the **GENES** it possesses, whether expressed in the **PHENOTYPE** or not.

genome sequencing, full genome sequencing The process of determining the complete DNA sequence of an organism's **GENOME** at one time. The genome of the Tasmanian devil (*Sarcophilus harrisii*) has been sequenced in the hope of discovering why the species has been decimated by a deadly contagious facial cancer (devil facial tumour disease) which is transmitted between individuals by bites. Results of the study will help conservationists select the best individuals for a captive breeding programme.

genomics The application of DNA sequencing and genome mapping to the production of biologically important molecules.

genotype
1. The genetic makeup of an organism. It may refer to the entire **GENOME** or the specific allelic composition of a particular gene or set of genes. *Compare* **PHENOTYPE**
2. To determine the genetic makeup by means of an **ASSAY**. *See also* **DNA FINGERPRINTING**

genus (genera *pl.***)** In **TAXONOMY**, a group of related **SPECIES**; a subdivision of a **FAMILY (1)** (Table C5).

geofence A virtual perimeter around a real geographical area. A wild animal may be fitted with an electronic device which monitors its location (a location-aware device (**GLOBAL POSITIONING SYSTEM (GPS)**) and notifies an appropriate authority when the animal crosses the 'fence boundary'. Used by conservationists to manage human–elephant conflict problems in Kenya.

Geographical Information Systems (GIS), Geographic Information Systems
1. '*A system for capturing, storing, checking, manipulating, analysing and displaying data which are spatially referenced to the earth*' (Anon., 1987).
2. A tool which allows the visualisation, analysis and interpretation of data in order to reveal relationships, patterns and trends, especially in the form of maps. GIS is widely used in wildlife conservation, especially to establish the home ranges, movements and migratory patterns of animals, and the relationship between the distribution of species and vegetation, human settlements and other components of the environment. Data collection analysis often involves the use of **REMOTE SENSING** techniques (e.g. the use of **SATELLITE IMAGES**) and data collected using the **GLOBAL POSITIONING SYSTEM (GPS)**.

geolocator A small electronic tracking device which may be attached to an animal. It includes an electronic calendar, a clock and a light sensor. The data recorded and stored by a geolocator can be used to determine where the animal was located at a particular point in time. Some devices are the size of a shirt button and may be attached to birds the size of a robin (*Erithacus rubecula*). *See also* **SATELLITE TAG**

geophagy The practice of eating soil. Occurs in a range of taxa including elephants and lemurs. May provide nutrients, aid digestion and assist in reducing the effects of gut parasites.

geotaxis A movement directly towards or away from the direction of gravity. Some species have sense organs capable of detecting gravity which assist them in maintaining an appropriate orientation with respect to the ground, e.g. the **MECHANO-RECEPTORS** in the ear of vertebrates.

geriatric

1. Relating to old age, e.g. geriatric diseases.
2. An old individual.

gestation The period of time between **CONCEPTION** (**FERTILISATION**) and birth in animals that give birth to live young. *See also* **DELAYED IMPLANTATION**, **VIVIPARITY**

gesture A deliberate movement (e.g. of the hand) intended to elicit a response from another animal. Field studies of wild chimpanzees (*Pan troglodytes*) have found that they use at least 66 distinct gestures to communicate with each other. Similar gestures are found in gorillas, orangutans, bonobos and humans. Bonobos (*P. paniscus*) have evolved a complex system of hand gestures to communicate instructions and intentions during sexual encounters. Gestures are also used by animal trainers when training their animals and by researchers studying communication in apes (especially using American Sign Language). *See also* **FACIAL EXPRESSION**, **LANGUAGE**, **POSTURE**

GI tract *See* **GASTROINTESTINAL TRACT**

giant panda breeding centres A group of specialist centres in the People's Republic of China dedicated to the breeding and conservation of giant pandas (*Ailuropoda melanoleuca*). They include the Chengdu Research Base of Giant Panda Breeding (founded in 1987), the Bifengxia Panda Centre (opened in 2002) and the Woolong Panda Centre (established in 1981 but closed in 2008 due to earthquake damage and being rebuilt to reopen late 2012), all located in Sichuan Province. These centres conduct research and promote educational awareness of the conservation needs of pandas.

giant panda loans The Chinese Government loans giant pandas to a number of zoos in other countries for a substantial fee, e.g. San Diego Zoo and Edinburgh Zoo. Any young born at these zoos remain the property of the Chinese Government. The money raised by this scheme is used for panda conservation in China.

giant panda sanctuaries The Sichuan Giant Panda Sanctuaries are home to more than 30% of the world's giant pandas (*Ailuropoda melanoleuca*) and are a UNESCO **WORLD HERITAGE SITE**. They cover 924500 ha with seven nature reserves and nine scenic parks in the Qionglai and Jiajin Moun-

tains and constitute the largest remaining contiguous habitat of the species.

giardiasis An intestinal infection caused by protozoans (*Giardia* spp.), which occur in humans and a range of animal species.

gill A respiratory organ which provides a large surface area for gaseous exchange in aquatic taxa, especially fish, but also in annelids and arthropods. In fish they may also be important in **OSMOREGULATION**.

gilt A young female pig, up to and including **PRIMIPAROUS** females.

gin trap A wire noose used for catching game, set as a trap or **SNARE**.

gingivitis Inflammation and possibly ulceration of the gums. May be caused by disease or the eruption of teeth.

GIS *See* **GEOGRAPHICAL INFORMATION SYSTEMS (GIS)**

giving set Apparatus for administering fluid to a patient from a plastic bag containing the material to be infused. Usually consists of clear flexible plastic tubes with a needle or **CATHETER**, and a chamber where the liquid pools to maintain a steady flow and prevent the formation of air bubbles. May also contain an injection port and a clamp. *See also* **BUTTERFLY NEEDLE**, **INFUSION**

gizzard A part of the anterior alimentary canal where food is broken into smaller particles prior to chemical digestion. It may contain grit to assist this process. Found especially in birds, but also in some annelids and insects.

gland A specialised cell or group of cells which secretes a particular substance. *See also* **ANAL GLANDS**, **ENDOCRINE GLAND**, **EXOCRINE GLAND**, **SCENT GLAND**

Global Animal Survival Plan (GASP) A management programme for a species bred in captivity, which involves a detailed demographic and genetic analysis, and management planning between regions, generally using an **INTERNATIONAL STUDBOOK** or several regional studbooks.

Global Biodiversity Information Facility (GBIF) An international organisation whose purpose is to make global biodiversity data accessible everywhere in the world via the internet. GBIF uses the 'Catalogue of Life' as the taxonomic basis for its web portal. *See also* **SPECIES 2000 AND ITIS CATALOGUE OF LIFE**

Global Captive Action Plan (CAP) A captive breeding plan which ensures that regional zoo organisations are co-operating fully in species management and can partition their captive space to allow for the conservation of as many species as possible.

global collection plan In relation to zoos, a plan which determines which species zoos should keep for captive breeding at a global level. *See also* **GLOBAL CAPTIVE ACTION PLAN (CAP)**

Global Federation of Animal Sanctuaries An organisation formed in 2007 to strengthen and support

G

the work of animal sanctuaries worldwide. It does not operate sanctuaries itself but encourages co-operation between organisations that do.

Global Positioning System (GPS) A satellite navigation system which provides information on location, time and altitude anywhere on Earth where there is line of sight to at least four satellites. Used for mapping, route tracing and monitoring the movements of animals (using GPS collars). *See also* GEO-GRAPHICAL INFORMATION SYSTEMS (GIS), REMOTE SENSING, SATELLITE IMAGE, SATELLITE TAG

glomerulus A narrow blood vessel in the BOWMAN'S CAPSULE of the NEPHRON of the kidney where waste materials and water are removed from the blood during the processes of EXCRETION and OSMOREGULATION.

glucagon A hormone secreted by the islets of Langerhans in the PANCREAS which stimulates the breakdown of glycogen to glucose in the liver, thereby increasing the blood sugar level. *Compare* INSULIN

glucometer An electronic device for measuring the concentration of glucose in blood. *See also* BLOOD SUGAR LEVEL

glucose A monosaccharide ($C_6H_{12}O_6$) which is the most common form of naturally occurring sugar. It is the main form in which the energy derived from CARBOHYDRATES is transported around the body in the blood and is used in AEROBIC RESPIRATION to produce ADENOSINE TRIPHOSPHATE (ATP). It is stored in the body as GLYCOGEN.

glycogen A highly branched polymer of GLUCOSE. The main form in which carbohydrate is stored in vertebrates, especially in the liver and muscles. *See* GLUCAGON

glycolysis The anaerobic biochemical pathway that converts glucose to lactic acid (in animal cells) or ethanol and carbon dioxide (in plant and fungal cells) with a net yield of two ADENOSINE TRIPHOSPHATE (ATP) molecules for each molecule of glucose. In most animal cells the end product of glycolysis is pyruvate which is then utilised by the KREBS CYCLE to yield more ATP.

GM Genetically modified. *See also* GENETICALLY MODIFIED ORGANISM (GMO)

GMO *See* GENETICALLY MODIFIED ORGANISM (GMO)

gnawing stick A wooden stick provided for some animals (e.g. rabbits, rodents, dogs, monkeys) as an enrichment.

goal The purpose towards which a particular behaviour is directed. When the goal is achieved the behaviour ceases. This may be CONSUMMATORY BEHAVIOUR which brings to an end a period of APPETITIVE BEHAVIOUR. For example, FORAGING BEHAVIOUR (appetitive behaviour) ends when food is found and eaten (consummatory behaviour).

goat There are eight species of goats including the domestic goat (*Capra hircus*) which occurs worldwide in association with people. Goats were among the first animals to be domesticated. They were kept by Neolithic farmers for their meat, milk, hair, bone, skin, sinew and dung (for fuel). Feral goats do considerable damage to vegetation in many locations around the world.

goat breeds There are over 300 breeds of goat. The BRITISH GOAT SOCIETY recognises 16 breeds (Table G1) including the British Toggenburg (Fig. G1) and Angora (Fig. G2).

Gobiesociformes An order of fishes: clingfishes and their allies.

gogga South African slang for an insect or other small crawling or flying animal.

Table G1 Goat breeds recognised by the British Goat Society.

Anglo Nubian	British Toggenburg
Angora	Cashmere
Bagot	English
Boer	Golden Guernsey
British	Harness
British Alpine	Pygmy
British Guernsey	Saanen
British Saanen	Toggenburg

Fig. G2 Angora goat ram.

Golgi body, Golgi apparatus, Golgi complex A cell organelle which consists of a network of smooth, membranous flattened sacs arranged in parallel and surrounded by vesicles. It is located near the **ENDO-PLASMIC RETICULUM (ER)** and involved in the packaging and transport of many of the products of cell metabolism, including many destined for the **CELL MEMBRANE**.

Gombe Stream National Park A protected area in western Tanzania where the chimpanzees have been studied for many decades as a result of the pioneering work of Dr. Jane **GOODALL**.

gonadotrophin Any of a group of hormones that regulate reproduction and are secreted by the **PITUITARY GLAND**, **PLACENTA** or the **ENDOMETRIUM**.

gonadotrophin-releasing hormone (GnRH), luteinising hormone-releasing hormone (LHRH), luliberin A hormone produced by the **HYPOTHALAMUS** which is involved in reproduction. In females of many mammal species GnRH is released in a regular cycle, thereby producing a cyclicity in the release of eggs. GnRH also controls **SPERMATOGENESIS** in males. *See also* **IMMUNO-CONTRACEPTION**, **OESTROUS CYCLE**

gonads The sex organs which produce the gametes (spermatozoa or ova): **TESTES** and **OVARIES**.

Gonorynchiformes An order of fishes: milkfish, beaked salmons and their allies.

Goodall, Jane (1934–) A British zoologist who is a world authority on chimpanzees. She is the author of *In the Shadow of Man* (1971) and *Through a Window* (1990). Goodall established the longest continuous field study of chimpanzees (*Pan troglodytes*) in the world in the **GOMBE STREAM NATIONAL PARK** in western Tanzania in 1960 and founded the Jane Goodall Institute in 1977 to support this research. Goodall holds a PhD from the University of Cambridge but never studied for a first degree. She is in very large part responsible for changing our attitude towards the treatment of chimpanzees in captivity. *See also* **GREYBEARD**

gooder A hybrid duck: goosander (*Mergus merganser*) × eider (*Somateria mollissima*).

goorie *See* **KURI**

goose
1. Any bird belonging to the tribe Anserini of the family Anatidae.
2. A domesticated goose. In Europe, most are descended from the greylag goose (*Anser anser*). Kept for meat and feather production and also to produce **FOIE GRAS**.

Gorillas in the Mist A book published in 1983 and a film released in 1988 about the work of Dian **FOSSEY** with the mountain gorillas of Rwanda.

Governing Council of the Cat Fancy A registration body for breeding and showing cats in the UK.

Government Program Population A captive breeding population program of the **ASSOCIATION OF ZOOS AND AQUARIUMS (AZA)** which focuses on species that are managed by state or federal programs (e.g. the manatee (*Trichechus manatus*)).

GPS *See* **GLOBAL POSITIONING SYSTEM (GPS)**. *See also* **PROXIMITY LOGGER**

Graafian follicle A mature **OVARIAN FOLLICLE** (fluid-filled vesicle) in a mammalian **OVARY** which contains an **OOCYTE** and produces **OESTROGEN**.

grade An animal whose parentage is unknown, e.g. a grade horse. *Compare* **PEDIGREE (1)**

graminivore An animal that feeds on grass or its seeds.

Grand National, The A steeplechase held at Aintree Racecourse in Merseyside, UK, which involves horses jumping 30 fences, including Becher's Brook, and running over 4.5 miles (7.2 km). It was first held in 1839 and is often described as the most famous horse race in the world. Horses regularly fall at the fences and fail to complete the course. Between 1839 and 2012, 69 horses died during the steeplechase or as a result of injuries sustained during the race.

granulation The first stage in the development of a scar, containing **COLLAGEN** and small blood vessels.

grass One of many hundreds of monocotyledonous plants used as food for animals on farms and in zoos and for decorative purposes in exhibits.

grass muzzle, grazing muzzle A cover strapped over a horse's mouth as a grazing management measure to restrict grass intake.

grass staggers, grass tetany, hypomagnesaemia A disease caused by a deficiency of **MAGNESIUM**. Common in lactating cows fed on grass. *See also* **LACTATION**

grass tetany *See* **GRASS STAGGERS**

gravid Pregnant.

gray wolf, grey wolf (*Canis lupus*) *See* **LOBO, MARTIN THE WOLF, WYOMING FARM BUREAU FEDERATION V. BABBITT (1997)**

grazing facilitation The phenomenon whereby some species of herbivores selectively graze plant species in the sward in such a way as to make food of suitable height or quality available to other herbivore species. **LIVESTOCK** grazing may facilitate grazing by geese and deer; buffalo (*Syncerus caffer*) graze grass to optimal heights for white rhino (*Ceratotherium simum*). *See also* **CONSERVATION GRAZING, RESOURCE PARTITIONING, SELECTIVE GRAZING**

grazing muzzle *See* **GRASS MUZZLE**

great ape A member of the family Pongidae (gorillas, chimpanzees, bonobos and orangutans). *Compare* **LESSER APE**

Great Ape Project
1. An international organisation, founded in 1994, of primatologists, anthropologists, ethicists and others who advocate a UN Declaration of the Rights of Great Apes that would give basic legal rights to great apes.

G

2. The title of a book edited by Peter **SINGER** and Paola Cavalieri who support the Great Ape Project.

great ape rights *See* **GREAT APE PROJECT**

Great Ape Trust A scientific research facility in Des Moines, Iowa, which is dedicated to understanding the origins and future of **CULTURE**, **LANGUAGE**, tools and **INTELLIGENCE**. It is home to seven bonobos (including *KANZI*) involved in studies of their cognitive and communicative capabilities, and two orangutans.

Great Apes Conservation Act of 2000 (USA) A US law which assists in the conservation of **GREAT APES** by supporting and providing financial resources for conservation programmes within their range states.

Greatest Show on Earth *See* **BARNUM AND BAILEY**

green exhibit A sustainable exhibit, especially in a zoo, which is constructed of recycled materials and is energy-efficient etc. The *Rhinos of Nepal* is a state-of-the-art sustainable exhibit at **WHIPSNADE ZOO** containing a herd of Asian greater one-horned rhinoceroses (*Rhinoceros unicornis*). It is the **ZOOLOGICAL SOCIETY OF LONDON (ZSL)**'s first fully 'green' exhibit. The building makes use of natural sunlight and utilises recycled and local materials such as recycled railway sleepers and local sandstone. The exhibit uses a rainwater utilisation system to supply water to the pool. The system collects rainwater in a 30 000 litre underground tank, but when there is insufficient rainwater the system automatically draws potable water from the mains supply. Wastewater from the enclosure is filtered through a reedbed system before it drains away. The pool water is heated with a solar thermal and an air heat exchanger which uses 75% less fossil fuel energy than a conventional gas boiler. The recycled water is filtered through a high-tech biological filter, saving 20 000 m³ of water per year. The exhibit cost £1 million to construct. *See also* **GREEN ROOF**

green gland *See* **ANTENNAL GLAND**

green paper A consultation document produced by the UK government and some other governments when they are considering introducing new legislation, e.g. *Preparing an Animal Health and Welfare Strategy for Great Britain*. Its purpose is to generate public debate in order to inform future decisions about new laws. Once government policy has been determined a **WHITE PAPER** may be produced. In the USA green papers have a similar purpose to those in the UK. The EU also produces discussion documents called green papers. *See also* **PUBLIC CONSULTATION**, **WHITE PAPER**

green plant A plant that has leaves, stems and other parts which contain the green pigment chlorophyll and are capable of producing food by **PHOTOSYNTHESIS**.

green roof, eco-roof, living roof A roof which is designed to support a plant community, sometimes used to camouflage buildings in rural areas and to

Fig. G3 Green roof, *Durrell* (Jersey Zoo).

provide useful habitat for birds, insects etc. in urban areas including zoos, e.g. the roof of the giraffe exhibit at Cincinnati Zoo (*Giraffe Ridge*), Karen Peck Katz Conservation Education Center at Milwaukee County Zoo, the Australasia Pavilion at Toronto Zoo. A green roof consists of various layers, typically, from bottom to top: the roof deck, a protection board, a waterproof membrane, an insulation layer, a drainage/water storage layer, filter fabric, growing medium, plants. These roofs absorb carbon dioxide, provide insulation for the building and store water, reducing run-off (Fig. G3).

green stick fracture An incomplete bone **FRACTURE** which appears similar to a green plant stem which has been partially broken.

Greenpeace An organisation founded in 1971 which investigates, exposes and documents the causes of environmental destruction, lobbies governments and promotes solutions for a greener future. It has led important campaigns against the testing of nuclear weapons and against commercial **WHALING**.

greenwash The process of disguising an activity or product to give the impression that it is environmentally friendly, is sustainable or in some other respect is beneficial to the environment, when it is not. *See also* **SUSTAINABILITY**

greeting ceremony **AFFILIATIVE BEHAVIOUR** which occurs between individuals of some species, especially mammals, when they have been apart for a period of time. For example, Asian elephants (*Elephas maximus*) vocalise, urinate, defecate and touch each other with their trunks.

gregariousness The tendency of animals to form social groups with **CONSPECIFICS** as a result of mutual attraction, often involving some type of **SOCIAL ORGANISATION**. *See also* **AGGREGATION**, **ASSOCIATION INDEX**

grey literature Printed material that is difficult to access because it not widely published, e.g. reports, newsletters etc. *Compare* **SCIENTIFIC PAPER**

grey matter Tissue in the CENTRAL NERVOUS SYSTEM (CNS) (1) that has a high density of nerve cell bodies, in the cortex of the CEREBELLUM and CEREBRUM and within the central part of the SPINAL CORD. *Compare* WHITE MATTER

grey wolf, gray wolf (*Canis lupus*) *See* LOBO, *MARTIN* THE WOLF, *WYOMING FARM BUREAU FEDERATION V. BABBITT* (1997)

Greybeard, David A male CHIMPANZEE named and studied by Dr. Jane GOODALL in what is now the GOMBE STREAM NATIONAL PARK in Tanzania. He was the first chimpanzee ever to be observed eating meat and the first observed making a TOOL by removing leaves from a twig and using it to extract termites from a termite mound. These observations changed the way we think about chimpanzees and our understanding of our genetic relationship to them.

Greyhound Board of Great Britain (GBGB) The Greyhound Board is the governing body for licensed greyhound racing in Britain. From 1 January 2009, the GBGB took on the functions of the British Greyhound Racing Board and the National Greyhound Racing Club. The GBGB promotes the interests of greyhound racing, approves changes to the RULES OF RACING, and formulates welfare policy and promotes the commercial interests of racing. *See also* GREYHOUND REGULATORY BOARD.

Greyhound Regulatory Board The Greyhound Regulatory Board is an independent body incorporated with the GREYHOUND BOARD OF GREAT BRITAIN (GBGB) which implements and manages the RULES OF RACING, licenses stadia and staff, identifies and registers dogs, investigates complaints and manages a drug sampling programme.

grief Some species appear to exhibit behaviours that would collectively be called grief in humans when a close relative or friend dies: LETHARGY, loss of appetite, weight loss, withdrawal from social activities. Jane GOODALL has argued that if we call this combination of behaviours grief in humans there is no reason why we should not call it grief in chimpanzees. *See also* ANTHROPOMORPHISM, EMPATHY

grooming
1. The action of one animal cleaning its own body (self-grooming) or the body of another animal of the same species (ALLOGROOMING). In some animal societies grooming is important in the maintenance of a DOMINANCE HIERARCHY.
2. The action of a person cleaning the body of an animal to maintain healthy skin or fur, or to prepare an animal for exhibition, e.g. in a dog show, an AGRICULTURAL SHOW etc. *See also* CURRY COMB, ROTATING CATTLE BRUSH

grooming claw *See* TOILET CLAW

gross assimilation efficiency (GAE), apparent dry matter digestibility (ADMD) A measure of the efficiency with which an animal converts food into its own body mass. Calculated as:

Table G2 Group living: benefits and costs.

Benefits	Costs
Reduced predation pressure	Increased risk of infection by parasites and disease
Improved foraging efficiency	Increased intraspecific competition, e.g. for mates, food, shelter
Improved defence against other groups of conspecifics	Increased risk of offspring being killed by conspecifics
Improved parental care/ co-operative breeding	Increased risk of exploitation of parental care by conspecifics

$$GAE(\%) = \frac{I - E}{I} \times 100$$

where I = the dry weight of food ingested, E = the dry weight of faeces egested, over a fixed period of time.

group fission The splitting of a social group. Results in DISPERSAL which is important in the spread, ISOLATION and SPECIATION of organisms.

group living Many species live in groups as a result of their gregarious nature. The benefits and costs of group living are listed in Table G2. *See also* ASSOCIATION INDEX, SOCIAL ORGANISATION

group selection A theory that suggests that natural selection will favour any individual whose behaviour lowers its own FITNESS if it benefits the group (or species) to which it belongs. It is difficult to explain in terms of our current understanding of evolutionary processes and is not widely accepted as a useful theory by evolutionary biologists. *See also* KIN SELECTION

grower diet A combination of foods formulated to promote growth and weight gain, especially in agricultural livestock in the GROWING PHASE. *See also* GROWER UNIT, PROTEIN CONCENTRATE

grower unit A facility where livestock are kept during their GROWING PHASE. *Compare* FINISHING UNIT

growing phase In agricultural animal production, the period of development of an animal when it grows continuously, gaining both height and weight. This occurs after the REARING PHASE and before the FINISHING PHASE. *See also* GROWER DIET, GROWER UNIT, GROWTH-PROMOTING HORMONES

growth
1. An increase in size of a cell or organism resulting from cell enlargement or cell division (MITOSIS). *See also* DETERMINATE GROWTH, GROWTH-PROMOTING HORMONES, INDETERMINATE GROWTH
2. An increase in the size of a population. *See also* POPULATION GROWTH

growth curve
1. A line representing the change in size of a population of living things (e.g. bacteria, elephants) with time. *See also* **POPULATION GROWTH**
2. A line representing the increase in size (e.g. weight, height) of an individual animal with time. Often compared to a standard line for the species and used to determine whether or not a young animal is growing normally.

growth hormone (GH), somatotrophin A hormone that controls growth and stimulates many aspects of metabolism. Secreted by the **ANTERIOR PITUITARY**. *See also* **GROWTH-PROMOTING HORMONES**

growth plate, epiphyseal plate, epiphysis Areas of developing cartilage at the ends of **LONG BONES** which produce new bone tissue and determine the length and shape of bones in the adult.

growth-promoting hormones Chemicals used to improve an animal's ability to utilise nutrients more efficiently and produce leaner meat; especially in cattle to produce cheaper beef. Originally given in feed but now generally as an implant at the base of the ear to reduce the risk of residues appearing in edible tissue. Use of these substances has been banned in the EU since 1988 except for veterinary purposes. Meat from animals where hormone implants have been used cannot be imported into the EU.

Gruiformes An order of birds: cranes, mesites, trumpeters, rails, sunbittern, bustards, kagus, finfoots, seriemas.

Grzimek, Bernard (1909–1987) Professor Bernard Grzimek was a German vet and was formerly Director of Frankfurt Zoo and President of the Frankfurt Zoological Society. Grzimek was instrumental in the establishment of the **SERENGETI** National Park in Tanzania. He was the author of *No Room for Wild Animals* and co-author, with his son Michael, of *Serengeti Shall Not Die*, both of which were released as films (in 1956 and 1959 respectively). Grzimek was editor-in-chief of the 13-volume *Grzimek's Animal Life Encyclopaedia* which was translated into English in 1975 and became a standard reference work.

guanine (G) A **NUCLEOTIDE** base that pairs with C (cytosine) in **DNA**.

guano Bird faeces. Used as an organic fertiliser in some parts of the world where it is collected in large quantities from bird colonies.

guide dog, seeing-eye dog An assistance dog trained to help a blind or visually impaired person move around obstacles. Suitable breeds include Labrador, German shepherd and golden retriever.

guillotine door A door that slides vertically between runners usually operated by steel cables running through a pulley system. Found between adjacent animal enclosures and usually operated from a **KEEPER AREA**.

guinea pig, cavy A medium-sized rodent (*Cavia porcellus*) that is a popular pet. Also used in medical research and farmed for meat in South America and Africa.

gun dog, bird dog A dog trained to assist hunters in finding and retrieving game, especially **GAME BIRDS** after they have been shot.

gustation
1. The act of tasting, achieved by the action of **CHEMORECEPTORS**.
2. The sense of taste.

gut loading The practice of rearing insects on a high-nutrient diet before feeding them to other animals. This is necessary because the insect itself is of low nutrient quality and when given as food the recipient receives not only the nutrients in the insect body but also any of the high-nutrient food that remains in the insect's gut. This is a method of supplementing the diet of, for example, amphibians when fed on crickets. *See also* **DUSTING (2)**

gut passage time, food passage time The time it takes food to pass through the gut from the time of **INGESTION** to the time it passes out of the body as faeces. May be measured using food dyes or small indigestible materials given in the food.

Gymnophiona An order of amphibians: caecilians.

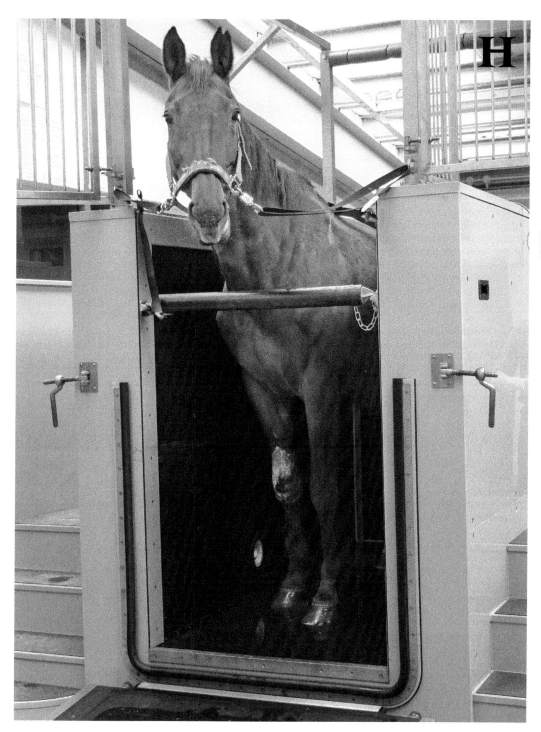

Fig. H1 H is for hydrotherapy. An equine hydrotherapy spa.

Dictionary of Zoo Biology and Animal Management: A guide to terminology used in zoo biology, animal welfare, wildlife conservation and livestock production, First Edition. Paul A. Rees.
© 2013 John Wiley & Sons, Ltd. Published 2013 by John Wiley & Sons, Ltd.

H1N1 virus, hog flu, pig flu, swine flu A human seasonal flu virus that also affect pigs and resulted in the term 'swine flu'.

H5N1 virus, bird flu A strain of the AVIAN INFLUENZA (bird flu) virus which may be deadly to humans.

habitat A place where organisms live; the place where a particular type of organism lives, e.g. tiger habitat. The UN CONVENTION ON BIOLOGICAL DIVERSITY 1992 (CBD) (Art. 2) defines a habitat as *'the place or type of site where an organism or population naturally occurs.'*

habitat fragmentation The process by which large areas of a particular habitat are broken up into smaller areas, or the result of this process. A forest may become fragmented as a result of CLEAR-FELLING or agriculture so that smaller, unconnected areas of forest remain. Fragmentation may cause local extinctions of particular species because the habitat fragments are too small to support viable populations and individuals are unable to disperse to nearby fragments due to a lack of suitable CONSERVATION CORRIDORS.

Habitats Directive Council Directive 92/43/EEC of 21 May 1992 on the conservation of natural habitats and wild fauna and flora (the Habitats Directive) promotes the maintenance of BIODIVERSITY in the EUROPEAN UNION (EU), while taking into account economic, social, cultural and regional requirements (Art. 2). It identifies certain priority species and habitats in need of special protection and recognises the transboundary nature of many of the threats to our natural heritage. The REINTRODUCTION of species listed in Annex IV (Animal and plant species of community interest in need of strict protection) is encouraged by Art. 22 (a), where this may contribute to the re-establishment of these species at a favourable conservation status. Annex IV species include species that are being captive bred in zoos such as the dormouse (*Muscardinus avellanarius*) and the sand lizard (*Lacerta agilis*). The Habitats Directive has led to the establishment of SPECIAL AREAS OF CONSERVATION (SACs) which make up a European network of protected areas known as Natura 2000. These include the SPECIAL PROTECTION AREAS (SPAs) established under the WILD BIRDS DIRECTIVES.

habituated group A group of animals which has become accustomed to the presence of humans and can therefore be followed and watched by scientists without their presence affecting the animals' behaviour. *See also* HABITUATION

habituation A type of LEARNING in which an individual becomes progressively less responsive to a particular STIMULUS and eventually ignores it altogether. This has an important adaptive value in the wild as it stops an individual from repeatedly reacting to a non-threatening stimulus and thereby wasting time and energy. Most animals living in zoos eventually habituate to the presence of visitors and may largely ignore their presence. *See also* HABITUATED GROUP

hack
1. An old, worn-out horse.
2. *See* HACKBOARD

hack aviary *See* AVIARY HACK

hackboard, hack An elevated board containing food used in the training of BIRDS OF PREY prior to release. The birds are taught to fly to the board from progressively increasing distances. *See also* HACKING (1)

hacking
1. A process that uses falconry techniques which allows a bird of prey (especially one that has been hand-reared) to develop hunting skills while food continues to be provided. Also called hacking back. *See also* HACKBOARD, KITE HACK, LURE HACK
2. Riding a horse for exercise or pleasure.

Haeckel, Ernst (1834–1919) A German zoologist who first coined the term 'ECOLOGY'.

haematology, hematology The scientific study of blood, blood diseases and the tissues where blood is formed.

haematoma, hematoma A swelling resulting from bleeding into the tissues.

haemoglobin, hemoglobin A respiratory pigment consisting of protein and an iron prosthetic group, found in ERYTHROCYTES in vertebrates and in solution in the blood of some invertebrates. It is responsible for the carriage of most of the oxygen in BLOOD. Oxygenated haemoglobin is called oxyhaemoglobin. *See also* OXYGEN SATURATION

haemolysis The rupture of red blood cells causing the release of HAEMOGLOBIN.

haemorrhage, hemorrhage A loss of blood (bleeding) from blood vessels either internally (internal haemorrhage) or externally (external haemorrhage). A cerebral haemorrhage is a bleed in the brain and it may result in death or serious disability.

haemostasis *See* BLOOD CLOTTING

Hagenbeck, Carl (1844–1913) A German animal trader and trainer who made a very significant contribution to the advancement of the design of enclosures and the exhibition of animals in zoos. Originally, he owned a travelling exhibition in which he displayed animals alongside people from different parts of the world, including Lapps, Nubians and Eskimos (Inuit), often with their traditional homes and domestic animals. This was extremely popular and on 6 October 1878 around 62 000 people visited Berlin Zoo to visit Hagenbeck's exhibition. In 1907 Hagenbeck founded his own zoo at Stellingen, near Hamburg (TIERPARK HAGENBECK). The zoo contained cleverly designed exhibits which appeared to house carnivores and herbivores together, but in reality they were separated by hidden moats. Prior to building these moated enclosures Hagenbeck investigated the

Fig. H2 Ha-ha.

Fig. H3 Hamster: Syrian or golden hamster (*Mesocricetus auratus*).

H

jumping abilities of the animals. As well as running his own zoo, he also supplied many well-known zoos with animals. Hagenbeck's zoo was destroyed by bombing in 1943, but was rebuilt after the end of the Second World War. *See also* **TRAVELLING MENAGERIE**

ha-ha A device used in landscape design (especially by Lancelot 'Capability' **BROWN**) consisting of a wall or fence placed in a ditch to hide it from view. Used in zoos to retain a wide range of large mammals, e.g. elephants, rhinos, antelopes (Fig. H2). Widely used by **HAGENBECK** in his enclosure designs. *See also* **DEPRESSED VERTICAL FENCE BARRIER**

hair A thread protruding from the **EPIDERMIS** of mammalian skin, made of a number of cornified cells. It develops from a **FOLLICLE (1)** at the base which is a useful source of **DNA** for analysis. Hair has a sensory function and in most mammals it is thick enough to have a thermoregulatory function.

hair follicle *See* **HAIR**

hair trap A device for collecting hair samples from mammals for DNA analysis, thereby allowing individuals to be distinguished. For small mammals the trap consists of a tube through which an animal must pass in order to reach food bait. As it passes through the tube it leaves hairs on a sticky pad. Such traps have been used on trees to count pine martens (*Martes martes*) in forests in Scotland. A scent lure and barbed wire fencing may also be used as a hair trap, e.g. for bears. Hair traps have also been constructed using strips of Velcro and roofing nails.

halal Relating to or denoting meat prepared according to Muslim law. *See also* **ISLAM AND ANIMALS, RELIGIOUS SLAUGHTER**

half-siblings Individuals that have only one parent in common, i.e. the same mother but a different father, or the same father but a different mother. Such individuals have only a quarter of their genes in common.

halitosis Bad-smelling breath. May be indicative of disease, e.g. mouth **ULCERS**.

Hamilton, William Donald (1936–2000) A British evolutionary biologist who was one of the founders of **SOCIOBIOLOGY** and developed a genetic basis for **ALTRUISM** and **KIN SELECTION** and also studied the evolution of sex.

hammock A suspended 'bed' provided for primates, often apes. Usually made of rope or fire hose and suspended from a climbing frame or other structure.

hamster A small rodent widely kept as a pet (Fig. H3). There are around 25 species, the most popular being the golden or Syrian hamster (*Mesocricetus auratus*).

Hancocks, David An architect and zoo historian who was director of **WOODLAND PARK ZOO**, Seattle (1976–1984), the **ARIZONA–SONORA DESERT MUSEUM**, Arizona (1989–1997) and Werribee Open Range Zoo (1998–2003). He is an advocate of revolutionising the design of zoos and is the author of *A Different Nature: The Paradoxical World of Zoos and Their Uncertain Future* (2001).

hand signals *See* **GESTURE, SIGN LANGUAGE**

handbag dog A miniature breed of dog (e.g. Chihuahua, pug, miniature smooth-haired dachshund, shih-tzu) small enough to carry in a designer handbag like a fashion accessory, imitating the behaviour of a number of female celebrities such as Paris Hilton and Britney Spears.

hand-reared Referring to an animal who has been reared by a keeper or other caretaker instead of their natural mother. Hand-rearing may be necessary because the animal was rejected at or around the time of birth, because it has been orphaned, or because eggs are removed from a adult bird to increase productivity as part of a **CAPTIVE BREEDING PROGRAMME**. It may involve the use of a **PUPPET MOTHER**. *See also* **SURROGATE (1)**

hanging stall apple A candy apple toy. An **ENVIRONMENTAL ENRICHMENT** device used in an equid stall.

haploid The condition in which a cell possesses only one CHROMOSOME of each pair which makes up the GENOME.

happiness A mental state of wellbeing. In most species it is difficult to assess whether or not an individual is happy and the concept of happiness is probably not very useful. Some species, e.g. howler monkeys (*Alouatta* spp.), have normal facial expressions which make them appear unhappy to humans. It is possible to use human measures of happiness with great apes because of our close evolutionary affinity. When scientists examined the relationship between subjective impressions of wellbeing ('happiness'), measured using questionnaires completed by keepers, and the longevity of orangutans living in zoos they found that happier individuals lived longer lives. This may be because undetected ill health makes an animal appear 'unhappy' or because 'happiness' makes it more likely that an individual will stay healthy. *Compare* GRIEF

hard and soft zoo environments In a zoo context, 'hard environments' refers to outdated concrete and metal structures and 'soft environments' to modern NATURALISTIC EXHIBITS.

hard release The release of an animal into the wild with no immediate possibility of further support from human caretakers, e.g. the release of a penguin back to the sea after recovery from an injury. *See also* PRECONDITIONING. *Compare* SOFT RELEASE

hard standing A hard surface (often concrete) where animals may stand or lie down. Provided as an alternative to standing on pasture which may be unhealthy for long periods if it is wet. Often located immediately outside the entrance to an animal house. It may help to wear down the hooves of UNGULATES and reduce foot problems.

hardpad disease *See* CANINE DISTEMPER

Hardy–Weinberg equilibrium In a closed population where the frequency of the DOMINANT ALLELE (A) is p and the frequency of the RECESSIVE ALLELE (a) of the same GENE is q (and $p + q = 1$) when random mating occurs the frequency of GENOTYPES in the next generation will be $p^2 + 2pq + q^2 = 1$ (where p^2 = the frequency of AA, $2pq$ = frequency of Aa and q^2 = frequency of aa). If $p = 0.5$ and $q = 0.5$, the frequency of the phenotypes in the next generation will be: AA = 0.25, Aa = 0.5, and aa = 0.25 (Fig. H4). This theorem assumes random mating, no SELECTION, no mutation, no IMMIGRATION and no EMIGRATION. *See also* GENE FREQUENCY

hare coursing The use of dogs to pursue and kill hares as a country 'sport'. Banned in England and Wales by the HUNTING ACT 2004.

harem A group of females especially when controlled by a male (e.g. occurs in some antelope species). *See also* POLYGYNY

Harlow, Harry Frederick (1905–1981) An American psychologist best known for his controversial experiments with rhesus monkeys which studied the effects of separating young from their mothers (maternal deprivation) and social isolation.

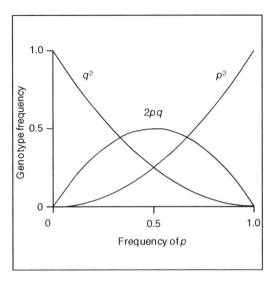

Fig. H4 Hardy–Weinberg equilibrium

harpoon A spear-like weapon used in the hunting of whales, walruses and other marine mammals. It may be a large device fired from a cannon on a whaling ship or thrown by hand by indigenous people from a small boat and is sometimes attached to floats to prevent the animal from diving.

Harrison, Ruth (1920–2000) An animal rights campaigner who was the first person to draw the attention of the public to the plight of animals in factory farms in her book *Animal Machines* (1964). She was influential in the development of laws to protect farm animals in the UK and in Europe, notably the Agriculture (Miscellaneous Provisions) Act 1968 and the EUROPEAN CONVENTION FOR THE PROTECTION OF ANIMALS KEPT FOR FARMING PURPOSES 1976.

Harvard mouse *See* ONCOMOUSE®

Harvard system A method of writing references to published and other work widely used in science. Its basic form for a scientific paper is: author's surname, followed by initials (repeated for each author if more than one); date of publication; title of paper; title of journal; volume of journal; page numbers. For a book the form is: author's surname followed by initials (repeated for each author if more than one); date of publication; title of book; edition (if more than one); town or city of publication; name of publisher. Sometimes the last two elements are reversed in order. *See also* ACADEMIC JOURNAL

harvesting of livestock The slaughtering of farm animals for market.

hatch room *See* HATCHERY

hatchability rate The proportion of eggs which hatch successfully.

hatchery, hatch room, hatching room A room, building or other place where eggs are hatched, especially those of birds, reptiles or fishes.

hatching The process of emerging from an EGG (2). Some birds raised in captivity may need assistance in hatching. *See also* ASSIST HATCH

hatching room *See* HATCHERY

hatchling In OVIPAROUS species, an animal that has very recently emerged from a hard-shelled egg. Generally applied to birds and reptiles.

Hawaiian goose *See* NĒ NĒ (*BRANTA SANDVICENSIS*)

hay Cut and dried grasses (e.g. ryegrass, timothy, fescue), legumes (e.g. LUCERNE) and herbaceous plants used for livestock fodder. Sometimes made from oats, barley or wheat. *See also* STRAW

hay net A net containing hay or similar fodder suspended from the roof of an animal house or pole. Sometimes use as a FEEDING ENRICHMENT. *See also* HAY RACK

hay rack A container made of widely spaced metal bars which contains hay or similar folder, and is fixed to a wall, usually at the head height of the animal being fed. Hay racks are sometimes used as a FEEDING ENRICHMENT device in an elephant house and fixed high on the walls so that the animals can only reach the food by extending their trunks. *See also* HAY NET

Hazardous Substances and New Organisms Act 1996 (NZ) Legislation in New Zealand which seeks, among other things, to control the release of NEW ORGANISMS to the environment.

head roll A repetitive behaviour consisting of rotational movements of the head exhibited in some captive animals, e.g. okapis, giraffes.

head starting The practice of captive breeding and subsequent release to the wild to alleviate population decline.

head-bobbing A STEREOTYPIC BEHAVIOUR which involves moving the head up and down repeatedly for no obvious reason.

Health and Safety Executive (HSE) An independent regulator in Great Britain that acts to reduce work-related deaths, serious injury and ill-health. Much of its work is preventative in nature but it also prosecutes employers where serious breaches of health and safety law occur. The HSE may become involved in serious accidents in zoos and on farms, and may bring prosecutions for breaches of health and safety law. *See also* COSHH REGULATIONS

Health of Animals Act 1990 (Canada) An Act of the Canadian Parliament concerned with the control of diseases and toxic substances that may affect animals or that may be transmitted by animals to persons, and the protection of animals.

hearing dog A dog which has been trained to recognise certain sounds and draw the attention of its hearing-impaired owner to them, e.g. a door bell, telephone ring tone, smoke alarm.

heart A muscular pump found in many taxa of animals, used to move blood around the body.

heart attack *See* CORONARY THROMBOSIS. *See also* CARDIAC ARREST

heart failure, congestive heart failure A physiological state in which the heart fails to provide sufficient blood to the body and lungs. The heart ventricles fail to pump blood out of the heart into the arteries at a sufficient rate to clear the blood entering through the veins, thereby causing increased venous pressure, possible tissue damage and breathlessness.

heart murmur An abnormal sound made by the heart which often indicates disease.

heart rate, heartbeat rate The number of HEART-BEATS per minute. A basic physiological measurement made by vets. An abnormal heart rate may indicate the presence of disease. *See also* BRADY-CARDIA, RESTING HEART RATE, TACHYCARDIA

heartbeat

1. The pulsation of the HEART which results from the alternate contraction and relaxation of the heart muscles (in the atria and ventricles) as it pumps blood around the body.

2. A single pumping action of the heart: a complete CARDIAC CYCLE of contraction and relaxation of the CARDIAC MUSCLE.

heartbeat rate *See* HEART RATE

heat (in) In OESTRUS; the time when female mammals are sexually active, i.e. during the oestrous phase of the oestrous cycle. The period immediately prior to OVULATION. *See also* BULLING

heat lamp *See* DULL EMITTER HEAT LAMP BULB, INFRARED LAMP

heat mat, heat pad An electrically heated mat used to keep animals such as reptiles warm. *See also* INFRARED LAMP

heat pad *See* HEAT MAT

Hediger, Heini (1908–1992) Professor Heini Hediger was a Swiss zoologist who conducted pioneering work in animal behaviour and was responsible for defining the concept of 'FLIGHT DISTANCE' in animals. He is considered to be the 'father of ZOO BIOLOGY' and was once the Director of Zurich Zoo. Hediger published a number of books on the biology of zoo animals including *Studies of the Psychology and Behaviour of Captive Animals in Zoos and Circuses* (1955), *Wild Animals in Captivity: an Outline of the Biology of Zoological Gardens* (1964), *Psychology and Behaviour of Animals in Zoos and Circuses* (1969), *Man and Animal in the Zoo: Zoo Biology* (1970) and he co-authored *Born in the Zoo* (1968).

hefting A traditional method of managing livestock (especially sheep or cattle) on large areas of common land or communal grazing land, without the use of fences. The animals learn which areas to use for grazing and this knowledge is passed from one generation to the next. Hefted flocks of sheep may only

H

be brought to the farm for lambing, dipping and shearing. *See also* **CONSERVATION GRAZING**

heifer A young female cow (over 1 year old) who has not had her first calf.

Heligoland trap, funnel trap A large, funnel-shaped trap made of netting and containing bushes and other vegetation, with a wide entrance and converging to a narrow section towards which birds are driven and eventually captured in a catching box. Used to capture and ring birds, especially on migration routes. Named after the site of the first such trap at the Heligoland Bird Observatory in Germany.

helminth A parasitic worm that lives inside its host, including **CESTODES** (tapeworms), trematodes (**FLUKES**) and **NEMATODES** (roundworms). *See also* **ANTHELMINTIC**

helper monkey, service monkey A monkey (capuchin (Cebinae)) trained to assist quadriplegics and others with mobility problems. It may be able to open bottles, microwave food, turn the pages of a book, operate knobs and switches etc.

hematology *See* **HAEMATOLOGY**

hematoma *See* **HAEMATOMA**

hemipenis (hemipenes *pl.*) One of a pair of intromittant erectile organs found in male snakes, lizards and worm lizards which protrudes through the **CLOACA (1)**.

hemoglobin *See* **HAEMOGLOBIN**

hemorrhage *See* **HAEMORRHAGE**

hemostasis *See* **BLOOD CLOTTING**

herbal veterinary medicine The use of medicines made from herbs e.g. dandelion as a **DIURETIC** for horses.

herbicide A weedkiller; a chemical which kills unwanted plants.

herbivore An animal that eats plants.

herd The collective term for groups of individuals of a wide range of species, e.g. elephants, moose, antelopes, deer, etc. *See* **HERDING**

herd book A book that records the **PEDIGREES** of a particular kind (breed) of animal, e.g. pigs, cattle, goats etc.

herding A type of **SOCIAL ORGANISATION** found in some herbivorous mammals. Herd organisation varies between species. It usually involves the control of females by competing males, and cooperative protection of the young, sometimes made easier by breeding synchronisation (**OESTRUS SYNCHRONY**), and the use of **PROTECTIVE FORMATIONS**. *See also* **HAREM**

heredity The transmission of characters from one generation to the next by **GENES**.

heritability The extent to which a trait is passed from one generation to another; the degree to which a **PHENOTYPE** is genetically influenced and subject to selection (Table H1).

hermaphrodite An individual which possesses both male and female sex organs simultaneously.

Table H1 Heritability: estimated heritability of selected cattle traits. Source: BREEDPLAN (http://breedplan.une.edu.au)

Trait	Heritability (%)
Milk	10
Weaning growth	20
Gestation length	22
Eye muscle area	25
400-day weight	30
Fat depth	30
Birth weight	40
Scrotal size	42

hernia A protrusion of part of an organ through the membrane that normally contains it. Commonly occurs in the stomach, bowel, bladder, liver, kidney, uterus.

herpetofauna Amphibian and reptilian species, especially those found in a particular geographical area.

herpetology The scientific study of amphibians and reptiles.

herps Abbreviation of **HERPTILES (1)**.

herptile
1. An amphibian or reptile.
2. Relating to amphibians and reptiles, e.g. herptile species.

heterodont dentition An assemblage of teeth of different types: **INCISORS, CANINES, MOLARS, PREMOLARS**. *See also* **DENTITION**. *Compare* **HOMODONT DENTITION**

Heterodontiformes An order of fishes: horn (bullhead) sharks.

heterogametic sex Within a species, that sex which has two different types of sex chromosome, e.g. XY. *See also* **SEX DETERMINATION**. *Compare* **HOMOGAMETIC SEX**

heterosexuality Sexual attraction or sexual behaviour between a male and a female of the same species. *Compare* **HOMOSEXUALITY**

heterosis, hybrid vigour The improvement in traits such as size, weight, growth rate, milk production, etc. resulting from the increased heterozygosity that occurs in F_1 generation crosses from two inbred lines. *See also* **HETEROZYGOUS, OUTBREEDING**

heterospecific An individual of another species to the one under discussion, i.e. one that is not a **CONSPECIFIC**.

heterotherm An animal that is intermediate between a **POIKILOTHERM** and a **HOMEOTHERM** in the manner in which it maintains its body temperature. For example, the counter-current heat exchange systems of some fishes allows them to maintain the muscles used for swimming at a temperature which is higher than that of the rest of the body. *See also* **ADAPTIVE HETEROTHERMY**

heterotroph An organism that obtains its energy and materials by feeding on other organisms, i.e. all animals and fungi, some bacteria and some flowering plants which catch and digest insects. *Compare* **AUTOTROPH**

heterotrophic Relating to a **HETEROTROPH**.

heterozygosity *See* **HETEROZYGOUS**

heterozygote An individual organism that possesses one **DOMINANT ALLELE** and one **RECESSIVE ALLELE** for a particular **GENE** (Aa). *Compare* **HOMOZYGOTE**

heterozygous The state of possessing two different **ALLELES** for a particular **GENE**. For example, with respect to gene A, having the genotype Aa. The term may be used to describe an individual organism or a particular **LOCUS**. A **POPULATION** in which a high proportion of the individuals are **HETEROZYGOTES** is said to exhibit high heterozygosity. *Compare* **HOMOZYGOUS**

Hexanchiformes An order of fishes: six- and seven-gill sharks.

hibernaculum, hibernarium A place, such as a cave, where an animal hibernates, e.g. bats, snakes, bears. At Ueno Zoological Gardens, in Tokyo, an exhibit has been created that allows visitors to observe Japanese black bears (*Ursus thibetanus japonicus*) during **HIBERNATION**. Strictly speaking, bears do not hibernate but undergo a period of **WINTER LETHARGY**.

hibernarium *See* **HIBERNACULUM**

hibernation A form of **DORMANCY** that occurs in some mammals in winter to avoid harsh conditions. It characteristically involves a lowering of body temperature and slowing down of metabolic processes. The animal lives on food reserves stored in the body during the summer. *See also* **HIBERNACULUM, TORPOR, WINTER LETHARGY**

hide
1. A small, collapsible, camouflaged shelter or a small building used for observing, photographing and filming wildlife, especially birds, usually through small windows. Used in the wild and in some zoo exhibits, especially for waterfowl. Often placed overlooking water bodies or shorelines. Sometimes built at an elevated location, often on stilts.
2. The skin of a large animal, especially mammals. The hides of some rare species are used to make products that are illegally traded, e.g. elephant hide is used to make boots, wallets and briefcases. *See also* **CITES, PELT**.

higher vertebrates Mammals and birds. *Compare* **LOWER VERTEBRATES**

Highland Wildlife Park A safari park and zoo which specialises in keeping past and present Scottish wildlife. It was opened in 1972 and has been run by the **ROYAL ZOOLOGICAL SOCIETY OF SCOTLAND (RZSS)** since 1986.

high-speed camera A video camera used to record behaviour at a higher frequency than that at which it will be viewed so that the action is slowed down. Used to slow down behaviour for analysis, e.g. the movement of wings during flight. *Compare* **TIME-LAPSE CAMERA**

high-tension wire fence A barrier made of a series of vertical or horizontal parallel strands of thin, high-tensile wire suspended between metal bars. Usually used for birds and small animals which do not climb. Visitors focus beyond the wire so that it appears almost invisible.

hill farming Extensive farming of animals, usually sheep (sometimes cattle), in upland areas in the UK, especially the Lake District and Scotland.

hindgut fermenter Herbivorous mammals (e.g. horses, rhinos and tapirs) that possess relatively simple guts in which the food is enzymatically digested in the stomach and small intestine. It then passes to the large intestine and the caecum where microbes ferment the cellulose in the food. *See also* **FERMENTATION**. *Compare* **BULK GRAZER, RUMINANT**

Hinduism and animals Most Hindus are vegetarian and no Hindu will eat beef: cows are sacred to Hindus. Some Hindu temples keep **SACRED ANIMALS** including elephants. Most Hindus believe that animals are lower forms of life than humans. Some Hindu gods take the form of animals (e.g. *Ganesh*) (Fig. A9).

Hiodontiformes An order of fishes: mooneyes.

hip dysplasia The result of a deformity of the hip joint that occurs during growth. Arthritic changes in the joint may result in lameness. Some breeds of dog are genetically predisposed to hip **DYSPLASIA**. It predominantly affects larger breeds such as the German Shepherd, St. Bernard and Labrador Retriever.

hippodrome
1. An arena for equestrian shows.
2. A building used for variety theatre or a **CIRCUS (1)**.
3. An open-air stadium consisting of an oval course used for horse and chariot races in ancient Greece and Rome. *See also* **CIRCUS MAXIMUS, COLOSSEUM**

His, bundle of Fibres in the heart that conduct **ACTION POTENTIALS** from the **ATRIOVENTRICULAR (AV) NODE** to the **VENTRICLES (2)** to initiate contraction of the ventricular muscles.

hispid Possessing short strong hairs or bristles.

histamine A substance that triggers the immune response. It is released by mast cells during an **ALLERGIC REACTION**. It causes the contraction of smooth muscle, increased permeability of capillaries (allowing white cells and proteins to enter the tissues) and dilation of blood vessels. *See* **ANTIHISTAMINE**

histology The scientific study of tissues.

histopathology The study of disease at the tissue level.

History of Animals A book written by **ARISTOTLE** in 350 BC in which he classified organisms into an

H

hierarchical system ('Ladder of Life') based on the complexity of their structure and functioning.

HLLE Head and lateral line erosion. A type of skin erosion that occurs in some captive fishes.

hoarding The storage of things, especially food, in a single cache (larder hoarding) or several caches scattered throughout the home range or territory (scatter hoarding). Larder hoarding is practised by some small mammals, birds and insects. Many mammals and birds practise scatter hoarding, e.g. crows, squirrels and foxes. Food is generally hoarded in order to overcome periods of low food availability or to protect food from scavengers.

hobble The action of tying together the legs of an animal with **HOBBLES**, or tying up a single foot, so that the animal has difficulty in walking. A temporary measure taken to prevent escape.

hobbles Straps or ropes used to bind together the two fore feet or two hind feet of an animal to prevent it from escaping. *See also* **FLIGHT RESTRAINT, TETHER**

hobby farm A small farm owned by someone whose primary occupation is not farming and who keeps it as a pastime, often while living and working elsewhere.

hog
1. A **PIG**.
2. A young **SHEEP** that has not yet been shorn.
3. The wool from such a sheep.
4. To trim a **HORSE'S** mane short so that it is bristly.

hog board *See* **PIG BOARD**

hog cholera *See* **CLASSICAL SWINE FEVER**

hog flu *See* **H1N1 VIRUS**

hog-dogging *See* **PIG-DOGGING**

holding pen An enclosure where animals are kept temporarily for some particular reason, e.g. prior to release to the wild.

holding power A measure of the amount of time a visitor spends viewing an **ANIMAL EXHIBIT**. This may be measured as the total number of seconds a person stops at an exhibit divided by the minimum number of seconds necessary to read and see the exhibit. *See also* **ATTRACTING POWER**

holistic Emphasising the whole of something and the interdependence of its composite parts. *See also* **HOLISTIC VETERINARY MEDICINE**

holistic veterinary medicine Animal care which is based on **HOLISTIC** principles (holism). This may include the use of **HOMEOPATHY, ACUPUNCTURE, HERBAL VETERINARY MEDICINE**, other types of natural medicine or a combination of any or all of these.

Holoarctic region *See* **FAUNAL REGIONS**

Holocephali A subclass of cartilaginous fishes: chimaeras.

holotype *See* **TYPE SPECIMEN**

holt A shelter made by otters.

home range The area routinely used by an animal while going about its normal activities. It does not normally include areas used during migration or dispersal: '. . . *that area traversed by an individual in its normal activities of food gathering, mating, and caring for the young*' Burt (1943). *Compare* **TERRITORY**

homeopathy, homoeopathy A system of treatment which involves giving the patient highly diluted substances to trigger the body's natural healing mechanisms. It is based on the principle that a substance which causes **SIGNS** if taken in large doses, can be used as treatment in small amounts for those signs. The technique has no scientific basis.

homeostasis, homoeostasis The tendency of an organism (or biological system) to maintain itself in a state of stable equilibrium.

homeotherm, endotherm, homiotherm An animal that maintains its body temperature physiologically, independent of fluctuation in the environment and within a relatively narrow range, i.e. mammals and birds. *Compare* **POIKILOTHERM**

homeothermic, homiothermic Of or relating to a **HOMEOTHERM**.

homing Returning to an area that is identified as 'home' after travelling elsewhere, for example after a foraging trip or a **MIGRATION (1)**. *See also* **COMPASS SENSE, HOMING PIGEON, NAVIGATION**

homing pigeon A pigeon which is used for racing or to carry messages (e.g. in time of war) because of its ability to return to its place of origin using a sophisticated navigation system.

homiotherm *See* **HOMEOTHERM**

homiothermic *See* **HOMEOTHERMIC**

homodont dentition **DENTITION** in which all the teeth are of the same type, e.g. as in crocodilians. *Compare* **HETERODONT DENTITION**

homoeopathy *See* **HOMEOPATHY**

homoeostasis *See* **HOMEOSTASIS**

homogametic sex Within a species, that sex which has two similar types of sex chromosome, e.g. XX. *See also* **SEX DETERMINATION**. *Compare* **HETEROGAMETIC SEX**

Homogocene An informal term used by some scientists to describe a new geological period in which the homogenisation of nature results from the loss of global biodiversity and the predominance of pest and weed species in many regions of the world. *See also* **ANTHROPOCENE**

homologous
1. An anatomical structure in one type of animal is said to be homologous with a structure in another type of animal when both have a similarity of form, position etc. due to descent from a common ancestor, even if the structures have different functions in the adult, e.g. the ear ossicles in mammals are homologous with bones involved in jaw attachment in fishes. *Compare* **ANALOGOUS**
2. The two **CHROMOSOMES** which form a pair in **DIPLOID** organisms are said to be homologous.

homosexuality A sexual preference for, or sexual behaviour performed with, members of the same

Fig. H5 Hoof pick.

sex. Exhibited by a wide range of species from bonobos (*Pan paniscus*) to elephants. *See also* **BULLING**. *Compare* **HETEROSEXUALITY**

homozygosity *See* **HOMOZYGOUS**

homozygote An organism that possesses two identical alleles for a particular gene, either two **DOMINANT ALLELES** (AA) or two **RECESSIVE ALLELES** (aa). *Compare* **HETEROZYGOTE**

homozygous The state of possessing two identical **ALLELES** for a particular **GENE**. For example, with respect to gene A, having the **GENOTYPE** AA or aa. The term may be used to describe an individual organism or a particular **LOCUS**. A **POPULATION** in which a high proportion of the individuals are **HOMOZYGOTES** is said to exhibit high homozygosity. *Compare* **HETEROZYGOUS**

hoof knife A specialised knife (often double-sided) used to trim the hooves of animals, especially horses, ungulates etc.

hoof pick A tool used for removing stones and other unwanted objects from the hooves of a horse (Fig. H5).

hoofstock A collective term for **EQUIDS** and **ARTIODACTYLS**.

hopper
1. A funnel-shaped device used to feed food or other material into a container, onto a surface or onto the ground. *See also* **SILO**
2. Immature stage of locusts, in which the wings are not fully developed. Used as food for captive lizards etc.
3. A grasshopper (especially in the USA).

horizontal fence An enclosure barrier made of mesh placed horizontally on the ground over a shallow pit.

hormone A substance which regulates physiological activity in the body. It is secreted by an **ENDOCRINE GLAND** and may act on all of the cells in the body or on cells distant from the gland by which it is produced. It is transported in the blood and active at low concentrations.

hormone assay *See* **ASSAY**

hormone implant A small quantity of a hormone placed under the skin which is gradually released into the blood to affect physiology, e.g. a **CONTRACEPTIVE**.

hormone therapy A veterinary treatment that adds, blocks or removes hormones to treat physiological imbalances in the body, including the use of hormones as **CONTRACEPTIVES** or to stimulate reproduction.

horn A pointed bony projection on the head of some species, covered with keratin, and used for fighting, digging etc. Usually present as a single pair but some wild and domestic species have two or more pairs. *Compare* **ANTLER**

Hornaday, William Temple (1854–1937) An American zoologist and conservationist who was the first director of the **SMITHSONIAN NATIONAL ZOOLOGICAL PARK** and later director of the New York Zoological Park (Bronx Zoo). He co-founded the American Bison Society.

hornbill ivory Material obtained from the **CASQUE** of the helmeted hornbill (*Rhinoplax vigil*). Also called hot-ting and golden jade. Valued as a carving material especially in Japan and China. *Compare* **IVORY**

horse
1. An individual of the species *Equus calabus*. Horses have been widely used as **BEASTS OF BURDEN** and for riding for pleasure, **HORSE RACING,** by the military (**WAR HORSES**) and as **POLICE HORSES**.
2. A generic term for a member of the family Equidae.
3. The **AMERICAN STUD BOOK** 2008 defines a horse as '*an* **ENTIRE** *male five years old or older*'.
4. In Great Britain, under s.6(4) of the **RIDING ESTABLISHMENTS ACT 1964** a horse includes '*any mare, gelding, pony, foal, colt, filly or stallion and also any ass, mule or jennet.*'

horse breeds There are several hundred horse breeds in existence today which have been bred for a variety of purposes from racing to pulling heavy loads. Some of the major breeds and their origin are listed in Table H2 (Fig. S1).

horse racing A competitive race between racehorses ridden by a person who controls their speed with a **WHIP**. *See also* **BRITISH HORSERACING AUTHORITY (BHA)**, **GRAND NATIONAL**, **WHIPPING REGULATION OF**

horse type A category of horse defined by its speed and temperament: **HOTBLOOD**, **WARMBLOOD**, **COLDBLOOD**, **PONY**.

horse whispering Techniques of natural horsemanship in which the horse whisperer builds a rapport with the horse and communicates with it using methods derived from observations of herds of free-ranging horses.

H

Table H2 Major horse breeds and their origin/homeland.

Breed	Origin/homeland	Breed	Origin/homeland
Akhal-Teké	Turkmenistan	Irish draught	Ireland
Altér Real	Portugal	Italian heavy draught	Italy
American Shetland	USA	Kabardin	Russia
Andalusian	Spain	Karabakh	Azerbaijan
Anglo-Arab	Britain	Karacabey	Turkey
Appaloosa	USA	Kladruber	Czech Republic
Arab	Arabian Peninsula	Knabstrup	Denmark
Ardennais	France	Konik	Poland
Australian pony	Australia	Latvian	Latvia
Australian stock horse	Australia	Lipizzaner	Austria
Azteca	Mexico	Lusitano	Portugal
Barb	North Africa	Malopolski	Poland
Bashkir	Ural Mountains, Russia	Mangalarga	Brazil
Brabant	Belgium	Mecklenburg	Germany
Breton	France	Métis trotter	Russia
Budyonny	Russia	Missouri foxtrotter	USA
Camargue	France	Morgan	USA
Carthusian	Spain	New Forest	Britain
Caspian	Iran	Nonius	Hungary
Clydesdale	Britain	Noriker	Austria
Connemara	Ireland	Norman cob	France
Criollo	South America	Northlands	Norway
Dales pony	Britain	Oldenburg	Germany
Danish warmblood	Denmark	Orlov trotter	Russia
Danubian	Bulgaria	Pasa Fino	Puerto Rico
Dartmoor pony	Britain	Percheron	France
Døle-Gudbrandsdal	Norway	Peruvian stepping horse	Peru
Don	Russia	Pony of the Americas	USA
Dutch draft	Holland	Przewalski's horse	Central Asia
Dutch warmblood	Holland	Quarter horse	USA
East Bulgarian	Bulgaria	Saddlebred	USA
East Friesian	Germany	Salerno	Italy
Exmoor pony	Britain	Schleswig heavy draft	Germany
Falabella	Argentina	Selle Français	France
Fell pony	Britain	Shagya Arab	Hungary
Fjord	Norway	Shetland pony	Shetland
Frederiksborg	Denmark	Shire	Britain
French Anglo-Arab	France	Standardbred	USA
French trotter	France	Suffolk Punch	Britain
Friesian	Holland	Swedish warmblood	Sweden
Furioso	Hungary	Swiss warmblood	Switzerland
Galiceño	Mexico	Tennessee walking horse	USA
Gelderlander	Holland	Tersky	Russia
Gidran	Hungary	Thoroughbred	Britain
Gotland	Sweden	Trakehner	Poland
Hackney	Britain	Turkmene	Iran/Turkmenistan
Haflinger	Austria	Ukrainian riding horse	Ukraine
Hanoverian	Germany	Welsh cob	Wales
Highland pony	Britain	Welsh mountain pony	Wales
Holstein	Germany	Welsh pony of cob type	Wales
Huçul	Poland	Westphalian	Germany
Icelandic	Iceland	Wielkopolski	Poland

horticulture The art or science of growing plants especially in gardens. Important in creating **NATURALISTIC EXHIBITS** in zoos and exhibits of interesting and rare plants.

host In **PARASITOLOGY**, an organism in, or upon which a particular stage of a **PARASITE** lives and from which it obtains food, shelter etc. *See also* **INTERMEDIATE HOST, PRIMARY HOST, VECTOR**

host-specific In relation to **PARASITES**, one that only targets certain species during particular stages of their life cycle.

hot desert *See* **DESERT**

hot wire *See* **ELECTRIC FENCE**

hotblood A **HORSE TYPE** consisting of **THOROUGHBREDS** and **ARABS**.

house rabbit A pet **RABBIT** that lives inside, and roams freely around, a dwelling house. Especially popular in the United States.

HPA axis The hypothalamic–pituitary–adrenal axis. Involved in the physiological response of the body to **STRESS**. *See also* **CORTISOL**

HSUS *See* **HUMANE SOCIETY OF THE UNITED STATES (HSUS)**

human exhibits, human zoo The exhibition of people for entertainment. The Romans exhibited dwarfs, giants, **HERMAPHRODITES** and deformed individuals and sold them as slaves. Peoples from different cultures were also exhibited at the World's Fair and Native Americans were exhibited by Wild Bill Cody's **WILD WEST SHOW** and by Carl **HAGENBECK** at his zoo. In January 2010 the bones of five Kawésqar tribesmen, who had been taken to Europe in 1881 by Hagenbeck to be exhibited at fairs and circuses, were repatriated to Chile. At one time they were exhibited at Berlin's zoo. Their bones were discovered in the Anthropological Institute of Zurich University. *See also* **OTA BENGA**

human–animal conflict A term used to describe the conflict that arises when animals and people compete for resources in the natural environment, e.g. when elephants raid crops and destroy houses and other property in Africa and Asia.

human–animal relationships (HAR) The associations and interaction between animals and people. These take many forms. We hunt them and farm them for food, fibres and leather. We conserve them and keep them as companion animals. We use them for clothing, entertainment and pleasure and to study life processes (investigating their behaviour and physiology as models of our own). Animals are also important in our ethical system. We consider their health and welfare, we regulate their exploitation with laws, we use them as symbols (in folklore, art and religion) and we value their diversity. *See also* **ANTHROZOOLOGY**

human–companion animal bond The connection and attachment formed between some people and their **COMPANION ANIMALS**. *See also* **CANINE SEPARATION-RELATED BEHAVIOUR (SRB)**.

humane education The promotion of compassion towards all animals and caring for the environment.

Humane Research Council A non-profit organisation based in the USA which is dedicated to conducting research in the fields of human–animal studies and critical animal studies. It produces a range of free resources for the public.

humane slaughter pistol A firearm designed for slaughtering animals humanely. *See also* **CAPTIVE BOLT, ELECTRIC STUNNER**

Humane Society of the United States (HSUS) An organisation, established in 1954, which works to prevent cruelty, exploitation and neglect of animals. Although much of its work concerns domestic and farm animals, it has also taken action against roadside zoos in America. The society operates a number of animal rehabilitation centres and shelters.

humane twitch *See* **TWITCH (2)**

humerus (humeri *pl.*) The single **LONG BONE** which forms the upper forelimb in tetrapod vertebrates.

humidity A measure of the quantity of water vapour in the atmosphere. Relative humidity is the ratio of the partial pressure of water vapour to the saturated vapour pressure of water (in the pertaining conditions), expressed as a percentage. In other words, the amount of water in the air compared with what it could hold at that temperature.

humidity pump A device which adds moisture to the atmosphere, thereby raising the humidity. May be used to control humidity automatically in egg incubators and **INTENSIVE CARE UNITS (ICUs) (1)**.

hummingbird feeder A feeding device designed to provide a sugary solution to hummingbirds through small apertures often simulating flowers.

hunt
1. To pursue game for food or sport. *See also* **AERIAL GUNNING, FOX HUNTING, HUNTING ACT 2004, INTERNET HUNTING, OTTER HUNTING, SPORT HUNTING, WILDFOWL**
2. A hunting expedition.
3. Those taking part in a hunting expedition, e.g. the Quorn Hunt.

Hunting Act 2004 A law in England and Wales which bans hunting a wild mammal with dogs except in certain specified conditions, e.g. hunting rats, flushing from cover with dogs for the purpose of enabling hunting with a bird of prey, recapturing a wild mammal that has escaped. The Act completely bans **HARE COURSING**, and effectively bans deer hunting and **FOX HUNTING** with dogs. In some situations dogs may be used for stalking, flushing wild mammals from cover and to protect game birds (exempt hunting). *See also* **BURNS REPORT, PROTECTION OF WILD MAMMALS (SCOTLAND) ACT 2002**

hunting concession An authority given to an individual or organisation to **HUNT (1)** animals on a particular area of land.

H

hunting park An area set aside for hunting (usually for large mammals), often exclusively for the use of royalty. *See also* ROYAL HUNTING GROUNDS

hunting quota The number of animals that may be legally killed by a hunter or hunters. Usually determined by scientific analysis of a population so that hunting has no effect on its future survival. *See also* CAMPFIRE, DUCK STAMP, FISHING QUOTA, MAXIMUM SUSTAINABLE YIELD (MSY)

Hurnik–Morris pig housing system A system of pig production which allows socially co-ordinated eating and resting, controlled and socially undisturbed feed intake and physical exercise. It provides housing for sows in small groups and regular access to boars. Food is provided in an individual, non-competitive environment using electronically controlled feeding compartments. *See also* EDINBURGH FAMILY PEN SYSTEM

husbandry The science, art and skill of raising animals, especially in farms, zoos and other captive situations.

husbandry guidelines Guidance on the keeping of particular taxa of animals in captivity, usually produced as a HUSBANDRY MANUAL. *See also* RSPCA WELFARE STANDARDS

husbandry manual In relation to zoos, a document, in printed or electronic form, which provides information and instructions relating to the HUSBANDRY of a specific TAXON of animals, e.g. tigers, elephants, kiwis. Usually produced by individual persons who have appropriate skills, knowledge and experience, or groups of such persons, including keepers, nutritionists and vets. Generally contains sections on natural history, distribution, nutrition, breeding, enclosure requirements, behaviour, enrichment, diseases, health and welfare problems. Some husbandry guidelines are available on-line, others are available through national or REGIONAL ASSOCIATIONS OF ZOOS.

Huxley, Sir Julian Sorrell (1887–1975) An influential British evolutionary biologist who was the Secretary of the ZOOLOGICAL SOCIETY OF LONDON (ZSL), the first Director of the UNITED NATIONS EDUCATIONAL, SCIENTIFIC AND CULTURAL ORGANIZATION (UNESCO) and a founding member of the WORLD WILDLIFE FUND (WWF).

hybrid
1. An individual which results from a cross between parents of the same species with different genotypes.
2. An individual which has been produced by the mating of organisms of two different SPECIES, varieties, RACES or BREEDS, e.g. a tigon is a cross between a male tiger and a female lion; a liger is a cross between a male lion and a female tiger. Some hybrids are fertile, others are infertile.
3. Any HETEROZYGOTE.
See also HETEROSIS, HYBRIDISATION

hybrid vigour *See* HETEROSIS

hybridisation The process of creating HYBRIDS as a result of the mating of individuals from different SPECIES or BREEDS. Hybridisation threatens the existence of some species in the wild. Hybrids are produced for the hobby market by some commercial fish breeders. *See also* BENGAL, DZO, FERAL CAT, GOODER, LIGER, MULE, RED WOLF, SAVANNAH CAT BREED, TIGON, TOYGER, WHOLPHIN

hydatid disease A disease caused by a small tapeworm, *Echinococcus granulosus*, in its cystic larval stage. The usual hosts are the dog and fox, but it can affect cattle, sheep, horses, wallabies and other species, including humans. Eggs released in the faeces of infected animals are eaten by grazing animals. Infection may also occur by drinking contaminated water or by exposure to wind-blown eggs. Swallowed eggs hatch in the intestines and migrate to the liver in the blood. Some remain there, forming hydatid cysts, while others may form cysts in the lungs, spleen, kidney, bone marrow cavity or the brain. Routine WORMING of animals is essential for the control of *Echinococcus*.

Hydracoidea An order of mammals: hyraxes

hydrocortisone The pharmaceutical name for CORTISOL: an anti-inflammatory drug.

hydrophone An underwater microphone, used to record sounds made by animals such as whales and dolphins when they are submerged.

hydroponic fodder Plant food (e.g. barley and legumes) grown in soil-less trays in a nutrient solution, especially for horses when unable to access pasture. Used for valuable animals, especially racehorses.

hydrostatic organ *See* SWIM BLADDER

hydrotherapy The treatment of disorders, injuries and diseases by the external use of water, especially the development of movement in water to treat disability (Fig. H1). *See also* HYDROTHERAPY SPA, UNDERWATER TREADMILL

hydrotherapy spa, hydrotherapy bath A vessel used for HYDROTHERAPY treatment (Fig. H1).

hydrotherapy water walker *See* UNDERWATER TREADMILL

hygiene
1. Sanitary practices and principles.
2. The practice or study of preserving health and preventing the spread of disease, especially by keeping bodies and the surroundings clean.

Hymenoptera A large order of insects: bees, wasps and ants.

hyperbaric oxygen therapy (HBOT) *See* OXYGEN THERAPY

hypercapnia, hypercapnea, hypercarbia An abnormally high level of carbon dioxide in the blood.

hyperkeratosis A thickening of the skin.

hyperplasia An increase in the size of an organ or tissue due to an increase in the number of cells.

hypersensitivity Excessive, undesirable, damaging and sometimes fatal reactions produced by the immune system. *See also* ALLERGIC REACTION, ANAPHYLACTIC SHOCK, HISTAMINE

hyper-sexual activity An abnormally high frequency of sexual behaviour.

hypertension High blood pressure. *Compare* **HYPOTENSION**

hyperthermia Elevated body temperature resulting from a failure of thermoregulation. May be fatal. *Compare* **HYPOTHERMIA**

hypertonic Of a liquid, having an **OSMOTIC PRESSURE** higher than a liquid with which it is being compared. *Compare* **HYPOTONIC, ISOTONIC**

hypertrophy An enlargement or overgrowth of a tissue or organ caused by an increase in cell size. Often occurs in one of a pair of organs when the other is removed, e.g. kidneys, ovaries. *Compare* **HYPOTROPHY**

hyperventilation Ventilation that exceeds metabolic demands. *Compare* **HYPOVENTILATION**

hypnosis A type of **TONIC IMMOBILITY** which may be induced in an animal, such as a bird, lizard or lamb, by holding it down on the ground for a short period. Similar to the **DEATH FEIGNING** that some animals exhibit when captured by a predator.

hypocapnea *See* **HYPOCAPNIA**

hypocapnia, hypocapnea, hypocarbia An abnormally low level of carbon dioxide in the blood.

hypocarbia *See* **HYPOCAPNIA**

hypodermic syringe A syringe (a hollow cylinder with a plunger) with a thin, hollow needle used for injecting drugs under the skin or into blood vessels, or for taking blood samples.

hypomagnesaemia *See* **GRASS STAGGERS**

hypotension Low blood pressure. *Compare* **HYPERTENSION**

hypothalamus The region of the brain located below the **THALAMUS** and above the **PITUITARY GLAND**. It acts as a control centre for the **AUTONOMIC NERVOUS SYSTEM (ANS)** and hormonal activity. It is involved in **THERMOREGULATION**.

hypothermia Excessively low body temperature resulting from a failure of **THERMOREGULATION**. May be fatal. *Compare* **HYPERTHERMIA**

hypothesis A statement or proposition which is assumed to be true for the sake of argument or as the basis for experimentation or investigation of the evidence; a provisional explanation. *See also* **NULL HYPOTHESIS (H_0), OCCAM'S RAZOR, SCIENTIFIC METHOD**

hypotonic Of a liquid, having an **OSMOTIC PRESSURE** lower than a liquid with which it is being compared. *Compare* **HYPERTONIC, ISOTONIC**

hypotrophy
1. The progressive degeneration of a tissue or organ as a result of a loss of cells.
2. Incomplete growth; atrophy.
3. Wasting of the body which may be the result of a nutritional deficiency.
Compare **HYPERTROPHY**

hypoventilation Ventilation that does not meet metabolic demands. *Compare* **HYPERVENTILATION**

hypoxia Insufficient oxygen in the blood or tissues. *See also* **ASPHYXIA**

hysterectomy Removal of the uterus due to disease or as a means of contraception.

H

Fig. I1 I is for iguana. Iguanas (*Iguana*) which had been rescued from inappropriate owners by a reptile group.

Dictionary of Zoo Biology and Animal Management: A guide to terminology used in zoo biology, animal welfare, wildlife conservation and livestock production, First Edition. Paul A. Rees.
© 2013 John Wiley & Sons, Ltd. Published 2013 by John Wiley & Sons, Ltd.

IATA Regulations *See* INTERNATIONAL AIR TRANSPORT ASSOCIATION (IATA)

ichthyology The scientific study of fishes.

ICU *See* INTENSIVE CARE UNIT (ICU)

ICZN *See* INTERNATIONAL COMMISSION ON ZOOLOGICAL NOMENCLATURE (ICZN)

identification guide A book or other document used to identify organisms and particularly to distinguish between similar SPECIES (or other TAXA). Usually consists of a series of questions to which there may be a limited number of possible responses. Each response leads to further questions until the organism is identified. Often contains diagrams or photographs to assist in identification. *See also* DICHOTOMOUS KEY, FIELD GUIDE

IFAW *See* INTERNATIONAL FUND FOR ANIMAL WELFARE (IFAW)

Illegal, Unreported and Unregulated Fishing Regulations 2008 Council Regulation (EC) No 1005/2008 of 29 September 2008 establishing a Community system to prevent, deter and eliminate illegal, unreported and unregulated fishing. It prohibits trade with the EU in fishery products stemming from IUU fishing and requires the introduction of a certification scheme applying to all trade in fishery products with the Community.

image intensifier An optical instrument used to create images in low light levels, either held to the eye like a telescope, or attached to a video camera.

imitation The copying of the behaviour of one animal by another. *See also* SOCIAL FACILITATION, TEACHING

immature
1. Generally, an animal which is not fully grown or developed.
2. A bird which is not yet an adult and is unable to breed. Some species pass through several distinct PLUMAGES during immaturity.

immediate environment The environment immediately outside the individual's body. *See also* BIOLOGICAL ENVIRONMENT, ENVIRONMENT. *Compare* PHYSICAL ENVIRONMENT

immersion exhibit An animal enclosure in which visitors become part of the exhibit, and feel as if they are entering the habitat of the animals, e.g. a walk-through aviary which simulates a tropical forest containing free-flying birds. *See also* NATURALISTIC EXHIBIT, WOODLAND PARK ZOO

immigration The movement of individuals into a POPULATION. *Compare* EMIGRATION

Immobilon, etorphine (hydrochloride), M99 An ANALGESIC (1) which is chemically similar to morphine. It is used as a tranquilliser for large animals and can produce CATATONIA at very low doses. It is extremely dangerous in even very small quantities and its use is regulated by law. The reversing agent is REVIVON.

immune reaction, immune response The reaction in the body which results from the recognition and binding of an antigen by its specific antibody or by a lymphocyte that has been previously sensitised. *See also* ALLERGIC REACTION, AUTOIMMUNE DISEASE

immune response *See* IMMUNE REACTION

immune system The tissues and cells of the body that neutralise and attack the ANTIGENS associated with disease organisms. *See also* LEUCOCYTES, LYMPHATIC SYSTEM

immunisation The protection of an animal from a disease by producing IMMUNITY to it by injecting the animal with an ANTISERUM or a treated ANTIGEN.

immunity The state of having sufficient body defences to prevent infection and disease, either naturally occurring or as a result of VACCINATION.

immuno-contraception A method of CONTRACEPTION which involves the administration of a vaccine that causes an animal to become temporarily infertile. This may be achieved by a number of methods, e.g. by causing immunity to GONADOTROPHIN-RELEASING HORMONE (GnRH) which is involved in gamete production in both sexes. Used to control wild elephants in South Africa and deer populations in the USA and Canada.

immunoglobulin *See* ANTIBODY

immunology The scientific study of IMMUNITY and the defence mechanisms used by the body to resist infection and disease.

immunotherapy The treatment of disease by enhancing or suppressing the patient's immune system. Used especially in the treatment of allergies and cancer.

implantation The process by which a zygote becomes attached to the uterine wall in a female mammal. *See also* DELAYED IMPLANTATION

import licence In relation to animals, a licence required by law (e.g. CITES) to import a species into a particular country. *Compare* EXPORT LICENCE

imprinting A type of LEARNING which occurs during a sensitive period in a young animal's development. A normal function of this behaviour is to identify the mother and form an attachment to her. A HAND-REARED mammal may imprint on its caretaker and follow her around. Young ducklings exhibit a 'following response' and will imprint on objects such as coloured boots or balloons and follow them instead of their mother if exposed to them during the sensitive period. If an animal imprints on a human it may not develop normal sexual behaviour and may be useless for breeding.

impure animal *See* UNCLEAN ANIMAL

in calf (in foal, in kid, in lamb, in pig) Pregnant, in cattle (in horses, goats, sheep, pigs). *Compare* EMPTY

In Defense of Animals (IDA) An international non-profit animal protection organisation based in California and founded in 1983. It is dedicated to ending the exploitation and abuse of animals by defending their rights, welfare and habitats.

I

inbreeding Reproduction between close relatives. The converse of **OUTBREEDING**. *See also* **INBREEDING DEPRESSION**

inbreeding coefficient The probability that two genes at any specific locus in an individual are identical, by descent, from the common ancestor(s) of the two parents.

inbreeding depression The lowered fitness, or vigour, of inbred individuals, compared with their non-inbred counterparts, which results in, for example, increased incidence of congenital diseases, reduced milk production in cattle. *See also* **HETEROSIS**. *Compare* **OUTBREEDING DEPRESSION**

incisor A sharp, chisel-edged tooth at the front of the mouth in mammals, used for biting and nibbling.

inclusive fitness The inclusive fitness of an organism is the sum of its **DIRECT FITNESS** (the contribution of its own offspring to future generations) and its **INDIRECT FITNESS** (the contribution to the next generation made by supporting related individuals). Altruistic behaviour directed towards close relatives may increase the inclusive fitness of an individual. *See also* **ALTRUISM**

incubate
1. To hatch eggs by keeping them warm, either naturally, e.g. by a bird sitting upon them, or artificially in an **INCUBATOR (2)**.
2. To encourage the development of microbes by creating favourable and controlled conditions, e.g. in a culture medium in a laboratory.
3. To maintain a newborn animal, especially one that is ill or born before the usual end of **GESTATION**, in a controlled environment (e.g. in relation to temperature, humidity and oxygen concentration) in order to provide optimal conditions for growth and development.
4. Experiencing the growth and reproduction of a disease-causing organism during the development of an infection from the time it enters the body until the time clinical **SIGNS (1)** and (in humans) **SYMPTOMS** appear. *See also* **INCUBATION TIME**

incubation medium Material placed in the bottom of a reptile incubator as a substitute for the sand or soil in which eggs might be laid in the wild, e.g. Vermiculite.

incubation time, latent period The time between infection with disease-causing organisms and the appearance of clinical **SIGNS** and (in humans) **SYMPTOMS**. Also called the latent period.

incubator
1. A container used to keep underdeveloped animals warm (particularly prematurely born mammals).
2. A machine, similar to an oven, used to keep eggs at an appropriate temperature for embryonic development and eventual hatching (Fig. I2). A still-air incubator requires the manual turning of eggs and allows the temperature and humidity to be quickly adjusted. A forced-draught incubator blows air past

Fig. I2 Incubator: reptile egg incubator.

the eggs at a set temperature and humidity and mechanically turns the eggs at intervals.

incus *See* **EAR OSSICLES**

independent variable A variable whose value determines that of other variables. These are referred to as **DEPENDENT VARIABLES** (Fig. D4). For example, temperature (the independent variable) may determine the frequency of stereotypic behaviour exhibited by an animal (the dependent variable).

Independent Zoo Enthusiasts' Society (IZES) Founded in 1995. It exists to foster interest in all good zoos and to counter misinformation about zoos. It has produced *The IZES Guide to British Zoos & Aquariums* and has a number of other publications.

indeterminate growth Growth which does not have a predetermined end point and continues indefinitely. Body growth is indeterminate in reptiles, most fish and many molluscs in the sense that they grow rapidly when very young and then continue to grow slowly. *Compare* **DETERMINATE GROWTH**

index case The earliest documented case in an epidemiological study of a disease outbreak.

Index of Exempted Dogs *See* **DANGEROUS DOGS ACT 1991**

indigenous species A species which naturally occurs in a particular area.

indirect fitness The component of **FITNESS** an animal gains by providing assistance to relatives in the production and support of their young rather than by producing its own offspring. *See also* **ALTRUISM**. *Compare* **DIRECT FITNESS, INCLUSIVE FITNESS**

individual distance The distance between two individuals of the same species at which aggression or **AVOIDANCE** behaviour occurs.

Indomalayan region *See* **FAUNAL REGIONS**

induced ovulator, reflex ovulator An animal that is stimulated by the process of **COPULATION** to release eggs from the ovary. This mechanism effectively synchronises **OVULATION** with the presence of sperm in the female's reproductive tract and increases the probability that **FERTILISATION** will occur. It occurs in cats, bears and many other carnivores, camelids and some rodents. Some desert rodents exhibit induced ovulation in response to the occurrence of green vegetation and associated changes in nutritional factors in the diet. Ovulation may be induced artificially by **HORMONES** to synchronise egg release with the timing of **ARTIFICIAL INSEMINATION (AI)**. *Compare* **SPONTANEOUS OVULATOR**

induction chamber *See* **ANAESTHESIA INDUCTION CHAMBER**

industrial melanism *See* **MELANISM**

infanticide The killing of a young animal by a parent or other animal. Sometimes occurs in inexperienced parents. Occurs in the wild in some species, e.g. lions, when males take over a pride, hamadryas baboons (*Papio hamadryas*). *See also* **CANNIBALISM, DECEPTIVE SWELLING, FOETICIDE, FRATRICIDE, PROLICIDE**

infection The invasion of the body by a disease-causing organism.

infectious
1. Of or relating to a disease caused by bacteria, viruses etc., capable of causing an infection.
2. Of or relating to an animal or person, capable of transmitting a disease to another.

inflammation A protective response of a **TISSUE** to the presence of an **ALLERGEN**, an injury, disease or infection. The affected area may become red, swollen and painful. Heat emitted from an inflamed area may be detected by a **THERMAL IMAGING CAMERA**.

influenza, flu Any of a number of viral infections of humans and other animals characterised generally by fever and respiratory signs. *See also* **AVIAN INFLUENZA, H1N1 VIRUS**

infradian rhythm A pattern in time that has a period of considerably more than 24 hours, e.g. a lunar cycle. *See also* **CIRCADIAN RHYTHM, CIRCANNUAL RHYTHM, LUNAR RHYTHM, ULTRADIAN RHYTHM**

infrared lamp A lamp used as a heat source to keep animals warm when kept in enclosures where the ambient temperature is too low, e.g. rock hyraxes (*Procavia capensis*), meerkats (*Suricata suricatta*), reptiles. *See also* **HEAT MAT**

infrared radiation, IR radiation Electromagnetic radiation of wavelengths between 700 nm and 1 mm, between the visible and microwave parts of the electromagnetic spectrum. 'Near infrared' light is closest in wavelength to visible light while 'far infrared' is closest to the microwave region. Far infrared radiation is thermal (i.e. experienced as heat) and can be used to measure the temperature of objects and animals (*see* **THERMAL IMAGING CAMERA**). Special cameras can be used to photograph and video animals at night using IR radiation.

Near infrared is used to activate electronic devices, e.g. in IR remote controllers. Passive IR sensors are used to activate **TRAIL MONITORS** and **CAMERA TRAPS**. IR is invisible to most species, but pit vipers (Crotalidae) possess sensory pits with which they can use IR to produce images of their warm-blooded (**HOMEOTHERMIC**) prey. *See also* **INFRARED LAMP**

infrared thermography *See* **THERMAL IMAGING CAMERA**

infrasound Sound waves at a frequency below the range of human hearing (lower than 20 Hz) and experienced as vibrations by humans, e.g. the low-frequency sounds elephants use to communicate. *Compare* **ULTRASOUND**

infusion
1. Something which is poured into or introduced into an animal, e.g. an intravenous infusion. *See also* **GIVING SET**
2. A solution produced by infusing (soaking to release flavour etc.).

infusion pump A device for moving liquid along a tube during infusion and capable of delivering a very accurate dose rate. *See also* **SYRINGE PUMP**

ingestion The taking in of food into the gut.

inhalation pneumonia, aspiration pneumonia **PNEUMONIA** caused as a result of the inhalation of a foreign substance, e.g. food, vomit, orally administered medicines.

injection The process of introducing something (e.g. a drug) into the body using a **HYPODERMIC SYRINGE**.

innate Inborn.

innate behaviour
1. Behaviour which is inborn, i.e. highly heritable.
2. Behaviour which is normal, natural and part of the animal's repertoire, but in this sense may be learned (for example, by **IMPRINTING**).

inner ear The innermost part of the vertebrate ear. In mammals this consists of the cochlea – which converts sound pressure from the bones in the middle ear into nerve impulses that allow the perception of sound by the brain – and the vestibular system (semi-circular canals), which is responsible for balance. *See also* **EAR OSSICLES, PINNA (1)**

inoculation
1. The introduction of a **VACCINE** or antigenic substance into the body to stimulate the body to produce **ANTIBODIES**, to create **IMMUNITY** to a particular disease.
2. The introduction of a microbe into a culture medium.

in-ovo Within the egg. *Compare* **IN-VITRO, IN-VIVO**

***in-ovo* feeding** Literally, feeding in the egg. Hatchability rates and chick quality may be improved in some species of birds (e.g. turkeys) by injecting a feeding solution into the amnion of later-term embryos a day before **INTERNAL PIPPING** occurs. It also increases **GLYCOGEN** reserves, advances gut

Fig. I3 Insect trap in cloud forest in Costa Rica.

development and promotes the development of muscles.

in-ovo **vaccination** Vaccination which occurs while a bird embryo is still inside the egg. Used in poultry production.

insect trap A device used for catching and killing insect pests (e.g. cockroaches) or for catching insects alive for the purpose of identification and ecological sampling (e.g. a moth trap (Fig. I3)).

Insecta A class of arthropods: insects.

insectarium A **VIVARIUM** containing insects or arachnids.

insecticide A naturally occurring or artificially manufactured chemical which is toxic to and capable of killing insects when applied in appropriate concentrations. Some insecticides, such as **DDT**, are concentrated as they pass along **FOOD CHAINS** and are toxic to **APEX PREDATORS**. *See also* **ORGANO-CHLORINE, ORGANOPHOSPHATE**

Insectivora An order of mammals: insectivores (hedgehogs, gymnures, golden moles, moles, tenrecs, solenodons, shrews, shrew moles, desmans).

insectivore

1. An animal that feeds on insects.

2. A member of the mammalian order **INSECTIVORA**.

insight learning A type of **LEARNING** in which a connection is made between various stimuli or events, all-of-a-sudden, in order to produce a novel response rather than as a result of trial and error. For example, a **CHIMPANZEE** arranging boxes on top of each other and climbing them in order to reach food suspended out of reach, or joining sticks together to make a long stick to reach food.

in-situ In its original position. *Compare* **EX-SITU**

in-situ **conservation**

1. Conservation which takes place in the wild.

2. The UN **CONVENTION ON BIOLOGICAL DIVERSITY 1992 (CBD)** (Art. 2) defines 'in-situ conservation' as *'the conservation of ecosystems and natural habitats and the maintenance and recovery of viable populations of species in their natural surroundings and, in the case of domesticated or cultivated species, in the surroundings where they have developed their distinctive properties.'*

Compare **EX-SITU CONSERVATION**

instantaneous scan sampling *See* **SCAN SAMPLING**

instar A stage in the development of arthropods between each moult. *See also* **ECDYSIS**

instinct An **INNATE** propensity to exhibit particular behaviours.

Institute for Animal Health (IAH) An institute of the Biotechnology and Biological Sciences Research Council (BBSRC) in the UK. The institute's Pirbright Laboratory is located in Surrey. It works to contain, control and eliminate viral diseases of animals and improve farm animal health in the UK by studying infectious diseases in cattle, sheep, pigs and poultry.

Institute for Conservation Research A research institute operated by the Zoological Society of San Diego which incorporates the **FROZEN ZOO®** (2).

Institute of Zoology (IoZ) A research institute which is part of the **ZOOLOGICAL SOCIETY OF LONDON (ZSL)** and is located in Regent's Park, London. It conducts a wide range of conservation-related research on animal species and their habitats in many different countries.

institutional collection plan (ICP) A collection plan produced by an individual zoo which defines which species it plans to keep and in what numbers. *See also* **REGIONAL COLLECTION PLAN (RCP)**

instrumental conditioning *See* **OPERANT CONDITIONING**

insulin A hormone which is produced by the islets of Langerhans in the **PANCREAS** and stimulates the utilisation of glucose by the cells thereby lowering the blood sugar level. Deficiency of insulin causes the condition **DIABETES MELLITUS**. *Compare* **GLUCAGON**

insurance population, assurance population A **POPULATION** of a species kept in captivity as insurance (assurance) against the species becoming extinct in the wild. The concept is used by zoos as a justification for keeping and breeding rare species.

intelligence Those aspects of behaviour which demonstrate a wide range of problem solving abilities

Fig. I4 Intensive care unit.

involving reasoning. *See also* **COGNITION, INSIGHT LEARNING**

intensive care brooder A container designed for use in rearing newly-hatched **ALTRICIAL** birds (e.g. parrots and birds of prey) in which air is filtered and temperature and humidity are controlled. See also **BROODER**.

intensive care unit (ICU)
1. A chamber in which small animals are kept when recovering from a surgical procedure, are unwell or for some other reason (e.g. premature birth), which allows the strict control of environmental variables, especially temperature, humidity and ventilation (Fig. I4).
2. A unit within a veterinary hospital, or other similar place, which has specialised equipment and is dedicated to providing high quality care for critically ill animals and those requiring frequent observations.

intensive farming The method of farming widely used in the developed world which utilises sophisticated machinery to produce food from large fields with high densities of animals, often kept in buildings. *Compare* **EXTENSIVE AGRICULTURE**

intention movement An incomplete behaviour that provides information about a behaviour that an animal is about to perform, e.g. a crouch that occurs before a leap.

inter alia A Latin term meaning 'among other things'. Often used in legal documents where the list of things given is not exhaustive. For example, the **WILD MAMMALS (PROTECTION) ACT 1996** prohibits, *inter alia*, the crushing, stabbing and burning of wild mammals.

interactive exhibit *See* **INTERACTIVE SIGN**

interactive sign, interactive exhibit A sign or other display with which a visitor may interact by pressing a button to activate a light or sound (e.g. a **VOCALISATION**), lifting a flap to expose the answer to a question etc. The exhibit may be a simple mechanical or electrical device, or it may be electronic.

inter-birth interval *See* **MEAN BIRTH INTERVAL**

interbreeding Breeding between two distinct groups or organisms, e.g. individuals from different **POPULATIONS** of the same species or between members of different **SPECIES** or **SUBSPECIES** (**HYBRIDISATION**).

intermediate host, secondary host An organism which harbours a stage in the development of a parasite for a short period of time during which some developmental (larval) stage is completed, before it is transmitted to the **PRIMARY HOST**. *See also* **FLUKE DISEASE**

intermediate neurone, relay neurone A **NEURONE** which connects a **MOTOR NEURONE** with a **SENSORY NEURONE**. It is located within the spinal cord.

internal fixator A device for supporting a skeletal break from inside the body; it may include bone plates, pins, screws and rods. *Compare* **EXTERNAL FIXATOR**

internal pipping In hatching chicks, the process of pushing the beak into the air cell in the egg. The air in this cell is then used to inflate the lungs. No outward signs of hatching are visible at this stage but internal pipping may be visible during **CANDLING** and chirping may be heard. *Compare* **EXTERNAL PIPPING**

International Air Transport Association (IATA) The global trade organisation for inter-airline co-operation in promoting safe, reliable, secure and economical air services. It produces regulations for the safe handling of animals on aircraft with which zoos and other organisations responsible for shipping animals must comply.

International Code of Zoological Nomenclature *See* **INTERNATIONAL COMMISSION ON ZOOLOGICAL NOMENCLATURE (ICZN)**

International Commission on Zoological Nomenclature (ICZN) An organisation founded in 1895 whose purpose is to provide and regulate a uniform system of zoological nomenclature to ensure that every animal has a unique and universally accepted **BINOMIAL NAME**. This is an important task because more than 2000 new genus names and 15 000 new species names are added to the zoological literature every year. The Commission publishes the *International Code of Zoological Nomenclature* which contains the rules for allocating scientific names to animals.

I

The Code regulates nomenclature only – the way names are created and published – not TAXONOMY (CLASSIFICATION). The Commission publishes a journal – *The Bulletin of Zoological Nomenclature* – which contains papers about problems related to the naming of animals which are resolved by the Commission. The Commission's work directly affects studies of BIODIVERSITY and CONSERVATION as it is essential that scientists can properly name and classify the animals with which they work.

International Congress of Zookeepers (ICZ) A global network of zookeepers and other professionals in the field of wildlife care and conservation which provides them with a means of sharing their experience and knowledge. *See also* KEEPER ASSOCIATION

international convention, international treaty A legally binding international agreement between at least two states, e.g. CITES. *See also* CONFERENCE OF THE PARTIES, INTERNATIONAL LAW, PROTOCOL (2)

International Convention for the Regulation of Whaling 1946 An international agreement whose purpose is to provide for the proper conservation of whale stocks and thereby facilitate the orderly development of the whaling industry. It established an International Whaling Commission (IWC) whose purpose is to encourage and organise research on whales and whaling, collect and analyse statistical information and study and disseminate information concerning methods of maintaining and increasing whale populations. The Commission also regulates the use of whale stocks by defining protected and unprotected species, establishing open and closed seasons, designating WHALE SANCTUARIES and determining the maximum catch in each season and the type of gear and apparatus that may be used.

International Fund for Animal Welfare (IFAW) An organisation founded in 1969 which saves individual animals, animal populations and habitats worldwide. It provides assistance to animals in need including companion animals, wildlife and livestock, and rescues animals in the wake of disasters. It also campaigns against cruelty, e.g. commercial whaling and seal hunts.

International Game Fish Association (IGFA) A not-for-profit organisation committed to the conservation of game fish and the promotion of responsible and ethical angling practices.

international law The law that regulates the way that states behave towards each other, largely by way of INTERNATIONAL CONVENTIONS, e.g. the UN CONVENTION ON BIOLOGICAL DIVERSITY 1992 (CBD). States are bound by international law by consent. This should not be confused with the NATIONAL LAWS of foreign countries, e.g. the laws of France or the laws of China. International law is difficult to enforce. In some cases states may impose TRADE SANCTIONS on others in order to encourage them to comply with their international legal obligations. *Compare* EUROPEAN LAW

International League for the Protection of Horses *See* WORLD HORSE WELFARE

International Primate Protection League (IPPL) A non-profit organisation dedicated to protecting primates. Founded in 1973, it works to expose primate abuse and combat international traffickers. IPPL operates a sanctuary for gibbons in South Carolina and supports primate rescue efforts worldwide.

International Society for Anthrozoology (ISAZ) A non-profit organisation that supports the scientific and scholarly study of HUMAN–ANIMAL RELATIONSHIPS. It publishes the academic journal *ANTHROZOÖS*.

International Species Information System (ISIS) A computer database, accessible via the internet, of the animal holdings of member institutions (i.e. most of the major zoos and aquariums in the world). It maintains information about the animals held by around 825 institutions in 76 countries. This is used to manage breeding programmes and includes data on sex, age, parentage, place of birth and cause of death. The ISIS central database contains information on over two million animals from almost 15 000 taxa and 10 000 species. ISIS was founded in 1973 by Drs. Ulysses SEAL and Dale Makey. Initially 51 zoos in North America and Europe contributed to the database, which was hosted by Minnesota Zoo for 15 years. ISIS currently distributes a number of different databases: ARKS4, (*see* ANIMAL RECORD KEEPING SYSTEM (ARKS)), EGGS, MEDICAL ANIMAL RECORD KEEPING SYSTEM (MEDARKS), SINGLE POPULATION ANALYSIS AND RECORD KEEPING SYSTEM (SPARKS), REGIONAL ANIMAL SPECIES COLLECTION PLAN (REGASP), ZOOLOGICAL INFORMATION MANAGEMENT SYSTEM (ZIMS).

International Stud Book Committee Formed in 1976 to develop and provide a consistent approach to accurately recording and publishing details of all THOROUGHBREDS, under the auspices of the International Federation of Horseracing Authorities. *See also* AMERICAN STUD BOOK

international studbook In relation to wild animal species, one of around 200 STUDBOOKS for an endangered or rare species which are kept under the auspices of the WORLD ASSOCIATION OF ZOOS AND AQUARIUMS (WAZA). They contain the most complete and accurate global data on the pedigree and demography *of ex-situ* populations. *See also* GLOBAL ANIMAL SURVIVAL PLAN (GASP), *INTERNATIONAL ZOO YEARBOOK (IZYB)*

international treaty *See* INTERNATIONAL CONVENTION

International Union for the Conservation of Nature and Natural Resources (IUCN) The IUCN was founded in 1948, originally as the International Union for the Protection of Nature (IUPN), but changed its name to the IUCN in 1956. From 1990 the organisation came to be known as the World Conservation Union, but this name is no longer commonly used. The IUCN is a partnership of states, government agencies and non-governmental organisations of over 1000 members and almost 11 000 volunteer scientists spread across over 160 countries. The IUCN seeks to assist in the conservation of **BIODIVERSITY** and ensure the responsible and equitable use of the world's natural resources. It has Official Observer Status at the UN General Assembly, and its headquarters are in Gland, near Geneva, Switzerland. The IUCN produces Red Data books which provide information about the conservation status of a wide range of species. *See also* **RED LIST**

International Whaling Commission (IWC) *See* **INTERNATIONAL CONVENTION FOR THE REGULATION OF WHALING 1946**

International Whaling Convention *See* **INTERNATIONAL CONVENTION FOR THE REGULATION OF WHALING 1946**

International Wildlife Rehabilitation Council (IWRC) The IWRC is the leading developer of professional training for wildlife care providers in North America and abroad. Its goal is to educate its members, colleagues and the public on issues relating to wildlife care and conservation. The IWRC publishes the *JOURNAL OF WILDLIFE REHABILITATION*.

International Zoo Educators Association (IZE) IZE is an organisation which is dedicated to expanding the educational impact of zoos and aquariums. It aims to improve zoo education programmes and provide access to the latest thinking, techniques and information in conservation education. It publishes the *International Zoo Educators Journal.*

International Zoo News (IZN) A journal which carries articles about zoos and animals kept in zoos which are of general interest to the zoo community, but which have not been peer-reviewed. Many of the articles are written by keepers and are concerned with husbandry techniques, enrichment and breeding records.

International Zoo Veterinary Group (IZVG) The largest and best-known independent freelance zoological veterinary practice in the world. It is based in Keighley, West Yorkshire, UK. *See also* **TAYLOR**

International Zoo Yearbook (IZYB) A book published by the **ZOOLOGICAL SOCIETY OF LONDON (ZSL)** approximately annually. It was first published in 1960 and has a different theme each year, e.g. New World Primates, Ungulates, Aquariums, Zoo Animal Nutrition, Amphibian Conservation. It contains reports of original research and review articles and lists **INTERNATIONAL STUDBOOKS**. It also contains a list of all of the major zoos in the world and provides information about the animals held (numbers of species and numbers of individuals),

annual **VISITOR ATTENDANCE**, area of the zoo, number of staff and the names of senior staff.

internet hunting, computer-assisted remote hunting, cyber hunting A system which allows a person to shoot an animal on a private game farm using a rifle fitted with a webcam which is remotely controlled via the internet. Animals are confined behind fences and often fed at feeding stations at predetermined times. Payment is made through a website and trophy mounts are prepared and sent to the customer. Supporters argue that the system allows disabled people to participate in hunting. Internet hunting is banned in many states in the USA. *See also* **CANNED HUNTING**

interobserver reliability *See* **BETWEEN-OBSERVER RELIABILITY**

INTERPOL The world's largest police organisation with 190 member countries. The word 'INTERPOL' is a contraction of 'international police'. *See also* **INTERPOL WILDLIFE CRIME WORKING GROUP**

INTERPOL Wildlife Crime Working Group The **INTERPOL** Wildlife Crime Working Group initiates and leads projects to combat the poaching, trafficking, or possession of legally protected flora and fauna, e.g. Operation RAMP was a worldwide operation involving 51 countries across five continents against the illegal trade in reptiles and amphibians. It resulted in arrests worldwide and the seizure of thousands of animals as well as products worth more than 25 million euros.

interpretation SIGNS (2) and other devices used to explain an exhibit. *See also* **INTERACTIVE SIGN**

interpreter *See* **PRESENTER**

interspecies communication Communication between different species. This has been achieved between humans and apes using **LEXIGRAM BOARD** and American Sign Language (ASL). *See also ALEX* **THE AFRICAN GREY PARROT, INTERSPECIES SEMANTIC COMMUNICATION,** *KANZI, KOKO, NIM CHIMPSKY*

interspecies embryo transfer The transfer of an embryo from one species to another **(SURROGATE (2))** species. Scientists at the Audubon Nature Institute's Center for Research of Endangered Species (ACRES) in New Orleans created an African wild cat (*Felis silvestris lybica*) by cloning in 2003. The African wildcat was born as a result of the world's first successful interspecies frozen/thawed **EMBRYO TRANSFER (ET)**. The embryo was transferred to a domestic cat. This was the first time a wild carnivore had been cloned. In the same year ACRES produced the world's first caracal (*F. caracal*) from a frozen embryo. It has since produced further cloned African wildcats and Arabian sand cats (*F. margarita*) by cloning. Surrogate species have now been successfully used to produce offspring by embryo transfer in a number of endangered species. For example, the eland (*Taurotragus oryx*) has been used as a surrogate for the bongo (*Tragelaphus euryceros*) and the

I

domestic horse has been used for **PRZEWALSKI'S HORSE** (*EQUUS FERUS PRZEWALSKII*) (Fig. P12).

interspecies semantic communication The ability of some species to understand the signals made by other species. For example, West African Diana monkeys (*Cercopithecus diana*) and Campbell's monkeys (*C. campbelli*) frequently form mixed-species associations in the wild and Diana monkeys understand the signals of male Campbell's monkeys when they make **ALARM CALLS** indicating the presence of two of their main predators: crowned eagles (*Stephanoaetus coronatus*) and leopards (*Panthera pardus*). *See also* **INTERSPECIES COMMUNICATION**

interspecific Occurring between individuals of different species. *Compare* **INTRASPECIFIC**

interspecific competition Competition between individuals of different species for resources, e.g. food, sheltering places. May occur in **MULTI-SPECIES EXHIBITS** in zoos. *Compare* **INTRASPECIFIC COMPETITION**

interstitial cell-stimulating hormone (ICSH) *See* **LUTEINISING HORMONE (LH)**

interstitial fluid, tissue fluid The fluid that occurs between the cells in the body. It consists of **PLASMA** without the large **PROTEINS** and is the basis of **LYMPH**.

intertidal zone The area of a beach which lies between the low watermark and the high watermark. Organisms that occur here are covered by seawater at high tide and exposed to the atmosphere during low tide, so they are adapted to survive **DESICCATION**.

intervertebral disc A disc-shaped piece of connective tissue (fibrocartilage) which occurs between adjacent vertebrae in the spine allowing movement and providing shock absorption. When the central part of the disc herniates through the outer fibrous rings of the disc the condition is referred to as an intervertebral (or spinal) disc herniation or 'slipped disc'.

intervertebral disc herniation *See* **INTERVERTEBRAL DISC**

intramuscular Within the muscle, e.g. an intramuscular injection.

intraspecific Occurring between individuals of the same species. *Compare* **INTERSPECIFIC**

intraspecific competition Competition between individuals of the same species for resources, e.g. food, mates. May occur in zoos in animals kept in large groups or in social groups where some individuals are dominant to others. *See also* **STOCKING DENSITY**. *Compare* **INTERSPECIFIC COMPETITION**

intraspecific variation Variation within a **SPECIES**; differences between the individuals which make up a species. This is affected by **NATURAL SELECTION** and is the raw material of **EVOLUTION**.

intravenous Into a vein, e.g. an intravenous injection.

intrinsic rate of natural increase The constant *r* in the exponential equation for population increase in an unlimited **ENVIRONMENT**; the rate of increase

per head of the population under specific physical conditions in an unlimited environment where the effects of increased population **DENSITY** do not need to be considered. The instantaneous rate of increase of such a population is given by:

$$\frac{dN}{dt} = rN$$

where N = the number of individuals present at a particular point in time, and t = time.
See also **EXPONENTIAL GROWTH**, **FECUNDITY**

introduced species

1. A species that has been accidentally or intentionally released into an area where it does not naturally occur, e.g. the European starling (*Sturnus vulgaris*) was introduced into the USA; the grey squirrel (*Sciurus carolinensis*) was introduced into the UK. Most countries have laws which prohibit the release of exotic species into the environment.

2. In Great Britain, s14(1) of the **WILDLIFE AND COUNTRYSIDE ACT 1981** makes the release of **EXOTIC (1)** species illegal – '. . . . . *if any person releases or allows to escape into the wild any animal which: (a) is of a kind not ordinarily resident in and is not a regular visitor to Great Britain in a wild state; or (b) is included in Part I of Schedule 9, he shall be guilty of an offence.'*
See also **INTRODUCTION**

introduction

1. An **INTRODUCED SPECIES**.

2. The act of releasing an introduced species.

3. In the US Executive Order 13112 of February 3, 1999 an introduction is defined as '. . . *the intentional or unintentional escape, release, dissemination, or placement of a species into an ecosystem as a result of human activity.'*
See also **POPULATION SUPPLEMENTATION**, **REINTRODUCTION**, **TRANSLOCATION**

intromission The insertion of the **PENIS** into the **VAGINA** during **COPULATION**.

intromittant organ A penis or similar organ for transferring sperm from the male to the female.

intubation The insertion of a tube into the lung through the mouth or nose to assist with breathing (by preventing the **COLLAPSE (1)** of the airway), or to supply oxygen or anaesthetic gases, or into a hollow organ or passageway for some other purpose.

invasive species

1. A species which spreads aggressively to new areas outside its natural range and is unwanted. It may come to dominate an area.

2. In US Executive Order, 13112 of February 3, 1999 an invasive species means '. . . *an alien species whose introduction does or is likely to cause economic or environmental harm or harm to human health.'*

invertebrate A general term used for an animal that does not possess a **VERTEBRAL COLUMN**, i.e. animals that are not **VERTEBRATES**. This is not a **TAXON**.

in-vitro In relation to biological processes or techniques, one performed outside a living organism in an artificial environment provided by scientific apparatus, e.g. a test tube or Petri dish. *Compare* IN-OVO, IN-VIVO

in-vitro **fertilisation** An ASSISTED REPRODUCTIVE TECHNOLOGY (ART) which involves the FERTILISATION of an ovum with sperm outside the female animal's body in culture medium and (usually) subsequent IMPLANTATION of the ZYGOTE in a female's uterus.

in-vivo In relation to biological processes or techniques, one performed inside a living organism. *Compare* IN-OVO, IN-VITRO

involuntary muscle *See* SMOOTH MUSCLE

involuntary nervous system *See* AUTONOMIC NERVOUS SYSTEM (ANS)

iodine An element which is a component of thyroid hormones that regulate tissue metabolism. Deficiency may occur in species feeding on vegetation growing on soil with low iodine concentration. Deficiency causes enlargement of the thyroid gland and retarded growth. Iodine occurs in high levels in some marine products.

IR radiation *See* INFRARED RADIATION

iron Important in the structure of haemoglobin, myoglobin and enzymes. Deficiencies are rare. Dietary iron may be important in animals that have suffered blood loss and are anaemic. Many mammalian milks are low in iron and hand-rearing using cow's milk may lead to deficiency.

irrigate
1. To wash, e.g. to pass sterile water over a wound in order to clean it.
2. To provide water to plants to encourage growth, particularly during drought conditions.

ischial callosity A thickened area of skin which overlies the posterior pelvis (ischial tuberosity) in OLD WORLD MONKEYS and some apes. May have evolved to allow them to sit comfortably and stably on thin branches.

ISIS *See* INTERNATIONAL SPECIES INFORMATION SYSTEM (ISIS)

Islam and animals Islam teaches respect for animals and that they exist for the benefit of humans. It requires that Muslims do not treat them cruelly or over-work them, engage in blood sports (unless hunting for food) or animal fighting. *See also* RELIGIOUS SLAUGHTER

island biogeography The study of the ecology and evolution of animals living on islands. Useful in studying SPECIATION and EXTINCTION. Studies of islands have helped in the design of nature reserves because often the species within them are isolated as if on an island.

isolation In the process of evolution, the separation of one population of a species from others which may eventually result in the formation of new species. *See also* SPECIATION

isolation ward *See* BARRIER NURSING

isotonic Of or relating to two solutions which have the same solute concentration. *Compare* HYPERTONIC, HYPOTONIC

iteroparity The condition of exhibiting multiple reproductive cycles during an animal's lifetime. *Compare* SEMELPARITY

IUCN *See* INTERNATIONAL UNION FOR THE CONSERVATION OF NATURE AND NATURAL RESOURCES (IUCN)

IUDZG International Union of Directors of Zoological Gardens; the forerunner of the WORLD ASSOCIATION OF ZOOS AND AQUARIUMS (WAZA).

IUU fishing Illegal, unreported and unregulated fishing. *See also* ILLEGAL, UNREPORTED AND UNREGULATED FISHING REGULATIONS 2008

iVET Birth Monitor An electronic device used to detect birth in cattle remotely. The device is inserted into the cow's vagina and is pushed out by the calf and amniotic sac when calving begins. The change in the sensor's position causes a text message to be sent to alert the farmer of the birth.

ivory Dentine. A hard white material which forms the TUSKS of elephants but also found in the walrus, hippopotamus, sperm whale and narwhal. Previously widely used to make ornaments and piano keys. International trade in elephant ivory is controlled by **CITES**. This has effectively banned international trade in ivory since 1989. In the same year President Daniel arap Moi ignited a 7-metre-high pile of ivory in Nairobi National Park to signal the Kenyan Government's determination to stop elephant poaching. *Compare* HORNBILL IVORY

IWC *See* INTERNATIONAL CONVENTION FOR THE REGULATION OF WHALING 1946

I

Fig. J1 J is for jaguar (*Panthera onca*).

Dictionary of Zoo Biology and Animal Management: A guide to terminology used in zoo biology, animal welfare, wildlife conservation and livestock production, First Edition. Paul A. Rees.
© 2013 John Wiley & Sons, Ltd. Published 2013 by John Wiley & Sons, Ltd.

Jacobson's organ *See* VOMERONASAL ORGAN

Jainism and animals Jainism is an Indian religion that teaches non-violence to all living things. Jains may not use leather or silk products. Jains operate animal shelters called panjrapoles, originally established for disabled cattle.

Jambo A male lowland gorilla at Jersey Zoo (now DURRELL) who stood guard over a small boy who fell into the gorilla enclosure until he was rescued. He has been credited with changing public attitudes to gorillas, who were previously considered by many to be aggressive.

Jane Goodall Institute UK A charity, founded by Dr. Jane GOODALL, to conduct research on chimpanzees, promote their conservation and educate the public about their needs.

Jardin des Plantes The Ménagerie du Jardin des Plantes is a botanical garden in Paris which contains a small zoo. When the VERSAILLES MENAGERIE closed, the animals were offered to the Jardin des Plantes. In 1793 the Jardin was incorporated into the new Muséum National d'Histoire Naturelle and in 1803 the zoologist George CUVIER assumed responsibility for the MENAGERIE. *See also* BOIS DE VINCENNES

jaundice A condition which causes a yellowing of the skin and whites of the eyes resulting from an excess of bilirubin – a breakdown product of HAEMOGLOBIN – in the blood which is often a sign of liver disease.

Jaws A fictional film, made in 1975, about a rogue great white shark (*Carcharodon carcharias*) who killed people and destroyed boats in the waters around a small seaside community in the United States. The film engendered negative attitudes towards sharks and portrayed them as highly dangerous, calculating and vengeful.

Jersey Zoo *See* DURRELL

jesses Thin straps (usually leather) used in falconry to tether a bird of prey by the legs (Fig. J2). Used when on the glove or training, and to secure the bird to its perch when outside the aviary.

jizz A term used by BIRDERS to indicate the distinctive character of a bird species that may assist with identification, e.g. lethargic or highly active.

JNCC *See* JOINT NATURE CONSERVATION COMMITTEE (JNCC)

Jockey Club (GB) An organisation founded in 1750 in London which established rules for HORSE RACING on Newmarket Heath which were eventually adopted all over Britain. The Jockey Club became the official governing body for racing in Britain. This function is now performed by the BRITISH HORSERACING AUTHORITY (BHA). The Jockey Club is made up of about 130 members, owns 14 racecourses and the NATIONAL STUD.

Jockey Club (USA) The breed registry for THOROUGHBRED horses in the United States, Canada and Puerto Rico. The organisation is dedicated to

Fig. J2 Jesses on a red-tailed hawk (*Buteo jamaicensis*).

the improvement of thoroughbred breeding and racing.

joey A young kangaroo, wallaby or koala.

Joint Management of Species Programme (JMSP) The structure under which captive breeding programmes were run in Britain and Ireland until 2006. They were overseen by a special group of experts set up by the BRITISH AND IRISH ASSOCIATION OF ZOOS AND AQUARIUMS (BIAZA) called the Joint Management of Species Committee (JMSC) and smaller, more specifically focused TAXON ADVISORY GROUPS (TAGs). BIAZA TAGs have now become TAXON WORKING GROUPS (TWGs).

Joint Nature Conservation Committee (JNCC) The public body that advises the UK Government on UK-wide and international nature conservation.

Jones & Jones Architects and Landscape Architects Ltd. A US company based in Seattle, Washington, that specialises in the design of zoo exhibits. Recent projects have included *Kaziranga Forest Trail* (Dublin Zoo), *Campo Gorilla Reserve* (Los Angeles Zoo), *Africa Live!* (San Antonio Zoo), ARCTIC RING OF LIFE (Detroit Zoo). *See also* WOODLAND PARK ZOO

journal *See* ACADEMIC JOURNAL

Journal of Applied Animal Welfare Science (JAAWS) An academic journal which publishes original studies on a wide range of welfare issues relating to domestic, farm and zoo animals. Published by the ANIMALS AND SOCIETY INSTITUTE.

J

Journal of Wildlife Rehabilitation A journal produced by the **INTERNATIONAL WILDLIFE REHABILITATION COUNCIL (IWRC)** since 1978 which provides reliable, relevant and useful information for wildlife rehabilitators and others involved in the care, treatment and conservation of wildlife.

Journal of Zoo and Aquarium Research An open-access online journal published by the **EUROPEAN ASSOCIATION OF ZOOS AND AQUARIA (EAZA)** whose aim is the rapid publication of a wide range of research relating to zoo and aquariums.

Journal of Zoo and Wildlife Medicine An academic journal that publishes original research findings, clinical observations and case reports in the field of veterinary medicine dealing with captive and free-ranging wild animals.

J-shaped growth A pattern of **POPULATION GROWTH** whereby numbers increase exponentially and then experience a sudden dramatic crash. Occurs when an organism invades a new area and over-exploits food resources. Also known as boom-and-bust growth (Fig. P8).

Judaism and animals Judaism teaches that animals should be treated with compassion and should not have pain inflicted upon them. However, it allows them to be used for food and clothing and it allows animal experimentation, but only if there is a clear benefit to humans and no unnecessary pain is caused. Animal fighting and hunting for sport are forbidden. *See also* **RELIGIOUS SLAUGHTER**

judicial review A legal process that challenges the manner in which a legal decision has been made (as opposed to the decision itself). In 2008 the **ROYAL SOCIETY FOR THE PREVENTION OF CRUELTY TO ANIMALS (RSPCA)** used the judicial review process to challenge the **DEPARTMENT FOR ENVIRONMENT, FOOD AND RURAL AFFAIRS (DEFRA)** regarding the legality of using **VENTILATION SHUTDOWN** as a method of killing poultry in England. In February 2012 the Badger Trust applied to the High Court for permission to seek judicial review of DEFRA's decision to carry out pilot badger culls in England in an attempt to reduce the incidence of bovine **TUBERCULOSIS**.

Jumbo A male African elephant who is undoubtedly the most famous zoo animal of all time. He was born wild in eastern Sudan, around Christmas 1860. In 1862 he was captured and sold to the **JARDINS**

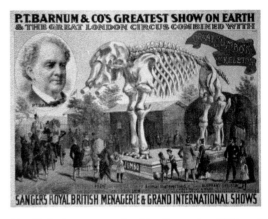

Fig. J3 *Jumbo*: a poster advertising Barnum's exhibition of *Jumbo*'s skeleton. (Source: Library of Congress Prints and Photographs Division, Washington DC.)

DES PLANTES in Paris. In 1865 he was sent to the **ZOOLOGICAL SOCIETY OF LONDON (ZSL)**. In 1882 *Jumbo* was sold by London Zoo to the **BARNUM AND BAILEY** Circus in America for £2000 (£138000 today). On 15 September 1885 he was hit by a train and killed while crossing a railway line in St. Thomas, Ontario, Canada. Barnum continued to make money out of the elephant by touring America with his stuffed body. On 4 April 1889 *Jumbo*'s mounted skin was delivered to Barnum's museum at Tufts College where it remained until 14 April 1975 when it was destroyed by fire (Fig. J3).

Jurassic Park A film released in 1993 about a fictional park where dinosaurs have been recreated using genetic engineering techniques. The story both fuelled concerns about the dangers of **GENETIC ENGINEERING** and acted as a prophecy of the possibilities of using this technology to save endangered species.

juvenile

1. Generally, an animal which has not yet developed into an adult.

2. In ornithology, a bird in its first plumage, before its first moult. The plumage of many juvenile birds lacks colour and is often brown, making it very difficult to identify the species to which they belong.

Fig. K1 K is for kangaroo. Eastern grey kangaroo (*Macropus giganteus*).

Dictionary of Zoo Biology and Animal Management: A guide to terminology used in zoo biology, animal welfare, wildlife conservation and livestock production, First Edition. Paul A. Rees.
© 2013 John Wiley & Sons, Ltd. Published 2013 by John Wiley & Sons, Ltd.

Kant, Immanuel (1724–1804) A late 18th century philosopher who argued that, to be moral, an act itself must have a pure intention behind it, regardless of the final consequences. His 'counter-utilitarian' idea would give greater consideration to the rights of the individual than the welfare of the group. This is the ethical basis for concern about the welfare of individual animals. *See also* **COUNTER-UTILITARIANISM**

Kanzi A male bonobo who was the subject of a language study using a **LEXIGRAM BOARD** conducted by Sue Savage-Rumbaugh at the Language Research Center of Georgia State University. *See also KOKO, NIM CHIMPSKY, WASHOE*

kaolin China clay. Used in some anti-diarrhoeal treatments. May absorb toxins etc. from the gut. Also used as a **POULTICE** for sprains etc. when mixed with glycerine and antiseptic.

keel *See* **CARINA**

keeper

1. A zookeeper. A person who works in a zoo, or similar facility, caring for animals by providing food, cleaning enclosures, keeping animal records, monitoring heath, providing information to visitors and performing other associated duties, sometimes including **TRAINING** animals. Legal definitions vary. *See also* **KEEPER-SCIENTIST**

2. Under section 6(3) of the **ANIMALS ACT 1971**, *'a person is a keeper of an animal if –*

    (a) he owns the animal or has it in his possession; or

    (b) he is the head of a household of which a member under the age of sixteen owns the animal or has it in his possession'.

3. Under the **DANGEROUS WILD ANIMALS ACT 1976** (s.7(1)) a *'person is a keeper of an animal if he has it in his possession; and if at any time an animal ceases to be in the possession of a person, any person who immediately before that time was a keeper . . .* [as defined here] *continues to be a keeper of the animal until another person becomes a keeper* [as defined here].'

4. Under s.2(1) of the **WELFARE OF FARMED ANIMALS (ENGLAND) REGULATIONS 2007** '*"keeper" means any person responsible for or in charge of animals whether on a permanent or temporary basis.*'

keeper area An area within an animal exhibit to which animals have no access and from which a **KEEPER (1)** is able to remotely operate doors between **ANIMAL ENCLOSURES**, prepare food, observe animals and perform other duties.

keeper association An organisation whose purpose is to provide for the interests of zoo keepers (Table K1). It may be concerned with a specific taxon (e.g. the **ELEPHANT MANAGERS ASSOCIATION (EMA)**) or for keepers in a particular geographical region (e.g. **ASSOCIATION OF BRITISH AND IRISH WILD ANIMAL KEEPERS (ABWAK)**). Keeper associations may promote the education of keepers in relation to animal welfare, husbandry

Table K1 Keeper associations.

American Association of Zoo Keepers (AAZK)
Animal Keepers Association of Africa (AKAA)
Association of British and Irish Wild Animal Keepers (ABWAK)
Australasian Society of Zoo Keeping Inc. (ASZK)
Berufsverband der Zootierpfleger (BdZ) – Union of zookeepers (Germany)
Elephant Managers Association (EMA)
European Elephant Keeper and Manager Association
International Congress of Zookeepers (ICZ)
International Rhino Keeper Association (IRKA)
La Asociación Ibérica de Cuidadores de Animales Salvajes (AICAS) – Iberian zookeepers association
Stichting de Harpij (The Harpy Foundation) – Organisation for Dutch and Belgian zoo professionals

and conservation. In some cases a keeper association may publish its own journal, e.g. *RATEL*.

keeper for a day A fund-raising scheme run by some zoos whereby a member of the public pays to work as a keeper for a day with an animal of their choice.

keeper-scientist The name given to keepers at *DURRELL* which acknowledges their role in monitoring animal health, designing **ANIMAL ENCLOSURES** and writing reports, in addition to the normal husbandry duties of a **KEEPER (1)**.

Keiko See FREE WILLY

kelp toy, artificial kelp Artificial kelp (seaweed) used as an enrichment for marine mammal pools. Constructed from long narrow sheets of thin plastic which are fixed to the bottom of a pool (Fig. K2).

kennel

1. A place where dogs are boarded or bred.

2. A small shelter for a dog.

3. To put or keep an animal in a kennel.

4. A place where **HUNT** hounds are housed and cared for by a **KENNELMAN**.

Kennel Club The UK's largest organisation dedicated to the health and welfare of dogs. It was founded in 1873. The Kennel Club provides dog owners with information and advice on dog welfare, health, training and breeding. *See also* **AMERICAN KENNEL CLUB**

kennelman Someone who cares for dogs, especially the hounds kept by a **HUNT (3)**.

keratin A tough, fibrous protein which is produced by the **EPIDERMIS** of vertebrates. It forms the dead outer layers of skin cells and the main component of hair, **NAILS**, **BILLS**, claws, **HORNS** and **FEATHERS**.

Kertesz, Peter Dental consultant to London Zoo who has performed dental operations on a wide range of species from elephants to pandas. He founded **ZOODENT INTERNATIONAL** in 1985 and is also dental consultant to the **INTERNATIONAL ZOO VETERINARY GROUP (IZVG)**. Kertesz is the

Fig. K2 Kelp: kelp toy, artificial kelp.

author of *A Colour Atlas of Veterinary Dentistry and Oral Surgery* (1993).

ketamine A general **ANAESTHETIC (1)** and **ANALGESIC (1)**.

ketonuria The presence of abnormally high levels of ketones in urine as a result of **KETOSIS**.

ketosis The excessive formation of acetone and other ketone bodies in the body. The result of incomplete oxidation of fats which occurs in **DIABETES MELLITUS**; it is also an indicator of starvation.

key *See* **DICHOTOMOUS KEY**

keystone species A term first used by Paine (1969) to describe a species which has a disproportionate effect on the diversity of a biological community for its size and abundance. In tropical forests ants, bees, bats and hummingbirds play keystone roles in pollination and seed dispersal.

kidney One of the paired abdominal organs in **VERTEBRATES** which removes **NITROGENOUS WASTE** from the body by producing **URINE**, and helps the body to maintain osmotic balance by releasing or reabsorbing water as necessary.

kidney failure The cessation of normal kidney function, often as a result of a urinary obstruction, an infectious disease, a physical injury or poisoning, but sometimes as a consequence of a congenital condition or old age. It may be an **ACUTE CONDITION** or a **CHRONIC CONDITION**.

kidney stone *See* **RENAL CALCULUS**

kill
1. To end the life of an organism.
2. The corpse of an animal that has been killed, especially one that has been taken in a hunt, e.g. a lion kill. Some zoos provide their carnivores with the carcasses of prey species, and even live prey, to allow them to exhibit normal feeding behaviour. *See also* **LIVE FEEDING IN ZOOS**

kin group *See* **BOND GROUP**

kin recognition The ability to recognise genetic relatives. This may be the result of **IMPRINTING** at an early age and there may also be a genetic effect. Recognition of relatives is important in the operation of **KIN SELECTION** in some species.

kin selection A form of **NATURAL SELECTION** which favours **GENES** which promote altruistic acts which benefit relatives. The increase in **FITNESS** achieved by the altruist is determined by the probability that his genes are carried by the recipient of his altruistic act, as calculated by the **COEFFICIENT OF (GENETIC) RELATEDNESS**. In some groups animals that live near to each other are likely to be related so it makes sense to help neighbours, for example, by emitting an **ALARM CALL** when a predator approaches even if it draws attention to yourself. **KIN RECOGNITION** is known in some species. *See* **GROUP SELECTION**

kinesis (kineses *pl.*) A movement which is proportional to the strength of a stimulus but the response is not directional. For example, woodlice are more active in dry atmospheres than in humid atmospheres. They are therefore more likely to remain in humid environments and more likely to move quickly through dry environments. The woodlice are not moving towards the areas of high humidity but encountering them by chance. In orthokinesis the speed of movement is proportional to the intensity of the stimulus. In klinokinesis the rate of turning (change of direction) is proportional to the stimulus.

King Kong A film about a fictional giant gorilla who was captured by explorers and taken from Africa to the United States to be exhibited to the public, escaped, destroyed property in New York and was eventually killed. The film portrayed gorillas as highly dangerous and destructive and engendered negative attitudes towards them. The original film was released in 1933 and remade in 2005. *See also* ***JAMBO***

kingdom A group of related **PHYLA**. *See also* **FIVE KINGDOM CLASSIFICATION** (Table C5).

kinship Family relationship; degree of genetic relatedness.

kinship group *See* **BOND GROUP**

kit
1. The young of some animals, e.g. ferret, fox, beaver, rabbit.
2. A slang term for a piece of equipment ('piece of kit'), especially if large and impressive.

K

K

Fig. K3 Knuckle walking: a male western lowland gorilla (*Gorilla gorilla gorilla*) standing with his knuckles supporting his forearms.

kite hack A technique used in the reintroduction of birds of prey to the wild in which the bird is flown to a kite to which food is attached. The food is released when struck by the bird. *See also* **HACKING (1)**, **LURE HACK**

knacker A person who buys old livestock and other animals unfit for human consumption and slaughters them for their hide or other products. *See also* **SLAUGHTERHOUSE**, **SLAUGHTERMAN**

knock-down box
1. A padded area within which a large animal (e.g. a horse) can be safely anaesthetised and allowed to collapse onto the ground.
2. A small transparent plastic chamber used for anaesthetising small animals into which anaesthetic gas is pumped through a tube (gas feed tube).

knock-out mouse A genetically engineered mouse in which a gene has been 'knocked out' (inactivated) in order to study its function. *See also* **ONCOMOUSE®**

knuckle walking A form of **LOCOMOTION** only seen in African **GREAT APES**, whereby the ape walks on all four limbs with the body weight partially supported on the middle **PHALANGES** (Fig. K3).

Koko A lowland gorilla who was taught to use American Sign Language by Dr. Penny Patterson. *Koko* learned over 1000 signs and even more spoken English in the only **INTERSPECIES COMMUNICATION** project using this species. *See also* **KANZI**, **NIM CHIMPSKY**, **WASHOE**

Kong A commercially produced plastic enrichment feeder in which food is hidden. Designed for pets, especially dogs, but may be used for small zoo animals, e.g. monkeys.

kopje A rocky outcrop with steep sides and sparse vegetation characteristically found in some African **SAVANNAS**. Sometimes built into naturalistic zoo exhibits. Part of the visitor centre in the **SERENGETI** National Park is constructed around a real kopje, and the hunting dog (*Lycaon pictus*) exhibit at Chester Zoo has been constructed to look like a kopje.

Krebs cycle, citric acid cycle, tricarboxylic acid cycle (TCA cycle) A biochemical cycle which takes place in the **MITOCHONDRION** in the presence of oxygen (although it does not use it directly) and produces **ADENOSINE TRIPHOSPHATE (ATP)** from the products of **GLYCOLYSIS**. *See also* **AEROBIC RESPIRATION**

Krebs Report A report by an independent scientific review group led by Prof. John Krebs on the link between bovine **TUBERCULOSIS** in cattle and badgers in the UK, published in 1997.

Kruskal–Wallis test A **NON-PARAMETRIC** statistical test which is used to compare three or more samples.

Kruuk, Hans (1934–) A zoologist, academic and field biologist who studied a wide range of species, including gulls, badgers and hyenas, and was co-founder of the Serengeti Research Institute in Tanzania. Dr. Kruuk has published several books including *The Spotted Hyena: A Study of Predation and Social Behaviour* (1972).

K-selected species A K-selected species, or K-strategist, is a stable species. It exhibits slow development, delayed reproduction, large body size and **ITEROPARITY**. It has low colonising ability but often a well-developed social structure. K-strategists exhibit density-dependent mortality and usually types I and II survivorship. They occur in constant, predictable environments. Such species are particularly prone to extinction because they are unable to evolve quickly in response to environmental change due to their long **GENERATION TIME**, e.g. elephants, beavers, dinosaurs. *Compare* **R-SELECTED SPECIES**

kuri, goorie A Maori term for a **MONGREL** dog in New Zealand.

K

Fig. L1 L is for lion (*Panthera leo*).

Dictionary of Zoo Biology and Animal Management: A guide to terminology used in zoo biology, animal welfare, wildlife conservation and livestock production, First Edition. Paul A. Rees.
© 2013 John Wiley & Sons, Ltd. Published 2013 by John Wiley & Sons, Ltd.

labor *See* **LABOUR**.

Laboratory Animal Science Association An organisation founded in the UK in 1963 by a consortium of industrial, university, ministry and research council representatives whose key aim is to advance knowledge of the care and welfare of laboratory animals and to promote the refinement of scientific procedures.

Laboratory for the Conservation of Endangered Species (LaCONES) A laboratory established by the **CENTRAL ZOO AUTHORITY (CZA)** in India at Hyderabad, which conducts **BIOTECHNOLOGY** research to assist in the conservation of endangered species.

labour, labor The process of giving birth in a mammal, especially from the point where contractions of the uterus begin. **OXYTOCIN** may be given to assist labour in cases of uterine inertia but not when the birth canal is obstructed or the foetus is oversized.

laceration A deep cut in a tissue.

Lacey Act of 1900 (USA) A US law which is concerned with the humane treatment of wildlife shipped to the United States. In addition the Act prohibits the importation, exportation, transportation, sale or purchase of wildlife or fish taken or possessed in violation of state, federal, tribal or foreign laws. It makes it illegal to import into the USA, without a permit, certain bird and mammal species that other countries have identified as requiring protection. The Act is an important tool in deterring the illegal trade and smuggling of wildlife, and it allows for the provision of federal assistance to the states and foreign governments in the enforcement of their wildlife laws.

Lack, David (1910–1973) A British evolutionary biologist and ornithologist who wrote many important books including *The Life of the Robin* (1943), *Darwin's Finches* (1947) and *The Natural Regulation of Animal Numbers* (1954). He began his career as a schoolmaster but eventually became Director of the Edward Grey Institute of Field Ornithology at the University of Oxford.

LaCONES *See* **LABORATORY FOR THE CONSERVATION OF ENDANGERED SPECIES (LaCONES)**

lactate meter A device for measuring the quantity of lactate in the blood. Lactate is a by-product of anaerobic metabolism and is produced by **GLYCOLYSIS**. Raised levels in the blood most commonly indicate shock (e.g. in colic in horses) or poor tissue perfusion, but may also indicate heart failure, liver problems, lung disease or sepsis.

lactation The production of **MILK** by a female **MAMMAL**, from **MAMMARY GLANDS** under the influence of **PROLACTIN** (which stimulates milk formation) and **OXYTOCIN** (which stimulates its release).

lactational amenorrhoea A period of postnatal infertility caused by the absence or suppression of **MEN-STRUATION** (ovulation) in some primates while nursing young. *See also* **AMENORRHOEA**

lactic acid fermentation A type of **ANAEROBIC RESPIRATION** which occurs in muscle cells during extreme exertion when the body cannot obtain enough oxygen for **AEROBIC RESPIRATION**. Glucose is converted to lactic acid, releasing energy. This leads to a build-up of lactic acid in the cells. The oxygen required to oxidise this lactic acid is the **OXYGEN DEBT**.

lagomorph A member of the order **LAGOMORPHA**.

Lagomorpha An order of mammals: pikas, rabbits, hares.

Laika The first animal to be launched into space. She was a 3-year-old husky-terrier cross and was launched into space aboard *Sputnik 2* on 3 November 1957 by the Soviet Union. She demonstrated that manned space flight was possible although she did not survive the flight. *See also* **CHIMPONAUT**

Lamarck, Jean-Baptiste (1744–1829) A French naturalist who proposed a theory of evolution in 1809 which suggested that change could take place by the inheritance of acquired characters, i.e. that an ability developed during an animal's lifetime can be passed on to the next generation. This came to be known as Lamarckism. *See also* **DARWIN**

Lamarckism *See* **LAMARCK**

lameness A failure to travel normally in a regular, even and sound manner using all limbs. May be caused by a number of conditions including **INFECTION**, injury to the **BONE**, **MUSCLE**, **TENDONS** or **LIGAMENTS (1)**, neurological disorders, **LAMINITIS**, compensation for back pain or injury etc.

laminitis A disease of the feet of ungulates characterised by **LAMENESS**. The cause is unclear but may be associated with obesity, stress, trauma, **TOXAEMIA**, some drugs and an excess of rich grass in the diet.

Lamniformes An order of fishes: mackerel sharks and their allies.

lamping
1. Hunting or fishing at night with torches. Illegal in some jurisdictions.
2. *See* **CANDLING**

Lampridiformes An order of fishes: oarfishes and their allies.

land bridge An area of land which connects two land masses. Land bridges have connected areas of the Earth in the past that are now separated by the sea as a result of **CONTINENTAL DRIFT**. These bridges have allowed the dispersal of species in ways that would now not be possible and explain why, for example, the animals present in Western Europe are similar to those in the British Isles.

land train A 'train' which travels on roads, made up of several carriages or trailers pulled by a powered vehicle. Often made to look like a locomotive but actually a road vehicle. Used to transport visitors around a large zoo or farm.

L

landmark A place or object in the environment used by an animal to aid NAVIGATION. Used by many birds, fishes and insects.

landscape immersion A technique used in zoo exhibit design which absorbs visitors into an environment which represents the natural habitat of the animals being exhibited. The concept of 'landscape immersion' developed in the 1970s. The first exhibit to have adopted a landscape immersion design is considered to have been the gorilla exhibit opened in 1978 at the Woodland Park Zoo in Seattle, Washington, designed by Grant Jones and Jon COE. *See also* JONES & JONES ARCHITECTS AND LANDSCAPE ARCHITECTS LTD, NATURALISTIC EXHIBIT

language A system of verbal or non-verbal abstract signals used to communicate information from one animal to another, usually of the same species. *See also* COMMUNICATION, INTERSPECIES COMMUNICATION, LEXIGRAM BOARD, NON-VERBAL LANGUAGE

laparoscope A surgical instrument consisting of a narrow, illuminated flexible tube that can be inserted into the body via a small incision especially in the abdominal wall or other cavity. May be fitted with a camera. Used for a variety of purposes including the sexing of birds and ARTIFICIAL INSEMINATION (AI).

laparoscopy An invasive procedure which uses a LAPAROSCOPE to examine the inside of an animal's body. May be used to determine the sex of birds.

large intestine The section of the gut located between the SMALL INTESTINE and the anus, and consisting of the colon, the CAECUM and the RECTUM. Water absorption occurs in the large intestine and VITAMINS are produced here by bacteria in some species.

large mammal Non-taxonomic term for taxa of large mammals, e.g. antelopes, elephants, giraffes etc. *Compare* SMALL MAMMAL. *See also* BIG FIVE

Lassie A fictional rough collie whose adventures were the subject of novels, several films (including *Lassie Come Home* (1943)) and several television series.

latent learning A form of learning in which an animal appears to make no response to a stimulus at the time the learning occurs (and therefore no REINFORCEMENT occurs) but demonstrates, by its behaviour, that learning has occurred at a later date. For example an animal returning to the location of a previously ignored food source when hungry.

latent period *See* INCUBATION TIME

lateral Relating to the side of an animal's body, e.g. a lateral fin on a fish.

lateral line system *See* ACOUSTIC-LATERALIS SYSTEM

Latin name *See* BINOMIAL NAME

latrine behaviour The repeated use of specific sites for defecation. *See also* MIDDEN

laufschlag *See* LEG BEAT

Lawton Review An independent review of England's wildlife sites and ecological network chaired by Prof. Sir John Lawton and published as *Making Space for Nature* in September 2010.

Laying Hens Directive An EU Directive that bans the use of conventional barren battery cages for hens in Member States from 1 January 2012 (1999/74/EC). *See also* BATTERY FARMING, ENRICHED COLONY CAGES

LD$_{50}$ test Lethal dose 50% test. A laboratory test used to determine the concentration or dose of chemical (e.g. a PESTICIDE) which will kill 50% of the individuals exposed to it. Unpopular with animal welfare organisations because of the large number of animals required for the test.

leaf litter The layer of decomposing leaves and fragments of plant material forming the upper layer of many soils.

League Against Cruel Sports A UK charity established in 1924 which investigates and campaigns against sports which involve cruelty to animals.

Leakey, Richard (1944–) A Kenyan paleoanthropologist, conservationist and politician. Son of the paleontologists Louis and Mary Leakey. Formerly Director of the Kenya Wildlife Service and the National Museum of Kenya. Dr. Leakey turned the wildlife service into a paramilitary organisation and adopted a shoot-to-kill policy to combat POACHING. Author of several books including *Wildlife Wars: My Battle to Save Kenya's Elephants* (2001).

learning A relatively permanent change in BEHAVIOUR made in response to a particular STIMULUS. *See also* AVERSION (2), AVOIDANCE, CLASSICAL CONDITIONING, HABITUATION, IMPRINTING, INSIGHT LEARNING, LATENT LEARNING, OPERANT CONDITIONING, SHAPING, SOCIAL FACILITATION, TRAINING

learning curve The curve that results when the time taken to perform a new task or learn a new skill (or the amount of material learnt) is plotted against trial number or time spent practising a new skill.

Least Concern (LC) A RED LIST category used by the INTERNATIONAL UNION FOR THE CONSERVATION OF NATURE AND NATURAL RESOURCES (IUCN): Least concern (LC) – The taxon does not qualify for CRITICALLY ENDANGERED (CR), ENDANGERED (EN), VULNERABLE (VU) or NEAR THREATENED (NT). Widespread and abundant taxa are included in this category.

leg beat, laufschlag An element of the COURTSHIP of the male of some antelope species (e.g. roan (*Hippotragus equinus*)) in which he checks if the female will allow him to mount by raising one of his fore feet and tapping one of her hind legs.

legal cases *See* ANIMAL RIGHTS LEGAL CASES, BAYLISS V. COLERIDGE (1903), CLIMATE CHANGE, *LEGAL EAGLE*, *LOCUS STANDI*, NORTHERN SPOTTED OWL (*STRIX OCCIDENTALIS CAURINA*), WYOMING FARM BUREAU FEDERATION V. BABBITT (1997)

legal definition The meaning of a word in the law. For any particular word this may vary from one law to another and between legal jurisdictions (*see* ZOO for example). The legal definition of a term may also be different from its scientific definition (*see* FISH for example).

Legal Eagle A publication of the ROYAL SOCIETY FOR THE PROTECTION OF BIRDS (RSPB) that is concerned with bird crime, especially cases brought to the courts.

legal instrument A law; an ACT OF PARLIAMENT, STATUTORY INSTRUMENT, EUROPEAN DIREC-TIVE, EUROPEAN REGULATION, INTERNA-TIONAL CONVENTION or other similar piece of law.

legal personality An entity to which rights and duties are attached under the law. Legal persons may be divided into natural persons (human beings) and artificial persons (e.g. an organisation such as a company, university or conservation organisation). Legal persons may bring an action in a court or be the subject of a lawsuit or a criminal prosecution. Lawyers have occasionally attempted to bring an action to a court on behalf of wild animals (as 'guardians') because they have no legal personality themselves and clearly cannot act on their own behalf. *See also* ANIMAL RIGHTS LEGAL CASES, *LOCUS STANDI*

leisure Non-essential activities which do not appear in an animal's repertoire when demands upon its time are severe, e.g. GROOMING, PLAY. *Compare* NEED

lek A place used as a communal mating ground held by males of some species as a type of TERRITORY, e.g. grouse, bowerbirds, some antelope species.

Leopold, Aldo (1887–1948) An American ecologist who worked for the US Forest Service. Credited with changing the way that Americans think about wildlife and wilderness areas. Author of *A Sand County Almanac* (1949). *See also* CARSON

leopon A hybrid formed by crossing a male leopard with a female lion.

Lepidoptera A large order of insects: butterflies and moths.

Lepidosireniformes An order of fishes: South American and African lungfishes.

Lepisosteiformes An order of fishes: gars.

leptospirosis A bacterial infection caused by species from the genus *Leptospira* which are found in surface water. It commonly occurs in cattle, horses, pigs, sheep, dogs and humans. It has also been found in wild mammals, including mice, rats, hedgehogs, voles, shrews. Spread may be partly via contamination of pasture with the urine of infected animals. Leptospires can be inhaled and can penetrate intact MUCOUS MEMBRANES and abraded skin. Signs may include generalised illness, JAUNDICE, KIDNEY FAILURE, fever, abortion and death. Treatment is by antibiotics (especially streptomycin) and vaccines are available. *Leptospira icterohaemor-rhagiae* causes jaundice in dogs and Weil's disease in humans.

lesion
1. An area of damage in a tissue or organ caused by disease.
2. A wound or injury.
3. A patch of skin that is infected or diseased.

lesser ape A gibbon (Hylobatidae). Gibbons are smaller than the GREAT APES, superficially resemble monkeys and exhibit low SEXUAL DIMOR-PHISM. *See also* APE

lethal temperature All species have an upper and lower lethal temperature which mark the boundaries of their temperature tolerance. *See also* HYPERTHERMIA, HYPOTHERMIA, THERMOREG-ULATION

lethargy A decrease in activity level. An unwillingness to take part in normal activities such as walking, eating and drinking, and an increase in sleeping. It may be a sign of a wide range of illnesses and conditions, e.g. infections, disorders of the digestive system, injury. *See also* WINTER LETHARGY. *Compare* RESPONSIVENESS

leucism A condition in which individuals have a white colour caused by a reduction in all types of skin pigmentation, e.g. white lions. *Compare* ALBINISM, MELANISM

leucocyte, white blood cell A blood cell found in vertebrates and invertebrates. There are three types: lymphocytes (that produce ANTIBODIES), monocytes (that ingest invading organisms) and polymorphs (phagocytic cells).

lexigram *See* LEXIGRAM BOARD

lexigram board A tool used in the study of communication between people and primates. The board consists of a range of abstract symbols (lexigrams) each of which represents an English word. Participants communicate by pointing at the symbols. *See also* GREAT APE TRUST, *KANZI*

liability for the actions of animals The legal responsibility for the action of animals when they cause injury, damage to property etc. This is defined in various laws and liability may be strict, e.g. under the ANIMALS ACT 1971 s2(1): '*Where any damage is caused by an animal which belongs to a dangerous species, any person who is a keeper of the animal is liable for the damage, except as otherwise provided by this Act.*' *See also* DANGEROUS DOGS ACT 1991, DANGER-OUS WILD ANIMALS ACT 1976, OCCUPIERS' LIABILITY, VICARIOUS LIABILITY, WORRYING LIVESTOCK

licence A document which permits an activity that is forbidden (illegal) without it, e.g. bird ringing, certain types of trapping, photographing rare birds at the nest, release of EXOTIC (1) animals to the wild, hunting, operation of a ZOO, import and export of animals, operation of a PET SHOP. *See also* CITES, EXPORT LICENCE, IMPORT LICENCE, WILDLIFE AND COUNTRYSIDE ACT 1981, ZOO LICENCE, ZOO LICENSING ACT 1981

L

Fig. L2 Life support system: a marine life support system for maintaining a marine aquarium.

licensing authority In relation to the activities of zoos, the organisation that issues a **ZOO LICENCE**. In the UK this is the local authority (i.e. local government). *See also* **ZOO LICENSING ACT 1981**

life history The sequence of events that make up an individual animal's life from birth to death. The pace is largely determined by the body size of the species, e.g. an individual from a large mammal species has a longer gestation, longer infancy and longer lifespan than one from a small mammal species.

life support, basic life support The artificial maintenance of the functioning of the respiratory and circulatory systems of the body of an animal, especially by giving mouth-to-nose resuscitation and chest compression over the area of the heart.

life support system Equipment that maintains constant and appropriate water chemistry for the aquatic organisms kept in an **AQUARIUM (2)** (Fig. L2). *See also* **INTENSIVE CARE UNIT (ICU) (1)**

life table A table of data showing the **MORTALITY** rates of different age classes within a **POPULATION** of organisms (Table L1). Used to produce a **SURVIVORSHIP CURVE**. Usually constructed separately for males and females. Static life tables are constructed by counting the number of animals in each age class present in a population at a single point in

Table L1 Theoretical example of a life table for a species with a maximum life span of less than 6 years.

Age (years) (x)	Survivors at start of age class x (lx)	Deaths between age class x and x+1 (dx)	Age-specific death rate qx (dx/lx)
0	100	100–87 = 13	0.130
1	87	87–54 = 33	0.379
2	54	54–34 = 20	0.370
3	34	34–20 = 14	0.412
4	20	20– 7 = 13	0.650
5	7	7– 0 = 7	1.000
6	0	0	–

time. Dynamic life tables follow the survival of a **COHORT** of animals born at the same time (e.g. in the same year). Most life tables are static because it is difficult to construct dynamic life tables for long-lived animals since they cannot be completed until all of the animals in the cohort have died. For some species this would take many decades. Static life tables suffer from the disadvantage that they assume environmental conditions for the individuals in all of the age classes have remained the same throughout their lives and that birth rates are stable from one year to the next. In zoo populations, poor survival in long-lived animals may be indicative of poor zoo conditions in the past rather than poor conditions now.

lifespan *See* **LONGEVITY**

ligament

1. A strip of connective tissue which connects adjacent bones.

2. A fold of **PERITONEUM** connecting two abdominal organs, e.g. the broad ligament which attaches the uterus, Fallopian tubes and ovaries to the pelvis.

ligature

1. The act of binding or tying a hollow structure, e.g. a blood vessel or other tube in the body, to close it off.

2. A wire, thread or cord (suture) used to create a ligature.

liger A hybrid produced when a male lion successfully mates with a female tiger. *Compare* **TIGON**

light The quality (wavelength) and quantity (duration) of light to which animals are exposed may be important in the welfare of some species. Visible light is electromagnetic radiation with a wavelength between approximately 400 nm and 700 nm. Some behaviours are cyclical and regulated by a photoperiod, e.g. reproduction in some species. In dairy cow sheds, maintaining the correct light levels can contribute to higher feed intakes, increase milk production and may increase the rate of mounting by bulls, and make **BULLING** easier to detect. In 2009 a farmer in West Yorkshire was prosecuted under the **ANIMAL WELFARE ACT 2006** and fined for failing to meet the psychological needs of a cow and her calf by not providing adequate lighting. *See also* **ULTRAVIOLET LIGHT**

lightbox A device for viewing **RADIOGRAPHS**, **POLYMERASE CHAIN REACTION (PCR)** plates, photographic materials (e.g. slides, negatives) etc. The material to be viewed is placed on a piece of frosted glass (or similar material) which is illuminated by light (similar to daylight) from underneath.

lignin A complex polymer which is a common component of plant cell walls and abundant in woody tissue. High lignin content adversely affects the **DIGESTIBILITY** of plant food.

limnology The scientific study of the biological, physical, geological and chemical aspects of **FRESHWATER** systems (lakes, ponds and rivers).

Lincoln index *See* **MARK–RECAPTURE TECHNIQUE**

Lincoln Park Zoo A zoo in Chicago which has one of the largest zoo-based conservation and science programmes in the USA. Its research centres include the Population Management Center, Urban Wildlife Institute, Lester E. Fisher Center for the Study and Conservation of Apes, Alexander Center for Applied Population Biology and Davee Center for Epidemiology and Endocrinology.

linear regression analysis A statistical method which produces a straight line through a series of points on

Fig. L3 Linnaeus.

L

a graph (line of best fit) which shows the relationship (**CORRELATION**) between two variables.

linear scale A scale on a graph, measuring instrument or elsewhere which is divided into equal divisions for equal values so, for example, the distance between 2 and 3 is the same as the distance between 5 and 6. *Compare* **LOGARITHMIC SCALE**

Ling Yu *See* **GARDEN OF INTELLIGENCE**

linkage *See* **GENE LINKAGE, SEX LINKAGE**

Linnaeus, Carolus (1707–1778) Linnaeus was a Swedish botanist and physician (Fig. L3). He devised the **BINOMIAL SYSTEM** of nomenclature used for assigning scientific names to organisms. This was first published in his *Systema Naturae* in 1735. In 1741 he was appointed Professor of Practical Medicine at the University of Uppsala and then in 1742 Professor of Botany, Dietetics and Materia Medica. Linnaeus named many thousands of animals and plants using his system, and although some have since been reclassified, many still retain his original names.

Linnean Society of London The world's oldest active biological society. It was founded in 1788, is named after Carolus **LINNAEUS** and is keeper of his botanical, zoological and library collections. The Society promotes the study of natural history and all aspects of the biological sciences, with particular emphasis on evolution, taxonomy, biodiversity and sustainability. It publishes several journals including the *Zoological Journal of the Linnean Society*. On 1 July 1858 papers describing their theory of evolution proposed by Charles **DARWIN** and Alfred Russel **WALLACE** were read in their absence at the Linnean Society. Both were Fellows of the Society.

Linzey, Andrew Revd Professor Linzey held the world's first academic post in Theology and Animal Welfare at Mansfield College, Oxford, between 1992 and 2000. He is currently the Director of the **OXFORD CENTRE FOR ANIMAL ETHICS**.

Lions of Longleat The first drive-through **SAFARI PARK** in the world. It was opened in 1966 by the Marquess of Bath in the grounds of Longleat House in Wiltshire. It began as a 100-acre (40.5-ha) lion reserve through which visitors could drive in their own cars. The park is still a major animal attraction. *See also AFRICA USA*, **CHIPPERFIELD**

lip smacking
1. Behaviour performed by many **OLD WORLD MONKEY** species, which may be a positive social communication and is possibly associated with social status. May have evolved into speech in humans.
2. A **STEREOTYPIC BEHAVIOUR** in some species.

lipase An enzyme produced by the **PANCREAS** which converts fats to fatty acids and glycerol.

lipid Lipids consist of a number of different types of molecules including fats, cholesterol and phospholipids. Fats are triglycerides, consisting of a backbone of glycerol linked to three fatty acids. The type of fat is determined by the types of fatty acids. Fats are stored in the body as **ADIPOSE TISSUE**. They act as long-term energy reserves and, in many animals, adipose tissue stored under the skin acts as an insulator, e.g. in marine mammals. Fats that are liquid at body temperature are called oils. Animals can synthesise most of the fatty acids they need. However, for most species there are some essential fatty acids that they cannot produce for themselves and must be present in the diet. Phospholipids are an important part of the **CELL MEMBRANE**. Cholesterol is also important in the structure of biological membranes and in the synthesis of **STEROID** hormones.

Lipizzaner, Lipizzan A breed of horse associated with the Spanish Riding School in Vienna where highly trained stallions demonstrate classical **DRESSAGE**.

listed building In the UK, a building which is protected because of its historical and/or architectural importance. Such buildings are graded based on their importance. In England and Wales the grades used are I, II* and II; in Scotland they are A, B and C; and in Northern Ireland they are A, B+ and B. Restrictions may be imposed on their alteration and even the colour they may be painted. This may severely constrain the use of a listed building intended to house animals if it is no longer considered to meet the animals' needs. Examples of listed buildings in zoos in England are the entrance to **DUDLEY ZOO**, and, at **LONDON ZOO**, the Penguin Pool, **RAVEN'S CAGE** and **MAPPIN TERRACES**.

Similar listing systems occur in other countries. *See also* **BÄRENGRABEN**, **NATIONAL REGISTER OF HISTORIC PLACES**

listlessness Being languid, indifferent, uninterested, inactive, apathetic, slow-moving.

litter
1. Material located on the floor of a cage or other enclosure, or used in a tray for a companion animal such as a cat (cat litter). For example the following may be used for hens: sand, gravel, wood shavings, wheat, **SPELT GLUMES**, rye straw, bark mulch, wood chips. *See also* **DEEP LITTER HOUSING SYSTEM**
2. *See* **LEAF LITTER**
3. All of the young born to a mother at the same time (especially in mammals).
4. Waste materials discarded in the environment which may be damaging to wildlife, farm animals and zoo animals, e.g. bottles, drinks cans and plastic bags.

littermates Individuals belonging to the same **LITTER (3)**.

littoral Relating to, or situated on, or near, the shore of a lake or (especially) a sea.

live cattle *See* **FAT CATTLE**

live feeding in zoos The process of providing living animals as food. Live invertebrates are widely fed to animals living in zoos, e.g. locusts and other insects are fed to lizards. Live prey are routinely fed to large predators in many zoos and wildlife parks in China, as a public spectacle. Big cats, hyenas and bears are fed a range of live prey including cattle, buffalo, horse, goat, rabbit, ostrich, duck, guinea fowl and chicken. In many countries, including the UK, **ANIMAL CRUELTY LAWS** would prohibit the feeding of live vertebrates to other species.

Live hard, die young A controversial report published by the **ROYAL SOCIETY FOR THE PREVENTION OF CRUELTY TO ANIMALS (RSPCA)** in 2002 on the welfare of **ELEPHANTS** in European zoos, based on a study by scientists from Oxford University.

live phytoplankton Commercially available phytoplankton used as food for **CORALS**, marine fishes etc. kept in aquariums. *See also* **LIVE ZOOPLANKTON, PLANKTON**

live transport *See* **TRANSPORT OF LIVE ANIMALS**

live zooplankton Commercially available zooplankton used as food for aquarium **CORALS** consisting of **COPEPODS** and rotifers. *See also* **LIVE PHYTOPLANKTON, PLANKTON**

liver A large gland in the abdomen of vertebrates which detoxifies the blood, produces bile salts used in the digestion of fats, stores sugar (as **GLYCOGEN**), **VITAMINS** and other foods, produces **ANTIBODIES** and various proteins, and removes wastes.

livestock
1. Animals which are kept or raised for use (as food or for other products) or pleasure, especially **FARM**

ANIMALS raised for profit. The term is defined in some legislation.

2. In Great Britain, the WILDLIFE AND COUN-TRYSIDE ACT 1981 (s.27(1)) defines livestock as including any animal kept . . . *'(a) for the provision of food, wool, skins or fur, (b) for the purpose of its use in the carrying on of any agricultural activity; or (c) for the provision or improvement of shooting or fishing.'*

3. Under the ANIMALS ACT 1971 (s11), *'. . ."livestock" means cattle, horses, asses, mules, hinnies, sheep, pigs, goats and poultry, and also deer not in the wild state and . . . while in captivity, pheasants, partridges and grouse'.*

livestock worrying *See* WORRYING LIVESTOCK

living collection *See* LIVING RESIDENT POPULATIONS

living modified organism (LMO) Any living organism that possesses a novel combination of genetic material obtained through the use of modern BIO-TECHNOLOGY (Art. 3, CARTAGENA PROTOCOL ON BIOSAFETY 2000).

living museum An outdated concept of a zoo as a collection of animals in small cages. *See also* MOD-ERNIST MOVEMENT. *Compare* UNZOO

Living Planet index A measure used in the *LIVING PLANET REPORT* which reflects changes in the state of the Earth's BIODIVERSITY, using trends in the size of 9014 populations of 2688 mammal, bird, reptile, amphibian and fish species from different BIOMES and regions. The index showed a decline of approximately 30% from 1970 to 2008 (Anon., 2012).

Living Planet Report The world's leading, science-based analysis of the health of the Earth and the impact of human activity. Produced by the WORLD WILDLIFE FUND (WWF) in collaboration with the ZOOLOGICAL SOCIETY OF LONDON (ZSL), the Global Footprint Network and the European Space Agency. *See* LIVING PLANET INDEX

living resident populations, living collection A term used by some zoo professionals, especially in the USA, to refer to an animal COLLECTION.

living roof *See* GREEN ROOF

LMO *See* LIVING MODIFIED ORGANISM (LMO)

loafing A behaviour of birds, especially ducks, whereby they laze around, usually communally, in loafing areas, which are separate from breeding and feeding areas, e.g. on shorelines, sandbars.

Lobo The alpha male of a pack of grey wolves (*Canis lupus*) who were killing LIVESTOCK in the Currumpaw Valley, New Mexico, in the 1890s. He evaded capture by the hunter and naturalist Ernest Thompson Seton but was eventually caught in 1894, and died soon thereafter. Seton deeply regretted *Lobo*'s death and wrote about this experience in his book *Wild Animals I have Known* (1898). This work is credited with changing American attitudes to wildlife, and leading to the establishment of new NATIONAL PARKS (NP) and environmental laws.

local extinction *See* LOCALLY EXTINCT

locally extinct, local extinction The extinction of a species from an isolated population in a particular locality although other populations of the same species may occur elsewhere, e.g. the grey wolf (*Canis lupus*) is extinct in the British Isles but present on the European mainland, North America and elsewhere.

locomotion Moving from place to place or having the capacity to do so. *See also* BRACHIATION, GAIT, KNUCKLE WALKING, MOTILE, SALTATION (3), TOBOGGANING

locus (loci *pl.*) In genetics, the place on a chromosome where a particular gene is located.

locus standi The right to bring a legal action or challenge a legal decision. This principle prevents individuals who have no legitimate interest in a particular legal case from bringing frivolous actions and wasting the time of the courts. GREENPEACE and other conservation NON-GOVERNMENTAL ORGANISATIONS (NGOs) have sometimes been granted *locus standi* in cases relating to the protection of wildlife and the environment. *See also* LEGAL PERSONALITY

lodge A shelter built by beavers.

logarithmic scale A scale used on a graph which plots the logarithm (usually to the base 10) of the values instead of the values themselves. For example the $\log_{10}$ of 10 is 1, the $\log_{10}$ of 100 is 2, and the $\log_{10}$ of 1000 is 3. This method is used to allow differences between low values to be distinguished more easily when the scale extends over a very wide range of values, and has the effect of making some curves appear as straight lines, or at least, straighter than if they had been plotted on a LINEAR SCALE.

logging The process of cutting down trees as a commercial activity. *See also* DEFORESTATION

logistic growth A type of POPULATION GROWTH in which numbers initially grow exponentially and then stabilise at the CARRYING CAPACITY of the environment (the asymptote). Also known as S-shaped growth. May occur when a species invades or is introduced to a new habitat and there are no significant competitors or predators (Fig. P8).

Loisel, Gustave Antoine Armand (1864–1933) A French physician, zoologist and zoo historian who became assistant professor of zoology at the Sorbonne, and was the author of the three-volume *Histoire des ménageries de l'antiquité à nos jours* (History of menageries from antiquity to present times) (1912).

London Wetlands Centre A large wetland nature reserve (100 acres; 40.5 ha) managed by the WILD-FOWL AND WETLANDS TRUST (WWT) in London, which was created from disused reservoirs and opened in 2000.

London Zoo, Regent's Park Zoo, ZSL London Zoo London Zoo (now ZSL London Zoo) is operated by the ZOOLOGICAL SOCIETY OF LONDON (ZSL). It is located in Regent's Park in central

L

London and opened in 1826 to Fellows of the Society. Paying visitors were first admitted in 1847. London Zoo was the first scientific zoo in the world and its first superintendent was Abraham Dee **BARTLETT**. The zoo opened the first reptile house (1849), the first aquarium – the **FISH HOUSE** (1853) – the first insect house (1881) and the first children's zoo (1938).

Lonesome George A male Pinta Island tortoise, who was the last remaining individual of a subspecies of Galapagos giant tortoise (*Chelonoidis nigra abingdonii*). He came to be known as the rarest animal in the world and was a symbol for conservation. He died on 24 June 2012.

long bone One of the major bones of the limbs, in which the body of the bone is longer than it is wide: **FEMUR, TIBIA, FIBIA, ULNA, HUMERUS, RADIUS**

longevity, lifespan The length of an animal's life between birth (or hatching) and death.

longitudinal study A study which follows the fate of all of the individuals in a population (or a sample) over a long period of time, possibly throughout their entire lives, e.g. a study of the longevity of a particular species in zoos which records the age of death of a cohort of individuals born in the same year. *Compare* **CROSS-SECTIONAL STUDY**

Longleat Safari Park *See LIONS OF LONGLEAT*

longline fishing A method of fishing which utilises a long length of line to which are attached baited hooks. Commercial longlines may be hundreds of kilometres long and consist of over 1000 bait hooks. Many seabirds, including albatrosses, dive for the bait and become entangled in the lines.

Longworth trap A trap designed to capture small mammals alive in field studies of their ecology, behaviour etc.

Lophiiformes An order of fishes: anglerfishes, goosefishes and frogfishes.

lordosis

1. An (excessive) inward curvature of the spine.
2. A posture which involves an arching downward of the back, exhibited by some female mammals (e.g. felids) during mating.

Lorenz, Konrad (1903–1989) Konrad Lorenz was an Austrian ethologist who studied the relationship between **INSTINCT** and learned behaviour, particularly in birds, in which he described the phenomenon of **IMPRINTING**. Lorenz is considered to be one of the founders of **ETHOLOGY**. In 1973 he shared the Nobel Prize for Physiology or Medicine with Nikolaas **TINBERGEN** and Karl von **FRISCH**. Lorenz published many influential books including *King Solomon's Ring* (1949) and *On Aggression* (1963).

Lovelock, James (1919–) A scientist and inventor who originated the **GAIA THEORY**.

lower vertebrates Reptiles, amphibians and fishes. *Compare* **HIGHER VERTEBRATES**

Lubetkin, Berthold (1901–1990) A Russian architect who formed a group called the **TECTON GROUP** with six other architects. *Tecton* designed many iconic zoo enclosures and buildings which were characterised by their sweeping curves and constructed from reinforced concrete. Many of these buildings are now protected, including the Gorilla House and Penguin Pool at **LONDON ZOO** and many buildings at **DUDLEY ZOO** (Fig. L4).

Fig. L4 Lubetkin: the entrance to Dudley Zoo is a Grade II* listed building. It was designed by Lubetkin and the Tecton Group.

lucerne, alfalfa Also called lucerne grass. A perennial forage legume.

luliberin *See* **GONADOTROPHIN-RELEASING HORMONE (GnRH)**

lumbar Relating to the abdominal section of the torso in a vertebrate, e.g. lumbar vertebrae.

lumen The space in the middle of a tubular structure, e.g. a blood vessel.

lumpy jaw A colloquial term used to describe an assortment of different conditions – including **ACTINOMYCOSIS** and **NECROBACILLOSIS** – involving facial bone abnormalities in **HOOFSTOCK** and other animals.

lunar rhythm A rhythmic behaviour influenced by the movement of the moon and tides which is important in affecting the seasonal behaviour of some marine animals. *See also* **BIOLOGICAL CLOCK, CIRCADIAN RHYTHM, CIRCANNUAL RHYTHM, INFRADIAN RHYTHM, ULTRADIAN RHYTHM**

lung
1. The respiratory organ of air-breathing **VERTEBRATES**, made up of tiny air sacs (alveoli) across which **GASEOUS EXCHANGE** occurs between the blood and the atmosphere.
2. Part of the mantle of terrestrial **MOLLUSCS** which is involved in gaseous exchange.

lungworm Any of a number of nematode parasites which infect the lower respiratory tract and sometimes the heart and the pulmonary circulation, e.g. *Angiostrongylus vasorum* and *Oslerus osleri* (domestic dogs and other canids), *Dictyocaulus viviparus* (cattle and deer) and *D. filaria* (sheep and goats), *Aelurostrongylus abstrusus* in cats. Slugs and snails act as **INTERMEDIATE HOSTS** for some species.

lupine Like or relating to a wolf.

lure
1. A device used to attract an animal to a trap, camera etc. Often used when training an animal to hunt. *See also* **CAMERA TRAP, LURE COURSING, LURE HACK, SOUND LURE**
2. Anything put on a line by an angler to induce a fish to bite.
3. In **FALCONRY**, a piece of meat attached to a bunch of feathers, swung around the head on a rope, that is used in training to encourage the bird to return to the falconer. *See also* **KITE HACK, LURE HACK**

lure coursing A **NATURALISTIC ENRICHMENT** whereby food or a **MANNEQUIN** is pulled at speed along a wire cable to simulate the movement of a prey animal. It may be used as an enrichment for cheetahs (*Acinonyx jubatus*) or other captive predators, especially when **PRECONDITIONING** them to release into the wild.

lure hack A technique used in the **REINTRODUCTION** of birds of prey to the wild after injury in which the bird is taught to fly to a lure to help build up its fitness and develop its skill. Most useful for mature birds with hunting experience.

Lush, Jay Laurence (1896–1982) A pioneering American geneticist who made important contributions to **LIVESTOCK** breeding.

luteinising hormone (LH), interstitial cell-stimulating hormone (ICSH) A **HORMONE** released by the **ANTERIOR PITUITARY** which stimulates the production of **TESTOSTERONE** in males and causes **OVULATION** in females, transforming the ruptured **GRAAFIAN FOLLICLE** into a **CORPUS LUTEUM**.

luteinising hormone-releasing hormone (LHRH) *See* **GONADOTROPHIN-RELEASING HORMONE (GnRH)**

luxator A very slim **ELEVATOR**.

Lyme disease A bacterial disease transmitted to humans via ticks which affects the nervous system, heart and joints. A natural reservoir of the disease occurs in mammals, especially **RODENTS** and **DEER**.

lymph A colourless liquid which is similar to **PLASMA** and contains lymphocytes and circulates in the **LYMPHATIC SYSTEM**. It drains from the interstitial spaces into the lymphatic system to lymph nodes and then into the blood. Lymph transports antigens (including bacteria) to the lymph nodes; metastatic cancer cells may be spread via the lymph. Fats are absorbed from the gut into the blood via the lymph.

lymph node A mass of **LYMPHOID TISSUE** to which antigens (including bacteria) are carried by the lymph and destroyed. Lymph nodes become enlarged in regions of the body where infection exists. *See also* **SPLEEN**

lymphatic system A branched, blind-ending system of tubes (lymphatic vessels) similar to veins which drains **LYMPH** from the tissues into the blood system. It also functions as part of the immune system.

lymphocystis A common infectious viral disease of freshwater and saltwater fishes that causes cell enlargement (**HYPERTROPHY**) usually on the skin and fins. *See also* **DYED FISH**

lymphoid tissue Part of the **IMMUNE SYSTEM** that is important for the **IMMUNE REACTION**. It is present throughout the body including the **LYMPH NODES, SPLEEN**, tonsils, and adenoids.

lysis The destruction of a cell – rupture of the **CELL MEMBRANE** – as a result of the action of an **ENZYME**, a **VIRUS** or an osmotic mechanism, e.g. when red blood cells are placed in water and undergo **HAEMOLYSIS**.

lysosome A membrane-bound organelle found in eukaryotic cells which contains hydrolase enzymes used to break down waste materials and debris in the cell.

L

Fig. M1 M is for macaw. Green-winged macaws (*Ara chloropterus*).

M99 *See* **IMMOBILON**.

macronutrient

1. The chemical compounds that animals consume in the largest quantities: **CARBOHYDRATES, PROTEINS** and **LIPIDS**.

2. The chemical elements that animals consume in the largest quantities. *Compare* **MICRONUTRIENT**

macropod Any **MARSUPIAL** belonging to the family Macropodidae, e.g. kangaroos, wallabies (Fig. K1).

Macroscelidea An order of mammals: elephant shrews.

mad cow disease *See* **BOVINE SPONGIFORM ENCEPHALOPATHY (BSE)**

maggot farm A place where maggots are produced commercially, especially as bait for fishermen.

magnesium An element which is involved in muscle contraction and nerve conduction, synthesis of proteins, fats, carbohydrates and nucleic acids. Deficiency may result in vasodilation, convulsions and calcification of soft tissues. Deficiency may occur in **RUMINANTS** grazing on spring pastures low in available magnesium.

magnetic orientation A directional sense conferred by the ability to detect the Earth's magnetic field. Magnetic materials have been found in the bodies of bacteria, bees and birds. *See also* **COMPASS SENSE, NAVIGATION**

magnetic resonance imaging (MRI) scanner A machine that uses a powerful magnetic field and radio frequencies to produce 2D and 3D images of the inside of the body. It is particularly useful in producing images of the brain, muscle, heart and cancers. *See also* **CT SCANNER, X-RAY**

mahout A person (usually a man) who works with, rides and cares for an **ELEPHANT** in the Indian subcontinent and SE Asia.

maiden

1. A racehorse that has never won a race.

2. According to the **AMERICAN STUD BOOK** (2008), '*a FILLY or MARE that has never been bred (mated).*'

maintenance of proximity index (MPI) This index measures the extent to which each animal in a pair (A or B) is responsible for maintaining their proximity. Possible values range from +1.0 (A totally responsible for maintaining proximity) to −1.0 (B totally responsible for maintaining proximity). A value of zero indicates A and B are equally responsible for maintaining their proximity. The extent to which individual A was responsible for maintaining proximity between itself and individual B is calculated as:

$$MPI = \frac{U_A}{U_A + U_B} - \frac{S_A}{S_A + S_B}$$

where U_A = the number of occasions on which the pair were united by A's movement; U_B = the number of occasions on which the pair were united by B's movement; S_A = the number of occasions when the pair were separated by A's movement; S_B = the number of occasions when the pair were separated by B's movement. *See also* **ASSOCIATION INDEX**

malaise A vague weakness, fatigue and bodily unease, often exhibited by an animal at the beginning of an illness.

malaria Any of a number of infectious diseases of humans and other animals (mammals, birds and reptiles) caused by protozoans from the genus *Plasmodium* which are transmitted by female mosquitoes of the genus *Anopheles* and are often fatal.

male

1. An organism capable of producing **SPERMATOZOA**.

2. A male animal.

See also **SEX DETERMINATION**. *Compare* **FEMALE**

male philopatry *See* **PHILOPATRY (1)**

malleus *See* **EAR OSSICLES**

malnutrition A state of poor nutrition. It may be undernutrition or over-nutrition, or a diet where the components are present in the wrong proportions.

Malthus, Thomas (1766–1834) An English historian who was professor of history and political economy at the East India Company's college in Haileybury, Hertfordshire. He was one of the founders of the Statistical Society of London. His most famous work is *An ESSAY ON THE PRINCIPLE OF POPULATION* (1798). His ideas on **POPULATION GROWTH** influenced the thinking of Charles **DARWIN**.

mammal A member of the class **MAMMALIA**.

Mammal Society An organisation, based in the UK, of amateur and professional biologists who are interested in mammals. It publishes *Mammal Review*.

Mammalia The mammals; a class of **CHORDATES**. Quadruped **VERTEBRATES** (and taxa which are secondarily evolved from **QUADRUPEDS**, e.g. cetaceans) which are **HOMEOTHERMIC** and whose bodies are covered with **HAIR**. They produce 'live' young as miniature adults which are initially fed on **MILK** from the mother's **MAMMARY GLANDS**. The lower jaw consists of a single bone (**DENTARY**) on each side.

mammalogy The scientific study of mammals. *See also* **MAMMALIA**

mammary glands The organs present in female mammals which produce **MILK** for their young.

manage *See* **MANÈGE**

mandible

1. The lower jaw of mammals. *See also* **DENTARY**. *Compare* **MAXILLA (1)**

2. The lower jaw and lower **BILL** in birds (or both upper and lower parts of the bill: upper and lower mandibles).

3. One of a pair of mouthparts used to seize and cut food in arthropods.

manège, manage

1. The actions, movements or paces of a trained horse.

2. The art of riding or training horses.

M

3. A school where horsemanship is taught and horses are trained. *See also* SAND SCHOOL

manganese An element which is involved in the development of the bone matrix, fat utilisation and gluconeogenesis. Higher levels are required for reproduction than for growth. Deficiency may cause ataxia in the newborn, neonatal death, loss of reproductive function, impaired growth and skeletal abnormalities. Some grains contain low levels of manganese.

mange A contagious skin disease caused by mites. The mites lay their eggs in the skin and the resulting larvae cause intense irritation. Attempts by the infected animal to relieve the discomfort causes damage to the skin. Different species of mite cause different types of mange: sarcoptic (scabies), psoroptic, chorioptic and demodectic. Treatment may involve the use of ivermectin, amitraz or doramectin. Treatment may be difficult where mites have penetrated deep within the skin.

manger A container for holding food for livestock which may be mounted on a wall, raised on a stand, or fixed to the ground behind a FEED BARRIER.

mangrove forest A tidal SALT MARSH in the tropics where the vegetation is dominated by trees and shrubs which have exposed roots at low tide and which are covered by salt water at high tide. The exposed roots are important in stabilising sediments.

M

Manifesto for Zoos A 2004 study by John Regan Associates Ltd. of the overall value of zoos to society in the UK. It was commissioned by a consortium of nine leading British zoos, facilitated by the BRITISH AND IRISH ASSOCIATION OF ZOOS AND AQUARIUMS (BIAZA) and aimed at persuading the government to work together with zoos on matters of mutual interest. The study examines the role of zoos in conservation, science and education. It also looks at the nature of zoo visitors, the economic environment in which zoos operate, their economic outputs and their potential role in regeneration policy. The report concluded that zoos have an enormous social, cultural, educational and economic impact on the British public and that they have the potential to do more.

manikin *See* MANNEQUIN

mannequin, manikin, mannikin An imitation body of a prey animal, e.g. an antelope, used to encourage large carnivores, such as big cats, to exhibit hunting and prey-dragging behaviour. It may be little more than a sack filled with straw.

mannikin *See* MANNEQUIN

Mann–Whitney U test A NON-PARAMETRIC statistical test that may be used in place of an unpaired (independent) T-TEST to test the NULL HYPOTHESIS (H_0) that two SAMPLES have been taken from the same POPULATION.

mantle An extension of the body wall in a MOLLUSC which encloses the mantle cavity and is covered by the shell in some species (e.g. snails) but not in others (e.g. octopuses).

manual pipping *See* ASSIST HATCH. *See also* EXTERNAL PIPPING, INTERNAL PIPPING

Maple, Terry (1946–) An American zoo director, research scientist, primatologist and academic who was the founding editor of the journal *ZOO BIOLOGY* and past president of the ASSOCIATION OF ZOOS AND AQUARIUMS (AZA). *See also* EMPIRICAL ZOO CONCEPT

Mappin Terraces An artificial mountain landscape at LONDON ZOO constructed from reinforced concrete in 1913–14. At different times in the past it has been home to a wide variety of species including polar bears, ibex, sloth bears and Hanuman langurs. The terraces currently house wallabies and emus in an Outback exhibit. The main structure cannot be substantially modified as it is a Grade II LISTED BUILDING.

mare

1. A female horse.

2. According to the AMERICAN STUD BOOK (2008) *'A female horse five years old or older.'*

marine Relating to the sea.

Marine and Coastal Access Act 2009 *See* MARINE CONSERVATION ZONE (MCZ)

Marine Conservation Society A UK charity which has been registered since 1983 and began as the Underwater Conservation Society. It promotes sustainable seafood, records and campaigns to protect marine life, promotes clean beaches and operates a network of volunteers who clean beaches (Beachwatch), and campaigns against marine pollution, especially from sewage.

Marine Conservation Zone (MCZ) A type of marine protected area created under the Marine and Coastal Access Act 2009 which may be designated in English and Welsh inshore and UK offshore waters. MCZs protect marine wildlife, habitats, geology and geomorphology which are nationally important.

marine mammal park *See* OCEANARIUM

Marine Mammal Protection Act of 1972 (USA) A US law under which a MORATORIUM on the taking and importation of marine mammals was established (including their parts and products). It defines federal responsibilities for marine mammal conservation and assigns management authority for the walrus, sea otter, polar bear, dugong and manatee to the UNITED STATES DEPARTMENT OF THE INTERIOR (DOI).

marine park

1. An area of sea, designated by a government, which is protected to preserve wildlife and/or a particular habitat, and for recreational use, e.g. the Great Barrier Reef Marine Park. *See also* MARINE PROTECTED AREA (MPA)

2. *See* OCEANARIUM

Marine Protected Area (MPA) A zone of the seas and coasts where wildlife is protected. In the UK a network of MPAs is being constructed which consists of **MARINE CONSERVATION ZONES (MCZ)**, **SPECIAL AREAS OF CONSERVATION (SACs)**, **SPECIAL PROTECTION AREAS (SPAs)**, **RAMSAR SITES** and **SITES OF SPECIAL SCIENTIFIC INTEREST (SSSIs)**.

Marine Turtles Conservation Act of 2002 (USA) A US law which aims to assist in the conservation of marine turtles and the nesting habitats of marine turtles in foreign countries.

Marineland *Marineland*, Florida, was the world's first **OCEANARIUM** and opened in 1938. The tanks were seascaped and contained coral reefs and a shipwreck. This was the first attempt to capture large sea animals, particularly dolphins, and sustain them in captivity. *Marineland* was originally called *Marine Studios* and was used as a set for a number of films.

mark An individual identification mark used in the field or in a zoo or farm (Fig. M2). *See also* **BIRD RINGING, EAR TAG, MICROCHIP, NECKLACE, V-NOTCHING**

mark test *See* **RED SPOT TEST**

marking
1. The application or attachment of a **MARK** to an animal for identification purposes.
2. A distinguishing mark or pattern on the body of an animal which may be used for identification of the species or a particular individual; the characteristic pattern of coloration, e.g. a pattern in the coat of a mammal or the feathers of a bird.
3. Territory marking. *See also* **SCENT-MARKING**

Markowitz, Hal (1934–2012) Formerly Emeritus Professor at San Francisco State University who pioneered the engineering of active environments for animals living in zoos. In 1982 he published a book entitled *Behavioral Enrichment in the Zoo*.

mark–recapture technique, Lincoln index A method of estimating the size of an animal **POPULATION** in the wild by marking a **SAMPLE**, releasing the marked animals and then capturing a second sample. The proportion of animals in the second sample which is marked should be the same as the

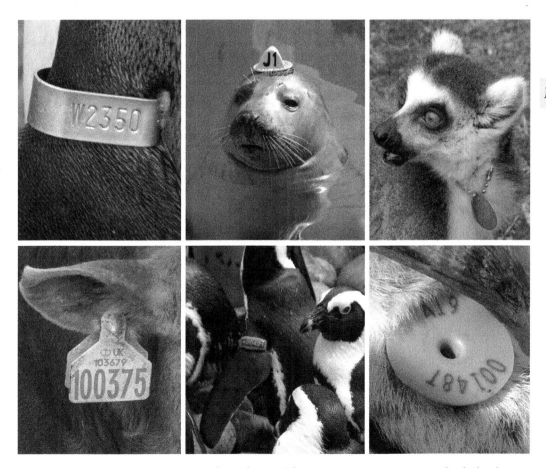

Fig. M2 Mark: metal wing tag on a penguin (Spheniscidae) (top left); temporary plastic cap on a rescued seal (Phocidae) (top middle); coloured necklace on a ring-tailed lemur (*Lemur catta*) (top right); DEFRA approved cattle ear tag (bottom left); silicone rubber wing tag (bottom middle); ear disc tag on a sheep (bottom right).

proportion marked in the entire population. The population size can be estimated as follows:

$$\text{Population estimate} = \frac{N_1 \times N_2}{R}$$

where N_1 = number captured and marked on first occasion; N_2 = number captured on second occasion; and R = number caught on both occasions (i.e. marked on first occasion and recaptured on second occasion). This particular method is known as the Lincoln index and it assumes that no immigration, emigration, births or deaths occur between the times that the two samples are taken. Other more sophisticated methods involve marking animals on more than one occasion. Mark–recapture methods are useful when animals are difficult to find. *See also* CENSUS

marrow *See* BONE MARROW

marsh A wetland area which is low-lying, poorly drained and sometimes covered with water. Occurs around the edge of lakes and streams and also on the coast (as SALT MARSH).

marsupial A pouched mammal. *See also* POUCH (1), MARSUPIALIA, METATHERIA

Marsupialia In some classifications, a mammalian order containing the marsupials, which give birth to underdeveloped young and raise them in a POUCH. Modern classifications divide marsupials into several orders: DASYUROMORPHIA, DIDELPHIMORPHIA, DIPROTODONTIA, MICROBIOTHERIA, NOTORYCTEMORPHIA, PAUCITUBERCULATA, PERAMELEMORPHIA. *See also* METATHERIA

Martha The last PASSENGER PIGEON (*Ectopistes migratorius*). Once extremely common in North America, the species was destroyed by hunting on a massive scale. *Martha* died in Cincinnati Zoo in 1914. She was stuffed and displayed in the SMITHSONIAN INSTITUTION and she has become a symbol of man's destructive effect upon animals.

Martin, Richard (1754–1834) A Member of Parliament for Galway, known as 'Humanity Dick' for his compassion for animals. He was a founding member of the SOCIETY FOR THE PREVENTION OF CRUELTY TO ANIMALS (now the ROYAL SOCIETY FOR THE PREVENTION OF CRUELTY TO ANIMALS (RSPCA)) and responsible for the passing of the first law in the world to protect animals from cruelty: MARTIN'S ACT.

Martin's Act An Act to Prevent the Improper Treatment of Cattle 1822, introduced by Richard MARTIN MP. The Act made it an offence to *'wantonly and cruelly beat, abuse or ill treat any Horse, Mare, Gelding, Mule, Ass, Ox, Cow, Heifer, Steer, Sheep or other Cattle . . .'* This Act was the first national legislation in the world which punished CRUELTY to animals.

Martin the wolf A grey wolf (*Canis lupus*) who achieved notoriety when he escaped death after a government order allowing the culling of a pack of ten wolves in Norway expired on Friday 6 April 2001. Animal rights groups 'offered' *Martin* 'political asylum' in Sweden if he could make it to the border. The cull was ordered because *Martin* and his pack had been killing livestock. However it caused an outcry in Norway and elsewhere; Norway had previously successfully reintroduced grey wolves in partnership with Sweden in the mid 1990s. The wolf was named after Martin Schanche, a veteran Norwegian rally driver who was one of the wolf's most ardent defenders. *See also* LOBO

mascots, animals as Some organisations keep animals as mascots to bring them good luck. They frequently accompany uniformed services on ceremonial occasions. Army regiments often have dogs, goats, ponies and sheep as mascots but other species have included a goose, a pelican, a mule, a baboon and an elephant.

Mason, Georgia A zoologist and expert in STEREOTYPIC BEHAVIOUR at the University of Guelph, Canada, who leads a research group studying how the housing of zoo, farm and laboratory animals affects their welfare and brain functioning.

mass extinction An event in geological time when the Earth experienced a sudden loss of macroscopic species over a wide geographical area. The five major extinction events were the Ordovician–Silurian, late Devonian, Permian–Triassic, Triassic–Jurassic and Cretaceous–Tertiary (K–T). *See also* SIXTH EXTINCTION

mass spectrometer An instrument which measures the masses and relative concentrations of atoms and molecules. It can be used to determine the elemental composition of a sample and the chemical structure of molecules. May be used in, for example, hormone ASSAYS of urine samples.

mass stranding The simultaneous beaching of a large number of CETACEANS. *See also* STRANDING

mastication The mechanical breakdown of food into smaller pieces by chewing with the teeth to aid swallowing and CHEMICAL DIGESTION. *See also* CUDDING

mastitis Inflammation of the mammary gland tissue, usually due to bacterial infection. Infected milk from cows and other livestock may be a health risk.

masturbation Stimulation of the genitals by means other than by sexual intercourse. A common and normal behaviour in many taxa.

mate guarding An individual animal remaining near its mate to prevent others from copulating with it. Mate guarding occurs in males and females in a range of species from shrimps to primates and elephants. *See also* SPERM COMPETITION, SPERM PLUG

mate selection The process by which an animal chooses a member of the opposite sex for reproduction. This process precedes COURTSHIP and involves the selection of an individual of the correct species (to avoid **HYBRIDISATION**), the correct sex and one who will make a good mate. In a captive environment it is often important to allow individual animals to select their own mates if they are to pair and mate successfully. *See also* **SEXUAL SELECTION**

mate suitability index (MSI) An index calculated by **MATERX** that indicates the relative genetic benefit or detriment to the population of breeding from a particular pair of animals. In calculating the MSI, the program considers the **MEAN KINSHIP** values of the pair, the difference in mean kinship values of the male and female, the **INBREEDING COEFFICIENT** of the offspring produced and the amount of unknown ancestry in the pair. In effect the MSI condenses everything known about the **GENETICS** of a pair of individuals into a single number.

maternal behaviour The behaviour exhibited by female animals (especially mammals) when caring for their young, consisting of providing food, shelter and protection etc. (Fig. P10).

MateRx A genetic software tool intended as an aid to population management. For each pair (male/female) in the population it calculates an index (**MATE SUITABILITY INDEX (MSI)**). MateRx was developed by staff at the **SMITHSONIAN NATIONAL ZOOLOGICAL PARK** and **LINCOLN PARK ZOO**.

matinal *See* **MATUTINAL**

mating system The aspect of **SOCIAL ORGANISATION** concerned with the manner in which males and females pair up in order to breed. There are essentially two types of system: **MONOGAMY** and **POLYGAMY**. *See also* **MATE SELECTION**, **MULTI-MALE/MULTI-FEMALE GROUP**

matriarch The dominant female in a group. She is important in some animal societies (e.g. elephants) as a controlling, stabilising influence and a repository of knowledge about the location of food sources etc.

matriline
1. A line of descent through the mother as opposed to the father (**PATRILINE**), i.e. daughter, mother, grandmother etc.
2. A social group which revolves around female **KINSHIP**.

matrilineal hierarchy, nepotistic hierarchy In a typical matrilineal hierarchy a mother's rank determines both her daughter's lowest rank (above lower-born females), and her daughter's highest rank (below the mother and higher-born females).

maturation
1. In relation to aquariums, the process by which the biological filter develops a population of bacteria sufficient to remove all of the ammonia and nitrite produced by the resident fishes and other animals. *See also* **BIOLOGICAL FILTRATION**, **FISHLESS CYCLING**

2. In animal behaviour, an irreversible part of development (**ONTOGENY**) which causes certain behaviours to appear at a particular age, independent of learning. For example, birds become able to fly at a particular point in their development without the need to practise.

matutinal, matinal Active at dawn or early morning. *Compare* **CATHEMERAL**, **CREPUSCULAR**, **DIURNAL (2)**, **NOCTURNAL**, **VESPERTINE**

Max Planck Institutes The Max Planck Society is a German research organisation, which was established in 1948 and consists of 80 research institutes. Konrad **LORENZ** conducted some of his research on animal behaviour while working as director of the Max Planck Institute for Behavioural Physiology which was founded in 1954 (now the Max Planck Institute for Ornithology). The Max Planck Institute for Evolutionary Anthropology in Leipzig, Germany, was founded in 1997. Its aim is to investigate the history of humankind. It conducts comparative research on different genes, cultures, cognitive abilities, languages and social systems of human populations from the past and present. It also studies primate taxa closely related to humans. The Max Planck Society also has research institutes for evolutionary biology, infection biology, human cognitive and brain sciences, and brain research.

maxilla (maxillae *pl.*)
1. The structure formed by the two fused bones of the upper jaw in humans and some other mammals (e.g. other primates). *Compare* **MANDIBLE (1)**
2. Each of a pair of mouthparts used in chewing in many arthropods.

maxillary Relating to a jaw or jaw bone, especially the upper jaw bone.

maximum avoidance of inbreeding (MAI) A designation applied to mating systems in which the least related individuals are mated. Defined by Sewall **WRIGHT**.

maximum sustainable yield (MSY) The largest theoretical yield (catch) that can be taken from a population of animals indefinitely, and which will maintain the population at the point of its maximum growth rate.

May, Robert M. (1938–) An Australian theoretical physicist who was the Chief Scientific Advisor to the UK Government and President of the Royal Society. Prof. Lord May holds professorships at Oxford University and Imperial College, London. He has been instrumental in developing the study of theoretical ecology and **BIODIVERSITY**. Prof. May used mathematical models to explain processes in population growth and disease transmission. He once wrote, '*It is a remarkable testament to humanity's narcissism that we know the number of books in the US Library of Congress on 1 February 2011 was 22 194 656 but cannot tell you – to within an order-of-magnitude – how many distinct species of plants and animals we share our world with*' (May, 2011).

M

Mayr, Ernst (1904–2005) An American evolutionary biologist and taxonomist who was an expert in **SPE-CIATION** and developed the **BIOLOGICAL SPECIES CONCEPT**. He became Director of the Museum of Comparative Zoology at Harvard University and was the author of many important books including *Populations, Species, and Evolution* (1970).

Mazuri® Zoo Feeds Mazuri® Zoo Feeds include a very wide range of products for exotic species, some of which are supplements, and others which are intended as the main diet. They include, for example: Mazuri Bear Diet, Mazuri Callitrichid High Fiber Diet, Mazuri Crocodilian Diet, Mazuri Insectivore Diet, Mazuri Ratite Diet and Mazuri Zebra Pellets. The company also manufactures products for animals that need special diets – such as Mazuri Callitrichid Diabetic Gel and Mazuri Ostrich Breeder (for very highly productive birds) – vitamin and mineral supplements and a **MILK SUBSTITUTE**. Each product is accompanied by a detailed diet sheet which provides information about the ingredients, nutrient content, mixing directions and feeding directions.

McKenna, Virginia (1931–) *See BORN FREE, RING OF BRIGHT WATER*

McMaster chamber A device similar to a microscope slide which is used for counting the eggs of parasites. It has two compartments, each with a grid etched onto the upper surface. When filled with a suspension of faeces in a flotation fluid, eggs float to the surface and those located under the grid are counted.

McMaster egg counting technique A technique for counting parasite eggs under a microscope using a counting chamber (**MCMASTER CHAMBER**) which enables a known volume of faecal suspension ($2 \times 0.15 \, \text{cm}^3$) to be examined.

MCZ *See MARINE CONSERVATION ZONE (MCZ)*

mean The arithmetic average of a set of values. The mean of the values 2, 5, 6, 3 = 16/4 = 4. *See also* **MEDIAN, MODE**

mean birth interval, inter-birth interval, mean calving interval The average time between successive births. In mammals this may be determined at autopsy by counting **PLACENTAL SCARS** and dividing this number into an estimate of the number of years the animal was likely to have been reproductively active.

mean calving interval *See* **MEAN BIRTH INTERVAL**

mean kinship A measure used to assess the genetic importance of an individual within a population by assessing the number of relatives it has in that population and the degree of relatedness. It is used to preserve genetic diversity in small populations and to avoid the negative consequences of inbreeding.

Meat & Livestock Australia (MLA) A producer-owned company which works with industry and the Australian government to achieve a sustainable and profitable red meat and livestock industry. It provides marketing, research and development services to cattle, sheep and goat producers.

Mech, L. David (1937–) An American wildlife biologist who is an expert on wolves. He has worked for the US Geological Survey and the University of Minnesota. Mech founded the International Wolf Center and published *The Wolf: The Ecology and Behavior of an Endangered Species* in 1970.

mechanical chicken harvester A large vehicle which is driven through a chicken house gathering up birds between rotating rubber 'fingers' and moving them onto a conveyor belt that puts them in shipping crates. A number of different designs exist. Studies have indicated that this method of catching chickens when they reach market weight causes less fear in the birds than the traditional method of capturing by hand and carrying them to the crates by the legs held upside-down.

mechanical digestion The physical breakdown of food into smaller pieces by **MASTICATION** and other activities of the **DIGESTIVE SYSTEM**, e.g. movement of the **STOMACH** wall. This aids **CHEMICAL DIGESTION** by providing a larger surface area upon which **ENZYMES** can act.

mechanical filtration, particulate filtration, physical filtration In relation to an aquarium or other aquatic exhibit, the process whereby a device that acts as a strainer removes particles when water is forced through filter media. This may be a sponge, filter floss, special filter pads, aquarium gravel, or a dense mass of air bubbles (only in salt water). *Compare* **BIOLOGICAL FILTRATION**

mechanoreceptor A type of **SENSORY RECEPTOR** located in various parts of the body that is sensitive to mechanical deformation, providing information to the **NERVOUS SYSTEM** about changes to the internal or external environment, e.g. **BLOOD PRESSURE (BP)**, **ORIENTATION (1)** of the body in relation to gravity (**BALANCE**) etc.

MedARKS *See* **MEDICAL ANIMAL RECORD KEEPING SYSTEM (MEDARKS)**

median The middle value when all of the values in a set of data are ranked from the highest to the lowest. The median of the values 7, 3, 6, 4, 1 is 4. If there is an even number of values the median is the mean of the two values either side of the middle. The median of the values 9, 3, 1, 4, 8, 2 is the mean of 3+4 = 3.5. This measure of central tendency is useful where the data set contains outliers as it prevents them from affecting the calculation. For example if the values were 1, 2, 5, 8, 11 the median would be 5. If the values were 1, 2, 5, 8, 169 the median would still be 5. *See also* **MEAN**, **MODE**

Medical Animal Record Keeping System (MedARKS) Software that supports the keeping of veterinary medical records and collection management within zoos.

medulla
1. The inner region of an organ (e.g. the vertebrate kidney) or a bone.
2. A shortened term for **MEDULLA OBLONGATA**.

medulla oblongata, myelencephalon The posterior region of the brain stem. It is concerned with many involuntary movements (e.g. breathing rate, heart rate), balance and hearing.

megafauna A community of large animal species, e.g. the large vertebrate species that inhabit African savannas. *See also* **CHARISMATIC MEGAFAUNA**

megazoo The concept which suggests that wild and captive populations of a species should be managed as a whole, suggested by Neesham (1990). *See also* **METAPOPULATION**

meiosis A type of nuclear division in which the daughter nuclei pass to the **GAMETES**, each of which has only one **CHROMOSOME** from each of the pairs present in the original cell. During this process **DIPLOID** cells produce **HAPLOID** gametes. *Compare* **MITOSIS**

melanic
1. Relating to **MELANISM**.
2. An individual animal who exhibits **MELANISM**.

melanin A dark brown pigment which gives a brown or yellow coloration to the hair, skin etc. It is often located in melanophores (a type of **CHROMATOPHORE**).

melanism An overdevelopment of dark pigmentation (**MELANIN**) in the skin. Some melanic forms are given characteristic vernacular names, e.g. a melanic leopard (*Panthera pardus*) is known as a black panther. In some industrial areas where pollution has darkened the surface of tree bark, walls and buildings some species have evolved melanic forms as a result of **NATURAL SELECTION** whereby they have been favoured over lighter forms which stand out against these dark surfaces. This 'industrial melanism' has occurred in the peppered moth *Biston betularia*. *See also* **ALBINISM, LEUCISM**

membrane *See* **CELL MEMBRANE, MUCOUS MEMBRANE, NICTITATING MEMBRANE, PATAGIAL MEMBRANE**

meme A term coined by Richard **DAWKINS** to represent the behavioural equivalent of a gene. A behaviour or an element of a **CULTURE** which may be transmitted from one individual to another by non-genetic means.

memorabilia *See* **ZOO MEMORABILIA**.

Memorandum of Understanding An agreement between two or more parties which indicates an intended common purpose but which is not legally binding, e.g. Chester Zoo has a Memorandum of Understanding with the Sichuan Forestry Department in China which allows its staff to work in forest reserves on conservation projects and with local communities.

memory The ability of an animal to store, retain and recall information and experiences. *See also* **MENTAL TIME TRAVEL (MTT)**

menagerie A collection of wild animals kept in cages for exhibition to the public. The animals may be housed in permanent buildings or in wagons as a **TRAVELLING MENAGERIE**. The term dates from the 18th century and is derived from the French word ménagerie, which originally meant the management of a household or domestic livestock (a farm). *See also* **BARNUM AND BAILEY, BOSTOCK AND WOMBWELL'S ROYAL MENAGERIE, MONTEZUMA II, TOWER OF LONDON MENAGERIE, TRAVELLING MENAGERIE, VATICAN MENAGERIE, WOODSTOCK**

menarche The first **MENSTRUATION**.

Mendel, Gregor Johann (1822–1884) An Augustinian monk who founded the science of **GENETICS** after conducting extensive experiments with pea plants (*Pisum sativum*). Mendel presented the results of his experiments at meetings of the Brünn Natural History Society in 1865. His ideas were largely ignored by the scientific community until the early 20th century. Mendel's laws of segregation and the independent assortment of characters now form the basis of modern genetics.

menses The period of time during which menstrual bleeding occurs. *See also* **MENSTRUATION**

menstrual cycle The type of reproductive cycle which occurs in humans, all **APES**, and some **MONKEYS**, and is characterised by a loss of blood from the **UTERUS** and **VAGINA** at the end of each cycle. This is regulated by changes in the levels of **SEX HORMONES** in the blood. Animals which exhibit a menstrual cycle are usually sexually receptive most of the time. *See also* **OESTROUS CYCLE**

menstruation The approximately monthly loss of blood and tissue from the uterus in humans, all **APES**, and some **MONKEYS** when the uterine lining (endometrium) breaks down at the end of the **MENSTRUAL CYCLE** if pregnancy does not occur.

mental map The internal representation of the geography of an area held within an animal's brain which allows it to find locations where it has previously found water, food, shelter etc.

mental state attribution The ability to ascribe a particular mental state to another individual; to know what another is thinking. For example, to know whether or not another individual can see a piece of partially concealed food. *See also* **THEORY OF MIND**

mental time travel (MTT) The capacity of some species to plan ahead. This involves accessing a past memory to fulfil a future need. In effect they are capable of 'planning ahead', e.g. the Eurasian jay (*Garrulus glandarius*) hoards food that will not be available in the future. Some psychologists believe that only humans are truly capable of MTT.

mesoderm The middle layer of cells in a triploblastic organism. It gives rise to muscles, blood vessels, some nervous tissues, many organs and the vertebrae.

M

messenger RNA (mRNA) A type of **RNA** which is produced during **TRANSLATION** and transfers the information contained in the **GENETIC CODE** of **DNA** to the **RIBOSOME** during **PROTEIN SYNTHESIS**.

meta-analysis A statistical technique used to combine the results of several independent but similar studies in order to create a larger data set.

metabolism The totality of all of the chemical reactions that occur within the cells of an organism when it is alive. May also refer to part of an organism, e.g. liver metabolism. *See also* **ANABOLISM**, **CATABOLISM**

metamorphosis The transformation of the larva into the adult form in some taxa such as insects, amphibians, e.g. a caterpillar metamorphoses into a butterfly; a tadpole metamorphoses into a frog.

metapopulation A regional group of populations that are spatially separated but interact at some level. Animals of a particular species kept by a number of zoos function as a metapopulation when they are managed and exchanged for breeding purposes as part of captive breeding programmes (e.g. **EUROPEAN ENDANGERED SPECIES PROGRAMME (EEP), SPECIES SURVIVAL PLAN® (SSP) PROGRAMS**). *See also* **MEGAZOO**

metastasis (metastases *pl.*)
1. The spread of a disease from one part of the body to another via the blood, lymph, or across the body cavity especially the cells of malignant tumours.
2. A secondary tumour resulting from the spread of malignant disease (**CANCER**).

Metatheria An infraclass of mammals; the marsupials and their extinct relatives *See also* **MARSUPIALIA**

metered dose inhaler (MDI) A device designed for the administration of drugs to the respiratory tract via small pressurised canisters. A single metered dose is administered through a mouthpiece by depressing the canister. The particles delivered by MDIs are larger than those created by nebulisation and, thus, do not penetrate as deeply into the respiratory tract. *See also* **NEBULISER**

methane A colourless, odourless, flammable gas (CH_4) which is released from organic waste, especially animal slurry, and may be explosive in a confined space. It is also produced by the guts of animals, especially **RUMINANTS**, as a product of digestion, and released largely through the mouth.

micro pig, mini pig, miniature pig Miniature pig breeds which have been bred for medical use and as pets.

microbe *See* **MICROORGANISM**

microbiology The scientific study of microorganisms.

Microbiotheria An order of mammals: Monito del Monte.

microchip A very small piece of semiconductor material which contains all of the components of an integrated circuit. Commonly used to **MARK** animals for identification purposes. *See also* **RADIO FREQUENCY IDENTIFICATION (RFID) TECHNOLOGY**

microcurrent treatment, microcurrent therapy The use of electrical stimulation with very low voltages to provide pain relief and facilitate faster healing of soft-tissue injuries. Used in the veterinary treatment of horses with a wide range of conditions including joint **INFLAMMATION**, **ABSCESSES**, **LAMINITIS**, **COLIC** and muscle spasms.

microhabitat A specialised **HABITAT** occupied by a species, which is small and narrowly defined, where environmental conditions are different from those of the surrounding area, e.g. a crack in a rock, a clump of grass, a cardboard box.

micronutrient Any essential nutrient required by an organism in minute quantities such as a **VITAMIN** or **TRACE ELEMENT**. *Compare* **MACRONUTRIENT**

microorganism, microbe An organism that is too small to see with the naked eye, especially **BACTERIA** and **PROTOZOANS** and some **FUNGI** and **ALGAE**, but not **VIRUSES** and **PRIONS**, which are considered to be non-living. Many microorganisms are involved in decomposition and some cause disease.

microsatellite marker, DNA microsatellite marker A type of genetic or **MOLECULAR MARKER** which consists of a specific sequence of **DNA** bases or **NUCLEOTIDE**s which contains mono (1), di (2), tri (3), or tetra (4) tandem repeats (Table M1). They are widely used for forensic identification of samples, relatedness testing and the identification of population fragmentation. Microsatellites are inherited in Mendelian fashion.

microvillus (microvilli *pl.*) Microscopically small finger-like projections protruding from the surface of a cell to increase its **SURFACE AREA:VOLUME RATIO**, thereby increasing its absorptive capacity, e.g. the epithelial cells covering the surface of the **VILLI** of the small intestine.

micturation Urination.

midden A pile of animal dung accumulated by the repeated use of the same place for defecation. May be used by some mammals to mark their territory, e.g. dik dik (*Madoqua* spp.), rhinoceros (Fig. T2).

Midgley, Mary (1919–) A moral philosopher who has written widely on animal rights. Author of *Animals and why they matter* (1983).

migration
1. The movement of animals from one place to another, especially over long distances along well-defined routes on a seasonal basis as a result of

Table M1 Microsatellite: examples of types of microsatellites.

Tandem repeat type	Base sequence	Abbreviation
mono-nucleotide	GGGGGGGGGG	$(G)_{10}$
di-nucleotide	CTCTCTCTCT	$(CT)_5$
tri-nucleotide	ACTACTACTACT	$(ACT)_4$
tetra-nucleotide	CTGACTGACTGA	$(CTGA)_3$

environmental changes. Baker (1978) defined migration in a very general sense as '*the act of moving from one spatial unit to another.*' *See also* **MIGRATORY SPECIES, NAVIGATION**

2. The movement of animals of a particular **SPECIES** from one **POPULATION** to another population. This is one of the factors that affects **POPULATION SIZE**.

3. An occurrence or act of migrating, e.g. a bird migration.

4. A group of organisms which are migrating, e.g. a migration of wildebeest.

Migratory Bird Hunting and Conservation Stamp Act of 1934 (USA) A US law also known as the Duck Stamp Act. It requires waterfowl hunters 16 years of age or older to possess a valid federal hunting stamp (duck stamp). Money raised from the purchase of these stamps is used to acquire migratory bird refuges and waterfowl production areas. The stamps are produced by the US Postal Service.

Migratory Bird Treaty Act of 1918 (USA) A US law which implements a number of treaties and conventions between the US and Canada, Japan, Mexico and the former Soviet Union aimed at the protection of migratory birds. It restricts the circumstances in which it is lawful to hunt, kill, capture, possess, buy, or sell any migratory species of bird, including its feathers, other parts, eggs, nests or products.

migratory species

1. A species that exhibits **MIGRATION (1)** (*compare* **RESIDENT SPECIES**). Biological and legal definitions differ.

2. Article I of the **CONVENTION ON THE CONSERVATION OF MIGRATORY SPECIES OF WILD ANIMALS 1979 (CMS)** defines 'migratory species' as '*the entire population or any geographically separate part of the population of any species or lower taxon of wild animals, a significant proportion of whose members cyclically and predictably cross one or more national jurisdictional boundaries.*' (The reference to 'national jurisdictional boundaries' is important in this context because the convention is concerned with protecting species that migrate from the territory of one state to that of another.)

milk A white or yellowish liquid produced by the **MAMMARY GLANDS** of female mammals to nourish their young. Contains carbohydrates, lipids, proteins, vitamins and minerals (especially calcium). The chemical composition varies considerably between taxa. *See also* **MILK SUBSTITUTE**

milk let-down The release of milk from the **MAMMARY GLANDS**. *See also* **OXYTOCIN**

milk substitute An artificially made **MILK** formulation used as a substitute for mother's milk when hand-rearing young mammals.

milker

1. A cow kept for its milk.

2. A person or machine that milks cows.

milking

1. The process of extracting milk from a mammal, for use as a food. Milking is either performed by hand, by squeezing the teats, or it is automated. *See also* **MILKING PARLOUR**

2. The process of removing venom from a snake. Often performed to make snakes used in entertainment safe to handle or in order to obtain **VENOM** which may be used to manufacture an **ANTIVENIN**.

milking parlour, milking parlor A place where cows or other animals are milked, often by **MILKING (1)** machines. The main types are: the side opening (tandem) parlour, the herringbone (fishbone) parlour, the parallel (side by side) parlour (Fig. M3), the swing (swing over) parlour and the rotary (carousel, turnstile) parlour. *See also* **ROBOTIC MILKING MACHINE**

mimicry The resemblance of one animal species (the mimic) to another animal or plant species (the model) such that they may be confused. *See also* **AUTOMIMICRY, BATESIAN MIMICRY, MÜLLERIAN MIMICRY**

mineral An element that is essential in the diet. The importance of different elements varies between species. Mammals need iron to make the blood pigment haemoglobin. However, in molluscs, crustaceans and some spiders the respiratory pigment is haemocyanin, and this contains copper in place of iron. Vertebrates need calcium for their skeletons and birds need it for egg shell production. While these nutrients are essential in trace amounts, most are highly toxic in large concentrations.

mineral lick A solid block of material containing minerals given as a nutrient supplement.

mini pig *See* **MICRO PIG**

miniature pig *See* **MICRO PIG**

miniature railway A narrow-gauge railway usually powered by steam engines used for giving rides to visitors. Many zoos operate such railways between or even within **ANIMAL ENCLOSURES**, giving visitors an interesting **VIEWPOINT**, e.g. **WHIPSNADE ZOO**.

mini-beasts A vernacular name for insects, spiders, small amphibians, small reptiles and other small species when kept in a zoo. Sometimes used for these taxa in the wild, especially with children.

minimum viable population (MVP) The minimum population size required to provide some specified probability that the population will survive for a given period of time. Some studies have suggested MVPs of more than 5000 are necessary for long-term persistence, regardless of the species and the environmental conditions. Recent studies have cast doubt on the general applicability of this figure and suggest it may not be useful for conservation planning. *See also* **DEMOGRAPHICALLY EXTINCT, FUNCTIONALLY EXTINCT**

mirror test *See* **RED SPOT TEST**

miscarriage *See* **ABORTION**

mist net A large fine net used for catching small birds and bats in the wild by extending it between trees or at cave entrances etc.

M

Fig. M3 Milking parlour: a parallel (side by side) parlour.

mite A small arachnid of the subclass Acarina, many species of which are of great economic importance because they are parasites of animals or plants. Some species cause allergic diseases such as hay fever, asthma and eczema, others cause sarcoptic mange and aggravate atopic **DERMATITIS**. May be controlled using an **ACARICIDE**.

mitochondrial DNA (mtDNA) DNA found in **MITO-CHONDRIA**. In **SEXUAL REPRODUCTION** mitochondria are inherited exclusively from the mother. Since mtDNA remains largely unchanged from generation to generation it is possible to trace maternal lineage far back in time, ultimately allowing the evolutionary history of a species to be traced.

mitochondrion (mitochondria *pl.*) A cell organelle which is approximately sausage-shaped and is involved in the production of **ADENOSINE TRI-PHOSPHATE** in **AEROBIC RESPIRATION** because it is the location of the **KREBS CYCLE** and the **ELECTRON TRANSPORT SYSTEM**. Mitochondria contain **MITOCHONDRIAL DNA (mtDNA)**

mitosis The process of nuclear division in which a single **NUCLEUS** splits to produce two identical daughter nuclei (which contain identical **CHROMO-SOMES** and hence **GENES**). This is normally fol-

lowed by cytokinesis (cell division) and is the normal method of division in **SOMATIC** cells during the growth and repair of tissues. *Compare* **MEIOSIS**

mixed-sex dyad An association between two animals (a **DYAD**) in which one is male and the other is female.

mixed-species exhibit *See* **MULTI-SPECIES EXHIBIT**

Mixiniformes An order of fishes: hagfishes.

moa Any of a group of extinct ostrich-like birds which were endemic to New Zealand.

moat A channel used to retain animals within an enclosure. It may be deep or shallow, dry or water-filled. Moats have been replaced by **HA-HAS** in many modern zoo enclosures. *See also* **DRY MOAT**, **WET MOAT**

mob
1. The collective term for a group of kangaroos or wallabies.
2. To attack as a group, e.g. crows mobbing a buzzard when it approaches their roost.

mobile field shelter A building used as an animal shelter, especially on farms. It is constructed on a metal frame and fitted with skids so that it may be towed across fields by a suitable vehicle.

modal value *See* **MODE**

mode, modal vale The value within a set of data which occurs most frequently. The mode of the values 3, 6, 7, 2, 6, 4 is 6. Modal values are useful where the calculation of a **MEAN** would not produce an integer. For example, it may be more useful to calculate the modal number of eggs produced by birds in a particular colony (e.g. 4) rather than calculate a mean (e.g. 3.8). *See also* **BIMODAL DISTRIBUTION**. *See also* **MEDIAN**

model In mathematics, a model is an equation or group of equations which allows events to be simulated, e.g. **POPULATION VIABILITY ANALYSIS (PVA)** allows theoretical changes in a population to be simulated. Some models are **DETERMINISTIC** and the outcome is the same each time the model is run. Others are **STOCHASTIC** and include random elements so the outcome is different each time. *See also* **MODEL SPECIES**

model species A relatively common species which may be kept in order that staff may develop expertise in the keeping of related rarer species, e.g. meerkats may act as model species for rarer mongoose species.

Modernist Movement A period in the history of zoos, up to the middle of the 20th century, when they were effectively living art galleries. *See also* **DISINFECTANT ERA**

moke
1. A **DONKEY**.
2. An Australian slang term for a horse, especially one which is old and worn out. *See also* **YARRAMAN**

molar A type of grinding tooth found at the rear of the jaw in most mammals.

molecular marker, genetic marker A fragment of **DNA** associated with a certain location within the **GENOME**. Used in molecular biology to identify a particular DNA sequence in a pool of unknown DNA. *See also* **DNA FINGERPRINTING, MICROSATELLITE MARKER**

mollusc A member of the phylum **MOLLUSCA**.

Mollusca A **PHYLUM** of animals. Molluscs are soft-bodied animals with a muscular 'foot' and often possess a shell, e.g. snails, slugs, octopuses, squids. The mantle is a sheet of specialised tissue which covers the viscera like a body wall. In most groups of molluscs it secretes a shell which protects the body. Within the mantle cavity are the gills (ctenidia) and the floor of the mouth possesses a radula which is used for scraping food. The cephalopods (octopuses, squids and their relatives) possess a well-formed head with a relatively large and advanced brain and eyes, along with arms and tentacles capable of manipulating objects.

molluscicide A chemical which kills molluscs.

molt *See* **MOULTING**

molting *See* **MOULTING**

mongrel An animal which has resulted from **INTERBREEDING**, especially a dog of undetermined or mixed breed.

monkey A primate belonging to the mammalian infraorder Simiiformes (simians) excluding the **APES**. Monkeys may be divided into **OLD WORLD MONKEYS** and **NEW WORLD MONKEYS**.

Monkey Forest An English zoo which consists of a single large walk-through naturalistic enclosure containing a group of Barbary macaques (*Macaca sylvanus*). The macaques live in a woodland, with no access to indoor accommodation.

monkey gap *See* **DIASTEMA**

Monkey World A specialist rescue centre in Dorset, UK, which campaigns to prevent the smuggling of primates and rehabilitates abused, neglected and confiscated animals in naturalistic groups.

monocotyledonous plant A flowering plant that has a single seed leaf and narrow leaves with parallel veins running along the long axis. They are important food plants for many grazers, and include grasses and cereal crops, e.g. wheat, barley, oats, bamboo. *Compare* **DICOTYLEDONOUS PLANT**

monoculture The practice of growing a single crop or plant species over a large area and for a number of consecutive years. Commonly used in intensive agricultural systems and forestry. Encourages the spread of disease, due to the genetic similarities of the individual plants, and the depletion of soil nutrients.

monoestrous Having a single breeding season (**OESTROUS CYCLE**) in a year, e.g. bears. *Compare* **POLYOESTROUS**

monogamous Having a single mate. *Compare* **POLYGAMOUS**

monogamy A social system (or mating system) in which one male and one female mate more or less exclusively over time. The group consists of the pair and its offspring.

monogastric Possessing a simple gut containing a stomach with a single chamber. *Compare* **RUMINANT**

monohybrid cross A genetic cross which considers the inheritance of a single gene which has two **ALLELES**. When both parents are **HETEROZYGOUS** for the gene (e.g. Aa × Aa), the **PHENOTYPES** produced in the F_1 generation occur in the ratio 3:1 (AA, Aa, Aa, aa). *Compare* **DIHYBRID CROSS**

monomorphic species
1. A species which has only one form.
2. A species in which there is only one genotype.
3. A species which has parts that exist in only one form. Often used to describe a species where the sexes superficially appear identical (i.e. they are sexually monomorphic). *Compare* **SEXUAL DIMORPHISM**

monorchid An individual animal in which only one testicle is apparently present, the other being absent, removed or undescended.

monosaccharide *See* **CARBOHYDRATE**

Monotremata An order of egg-laying mammals: echidnas and the duck-billed platypus (*Ornithorhynchus anatinus*).

M

monotreme A mammal belonging to the order **MONOTREMATA**.

monotypic
1. In **TAXONOMY**, relating to a **TAXON** which contains only one subgroup at the next lowest taxonomic level, e.g. a genus with just one species, a family with a single genus.
2. A species is said to be monotypic if it exhibits no geographical variation, i.e. has no recognised **SUBSPECIES**. Especially used with reference to birds. *Compare* **POLYTYPIC**

montane Relating to or living in a mountainous region, e.g. montane forest.

Montezuma II An Aztec ruler who kept a large menagerie in his palace at Tenochtitlán, Mexico. The Spanish conqueror Hernando Cortés discovered it in 1519. The menagerie was reputed to have had 600 keepers.

moose farm A farm where moose (*Alces alces*) are reared as domesticated animals to produce milk and cheese, and kept as a tourist attraction, e.g. in Sweden and Russia.

morality
1. The quality of exhibiting good (virtuous) conduct. Some species appear to exhibit a moral sense, e.g. captive chimpanzees (*Pan troglodytes*) will help other chimps to obtain food when it is out of their reach.
2. A system of ideas which distinguishes right from wrong; a system of moral principles.
3. '. . . *a suite of interrelated other-regarding behaviors that cultivate and regulate complex interactions within social groups. These behaviors relate to well-being and harm, and norms of right and wrong attach to many of them*' (Bekoff and Pierce, 2009).
See also **ANIMAL RIGHTS**

moratorium An agreed temporary cessation of an activity, e.g. hunting. *See* **WHALING MORATORIUM**

morbid Indicative of disease; of the nature of disease. *See also* **MORBIDITY**

morbidity
1. The state of being **MORBID**.
2. Morbidity rate; rate of incidence of a disease.

morbidity rate *See* **MORBIDITY (2)**

morph
1. A particular distinct morphological form of an organism or species; a phenotypically distinct form.
2. Any one of the particular forms of individual found in a **POLYMORPHIC SPECIES**.

morphological species A group of individuals which are considered to represent a single species by virtue of similarities in their anatomy. A species concept widely used by palaeontologists when classifying fossil forms where genetic evidence is not available. *See also* **SPECIES**

morphology
1. The physical form and structure of an organism, especially its external features.
2. The scientific study of the physical form of organisms.

Morris, Desmond (1928–) Dr. Desmond Morris is a zoologist, TV presenter and artist who was formerly Curator of Mammals at **LONDON ZOO**. He is the author of many academic and popular books on zoology including *The Naked Ape* (1967), which described human behaviour from the point of view of a zoologist, *The Human Zoo* (1969) and *The Animal Contract* (1990). Dr. Morris was the first presenter of *Zoo Time*, the first television wildlife series in the world aimed at children. A total of 331 weekly episodes was transmitted between 1956 and 1968, broadcast from a specially built TV studio in London Zoo. *See also* **TEN COMMANDMENTS**

Morris, Johnny (1916–1999) *See* **ANIMAL MAGIC**

mortality
1. The death rate of organisms in a population, often expressed as deaths per 1000 individuals per year.
2. The number of deaths.
3. Death, especially on a large scale.

mortality rate *See* **MORTALITY (1)**

moth trap A device for catching moths by attracting them with a light (often **ULTRAVIOLET LIGHT**) (Fig. I3).

motile Relating to animals or their gametes, moving or having the capacity for spontaneous movement. *Compare* **SESSILE**

motivation In animal behaviour, the underlying cause of a behaviour. This may be affected by both internal changes (in the body, e.g. hunger) and external changes (in the outside environment, e.g. a change in temperature).

motor co-ordination *See* **CO-ORDINATION**

motor neurone A **NEURONE** located in the **CENTRAL NERVOUS SYSTEM** **(CNS)** whose **AXON** projects beyond the CNS and controls muscles. *See also* **INTERMEDIATE NEURONE**, **SENSORY NEURONE**

moult *See* **MOULTING**.

moulting, molting
1. The shedding of the skin or outer layer of the body during growth in insects, other arthropods and reptiles. *See also* **ECDYSIS**, **INSTAR**
2. The seasonal loss of fur in mammals or feathers in birds. *See also* **CATASTROPHIC MOULT**

mountain chicken frog (*Leptodactylus fallax*) A large frog that is only found in Dominica and Monserrat, in the Caribbean. Its wild populations have suffered from a dramatic decline due to **CHYTRIDIOMYCOSIS**. As part of a rescue effort for this species scientists from **LONDON ZOO** and elsewhere have taken several individuals into captivity.

mouth gag A metal device for separating the jaws of a large animal (e.g. a horse) during dental examination and treatment.

mouthbrooder An animal that protects its young by holding them in its mouth for an extended period. Mouthbreeding is also called oral incubation or **BUCCAL** incubation. Occurs in some fishes and frogs. *See also* **INCUBATE (1)**

movement document A document which must be completed when animals are moved between farms. For example the movement document required under the Sheep and Goats (Records Identification and Movement) (England) Order 2009, which indicates the point of departure, destination, date and the individual identification numbers of the animals being moved. *See also* CATTLE PASSPORT, PET PASSPORT

movement restrictions Restrictions placed by law upon the transportation of LIVESTOCK, especially during disease outbreaks, e.g. the Disease Control (Interim Measures) (Scotland) Order 2002.

Mr. Ed A 1960s American television situation comedy that starred a 'talking' horse (*Mr. Ed*) whose voice could only be heard by his friend Wilbur Post.

MRI scanner *See* MAGNETIC RESONANCE IMAGING (MRI) SCANNER

mRNA *See* MESSENGER RNA (mRNA)

mtDNA *See* MITOCHONDRIAL DNA (mtDNA)

mucking out A colloquial term for removing waste (dung, straw, waste food etc.) from ANIMAL ENCLOSURES and animal houses.

mucosa *See* MUCOUS MEMBRANE

mucous (*adj.*) Of or relating to MUCUS.

mucous membrane, mucosa A term applied to the EPITHELIUM (which contains MUCUS-secreting goblet cells) forming the lining of the gut, urogenital tract and other surfaces in vertebrates, along with the immediately underlying CONNECTIVE TISSUE.

mucus A slimy secretion produced by goblet cells in the MUCOUS MEMBRANES of vertebrates. It contains mucins (a type of glycoprotein).

mucus plug *See* SPERM PLUG

Muir, John *See* SIERRA CLUB

mule The sterile offspring of a male DONKEY and a female horse. Valued as a BEAST OF BURDEN (Fig. M4).

mulesing A method of reducing FLY STRIKE in sheep in Australia by the removal of the skin around

Fig. M4 Mule: a mule working as a draught animal in Marrakesh, Morocco.

the hindquarters to produce a smooth, wool-free area which is less attractive to flies. *See also* TAIL DOCKING

Müllerian mimicry The similar appearance of two species where both are distasteful to predators so both species are avoided by a predator once it has tasted just one of them. *Compare* AGGRESSIVE MIMICRY, AUTOMIMICRY, BATESIAN MIMICRY

multicellular Composed of many cells. A characteristic of all animals. Almost all multicellular animals have CELLS organised into discrete TISSUES, apart from sponges (PORIFERA).

multi-male/multi-female group A social group consisting of many sexually mature males and sexually mature females, along with their young of various ages. *See also* MATING SYSTEM

multipara A MULTIPAROUS female.

multiparous
1. Having given birth more than once.
2. Producing more than one offspring at each birth. *Compare* NULLIPAROUS, POLYPAROUS, PRIMIPAROUS, UNIPAROUS

multiple ocular coloboma (MOC) A congenital eye malformation that occurs in snow leopards (*Uncia uncia*) and some other species, including humans. The malformation affects the upper eyelid, retina and optic nerve, but the cause is not fully understood. There may be a genetic link or the condition may arise in offspring following a nutritional deficiency or other problems during pregnancy.

multi-species exhibit, mixed-species exhibit An exhibit which contains more than one species kept in the same enclosure, e.g. *African Plains*, Dublin Zoo, which contains giraffes, ostriches, zebra and scimitar-horned oryx. The best multi-species exhibits contain species which would be found together in the same ecosystem in the wild. However, some zoos house compatible species from different ecosystems together for convenience.

mural A painting on a wall, for example, in a zoo or museum exhibit. Usually a scene of a forest, desert, or some other habitat. Often used on the sides and back of a VIVARIUM or cage to create the illusion of a much larger space. Sometimes painted on the walls of the visitor area inside an animal house to create an immersion effect. *See also* IMMERSION EXHIBIT

murine Relating to members of the RODENT family Muridae: rats, mice and their relatives.

Murray Valley encephalitis (MVE) *See* ANIMAL SENTINELS

muscle A contractile tissue composed mainly of the proteins actin and myosin which slide over each other during contraction. There are three basic forms: SKELETAL MUSCLE, SMOOTH MUSCLE and CARDIAC MUSCLE. Most muscles are controlled by the CENTRAL NERVOUS SYSTEM (CNS).

muscle system, muscular system An organ system consisting of SKELETAL MUSCLES, SMOOTH MUSCLES and CARDIAC MUSCLES, which

M

controls movements, maintains posture and circulates blood.

museum An institution of education and research that displays exhibits to the public which may include preserved animals. Some museums also contain collections of live animals, especially those that may be kept in **AQUARIUMS (2)** or **VIVARIA**. *See also* **ARIZONA–SONORA DESERT MUSEUM, NATURAL HISTORY MUSEUM (LONDON), SMITHSONIAN INSTITUTION**

music as enrichment *See* **AUDITORY ENRICHMENT**

musk gland A gland, usually an **ANAL GLAND**, that secretes a highly odorous substance called musk. It occurs in a number of species including Siberian musk deer (*Moschus moschiferus*) and the house musk shrew (*Suncus murinus*).

mustelid A member of the family **MUSTELIDAE**.

Mustelidae A mammalian family containing weasels, **FERRETS**, otters and their relatives (Fig. F1).

musth A condition associated with sexual activity which occurs seasonally in adult bull elephants in which they undergo dramatic increases in the levels of **TESTOSTERONE** and become particularly restless and aggressive. A secretion is released from the **TEMPORAL GLAND** and they continually dribble urine. Some elephant keepers are able to prevent bulls from coming into musth by dominating them. Risks to keepers can be reduced by handling bulls using **PROTECTED CONTACT** techniques. Musth also occurs in camels.

mutagen An agent that is capable of causing a mutation or increasing mutation frequency, e.g. X-rays, colchicine, viruses.

mutation *See* **CHROMOSOME MUTATION, GENE MUTATION, MUTAGEN**

mutton busting An event held at **RODEOS** in the USA whereby children ride **SHEEP** in the same manner as men ride steers or unbroken horses in a rodeo, until they are thrown off. The object of the contest is to remain on the sheep for as long as possible.

mutualism An association between two different **SPECIES** in which both gain some benefit to their **FITNESS**. For example, many species of plants provide food for **HERBIVORES** who in turn disperse their seeds. *Compare* **INTERSPECIFIC COMPETITION, PARASITE**

muzzle
1. The snout; the projecting forward part of the head of certain animals (e.g. canids), including the nose, mouth and jaws.
2. A cover strapped over an animal's mouth (especially a dog) to prevent it from biting or to prevent feeding. *See also* **GRASS MUZZLE**
3. The action of applying a muzzle.

mycotoxin A toxic substance produced by a fungus which may spoil animal feed, especially when stored.

Myctophiformes An order of fishes: lanternfishes and their allies.

myelencephalon *See* **MEDULLA OBLONGATA**

myelin sheath A covering around the outside of the **AXON** in some **NEURONES** (myelinated nerve fibres) which increases the speed of propagation of the nerve impulse. It is composed mostly of lipid and protein.

myiasis *See* **FLY STRIKE**

Myliobatiformes An order of fishes: stingrays and their allies.

myosin *See* **MUSCLE**

Myriapoda A subphylum of arthropods containing centipedes, millipedes and their relatives.

Myxini A class of fishes: hagfishes.

myxomatosis A fatal disease of rabbits caused by the myxoma virus which occurs in many countries. It was introduced into Australia in 1950 in an attempt to control the introduced rabbit population. A vaccine is available to protect pet rabbits.

M

Fig. N1 N is for Natural History Museum (London).

Dictionary of Zoo Biology and Animal Management: A guide to terminology used in zoo biology, animal welfare, wildlife conservation and livestock production, First Edition. Paul A. Rees.
© 2013 John Wiley & Sons, Ltd. Published 2013 by John Wiley & Sons, Ltd.

nace The 'pointed' end of a bird's egg.

Nagoya Protocol 2010 The Nagoya Protocol on Access to Genetic Resources and the Fair and Equitable Sharing of Benefits Arising from their Utilisation is a **PROTOCOL** to the UN **CONVENTION ON BIOLOGICAL DIVERSITY 1992 (CBD)**. Its aim is to ensure the sharing and equitable distribution of the benefits arising from the use of genetic resources by the appropriate transfer of technologies and funding, taking into account all rights over such resources, thereby promoting the conservation of biological diversity and the sustainable use of its components.

nail A flattened covering of horn-like material (**KERATIN**) on the dorsal aspect of the terminal **PHALANGES** of fingers and toes. Nails are found in most **PRIMATES** (except marmosets and tamarins) and are important in providing a hard backing to the fingertips. They also occur in a small number of other mammalian species, e.g. elephants and manatees.

nares Nostrils.

nasal discharge A liquid released from the nose which may be indicative of the presence of a respiratory infection or some other disease.

natal Relating to birth, e.g. natal group.

natal coat The fur of a newborn mammal, especially one which is coloured differently from that of an adult of the same species, and which develops the adult coloration as it ages. Spotted natal coats occur in some artiodactyls (e.g. peccaries) and felids (e.g. pumas) and cream, black or ostentatious coats occur in some primates (e.g. gibbons).

natal group The social group into which an individual animal is born. *See also* **PHILOPATRY**

natality *See* **BIRTH RATE**. *See also* **FECUNDITY**

natality rate *See* **BIRTH RATE**

National Animal Identification and Tracing (NAIT) An IT system which is used to tag individual livestock and deer with electronic ear tags in New Zealand. In the event of a biosecurity alert, the system will allow infected animals to be identified and traced by their individual tag number. Established under the National Animal Identification and Tracing Act 2012.

National Animal Identification System (NAIS) The identification and traceability system for livestock in the USA, established to facilitate animal health surveillance by tracking movements of individual animals from birth to slaughter. A premises registration system and a system to administer animal identification numbers is maintained by the **UNITED STATES DEPARTMENT OF AGRICULTURE (USDA)** (administered by the **ANIMAL AND PLANT HEALTH INSPECTION SERVICE (APHIS)**) and the individual states (by animal health boards). The system covers most livestock species (including some fishes).

National Audubon Society, Audubon An American conservation organisation, named after John AUDUBON, whose mission is to conserve and restore natural ecosystems, focusing on birds, other wildlife, and their habitats for the benefit of humanity and the Earth's biological diversity. *See also* **INTERSPECIES EMBRYO TRANSFER**

National Beef Association An organisation which promotes the preservation and improvement of beef cattle in the UK and elsewhere, by promoting the breeding of beef cattle, and improving the standards of management, transport, slaughter and treatment of cattle by education.

National Canine Defence League *See* **DOGS TRUST**

National Cattle Association (Dairy) An umbrella organisation for the dairy cattle breed societies in the UK.

National Centre for the Replacement, Refinement and Reduction of Animals in Research (NC3Rs) An independent scientific organisation, tasked by the UK Government with supporting the country's science base through the application of the 3Rs (replacement, refinement and reduction) to research.

National Fallen Stock Company (NFSCo) A not-for-profit organisation that provides a fallen livestock collection and disposal service to farmers in the UK. It was established by the **DEPARTMENT FOR ENVIRONMENT, FOOD AND RURAL AFFAIRS (DEFRA)** and the devolved administrations of Wales, Scotland and Northern Ireland, but is not a government organisation. *See also* **FALLEN STOCK**

National Farmers' Union (NFU) An organisation which represents the interests of British farming and provides professional representation and services to its members.

national flock *See* **NATIONAL HERD**

National Geographic Association A large non-profit scientific and educational institution, based in Washington DC, whose interests include geography, archaeology, natural science, environmental and historical conservation, world culture and history. It has published its magazine *National Geographic* (originally *The National Geographic Magazine*) since 1888, the year the organisation was founded.

national herd, national flock The population of individual **LIVESTOCK** animals of a particular species or breed which is resident within a particularly country, e.g. the national herd of dairy cattle in the USA, the UK national alpaca herd, the national sheep flock of Australia (Table N1).

National Heritage Animal Status afforded to certain animals in India to focus attention on the need for their conservation: tigers and elephants.

national law, domestic law The law made by, and in force in, a particular state, e.g. English law, French law, Kenyan law. *Compare* **EUROPEAN LAW**, **INTERNATIONAL LAW**

National Livestock Identification System (NLIS) The identification and traceability system for livestock in Australia. It allows animals to be identified and tracked from their property of birth through to

Table N1 National herds and flocks in England, June 2011 (Anon., 2011b).

Livestock type	Total animals
Poultry	120 000 000
Sheep	14 300 000
Cattle	5 400 000
Pigs	3 600 000
Horses	217 000
Goats	79 000
Deer	21 000

slaughter and vice versa. It is based on a property registration system for anyone keeping cattle, pigs, sheep or goats.

National Marine Fisheries Service (NMFS) A US federal agency which is responsible for the management, conservation and protection of living marine resources within the United States' Exclusive Economic Zone (water three to 200 miles offshore) and is jointly responsible (with the **UNITED STATES FISH AND WILDLIFE SERVICE (USWFS)**) for implementing the **ENDANGERED SPECIES ACT OF 1973 (USA)**.

National Nature Reserve (NNR) A nationally important site for nature conservation in the UK designated in Great Britain under the **NATIONAL PARKS AND ACCESS TO THE COUNTRYSIDE ACT 1949** and the **WILDLIFE AND COUNTRYSIDE ACT 1981** and in Northern Ireland under the Amenity Lands Act (Northern Ireland) 1965. All NNRs are protected as **SITES OF SPECIAL SCIENTIFIC INTEREST (SSSI)**.

national park (NP) Under the Global Protected Areas Programme of the **INTERNATIONAL UNION FOR THE CONSERVATION OF NATURE AND NATURAL RESOURCES (IUCN)**, national parks are category II protected areas and defined as '*large natural or near natural areas set aside to protect large-scale ecological processes, along with the complement of species and ecosystems characteristic of the area, which also provide a foundation for environmentally and culturally compatible spiritual, scientific, educational, recreational and visitor opportunities.*' Some zoo exhibits use the name of important national parks in their titles: *Kaziranga Forest Trail*, the Asian elephant exhibit at Dublin Zoo (based on Kaziranga NP in Assam, India); *Tsavo Bird Safari*, **CHESTER ZOO** (based on **TSAVO** NP, Kenya); *Lions of the Serengeti*, **WHIPSNADE ZOO** (based on the **SERENGETI** NP, Tanzania). *See also* **NATIONAL PARKS AND ACCESS TO THE COUNTRYSIDE ACT 1949**, **UNITED STATES NATIONAL PARKS SERVICE**, **YELLOWSTONE NATIONAL PARK**

National Parks and Access to the Countryside Act 1949 An Act in Great Britain which made provision for **NATIONAL PARKS (NP)** and the establishment

of a National Parks Commission and created powers for the establishment and maintenance of nature reserves.

National Parks Service *See* **UNITED STATES NATIONAL PARKS SERVICE**

National Pig Association The representative trade association for British commercial pig producers.

National Register of Historic Places A list of protected buildings of historical interest in the USA, authorised under the National Historic Preservation Act of 1966. It includes structures in some zoos, e.g. Detroit Zoo, Pueblo City Park Zoo (Colorado). *See also* **LISTED BUILDING**

National Sheep Association The trade association in the UK that represents sheep farmers.

National Trust (NT) A conservation charity in the UK which protects historic places and green spaces including forests, woods, fens, beaches, farmland, downs, moorland, islands and nature reserves.

National Wild Horse and Burro Program *See* **BUREAU OF LAND MANAGEMENT (BLM)**

National Wildlife Health Center *See* USGS **NATIONAL WILDLIFE HEALTH CENTER (NWHC)**

National Wildlife Refuge System A system of public lands and waters set aside to conserve America's fish, wildlife and plants which is managed by the **UNITED STATES FISH AND WILDLIFE SERVICE (USFWS)**. The first wildlife refuge, Pelican Island, Florida, was designated in 1903 by President Theodore Roosevelt.

National Zoological Association of Great Britain A rival organisation to the **FEDERATION OF ZOOLOGICAL GARDENS OF GREAT BRITAIN**, which was founded in 1972 but is no longer in existence. *See also* **BRITISH AND IRISH ASSOCIATION OF ZOOS AND AQUARIUMS (BIAZA)**

native species
1. A species which occurs naturally in a particular area, especially a particular region or country.
2. The US Executive Order 13112 of February 3, 1999 (s.1(g)) defines native species as '*. . .with respect to a particular ecosystem, a species that, other than as a result of an introduction, historically occurred or currently occurs in that ecosystem.*'
See also **FARMING OF NATIVE SPECIES**. *Compare* **ENDEMIC (1)**, **NATURALISED SPECIES**

Natura 2000 A network of protected areas in Europe established by the **HABITATS DIRECTIVE**, consisting of **SPECIAL AREAS OF CONSERVATION (SACs)** and **SPECIAL PROTECTION AREAS (SPAs)** (designated under the **WILD BIRDS DIRECTIVES**).

natural behaviour Those species-specific behaviours that are usually observed in the wild. This may vary between populations due to the development of **CULTURE**. *See also* **ABNORMAL BEHAVIOUR**, **NORMAL BEHAVIOUR**

Natural England The statutory nature conservation agency for England. Formerly called English Nature,

N

and before that the Nature Conservancy Council (which was the statutory body for Britain as a whole).

Natural Environment Research Council (NERC) A UK government organisation that funds environmental research in British universities and its own research centres, which include the **BRITISH ANTARCTIC SURVEY (BAS)**, the British Geological Survey, the Centre for Ecology and Hydrology and the National Oceanography Centre.

natural history
1. The popular study of animal, plants and their environments.
2. The animals, plants and other organisms, geology, climate, etc. associated with a particular place. Often taken to refer only to the living components of an **ECOSYSTEM**.
3. In relation to a species, its ecology, behaviour, life cycle and other aspects of its biology.

Natural History and Antiquities of Selborne, The A book published in 1789 by Gilbert **WHITE** which was a compilation of his letters to Thomas Pennant and Daines Barrington, both leading naturalists of the day. He discussed his observations and theories about the local fauna and flora and their inter-relationships. The book has never been out of print.

Natural History Museum at Tring *See* **ROTHSCHILD**

Natural History Museum (London) Formerly the British Museum (Natural History). A major natural history museum in London which incorporates the collections of **ROTHSCHILD** and the Darwin Centre (Fig. N1).

natural selection The process that occurs in nature whereby some individuals survive and breed and pass on their traits to future generations while others are prevented from doing so due to the inter-action between their **GENOMES** and adverse factors in the **ENVIRONMENT**, e.g. severe weather conditions, predation, food shortage, disease. *Compare* **ARTIFICIAL SELECTION**

naturalised species A species that has been introduced into the wild in an area which is not part of its natural range and which has become adapted to its new environment. Such species may become **INVASIVE SPECIES**. *Compare* **ENDEMIC (1)**, **NATIVE SPECIES**

naturalistic enrichment An enrichment device that is visually and functionally compatible with a **NATURALISTIC EXHIBIT**, e.g. a root feeder for babirusa (*Babyrousa* spp.), a **SWAY BRANCH** or **SWAY FEEDING POLE**, an artificial 'river' for elephants. Such devices might be found in the **ON-SHOW** areas but the appearance of enrichment devices is less important in **OFF-SHOW** areas, or in exhibits which are not naturalistic. *See also* **LURE COURSING**

naturalistic exhibit An exhibit in a zoo or elsewhere which presents animals in a habitat that has been designed to resemble closely their natural habitat.

See also **IMMERSION EXHIBIT**, **WOODLAND PARK ZOO**

naturalistic grazing The restoration of grazing by large wild herbivores. It improves natural biodiversity on farmland and was pioneered in the Oostvaardersplassen, a nature reserve in the Netherlands. *See also* **CONSERVATION GRAZING**, **HEFTING**, **REWILDING (1)**

nature
1. The totality of the natural, physical and material world, including the forces which govern its activity.
2. Living things, geology (landscape) and weather.

Nature A prestigious weekly interdisciplinary science journal which publishes original research across a wide range of fields. It has been published since 1869 and has produced many landmark papers including one reporting the first cloning of a mammal in 1997 (*DOLLY* **THE SHEEP**).

nature deficit disorder A term used to describe the alienation of children from nature by virtue of a lack of experience of outdoor activities due to concerns about safety, urban lifestyles, increased time spent playing computer games etc., which appears to be associated with an increase in attention disorders, obesity and depression in children. It is not a recognised medical or psychological disorder. The term was first coined by Richard Louv, author of *Last Child in the Woods: Saving Our Children From Nature-Deficit Disorder*, published in 2005. *Compare* **BIOPHILIA**

nature reserve A general term for an area set aside for wild animals and plants. It may or may not be accessible to the public and may or may not have legal status. *See also* **GAME RESERVE (1)**, **NATIONAL NATURE RESERVE (NNR)**, **NATIONAL PARK (NP)**

navigation A form of **ORIENTATION (1)** exhibited by an animal in which it travels long distances, moving from location to location, towards a goal, e.g. during **MIGRATION (1)**. *See also* **COMPASS SENSE**, **HOMING**, **LANDMARK**, **MENTAL MAP**

Nē Nē (*Branta sandvicensis*) The Hawaiian goose, which was saved from extinction by captive breeding by Peter **SCOTT** at **SLIMBRIDGE** and elsewhere.

Near Threatened (NT) A **RED LIST** category used by the **INTERNATIONAL UNION FOR THE CONSERVATION OF NATURE AND NATURAL RESOURCES (IUCN)**: Near threatened (NT) – the taxon does not qualify for **CRITICALLY ENDANGERED (CR)**, **ENDANGERED (EN)** or **VULNERABLE (VU)** now, but is likely to qualify for a **THREATENED** category in the near future.

Nearctic region *See* **FAUNAL REGIONS**

nebuliser An aerosol delivery system which converts a drug into a fine mist for inhalation. Nebulised liquid may be administered to an animal by face mask, by tent, in a closed aquarium-type tank into which the animal is placed, or through an **ENDOTRACHEAL TUBE** or **TRACHEOSTOMY TUBE**. *See also* **METERED DOSE INHALER (MDI)**

neck collar A means of marking large birds, consisting of a large coloured plastic ring bearing a number. The **WILDFOWL AND WETLANDS TRUST (WWT)** has attached neck collars to pink-footed geese (*Anser brachyrhynchus*) which can be read from 300 m with a telescope in order to follow their migrations and study their **POPULATION DYNAMICS**. *See also* **MARK, RING**

necklace A device worn around the neck of some captive animals (e.g. lemurs) for the purpose of individual identification, especially in zoos (Fig. M2).

neck-twisting A **STEREOTYPIC BEHAVIOUR** which involves an unnatural twisting and rolling of the neck, sometimes bending the neck back, or flicking the head around. Exhibited by some giraffes, llamas and monkeys.

necrobacillosis A disease or lesion associated with the presence of the anaerobic soil bacterium *Fusobacterium necrophorus*, including footrot in cattle, foot abscesses in sheep, necrotic rhinitis in pigs and **LUMPY JAW** in **MACROPODS** and **ARTIODACTYLS**.

necropsy *See POST-MORTEM* (2)

necrosis The death of tissue, especially where the supply of blood has been interrupted.

necrotic Relating to **NECROSIS**.

need In animal welfare, a need is a deficiency in an animal that can be remedied by acquiring a particular resource (e.g. food), or by responding to a particular environmental or bodily stimulus. A need is a consequence of behaviour that is necessary for an individual's survival; an essential behaviour such as **FORAGING BEHAVIOUR** (because food is essential for survival), as opposed to a **LEISURE** activity. *See also* **BASIC NEEDS TEST, NEGLECT**

negative buoyancy The inability to float in water. Occurs when the **BUOYANCY** force is lower than the gravitational pull acting on a body. Some fish need to swim continually to maintain their vertical position in water, e.g. sharks. Fish which exhibit negative buoyancy can remain on the sea floor. *See also* **NEUTRAL BUOYANCY**

negative correlation *See* **CORRELATION**

negative reinforcement A situation in **OPERANT CONDITIONING** whereby behaviour is strengthened (reinforced) because it removes or prevents an aversive stimulus (negative reinforcer), e.g. a shock from an electric fence. Training by using punishment. *See also* **LEARNING, REINFORCEMENT**. *Compare* **POSITIVE REINFORCEMENT**

negative reinforcer *See* **NEGATIVE REINFORCEMENT**

neglect The persistent failure to provide for the **NEEDS** of an animal.

negligence In law, the breach of a duty to take reasonable care or exercise reasonable skill, e.g. a zoo might be accused of negligence if it allowed a dan-

gerous animal to escape (by leaving a cage door open) and injure a visitor. *See also* **DUTY OF CARE**

Nematoda A **PHYLUM** of animals. Nematodes are free-living and parasitic roundworms, with a tough outer cuticle. They are extremely numerous but rarely seen because most are very small. Some species live in the soil and the bottom of lakes and streams, while others are **PARASITES** and live inside plants or other animals. They are worm-like in appearance and all species share a general form.

nematode A member of the phylum **NEMATODA**.

neonatal Relating to a **NEONATE**.

neonate A newborn animal.

neophilia A tendency to be attracted to new things. *Compare* **NEOPHOBIA**

neophilic Attracted to novel situations; enjoys novelty. *Compare* **NEOPHOBIC**

neophobia The fear of new things. May be a cause of stress in some zoo animals. *Compare* **NEOPHILIA**

neophobic Fears novel situations. *Compare* **NEOPHILIC**

neoplasm *See* **TUMOUR**

Neotropical region *See* **FAUNAL REGIONS**

nephrolithiasis *See* **RENAL CALCULUS**

nephron The basic functional unit of the **KIDNEY**. A microscopic tubule which filters waste material from the blood in the **BOWMAN'S CAPSULE**, producing **URINE**, and reabsorbs some water.

nepotistic hierarchy *See* **MATRILINEAL HIERARCHY**

NERC *See* **NATURAL ENVIRONMENT RESEARCH COUNCIL (NERC)**

nerve A bundle of **AXONS** from different **NEURONES** which carries information relatively long distances in the body.

nerve impulse The passage of an **ACTION POTENTIAL** along a **NEURONE**.

nervous system The organ system which is made up of specialised cells called **NEURONES** which coordinate the functioning and actions of the body. It receives information from the environment and controls the body's response to it. In most animals it is divided into the **CENTRAL NERVOUS SYSTEM (CNS)** – the **BRAIN** and **SPINAL CORD** – and the **PERIPHERAL NERVOUS SYSTEM (PNS)** (everything else).

nest A structure used by an animal for shelter and especially for breeding, notably in birds and small mammals. Often constructed from dead vegetation. Adélie penguins (*Pygoscelis adeliae*) build nests from small rocks. *See also* **SCRAPE**

nest box A container used by animals for sheltering and especially breeding, particularly by birds. It usually has a removable roof to allow access by keepers etc. for cleaning, egg removal. Nest boxes designed for birds need to have an access hole of a suitable size for the intended species. Boxes are sometimes fitted with a **CCTV** camera so that **BEHAVIOUR** and breeding can be monitored.

nestling A young bird that is not old enough to leave the nest.

N

net

1. A bag or sheet of material made of thread or cord worked into a meshed fabric and used for catching fish, birds, bats, insects and other animals. Nets may also be used to exclude animals from particular areas (*see* **SHARK NET**) or to prevent escape from the top of zoo enclosures. *See also* **CANNON NETTING, MIST NET, SEINE NET**. Nets may be defined by law.

2. In Ohio, in the Ohio Revised Code, Title [15] XV, Conservation of Natural Resources, Chapter 1531.01 (PP) '"*Net*" *means fishing devices with meshes composed of twine or synthetic material and includes, but is not limited to, trap nets, fyke nets, crib nets, carp aprons, dip nets, and seines, except minnow seines and minnow dip nets.*'

neural Relating to **NERVES** or the **NERVOUS SYSTEM**.

neurohypophysis *See* **POSTERIOR PITUITARY**

neurology The scientific study of the nervous system and its disorders and diseases.

neuron *See* **NEURONE**

neurone, neuron A nerve cell. *See also* **INTERMEDIATE NEURONE, MOTOR NEURONE, SENSORY NEURONE**

neurotoxin A toxin which adversely affects the nervous system.

neurotransmitter A substance that transmits a **NERVE IMPULSE** across a **SYNAPSE**, i.e. from one **NEURONE** to another or from a neurone to a **MUSCLE** or **GLAND**. It may be excitatory or inhibitory. Examples include **ADRENALINE, ENDORPHINS, NORADRENALINE, SEROTONIN**.

neuter(ed) Sterilise(d).

neutering The sterilisation of an individual so that it cannot reproduce or to prevent sexual or aggressive behaviour. Used to prevent animals from breeding where their genes are over-represented in a captive population to prevent **INBREEDING**. Also used to prevent interbreeding between closely related species in the wild. In Scotland domestic cats and Scottish wildcat–domestic cat hybrids are being neutered to protect the gene pool of pure-bred wildcats (*Felis sylvestris grampis*).

neutral buoyancy In relation to water, the condition in which an animal's mass displaces an equal mass of water so that it neither rises nor sinks. In fish, the swim bladder is used to alter **BUOYANCY** by changing the amount of air and water it contains. *See also* **NEGATIVE BUOYANCY**

neutral pH Neither **ACID**, nor **ALKALI**; having a pH of 7.0. See **pH SCALE**

neutral stimulus In animal behaviour, a stimulus which does not normally elicit a response, e.g. a background sound. *See also* **ASSOCIATION (1), HABITUATION**

new organism In New Zealand, under the s2A(1) of the **HAZARDOUS SUBSTANCES AND NEW ORGANISMS ACT 1996 (NZ)** a new organism is defined as:

'(a) *an organism belonging to a species that was not present in New Zealand immediately before 29 July 1998;*

(b) *an organism belonging to a species, subspecies, infrasubspecies, variety, strain, or cultivar prescribed as a risk species, where that organism was not present in New Zealand at the time of promulgation of the relevant regulation;*

(c) *an organism for which a containment approval has been given under this Act;*

 (ca) *an organism for which a conditional release approval has been given;*

 (cb) *a qualifying organism approved for release with controls;*

(d) *a genetically modified organism;*

(e) *an organism that belongs to a species, subspecies, infrasubspecies, variety, strain, or cultivar that has been eradicated from New Zealand.*'

Compare **NEW SPECIES**

new species A type of organism may be described as a 'new species' because:

1. During the process of **EVOLUTION**, it has developed from a pre-existing species by **SPECIATION**.

2. It is a newly recognised species which has only just been discovered in the wild.

3. It was previously considered to be a **SUBSPECIES** but has been given species status in the light of information from **DNA FINGERPRINTING** or for some other reason.

See also **CRYPTOZOOLOGY**. *Compare* **NEW ORGANISM**

new tank syndrome The condition in which a new aquarium tank accumulates the waste produced by fishes but is incapable of removing it because the bacterial population in the biological filter has not had sufficient time to grow. This results in fish death. *See also* **MATURATION (1)**

New World Relating to the Western Hemisphere: the Americas, e.g. **NEW WORLD MONKEY**. *Compare* **OLD WORLD**

New World monkey A monkey which naturally occurs in the Americas, e.g. spider monkeys, marmosets, howler monkeys, capuchins. *Compare* **OLD WORLD MONKEY**

New York Aquarium *See* **WILDLIFE CONSERVATION SOCIETY (WCS)**

New York Central Park Zoo *See* **WILDLIFE CONSERVATION SOCIETY (WCS)**

New York Zoological Society *See* **WILDLIFE CONSERVATION SOCIETY (WCS)**

Newcastle disease A notifiable disease caused by a **PARAMYXOVIRUS** that affects birds and may be contracted by humans. It may be spread by the wind. The disease may cause a reduction in egg production and soft-shelled eggs. Infected birds may exhibit respiratory signs (breathing difficulties) or nervous signs (e.g. paralysis of wings or legs), but rarely both. In mild cases the only sign may be black **DIARRHOEA**. Live and inactivated vaccines are available.

NFU *See* **NATIONAL FARMERS' UNION (NFU)**

NGO *See* **NON-GOVERNMENTAL ORGANISATION (NGO)**

Ngorongoro Conservation Area (NCA) A multi-land use area in northern Tanzania where people (the Masaai) live alongside protected wildlife. The NCA includes Ngorongoro Crater (Fig. R2) which is an important wildlife area supporting approximately 25 000 large mammals, mostly wildebeest, zebra, buffalo, Grant's gazelle and Thomson's gazelle, along with high densities of lion and hyena and a small population of black rhino.

Ngorongoro Crater *See* **NGORONGORO CONSERVATION AREA (NCA)**

niche *See* **ECOLOGICAL NICHE, FUNDAMENTAL NICHE**

niche separation The phenomenon whereby closely related species that utilise the same habitats (**SYMPATRIC** species) avoid **INTERSPECIFIC COMPETITION** by evolving so that they occupy different niches, e.g. feeding on different species (**RESOURCE PARTITIONING**) or in different places. *See also* **DISRUPTIVE SELECTION**

nictitating membrane A thin membrane which acts as a third (inner) eyelid that closes to protect the eye in birds, reptiles and some mammals. It may be transparent, translucent or opaque, depending upon the species.

nidicolous Referring to a young bird that remains in the nest until able to fly. *Compare* **NIDIFUGOUS**

nidifugous Referring to a **PRECOCIAL** young bird that leaves the nest soon after hatching. *Compare* **NIDICOLOUS**

night faeces *See* **COPROPHAGY**

Night Safari The world's first wildlife park for **NOCTURNAL** animals, opened in 1994. Located in Singapore, the attraction holds around 115 species of animals. The site covers 40 hectares and exhibits represent eight geographical zones.

night scope *See* **NIGHT VISION EQUIPMENT**

night shine *See* **EYE SHINE**

night sight *See* **NIGHT VISION EQUIPMENT**

night vision equipment, night scope, night sight An optical device, e.g. binoculars, for seeing in low light levels. Some use ambient light, others use infrared. *See* **IMAGE INTENSIFIER, THERMAL IMAGING CAMERA**

Nim Chimpsky A chimpanzee who was the subject of a long-running study of animal language acquisition at Columbia University (Project Nim). He was raised by surrogate human parents. His name was a play on words on the name of the famous language theorist Noam Chomsky. *See also* **KANZI, KOKO, WASHOE**

nipple drinking system A system which supplies water continuously through a system of pipes via 'nipples' which release the water when touched. Used in intensive poultry rearing systems.

nitrogen cycle The cyclic movement of carbon between the biological and physical components of the environment. Inorganic nitrogenous compounds (mainly nitrates) are absorbed by plants and synthesised into organic compounds (e.g. proteins and nucleic acids). These compounds are passed to animals when plants are eaten and nitrogen is returned to the soil and water when organisms die and by the process of excretion. Nitrogen fixation by soil microbes is important in making atmospheric nitrogen available to plants. Anaerobic soil bacteria release nitrogen to the atmosphere by a process called denitrification, thereby reducing soil fertility.

nitrogenous waste Waste substances which are high in nitrogen and produced by animal bodies as a by-product of protein metabolism. The chemicals released from the body vary between taxa and are largely determined by the availability of water. Fishes release nitrogenous waste into water as ammonia (which is toxic). Amphibians metabolise ammonia into the less toxic urea but their larvae excrete ammonia. Mammals also produce urea but it leaves the body diluted in water as urine. Birds produce uric acid.

NMFS *See* **NATIONAL MARINE FISHERIES SERVICE (NMFS)**

NNR *See* **NATIONAL NATURE RESERVE (NNR)**

nocturnal

1. Relating to the night; occurring at night.

2. Of animals, active during the night, e.g. a nocturnal species (as opposed to a **DIURNAL** species). *See also* **NOCTURNAL HOUSE.** *Compare* **CATHEMERAL, CREPUSCULAR, MATUTINAL, VESPERTINE**

nocturnal house A zoo building, open to visitors, where **NOCTURNAL** animals are kept in individual enclosures usually behind glass. It usually contains small mammals, e.g. bushbabies (Galagidae), bats, aye-ayes (*Daubentonia madagascariensis*). Day and night are reversed by operating lighting using a timer so that visitors are able to see animals when they are most active. *See also* **BAT HOUSE**

noise and stress Noise may cause **STRESS** in some animals. In the wild, noise from shipping causes stress in whales. Some animals exposed to noise in zoos may suffer chronic stress. *See also* **AUDITORY ENRICHMENT**

nominate subspecies, nominotypical subspecies When a **SPECIES** is divided into **SUBSPECIES** the originally described population is called the nominate subspecies and its scientific name repeats the species name to form a **TRINOMIAL** name, e.g. *Loxodonta africana africana*.

nominotypical subspecies *See* **NOMINATE SUBSPECIES**

non-governmental organisation (NGO) An organisation which is independent of any government. An NGO may be a professional association, a charity, a business or simply a group of people with a common interest. There is no generally accepted legal definition of an NGO.

N

non-human primate Any primate other than man. Used to exclude humans from references to primates in general, e.g. a study of the effect of visitors on the behaviour of non-human primates in zoos.

non-parametric In statistics, not based on a particular statistical distribution, e.g. a non-parametric test. *Compare* PARAMETRIC

non-striated muscle *See* SMOOTH MUSCLE

non-verbal language A form of COMMUNICATION which is not spoken and does not involve words. May involve FACIAL EXPRESSIONS or DISPLAYS in some species. Studies of language in great apes have used LEXIGRAM BOARDS.

noradrenaline, norepinephrine A hormone produced by the ADRENAL MEDULLA which has similar effects to ADRENALINE. It stimulates the breakdown of glycogen to glucose and thus raises blood sugar levels. It also functions as a neurotransmitter in the SYMPATHETIC NERVOUS SYSTEM (SNS) and raises blood pressure in HYPOTENSION.

norepinephrine *See* NORADRENALINE

normal behaviour The behaviour that an animal usually exhibits, especially in the wild, in the absence of disease or behavioural problems. This may vary within a species between the sexes and at different ages, as a result of MATURATION. It may also vary between groups of the same species as a result of a difference in CULTURAL BEHAVIOURS. *See also* ABNORMAL BEHAVIOUR, NATURAL BEHAVIOUR, STEREOTYPIC BEHAVIOUR

normal distribution, Gaussian distribution A bell-shaped distribution of values, where the MEAN, MODE and MEDIAN all occur at the middle. It occurs widely in nature, especially in body measurements such as height, mass etc. It occurs when a particular characteristic is influenced by a number of genes together with various factors in the environment, e.g. nutrition.

northern spotted owl (*Strix occidentalis caurina*) An owl species that is native to the north-west USA. By the early 1980s there was clear evidence that this owl was endangered in the USA, but the UNITED STATES FISH AND WILDLIFE SERVICE (USFWS) refused to list the species under the ENDANGERED SPECIES ACT OF 1973 (USA) because of concerns about the effect on the timber industry in the Pacific Northwest. Thirty organisations petitioned the USFWS to list the owl. In 1987 the petitions were denied. In 1988 the SIERRA CLUB Legal Defense Fund filed a suit in the federal district court in Seattle. They claimed that the decision not to list was based on economic, not scientific evidence. The court ordered the USFWS to list the owl (*Northern Spotted Owl v. Hodel* (1998)). In a second case the court ordered the FWS to designate critical habitat (*Northern Spotted Owl v. Lujan* (1991)).

***Northern Spotted Owl v. Hodel* (1998)/*Northern Spotted Owl v. Lujan* (1991)** *See* NORTHERN SPOTTED OWL (*STRIX OCCIDENTALIS CAURINA*)

nose fill drinker A wall-mounted drinking water bowl fitted with a valve operated by the animal's nose. This design ensures that fresh water is piped to the bowl each time the animal drinks and water does not become stagnant. Often used for horses and other large livestock. *See also* AUTOMATIC WATER DRINKER

Not Evaluated (NE) A RED LIST category used by the INTERNATIONAL UNION FOR THE CONSERVATION OF NATURE AND NATURAL RESOURCES (IUCN): Not Evaluated (NE) – The taxon has not yet been evaluated against the IUCN Red List criteria.

Notacanthiformes An order of fishes: spiny eels and their allies.

notifiable disease A disease which is considered serious – either because it threatens human health, or because it is of great economic importance – that must, by law, be notified to the state veterinary authorities in the countries where it occurs. Notifiable diseases vary from country to country. In the UK they include ANTHRAX, BRUCELLOSIS, BOVINE SPONGIFORM ENCEPHALOPATHY (BSE), FOOT AND MOUTH DISEASE (FMD), NEWCASTLE DISEASE and SCRAPIE.

Notoryctemorphia An order of mammals: marsupial 'mole'.

NSAIDs Non-steroidal anti-inflammatory drugs.

nuclear transfer, somatic cell nuclear transfer (SCNT) A method of CLONING whereby the GENETIC MATERIAL from a body cell (SOMATIC CELL) from one individual is transferred to an egg cell from a different individual from which the genetic material has previously been removed. This cell is then stimulated with an electric current to make it start dividing like a normal EMBRYO. It is then implanted into the UTERUS of a SURROGATE (2) mother who later gives birth normally. The first mammal to be cloned from an adult somatic cell (from a sheep's udder) was *DOLLY* THE SHEEP.

nucleic acid A polymer made from nucleotides; a polynucleotide. The naturally occurring forms are DNA and RNA.

nucleolus (nucleoli *pl.*) A non-membrane bound structure found in the NUCLEUS where ribosomal RNA (rRNA) is transcribed and assembled.

nucleotide One of the structural units of **RNA** and **DNA**, composed of a five carbon sugar (ribose in RNA or deoxyribose in DNA), a phosphate group and a nitrogenous base. Each contains either a purine base (adenine or guanine) or a pyrimidine base (thymine or cytosine). In RNA uracil replaces thymine. In the structure of DNA adenine pairs with thymine, and guanine pairs with cytosine.

nucleus A large membrane-bound ORGANELLE found in EUKARYOTES which contains CHROMOSOMES and controls the functioning of the cell by regulating GENE EXPRESSION. *See also* MEIOSIS, MITOSIS, NUCLEOLUS

Fig. N2 Nursery: the duck nursery at Martin Mere (Wildfowl and Wetlands Trust (WWT)), UK.

N

null hypothesis (H₀) An **HYPOTHESIS** that a scientist attempts to disprove (reject or refute). It is generally paired with an alternative hypothesis (H_1). In statistics often the purpose of a test is to try to prove the null hypothesis wrong, or reject or refute the null hypothesis. For example, if we were studying the effect of visitor numbers on aggression in chimpanzees we could formulate the following hypotheses:

H_1 = High visitor numbers increase the frequency of aggressive behaviours in chimpanzees.

H_0 = High visitor numbers do not increase the frequency of aggressive behaviours in chimpanzees.

In attempting to establish that H_1 is true we must obtain evidence from a statistical test that allows us to reject H_0. *See also* **HYPOTHESIS**

nullipara A **NULLIPAROUS** female.

nulliparous Describing a female that has not borne any offspring. *See also* **MULTIPAROUS, POLYPAROUS, PRIMIPAROUS, UNIPAROUS**

nursery
1. A nest, burrow or other place where young are born and raised away from an animal's main living quarters.
2. A facility within a zoo or similar institution where young animals are reared when rejected by the mother or as part of a captive breeding programme (Fig. N2).

nutrient Any food material that provides energy or raw materials for growth, tissue repair and reproduction in an organism. *See also* **MACRONUTRIENT, MICRONUTRIENT**

nutrient enrichment The addition of nutrients to the environment of an organism, especially water or soil (in relation to plants), which generally results in increased growth. *See also* **EUTROPHICATION**

nutrition
1. The process by which an organism obtains and assimilates food for its energy needs, and uses it for growth, the replacement of tissues and reproduction.
2. Food.
3. The scientific study of food, including food composition, diet and the role of various nutrients in health.

nutritional enrichment The presentation of varied or novel food types to animals or varying the method of delivery as an enrichment. *See also* **SCATTER FEED, SWAY BRANCH, SWAY FEEDING POLE**

nutritional wisdom *See* **FOOD SELECTION**

nutritionist A person who is qualified in the science of nutrition, and has knowledge of foods and the nutritional requirements of animals.

Fig. O1 O is for owl. Great grey owl (*Strix nebulosa*).

Dictionary of Zoo Biology and Animal Management: A guide to terminology used in zoo biology, animal welfare, wildlife conservation and livestock production, First Edition. Paul A. Rees.

obedience training Behavioural training, generally for dogs, in which they learn to obey basic commands. *See also* **DOG AGILITY TRAINING**

obesity The possession of excessive weight due to the accumulation of body fat, generally resulting from an energy intake that greatly exceeds the energy requirements of the body. Associated with health problems in many species, including **DIABETES MELLITUS**, **ARTHRITIS**, cardiovascular disease and respiratory disease. The amount of subcutaneous body fat may be estimated in some species by measuring skinfold thickness. In some taxa (e.g. monkeys) the measurement of body mass and height may be used to calculate the **BODY MASS INDEX (BMI)** which serves as a proxy for measuring body fat. Obesity is a problem in some animals living in zoos (e.g. chimpanzees and elephants) and in some companion animals (e.g. cats and dogs). *See also* **ABDOMINAL SKINFOLD**, **BODY CONDITION SCORE**

object licking The repetitive licking of non-food objects. A common **STEREOTYPIC BEHAVIOUR** of cattle, sheep and sows.

Occam's razor, Ockham's razor A principle that argues that the simplest explanation of a phenomenon – one which makes the fewest assumptions – is the most plausible until evidence is provided to the contrary, i.e. it favours simple explanations over complex ones.

occupational enrichment Enrichment that includes both psychological enrichment (devices which give the animal control over its environment or mental challenges) and enrichment that encourages physical exercise.

occupiers' liability A concept in the law which makes an owner or occupier of land or premises liable for the safety of those who enter their land or premises. A **DUTY OF CARE** is owed to legitimate visitors and often to trespassers. Great care must be taken to provide for the safety of children on premises where children are routinely present, e.g. in **ZOOS** or **FARM PARKS**.

Ocean Project A network of organisations (including zoos, aquariums, museums, conservation organisations and agencies) that work together to conserve the oceans using education, action and networking.

oceanarium, marine mammal park, marine park, seaquarium A large-scale seawater aquarium, simulating the ocean, in which large marine animals, especially seals, dolphins and whales, are kept for research or display to the public. The world's first oceanarium was *MARINELAND* in Florida, which opened in 1938. Tanks were 'seascaped' and included an artificial coral reef and a shipwreck. The Miami Seaquarium, also in Florida, opened in 1955.

oceanography The scientific study of the oceans, including their chemistry, physics and biology.

Ockham's razor *See* **OCCAM'S RAZOR**

ocular Relating to the eye.

ocyte *See* **OOCYTE**

odd-toed ungulate A member of the mammalian order **PERRISODACTYLA**. *Compare* **EVEN-TOED UNGULATE**

oedema, edema An abnormal accumulation of fluid around the **CELLS** or **TISSUES**.

oesophagus, esophagus A muscular tube which carries food from the **BUCCAL CAVITY** to the **STOMACH** by **PERISTALSIS**.

oestrogens, estrogens A group of **STEROID** hormones produced by the ovaries. In mammals they regulate fat deposition and the growth of the endometrium, promote the widening of the pelvic girdle, and control **OVULATION** by causing a surge in **LUTEINISING HORMONE (LH)**. They may also affect **MATERNAL BEHAVIOUR**. Oestrogens are major components of **CONTRACEPTIVE** drugs.

oestrous, estrous (*adj.*) Relating to **OESTRUS**, e.g. oestrous female.

oestrous cycle, estrous cycle, ovarian cycle The cycle of change in the physiology and anatomy of female mammals (except most primates) during which an ovum (or ova) are released from the ovary (or ovaries). The cycle is controlled by hormonal changes. *See also* **ANOESTROUS**, **FOLLICLE-STIMULATING HORMONE (FSH)**, **LUTEINISING HORMONE (LH)**, **MENSTRUAL CYCLE**, **OESTROGEN**, **OESTRUS SYNCHRONY**, **OVULATION**

oestrus, estrus A restricted but regularly occurring period of sexual receptivity exhibited by most mammal species. Females in oestrus are referred to as being 'in heat'. *See also* **FLEHMEN**, **OESTROUS CYCLE**, **OESTRUS SYNCHRONY**

oestrus synchrony, estrus synchrony The synchronisation of the **OESTROUS CYCLE** in a group of females of the same species living together. Occurs in nature, especially in herd animals (e.g. wildebeest (*Connochaetes taurinus*)) as a protection against predators. It may be artificially induced in farm animals to improve herd management by allowing an intensive period of **ARTIFICIAL INSEMINATION (AI)**, the scheduling of breeding and calving, and the production of young at an appropriate time (season). *See also* **REPRODUCTIVE SYNCHRONY**

offal Internal organs of animals (e.g. liver, pancreas, kidneys) sometimes given as food to carnivores in captivity. Includes most internal organs apart from muscle and bone.

off-exhibit *See* **OFF-SHOW**

off-show, off-exhibit In relation to animals that are normally exhibited (**ON-SHOW**) in a zoo or similar facility, not on display to the public.

off-show area A part of an **ANIMAL ENCLOSURE** that is separate from the **ON-SHOW AREA** and is designed so that animals are not visible to the public.

Old World Relating to the Eastern Hemisphere, especially Africa and Eurasia, e.g. **OLD WORLD MONKEYS**. *Compare* **NEW WORLD**

O

Old World monkey A monkey which naturally occurs in Africa or Asia, e.g. baboons, langurs, macaques. *Compare* **NEW WORLD MONKEY**

olfaction The sense of smell.

olfactory communication The perception of molecules passed from one animal to another through air or water which results in an alteration of behaviour in the receiver animal. This may occur, for example, via chemicals produced by **SCENT GLANDS** or in urine or faeces. It is important in communicating reproductive state and the position of **TERRITORY** in some species. *See also* **PHEROMONE**

olfactory enrichment An enrichment technique that enriches an animal's environment by stimulating its sense of smell, e.g. exposure to the faeces of a predator, **CONSPECIFIC** or prey organism. A number of scents are available commercially (e.g. **FELIWAY**).

olfactory system The system of the body that detects odours and provides the sense of smell (olfaction). *See also* **OLFACTORY COMMUNICATION**

omasum The third stomach of a **RUMINANT**.

omnivore An animal that eats both animals and plants.

oncogene A gene that makes an organism particularly susceptible to cancer. *See also* **ONCOMOUSE®**

OncoMouse®, Harvard mouse A genetically modified laboratory mouse that carries an introduced gene which makes it particularly susceptible to cancer and therefore useful in cancer research. The mouse was one of the first **TRANSGENIC ORGANISMS** and was produced by researchers at the Harvard Medical School in the 1980s. 'OncoMouse' is a registered trademark and has been patented in a number of countries. *See also* **GENETICALLY MODIFIED ORGANISM (GMO)**, **KNOCK-OUT MOUSE**

one-way window barrier A window forming part of an exhibit or **ANIMAL ENCLOSURE** which is only transparent in one direction. This allows visitors and staff to observe the animals through the window without disturbing them.

on-exhibit *See* **ON-SHOW**

one-zero sampling A sampling method use in behaviour studies. The action of the **FOCAL ANIMAL**(s) is recorded between specific points in time. Behaviour is recorded as either occurring during this period (one), regardless of the frequency of occurrence, or not occurring (zero). *See also* **FOCAL SAMPLING**. *Compare* **AD LIB SAMPLING**, **SCAN SAMPLING**

on-line journal A journal which is available only on the internet. *See also* **ACADEMIC JOURNAL**

on-show, on-exhibit In relation to animals, on display to the public. *Compare* **OFF-SHOW**

on-show area A part of an **ANIMAL ENCLOSURE** that is separate from the **OFF-SHOW AREA** and is designed so that animals are visible to the public.

ontogeny Development; the history of the development of an organism from **CONCEPTION** until death, including the influences of **GENETICS**, **MATURATION** and **LEARNING** on its behaviour.

oocyst A cyst containing a **ZYGOTE** produced by some protozoan parasites which may be transmitted to a new host, e.g. *Toxoplasma* spp.

oocyte, ocyte A cell that forms an **OVUM** when it divides by **MEIOSIS**.

oogenesis The process by which eggs (ova) are formed by **MEIOSIS** and the formation of the egg membranes and yolk.

open circulatory system A simple system in which 'blood' (haemolymph) circulates around a body cavity (haemocoel) bathing the organs directly with nutrients and oxygen. A heart pumps out haemolymph which returns via ostia (pores) but there is no network of blood vessels. Occurs in invertebrates such as arthropods and molluscs. *Compare* **CLOSED CIRCULATORY SYSTEM**

open range zoo *See* **SAFARI PARK**

open season A period during the year when the hunting of particular types of animals is allowed by law, e.g. deer, wildfowl. *Compare* **CLOSE SEASON**

operant conditioning, instrumental conditioning, trial and error learning, type II conditioning A type of conditioning in which a particular response is instrumental in receiving a reward (**POSITIVE REINFORCEMENT**, e.g. food) or punishment (**NEGATIVE REINFORCEMENT**, e.g. an electric shock). For example, a pigeon may associate pushing a button with receiving food, or a monkey may associate receiving an electric shock with touching an electric fence. This type of conditioning is used in animal **TRAINING**. *See also* **CLASSICAL CONDITIONING**, **LEARNING**

Operation Oryx *See* **ARABIAN ORYX (*ORYX LEUCORYX*)**

operculum
1. A hard, bony plate that covers and protects the gills of a fish, or a similar structure in an amphibian.
2. A small 'lid' which covers the opening in the shell of many snails, protecting the soft parts of the animal when they are retracted.
3. Any of several parts of the **CEREBRUM** of the brain.

Ophidiiformes An order of fishes: cuskeels, pearlfishes and their allies.

ophthalmologist A person who examines, diagnoses and treats injuries to, and diseases of, the eye.

ophthalmology The scientific study of the anatomy, physiology and diseases of eyes.

opioids A group of pain-relieving drugs.

Opisthocomiformes An order of birds: hoatzin.

opportunistic
1. In animal behaviour, relating to the ability of an animal to exploit newly available resources, e.g. foods, habitats, mates etc.
2. In ecology, describing a species which is able to adapt to exploit newly available habitats. Such species are r-selected. *See also* **R-SELECTED SPECIES**

opportunistic sampling *See AD LIB SAMPLING*

opposable thumb A thumb which can be turned so that its tip may come into contact with the fingertips of all of the other digits. This allows the hand to grasp objects such as food, weapons and tools. **OLD WORLD MONKEYS**, all **APES** and some lemurs and lorises possess opposable thumbs.

optic Relating to the eye, e.g. optic nerve.

oral Relating to the mouth and **BUCCAL CAVITY**, e.g. oral hygiene. *Compare* **ABORAL**

orbit Eye socket: a bony cavity in the skull which contains an eye.

order
1. In **TAXONOMY**, a group of related **FAMILIES**; a subdivision of a **CLASS** (Table C5).
2. In English law, a type of delegated legislation (a **STATUTORY INSTRUMENT**) made under the authority of a parent Act, e.g. the Spring Traps Approval Order 1995, made under the Pests Act 1954.

Orectolobiformes An order of fishes: carpet sharks and their allies.

Orf disease A viral disease of sheep which causes scabby **LESIONS (1)** around the teats of nursing ewes and around the mouth and nostrils of lambs. It also results in poor growth.

organ A part of the body which has a specific function, or functions, and is made up of specialised tissues, e.g. the **BRAIN, STOMACH, PANCREAS** etc.

organ system A collection of **ORGANS** which function together to perform a particular physiological function, e.g. digestion: the **DIGESTIVE SYSTEM**.

organelle A structural component of a **CELL** which has a specific biological function. *See also* **ENDOPLASMIC RETICULUM (ER)**, **GOLGI BODY**, **LYSOSOME**, **MITOCHONDRION**, **NUCLEOLUS**, **NUCLEUS**, **RIBOSOME**

organic
1. Having its origin in a living or once-living organism.
2. Of or relating to a bodily organ or organs.
3. Of or relating to carbon-based chemicals of biological origin.
See also **ORGANIC CERTIFICATION**, **ORGANIC FARMING**, **ORGANIC PEST AND WEED CONTROL**

organic certification The process of certifying that organic food has been produced to a particular standard, depending upon the country where it is sold. In the EU these standards are laid down in Council Regulation (EC) No 834/2007 of 28 June 2007 on organic production and labelling of organic products and repealing Regulation (EEC) No 2092/91. *See also* **ORGANIC FARMING**

organic farming The production of food from plants and animals without the use of chemicals such as fertilisers, **HERBICIDES**, **INSECTICIDES** and other artificial agents. *See also* **ORGANIC CERTIFICATION**, **SOIL ASSOCIATION**

organic feed Livestock feed which has been produced organically, i.e. without the use of chemicals.

organic pest and weed control Methods of pest and weed control which do not use artificially produced chemicals, e.g. the use of natural predators to kill animal pests (**BIOLOGICAL CONTROL**).

organism A living thing; an individual animal, plant or microbe.

organochlorine An organic compound with at least one covalently bonded chlorine atom, including many chemicals which are harmful to humans and animals, e.g. dioxins and the insecticide **DDT**. *See also* **CARSON, ORGANOPHOSPHATE**

organophosphate An ester of phosphoric acid, including **DNA**, **RNA** and old types of **INSECTICIDE** and **HERBICIDE** which are highly toxic to many animals, including humans. *See also* **ORGANOCHLORINE**

Oriental region *See* **FAUNAL REGIONS**

orientation
1. In animal behaviour, the positioning of the body, or parts of the body, in relation to the environment.
2. In relation to a visit to a zoo, or other visitor attraction, appreciating one's location in that place. *Compare* **WAYFINDING**

Origin of Species, The Abbreviated title of *On the Origin of Species by Means of Natural Selection* published by Charles **DARWIN** in 1859. This work forms the basis of our modern-day understanding of evolutionary processes and the formation of **NEW SPECIES** by **NATURAL SELECTION**.

ornamental fishes Any species of fish which is bred for the purpose of display and not for human consumption. *See also* **DYED FISH**

ornithology The scientific study of birds. Also, a popular pastime with amateur naturalists. *See* **BIRDING, TWITCHING**

ornithosis *See* **PSITTACOSIS**

orthokinesis *See* **KINESIS**

orthopnoea An increased **RESPIRATORY DISTRESS** when the patient is lying down or when the chest is compressed.

Oryx An academic journal published by **FAUNA AND FLORA INTERNATIONAL (FFI)** which contains reports of original research relating to the **CONSERVATION** of animals and plants.

os penis, baculum, penile bone, penis bone The bone supporting the penis in some mammals. It functions as an aid to copulation.

Osborn, Henry Fairfield (1857–1935) An American palaeontologist who was president of the American Museum of Natural History for 25 years (1908–1933), a professor at Columbia University and president of the New York Zoological Society (1909–1924). He was one of the foremost zoologists of his day and published more than a thousand works, including the two-volume work *Proboscidea. A Monograph of the Discovery, Evolution, Migration and Extinction of the*

O

Mastodonts and Elephants of the World, which was published posthumously (1936–1942).

osmoregulation The control of solute concentration in cells and/or total water volume in the body. *See also* **KIDNEY**

osmosis The movement of water molecules (or other solvent) from a weak solution to a more concentrated solution through a semi-permeable membrane (e.g. the **CELL MEMBRANE**). When there is no net movement of water across the cell membrane the cell is in osmotic balance.

osmotic balance *See* **OSMOSIS**. *See also* **KIDNEY**

osmotic pressure The pressure required to prevent the inward passage of water across a semi-permeable membrane, i.e. the pressure required to counteract **OSMOSIS**.

Osteichthyes A class of fishes containing the bony fishes.

osteoarthritis, degenerative joint disease A progressive disease in which there is degradation of the articular cartilage and the development of bony outgrowths at the margins of the joint.

osteodystrophia fibrosa *See* **ANGEL WING**

Osteoglossiformes An order of fishes: bonytongues, elephantfishes and their allies.

osteopath A person who practises **OSTEOPATHY**.

osteopathy A system of treatment which is based on the theory that disturbances in the musculoskeletal system affect other parts of the body. Many of these disorders (e.g. gait problems) may be corrected by manipulative techniques used in conjunction with conventional treatments.

osteoporosis A disease in which the bones lose density and become extremely porous, making them prone to fracture and slow to heal. Typically occurs in older animals.

ostrich farm A place where ostriches (*Struthio camelus*) are reared commercially for their meat and other products. May also be a tourist attraction.

ostrich racing Competitions between people riding ostriches (*Struthio camelus*), using special saddles and reins, which occur in some countries in Africa and in parts of the USA.

Ota Benga A young bushman who was brought from central Africa by Prof. S. P. Verner. Verner gave Ota Benga to the New York Zoological Society and they put him on display in a cage in the Bronx Zoo in 1906, alongside orangutans and monkeys. On 9 September 1906 *The New York Times* carried an article entitled 'Bushman shares a cage with Bronx Park apes'. Ota Benga had previously been part of an exhibit of pygmies at the World's Fair 1904 held at St. Louis, Missouri. *See also* **HUMAN EXHIBITS**

otter hunting The hunting of otters (Lutrinae) using specially bred **OTTERHOUNDS**. Banned in the UK in 1978, at which time there were only two packs left. Some otter hunts switched to coypu (*Myocastor coypus*) in East Anglia and later to mink (Mustelinae).

otterhound An old breed of British dog which is large, strong with a thick coat, and was used for **OTTER HUNTING**.

OUTBREAK Modelling software which is used for wildlife disease **RISK ASSESSMENT**. Designed to be used with **VORTEX** to provide a more extensive assessment incorporating a consideration of disease, genetic change, demographic stochasticity, environmental variation and management actions.

outbreeding The breeding of distant relatives or unrelated animals. The converse of **INBREEDING**.

outbreeding depression A lowering of fitness in the offspring produced by crossing individuals from different populations compared with that of offspring produced from crosses between individuals from the same population. This is effectively the converse of **INBREEDING DEPRESSION**. For example, population A may possess large body size and be adapted to an environment where large body size confers an advantage. Population B may possess small body size and be adapted to an environment where small body size is an advantage. If an individual from population A crosses with one from population B an intermediate body size may result which is not adapted to either environment. *See also* **HETEROSIS**

outreach programme A programme taken out into the community, with the purpose of increasing community engagement, e.g. a conservation project, an **ENVIRONMENTAL EDUCATION** project.

ovarian acyclicity The condition in which a female animal does not exhibit a normal **OESTROUS CYCLE**, possibly due to sexual immaturity, disease or because she is in the post-reproductive phase of life.

ovarian cycle *See* **OESTROUS CYCLE**

ovarian follicle A roughly spherical aggregation of cells found in the ovary which contains a single egg (oocyte). Follicles periodically grow under hormonal control and release the egg during a process called **OVULATION**. *See also* **GRAAFIAN FOLLICLE**

ovariectomy Removal of the ovary or ovaries due to disease or as a method of contraception.

ovary The female gonad, which produces ova.

overdriving Driving an animal too hard; overworking an animal, especially livestock. A type of **CRUELTY** identified in some legislation.

overfeeding The provision of too much food, especially for livestock. In sheep it may cause **DYSTOCIA**. *Compare* **UNDERFEEDING**

overfishing The over-exploitation of fish resources, especially in the sea. *See also* **FISH DISCARDS, FISHING QUOTA**

overgrooming Abnormal excessive **GROOMING** (e.g. pulling at **HAIR** or **FEATHERS**) resulting in bald patches.

overhunting The over-exploitation of wildlife by hunters. This is not a recent phenomenon and there is fossil and archaeological evidence of mass killings of some large herbivore species. At sites known as 'buffalo jumps' native American Indians drove bison (*Bison bison*) over the edge of rock formations to their deaths in large numbers. In Syria over 6000 years ago, migrating gazelles were herded into long stone structures called desert kites (so called because of their appearance from the air) and slaughtered.

overmarking SCENT-MARKING over the marks of CONSPECIFICS.

overwintering The practice in some species (especially birds) of spending the winter in a particular locality.

oviparity Reproduction in which eggs are laid and embryos mature and hatch after being expelled from the mother's body. Oviparous taxa include most invertebrates, birds, most reptiles, amphibians, fishes and MONOTREMES. *Compare* OVOVIVIPAR-ITY, VIVIPARITY

oviparous Relating to or exhibiting OVIPARITY.

ovoviviparity Reproduction in which the embryo develops within the mother from which it may derive nutrition, but from which it is separated for all or almost all of development by egg membranes. Occurs in many insects, snails, some fishes and some reptiles. *Compare* OVIPARITY, VIVIPARITY

ovoviviparous Relating to or exhibiting OVOVIVI-PARITY.

ovulation The formation and release of an OVUM from an OVARY. In some species ovulation only occurs after mating (i.e. in INDUCED OVULA-TORS). *See also* SPONTANEOUS OVULATOR

ovum (ova *pl.***)** An unfertilised egg cell.

owl pellet A bolus of material regurgitated by an owl which contains the undigested remains of its prey, e.g. the bones and hair of small mammals. May be used to detect the presence of owls and to determine their diets. May also be used to monitor indirectly changes in the relative abundance of small mammal species.

Oxford Centre for Animal Ethics An independent education company founded in 2006 by Prof. Andrew LINZEY. It is the first centre in the world dedicated to pioneering ethical perspectives on animals through academic research, teaching and publication.

oximeter *See* PULSE OXIMETER

oxygen A colourless gas present in the atmosphere which is essential for AEROBIC RESPIRATION and is produced by green plants during PHOTOSYN-THESIS. *See also* OZONE (O_3)

oxygen debt The amount of oxygen required to metabolise the lactic acid which accumulates in the muscles after extreme exertion as a result of their obtaining energy anaerobically by LACTIC ACID FERMENTATION. *See also* CAPTURE MYOPATHY

oxygen saturation

1. A relative measure of the amount of oxygen dissolved or carried in a medium, e.g. the blood. It can be measured non-invasively using a PULSE OXIMETER.

2. The ratio of oxyhaemoglobin to the total concentration of HAEMOGLOBIN present in the blood.

oxygen tent A tent-like structure which can be erected around a patient and supplied with oxygen to aid breathing.

oxygen therapy, hyperbaric oxygen therapy (HBOT) A procedure used in emergency and critical care veterinary medicine where the animal has problems breathing or absorbing oxygen and where HYPOXIA may occur. It involves the optimisation of the oxygen supplied to the tissues. The animal enters a chamber into which 95% oxygen is pumped at raised pressure (above normal atmospheric pressure). Oxygen therapy may be useful in wound healing and the treatment of burns, skin grafts, fractures, cardiac disease, strokes and a number of other conditions.

oxygenated Carrying oxygen, e.g. oxygenated blood, oxygenated water.

oxytocin A hormone in mammals which is secreted by the POSTERIOR PITUITARY. It induces contractions of the uterus during LABOUR and stimulates the flow of milk from the breasts during suckling. A synthetic oxytocin may be given artificially to assist labour, but only if there is no obstruction to the birth or foetal oversize.

oyster farm A facility which rears oysters commercially as food, usually in tanks and lagoons.

ozone (O_3) A form of oxygen in which three oxygen atoms are joined together. It may be used to disinfect water. *See also* OZONISER

ozoniser An ozone-generating device used in aquariums and other aquatic exhibits to kill bacteria, viruses and other pathogens. Also improves the efficiency of BIOLOGICAL FILTRATION and PROTEIN SKIMMERS by speeding up the breakdown of ammonia and nitrites.

O

Fig. P1 P is for python. Reticulated python (*Python reticulates*).

Dictionary of Zoo Biology and Animal Management: A guide to terminology used in zoo biology, animal welfare, wildlife conservation and livestock production, First Edition. Paul A. Rees.
© 2013 John Wiley & Sons, Ltd. Published 2013 by John Wiley & Sons, Ltd.

Fig. P2 Paca (*Agouti paca*) farming.

PAAZAB The African Association of Zoos and Aquaria. *See also* **KEEPER ASSOCIATION**

paca farming A paca (*Agouti paca*) is a species of agouti which is farmed for its meat on a small scale in Costa Rica (Fig. P2). This activity helps to reduce the taking of wild pacas from the forest and some individuals are also released back into the wild.

pachyderm Originally, a member of an obsolete order of mammals, the Pachydermata. Now used to refer to large, thick-skinned mammals: elephant, rhinoceros, hippopotamus. Some zoos have historically housed such taxa together in a Pachyderm House.

Pachyderm The journal of the African Elephant, African Rhino and Asian Rhino **SPECIALIST GROUPS**.

pacing STEREOTYPIC BEHAVIOUR which involves repetitively walking forwards and backwards, often as a result of **PICKETING**. *See also* **STEREOTYPED ROUTE-TRACING**

pack Collective noun for a group of certain species, e.g. wolves, hunting dogs.

pack animal
1. An animal used for carrying loads, e.g. horse, mule, donkey. *See also* **BEAST OF BURDEN**
2. A species that normally lives in a **PACK**.

pain
1. A distressing or uncomfortable sensation caused by the stimulation of specialised nerve endings which are sensitive to heat, cold, pressure and other strong stimuli. Pain is impossible to measure scientifically and, although it may be experienced by some animals, especially mammals, in the same way as it is in man, non-mammals exhibit physiological responses to painful stimuli which are different from those of humans.
2. **SUFFERING** caused as a result of **EMOTION**.

painted fish *See* **DYED FISH**

pair bond A temporary or permanent association formed between a male and a female of a species which usually results in mating. This may imply a lifelong monogamous relationship, or a stage in the mating interaction in socially monogamous species. **VASOPRESSIN** and **DOPAMINE** appear to be important in the formation of pair bonds. Same-sex pair bonding occurs in some species, e.g. the zebra finch (*Taeniopygia guttata*). *See also* **CONSORTS**

Palaearctic region *See* **FAUNAL REGIONS**

palaeontology The scientific study of fossils.

palatability The quality of food that makes it acceptable to an animal as food, which may be related to its taste, smell, appearance etc.

palatable species
1. Organisms that taste pleasant, are not harmful and are suitable as food for other organisms.
2. In relation to herbivores, plant species that are preferred by grazers or browsers because their taste is agreeable, because they are less fibrous than others or for some other reason.

palliative care The care given to an animal to keep it comfortable when it is seriously ill and unlikely to recover.

palmar, volar Relating to the palm of the hand or sole of the foot; the back (caudal) surface of the forelimb below the carpus or wrist. *Compare* **PLANTAR**

paludarium A **VIVARIUM** that simulates a swamp or **RAINFOREST** habitat, and contains underwater areas. A mixture of a terrarium and an aquarium. *See also* **RIPARIUM**

pancreas A glandular abdominal organ in vertebrates that secretes digestive enzyme (including **AMYLASE** and **TRYPSIN**) into the small intestine and contains the islets of Langerhans, which secret **INSULIN** and **GLUCAGON**.

Pantanal A tropical wetland in South America which is high in biodiversity and is home to a large number of rare species including the giant anteater (*Myrmecophaga tridactyla*), jaguar (*Panthera onca*), maned wolf (*Chrysocyon brachyurus*), bush dog (*Speothos venaticus*), giant river otter (*Pteroneura brasiliensis*), South American tapir (*Tapirus terrestris*) and hyacinth macaw (*Anodorhyncus hyacinthinus*).

panther An alternative name for a leopard (*Panthera pardus*). A black panther is a **MELANIC (1)** leopard.

PanTHERIA An on-line database of information on extant and recently extinct mammals, including their ecology, distribution, life history and phylogeny. *See also* **YOUTHERIA**

panting A method of losing body heat by evaporating water in the warm air passed out of the lungs during expiration, used by some mammals, e.g. canids and reindeer.

paradeisos A walled park in which animals were kept for the enjoyment of a monarch (a royal 'paradise'). They provided animals for royal hunts and processions, and housed animals received as gifts from foreign rulers. The earliest paradeisos was established by the Chinese Emperor Wen Wang around 1150 BC. Similar parks were established in Assyria, Babylonia and Egypt.

paralysis A loss of muscular function or sensation in a part of the body. It may be temporary or permanent, and is usually caused by damage to the **NERVOUS SYSTEM** as a result of injury or disease. *See also* **PARAPLEGIA**

parametric In statistics, relating to a particular distribution, e.g. normal, Poisson. *Compare* **NON-PARAMETRIC**

paramyxoviruses An important group of viruses that cause diseases such as **NEWCASTLE DISEASE**, equine respiratory diseases, **CANINE DISTEMPER**, phocine distemper (in seals) and **RINDERPEST**.

parapatric Referring to species whose geographical ranges do not overlap but are immediately adjacent to each other with no physical barriers between them. *Compare* **ALLOPATRIC**, **SYMPATRIC**

parapatric speciation Speciation that occurs despite a minor flow of genes between **DEMES**. *Compare* **ALLOPATRIC SPECIATION**, **SYMPATRIC SPECIATION**

paraplegia **PARALYSIS** of the hind legs. Often associated with spinal injuries.

parasitaemia, parasitemia The condition in which parasites are present in the blood.

parasite An organism that lives on (**ECTOPARASITE**) or in (**ENDOPARASITE**) another organism (the **HOST**) and benefits by deriving nutrients (and often shelter) at the host's expense.

parasitism An association between two species where one (the **PARASITE**) gains some benefit at the expense of the other (the **HOST**).

parasitology The scientific study of **PARASITES** and parasitism. It involves the study of parasites, their hosts and **VECTORS**, the relationship between them and the diseases they cause.

parasympathetic nervous system (PNS) Part of the **AUTONOMIC NERVOUS SYSTEM (ANS)**. Acts antagonistically to the **SYMPATHETIC NERVOUS SYSTEM (SNS)**. It tends to control physiology in 'calm' conditions. For example, it constricts the pupils, stimulates **SALIVATION**, constricts the airways, slows the **HEART RATE**, stimulates digestion, the uptake of glucose and glycogen synthesis.

parent cell In **CELL DIVISION**, a **CELL** which divides to produce two **DAUGHTER CELLS**.

parental care The provision of food, shelter, protection and other materials and services by parents to their offspring, which is a form of **ALTRUISM**. Characteristic of higher vertebrates, especially mammals, but also present in birds, some reptiles, fishes, amphibians and even insects (Figs. P10, S12). *See also* **PARENTAL INVESTMENT**, **PARKING**

parental investment Any investment made by a parent in an offspring which increases the chance of that offspring surviving and reproducing at the expense of the parent's ability to invest in any other offspring. *See also* **PARENTAL CARE**, **PARKING**

parenteral In relation to the administration of a drug or other substance, a route other than by mouth. Usually means by **INJECTION**.

Paris Zoo *See* **BOIS DE VINCENNES**

parking In mammals, especially monkeys and prosimians, the action of a mother temporarily leaving an infant alone in a nest, tree or elsewhere, possibly to reduce the energetic cost of carrying it and to reduce mortality in carried infants. *See also* **PARENTAL CARE**

P

parrot disease *See* PSITTACOSIS

parrot fever *See* PSITTACOSIS

parthenogenesis The development of an individual from an egg which has not been fertilised. It occurs in some taxa where males are absent, e.g. leeches and flatworms. Komodo dragons (*Varanus komodoensis*) normally reproduce sexually. However, Watts *et al.* (2006) used **DNA** FINGERPRINTING to identify parthenogenetic offspring produced by two female Komodo dragons that had been kept at separate zoos and isolated from males.

particulate filtration *See* MECHANICAL FITRATION

PARTINBR Software which can be used to calculate INBREEDING COEFFICIENTS and partial inbreeding coefficients from PEDIGREES.

***Partula* snails** A genus of rare small tree snails endemic to the islands of French Polynesia for which captive breeding programmes in zoos (e.g. LONDON ZOO, Bristol Zoo, *DURRELL*, St. Louis Zoo and WOODLAND PARK ZOO) and reintroduction programmes exist. Often quoted as a captive breeding success story.

parturition The process of delivering an offspring and the PLACENTA from the UTERUS to the outside via the VAGINA in a MAMMAL; giving birth.

parvoviruses A group of small viruses which cause a range of diseases in animals including canine parvovirus, feline parvovirus (FELINE PANLEUCOPENIA), mouse parvovirus, porcine parvovirus and Aleutian disease (in mustelids).

passage migrant A bird that occurs in a particular geographical location for a brief period while travelling from its origin to its destination during MIGRATION (1).

passenger pigeon (*Ectopistes migratorius*) An extinct bird species which was once very common in North America but was exterminated by hunters. Often used as an example of the devastating effect human hunting has had in causing the extinction of once common species. *See also* MARTHA

Passeriformes An order of birds: broadbills, pitas, ovenbirds, woodcreepers, tyrant flycatchers, and songbirds, e.g. lyrebirds, pipits, wrens, thrushes, sparrows, titmice, weavers, drongos, crows, bowerbirds, birds of paradise.

passerine A member of the order PASSERIFORMES.

passport *See* CATTLE PASSPORT, PET PASSPORT

pasted vent A condition in birds (especially POULTRY (1 and 2)) in which the CLOACA (1) becomes matted with faeces. It is caused by DIARRHOEA or excessive excretion of urates.

Pasteur, Louis (1822–1895) A French scientist who performed experiments that supported the germ theory of disease and created the first VACCINES for rabies and anthrax. He also invented the process of pasteurisation used to treat milk.

pasteurellosis An infection caused by a bacterium from the genus *Pasteurella*. Occurs in cats, cattle, pigs, sheep, ducks and many other animals including humans.

pastoralist A nomad who raises livestock on natural pastures, e.g. cattle, sheep, goats, llamas, yaks etc.

patagial membrane, patagium
1. A thin membrane which extends between a limb and the body of an animal to form a wing or wing-like extension as seen in bats and flying squirrels.
2. An expandable membranous skin fold located between the wing and body of a bird.

patagium *See* PATAGIAL MEMBRANE

paternity certainty The situation in which a male animal is reasonably sure that the young he is caring for are his own. This knowledge is important in social systems in which males may provide considerable PARENTAL CARE if young are to survive.

pathogen A disease-causing agent or organism, e.g. BACTERIUM, FUNGUS, PRION, VIRUS.

pathological Relating to disease.

pathology The scientific study of DISEASE.

patient warming unit A device for keeping a patient warm during a veterinary procedure or recovery using warm air or heat generated from electrical resistance.

patriarch An alpha male; the most senior male within a social group who exercises leadership functions.

patriline
1. A line of descent through the father, i.e. son, father, grandfather etc.
2. A social group which revolves around male kinship.
Compare MATRILINE

Paucituberculata An order of mammals: 'shrew' opossums.

Pavlov, Ivan Petrovich (1849–1936) A Russian physiologist who received the Nobel Prize for his work on CLASSICAL CONDITIONING.

Pavlovian conditioning *See* CLASSICAL CONDITIONING

PCR *See* POLYMERASE CHAIN REACTION (PCR)

PDA *See* PERSONAL DIGITAL ASSISTANT (PDA)

PDSA *See* PEOPLE'S DISPENSARY FOR SICK ANIMALS (PDSA)

peck order *See* PECKING ORDER

pecking order, peck order A DOMINANCE HIERARCHY found in gregarious birds whereby dominant birds peck SUBORDINATE INDIVIDUALS when they approach too near. The hierarchy becomes established as birds learn to recognise each other. First studied in domestic fowl (*Gallus gallus domesticus*), but the term has now been extended to apply to similar relationships in other taxa when actual pecking does not occur.

pectoral girdle, shoulder girdle The set of bones in vertebrates which is part of the APPENDICULAR SKELETON and connects the forelimbs or fins to the AXIAL SKELETON providing them with

P

support. In vertebrates this includes the shoulder blades (scapulae). *See also* **PELVIC GIRDLE**

pedicure Treatment of the feet and toe nails.

pedigree
1. An animal's line of descent, which may be used as evidence of pure breeding. *Compare* **GRADE**
2. A family tree showing this.
See also **STUDBOOK**

pedigree animal
1. An animal which is pure-bred; descended from a long line of known ancestors of the same breed.
2. In Great Britain the **PET ANIMALS ACT 1951** (s.7(3))defines a pedigree animal as '*an animal of any description which is by its breeding eligible for registration with a recognised club or society keeping a register of animals of that description*'.

pedigree register A document or database which records the parentage of a particular type of pedigree animal.

peer review In relation to a **SCIENTIFIC PAPER**, the process by which the work of scientists and others is verified by other experts for its accuracy and usefulness before publication in an **ACADEMIC JOURNAL**. The process is intended to ensure that only high-quality work is published in academic journals.

pelage The coat of a **MAMMAL**; its fur, hair, wool or other soft covering.

pelagic
1. Relating to the coat of a mammal (its **PELAGE**).
2. Relating to occurring in the upper waters of open seas and oceans, as opposed to waters near land or the seabed, e.g. pelagic fishes, pelagic birds, pelagic whaling. This area of the sea is the pelagic zone.

Pelecaniformes An order of birds: pelicans, tropic-birds, cormorants, gannets, boobies, frigatebirds.

pellet
1. A food pellet; a food supplement containing vitamins, minerals etc.
2. A ball of waste regurgitated by a bird of prey containing hair, bones etc. of prey, e.g. **OWL PELLET**. May be used to identify prey organisms in field studies.

pelt The skin or untanned hide of an animal, especially the fur of a mammal. The pelts of some rare species are highly prized and illegally traded, especially those that have distinctive markings, e.g. tigers, leopards and other cats, zebras. *See also* **CITES, HIDE (2)**

pelvic girdle, pelvis A ring-like structure of bones at the lower end of the trunk in vertebrates, which is fused to the sacral vertebrae and supports the hindlimbs. It is part of the **APPENDICULAR SKELETON**. *See also* **PECTORAL GIRDLE**

pelvis *See* **PELVIC GIRDLE**

penile bone *See* **OS PENIS**

penis An intromittant organ found in mammals, some reptiles and a few birds. It is used to transfer semen from the male to the reproductive system of

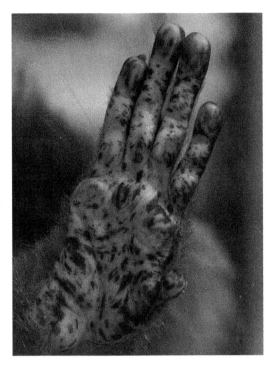

Fig. P3 Pentadactyl limb. Hand of a Javan langur (*Trachypithecus auratus auratus*).

a female. In mammals it also contains the terminal part of the urethra and is used for urination. A bone (**OS PENIS**) is present in some mammal species. *See also* **HEMIPENIS**

penis bone *See* **OS PENIS**

pentadactyl limb A limb possessing five digits or one which has evolved from such a limb, e.g. a hoof. Characteristic of all living **TETRAPODS** (Fig. P3).

pentamerous (radial) symmetry A special type of radial symmetry in which there are five lines of symmetry about a central axis. Adult echinoderms exhibit this type of symmetry, e.g. the five arms of a starfish. *Compare* **BILATERAL SYMMETRY, RADIAL SYMMETRY**

People for the Ethical Treatment of Animals (PETA) Foundation A UK-based charity dedicated to establishing and protecting the rights of all animals through public education, research, legislation and protest campaigns.

People's Dispensary for Sick Animals (PDSA) A charity in the UK that provides free veterinary care for pets whose owners cannot afford to pay for it. Founded in 1917 by Maria **DICKIN**. *See also* **DICKIN MEDAL**

People's Trust for Endangered Species A UK charity which helps to protect endangered species and their habitats around the world. It was founded in 1977.

pepper spray A spray that contains a chemical which irritates the eyes and may cause temporary blindness and is used in the control of some animals. Carried by keepers in some zoos that work with dangerous animals (e.g. carnivores and large primates). Carried by some hikers in the USA as a deterrent in case of attack by bears. Also used against dogs.

per os (PO) By mouth, as in the administration of a drug.

peracute In relation to disease, a very acute and violent form. *Compare* ACUTE CONDITION, CHRONIC CONDITION, SUBACUTE

Peramelemorphia An order of mammals: bandicoots.

perception The appreciation of the world by an animal as detected by, and limited by, its senses. This depends upon the type and sensitivity of the sense organs, e.g. an animal may possess an ability to hear a narrow range of sound frequencies or a wide range; it may have eyes that see in colour or only in black and white.

perceptual barrier *See* PSYCHOLOGICAL BARRIER

perchery system *See* BARN SYSTEM

Perciformes An order of fishes: perches and their allies.

Percopsiformes An order of fishes: troutperches and their allies.

Père David's deer (*Elaphurus davidianus*) A deer species originally native to China which was discovered by a French missionary, Father Armand David (Père David) in 1865. It was hunted to extinction in the wild but was saved by CAPTIVE BREEDING, notably by the Duke of Bedford at Woburn Abbey in England. The species has since been reintroduced into China.

perennial plant A plant that lives for more than 2 years. *See also* ANNUAL PLANT

Performing Animal Welfare Society (PAWS) A society founded in the USA in 1984 dedicated to the protection of performing animals. It provides sanctuary to abused, abandoned and retired captive wildlife, campaigns for changes to, and enforcement of, animal laws and engages in public education. *See also* CAPTIVE ANIMALS' PROTECTION SOCIETY (CAPS), CIRCUS, TRAVELING EXOTIC ANIMAL PROTECTION ACT (USA)

Performing Animals (Regulation) Act 1925 An Act to regulate the exhibition and training of performing animals in Great Britain. It requires the registration of persons who train or exhibit performing animals with the relevant local authority, and prohibits the use of cruelty. The Act does not apply to the training of animals for military, police, agricultural or sporting purposes or the exhibition of animals so trained.

perinatal Relating to the period immediately before and after birth. The length of the period to which this term applies is variable.

perinatal mortality The total number of animals (FOETUSES and NEONATES) that die as a proportion of the total at risk during the PERINATAL period.

Fig. **P4** Perineal tumescence in a female Sulawesi crested macaque (*Macaca nigra*).

perineal tumescence, sexual swelling A swelling in the area of the pelvic region occupied by the urogenital passages and rectum exhibited by females of some primate species as an indication of sexual receptivity (Fig. P4). *See also* DECEPTIVE SWELLING

periparturient, peripartum Relating to any condition which occurs in the mother just before or just after birth.

peripheral nervous system (PNS) In vertebrates, the nerves of the SOMATIC NERVOUS SYSTEM (SNS) which are responsible for carrying motor and sensory information to and from the CENTRAL NERVOUS SYSTEM (CNS) (1) (via the nerves that connect to the skin, sense organs and skeletal muscles), and those of the AUTONOMIC NERVOUS SYSTEM (ANS), which supply the intestines and other internal organs.

Perissodactyla An order of mammals: odd-toed ungulates (horses, zebras, asses, tapirs, rhinoceroses).

peristalsis A wave of constriction that passes along a tubular organ (e.g. the oesophagus) as the result of the sequential contraction of muscles, the purpose of which is to move material through the tube. As a bolus of food passes down the oesophagus the circular muscles behind it contract (constricting the lumen and pushing the food down) and the longitudinal muscles relax (allowing the lumen to dilate).

peritoneal cavity A fluid-filled potential space between the organs of the ABDOMEN (1) and the abdominal wall. In normal healthy animals there is very little actual space. It may be used as an INJECTION site.

peritoneum A thin membrane which lines the abdominal and pelvic cavities and covers most of the abdominal viscera.

permanent teeth Adult teeth. The second set of teeth found in most mammal species. *Compare* TEMPORARY TEETH

permit to work A document authorising a person or persons to work in a dangerous or potentially

P

dangerous area, such as inside the enclosure of a category 1 animal in a zoo. *See also* **CATEGORY 1, 2 AND 3 SPECIES**

person
1. In general usage, a human being, but some people think that certain animals (e.g. chimpanzees) should be treated as persons.
2. *See* **LEGAL PERSONALITY**

personal digital assistant (PDA) A hand-held computer that may be used for collecting data that can later be downloaded to a computer database, or to receive information wirelessly from a transmitter. *See also* **RADIO FREQUENCY IDENTIFICATION (RFID) TECHNOLOGY**

personality *See* **ANIMAL PERSONALITY, LEGAL PERSONALITY**

pest An unwanted species of insect, rodent, bird or other taxon which usually causes some harm to other species or the environment. Many pest species, e.g. rats and pigeons, may transmit diseases or parasites to farm or zoo animals and must therefore be controlled.

pest repeller A device which scares away animals (e.g. ants, cockroaches, mice, rats, squirrels, deer, foxes, cats) with bursts of ultrasound, and sometimes flashing strobe lights, activated by a **PIR SENSOR**.

pesticide A substance used to kill pest species. *See also* **DDT, INSECTICIDE, MOLLUSCICIDE, ORGANIC PEST AND WEED CONTROL, ORGANOCHLORINE, ORGANOPHOSPHATE, RODENTICIDE**

pet An animal kept by a person for companionship. The term '**COMPANION ANIMAL**' is preferred by some advocates for animals and ethicists.

Pet Animals Act 1951 An Act in Great Britain which requires the licensing and inspection of pet shops by the relevant local authority. All commercial selling of vertebrate animals, whether from business premises, private dwellings or over the internet, must be licensed by the local authority within which the premises are located. Vertebrate animals may not be sold to children under the age of 12 years, or in any street, road or public place or in a stall or barrow in a market. Local authorities have powers to inspect premises where pet animals are sold.

Pet Obesity Task Force A group launched in conjunction with the **ROYAL SOCIETY FOR THE PREVENTION OF CRUELTY TO ANIMALS (RSPCA)** at the European Pet Obesity Conference in 2008 in response to an epidemic of **OBESITY** among many companion animal species.

pet passport A document issued as part of the **PET TRAVEL SCHEME (PETS)** that records information relating to a particular animal, including its **MICROCHIP** or tattoo number and a rabies vaccination certificate. *See also* **CATTLE PASSPORT, MOVEMENT DOCUMENT**

pet shop A retail outlet that sells small companion and exotic animals, animal food, equipment and other related items to the public. Pets shops in Great Britain must be licensed. *See* **PET ANIMALS ACT 1951**

Pet Travel Scheme (PETS) A system that allows pet dogs, cats and ferrets to enter the UK without **QUARANTINE** as long as they comply with certain rules (which vary depending upon the country from which the animal is travelling). It also allows people in the UK to take these animals to other countries and return with them to the UK without the need for quarantine. The basic requirements are that the animal is microchipped, vaccinated against rabies and then blood tested at least 30 days later to make sure the vaccine has worked, has appropriate travel documents, has tapeworm treatment (dogs only) and travel is arranged on an authorised route with an approved transport company. Similar schemes operate in some EU countries, the USA, Canada, Australia and New Zealand. *See also* **PET PASSPORT**

pet-facilitated psychotherapy The use of companion animals (especially cats or dogs) to treat people with psychiatric illnesses. *See also* **ANIMAL-ASSISTED THERAPY, DOLPHIN THERAPY**

PETA *See* **PEOPLE FOR THE ETHICAL TREATMENT OF ANIMALS (PETA) FOUNDATION**

petri dish A shallow, clear glass or plastic circular dish with a lid used in **MICROBIOLOGY** to grow cultures.

Petromyzontiformes An order of fishes: lampreys.

pets' corner, petting zoo *See* **CHILDREN'S ZOO**

pH scale A scale used to measure acidity or alkalinity; 1.0 = very acid, 7.0 = neutral, 14.0 = very alkaline. The pH value is the reciprocal of the hydrogen ion concentration in moles per litre.

phalange, phalanx One of the bones which make up the digits.

phase feeding A method of livestock feeding that involves altering the nutrient concentrations in a series of diets formulated to meet the animal's nutrient requirements more precisely at each stage of growth or production.

Phase-In Population An **ASSOCIATION OF ZOOS AND AQUARIUMS (AZA)** population management level consisting of a species which is being phased into AZA-accredited institutions and will be transferred to another population management level as appropriate.

Phase-Out Population An **ASSOCIATION OF ZOOS AND AQUARIUMS (AZA)** population management level consisting of a species that is being phased out of AZA-accredited institutions by a **MORATORIUM** on breeding.

phenotype The sum total of all of the structural and functional characteristics of an organism, including morphological features and those that cannot be directly observed, e.g. **BLOOD GROUPS**. *Compare* **GENOTYPE**

pheromone A chemical secreted by an organism and released into the environment which is capable of altering the behaviour of a **CONSPECIFIC**. In vertebrates they are detected by the olfactory system. Pheromones may be important as sexual attractants (e.g. in moths) and in **SCENT-MARKING**. *See also* **FELIWAY**

Philodota An order of mammals: pangolins.

philopatry

1. The tendency of an animal to remain in or return to its home area or birth place. Sometimes called natal philopatry. The opposite of **DISPERSAL**. In bird species philopatry is prevalent in males (male philopatry), whereas in mammal species it is prevalent in females (female philopatry).

2. In behavioural ecology, the behaviour of elder offspring sharing the burden of rearing their siblings with their parents.

Phoenicopteriformes An order of birds: flamingos.

phonotaxis Movement of an animal in relation to a source of sound. Positive phonotaxis occurs in some territorial frog species which move towards the calls of other males of the same species.

phosphorus An important element required by organisms. Phosphorus plays a role in the utilisation of carbohydrates and fats in the body and in **PROTEIN SYNTHESIS**. In combination with calcium it forms the mineral portion of bones and teeth, and it is required for the production of **ADENOSINE TRIPHOSPHATE (ATP)**. It assists in muscle contraction, kidney function, nerve impulse conduction and in the regulation of the heartbeat. The maintenance of an appropriate calcium: phosphorus ratio is essential for good health.

photography and animal exploitation Animals are widely exploited to produce souvenir photographs for tourists and close-up images of exotic species. In many tourist destinations, particularly in developing countries (e.g. Morocco) but also in some European countries (e.g. Spain), animals, especially primates, are exploited by photographers. They take and sell photographs of tourists with their animals, or allow tourists to take their own photographs for a fee (Fig. P5). In some cases primates have their teeth removed so they cannot inflict a serious bite. Some animal welfare organisations actively work to put an end to this use of animals and place animals in sanctuaries. *MONKEY WORLD* has confiscated a number of illegally held apes from photographers. Animals are kept by some organisations to provide subjects

P

Fig. P5 Photography and animal exploitation: tourists paying to be photographed with Barbary macaques (*Macaca sylvanus*) in Marrakesh, Morocco.

for professional or amateur photographers. They provide opportunities for producing still photographs or video of animals that would be difficult to approach in the wild, e.g. felids, birds of prey etc. Some of these organisations provide trained exotic animals for use by the television and film industry.

photoperiod The length of time an organism is exposed to sunlight each day. This varies with the season and acts as a trigger for some seasonal changes in behaviour and physiology, e.g. reproduction.

photoreceptor A sensory receptor that is sensitive to light.

photosynthesis A biochemical process by which GREEN PLANTS produce glucose and oxygen from carbon dioxide and water using the energy from sunlight.

phototaxis A movement (TAXIS) in relation to a light source: positive phototaxis is movement towards light, negative phototaxis is movement away from light.

photo-trap *See* CAMERA TRAP

phylogenetic Relating to the sequence of changes that have occurred during the course of the EVOLUTION of a particular organism or TAXON.

phylogenetic systematics *See* CLADISTICS

phylogeny The evolutionary relationships within and between groups of organisms.

phylum In taxonomy, a group of related classes; a subdivision of a kingdom (Table C5).

physical enrichment *See* STRUCTURAL ENRICHMENT

physical environment, abiotic environment The non-biological components of the surroundings of an organism, e.g. air, water, rock, temperature, chemical ions, oxygen, humidity. *See also* ENVIRONMENT, IMMEDIATE ENVIRONMENT. *Compare* BIOLOGICAL ENVIRONMENT

physical filtration *See* MECHANICAL FILTRATION

physiology
1. The scientific study of the functions and activities of living organisms and their parts, including all physical and biochemical processes.
2. The biological processes or functions which occur in an organism or in any of its parts.

phytoplankton *See* PLANKTON. *See also* LIVE PHYTOPLANKTON

PIC *See* PIG IMPROVEMENT COMPANY (PIC)

pica
1. The eating of non-food substances, e.g. cats have been reported to eat rubber, fabric, electrical cables and wool.
2. Alternative outdated spelling of pika: a small LAGOMORPH of the genus *Ochotona*.

Piciformes An order of birds: woodpeckers, barbets, jacamars, toucans, honeyguides, puffbirds.

picketing The practice of restraining large animals (especially elephants) by tethering them (with ropes or chains) to a stake or SCREW PICKET in the ground, or to a metal ring fixed in a concrete floor. A common practice in travelling circuses. Elephants are traditionally chained by one foreleg and the diagonally opposite hind leg so they are able to take just one step forward and one step backwards. Used by circuses and some zoos to restrain elephants while washing, inspecting the body and for other reasons. *See also* PROTECTED CONTACT, TETHER

Pidcock's Wild Beast Show Possibly the first recorded TRAVELLING MENAGERIE to take to the road in the UK, in 1708.

piedmont An area at the base of mountains.

pig, hog, swine
1. A pig is any of the species in the genus *Sus* (Suidae).
2. One of the oldest domesticated animals, it was kept by Neolithic farmers. Modern pig breeds are descended from the wild boar (*Sus scrofa*) which is still kept by some farmers. It was once common in the wild in the UK and still occurs in the wild in Europe. *See also* BACONER, MICRO PIG, PIG BREEDS, SAUSAGE PIG

pig board, hog board A wooden board, with handles, which is held vertically by a keeper or farmer and used to guide the movement of an animal by blocking its path and vision. Used to transfer animals between pens, load them onto vehicles, CORRAL (2) them for capture etc. Effective because LIVESTOCK tend to move away from walls and into open spaces.

pig breeds There are hundreds of breeds of domestic pig. Some notable examples are listed in Table P1 and illustrated in Fig. P6.

pig flu *See* H1N1 VIRUS

Pig Grading Information System (PiGIS) A system which provides pig producers with information about their animals from the abattoir including weight details, grading, changes in carcass quality over time, and the level of condemnations.

Pig Improvement Company (PIC) A company which breeds and supplies genetically superior pig breeding stock and technical support for farmers. *See also* DANBRED INTERNATIONAL (DBI)

Table P1 Major pig breeds.

American Yorkshire	Large Black
Bentheim Black Pied	Large White
Berkshire	Mangalitza
British Lop	Meishan
British Saddleback	Middle White
Chester White	Oxford Sandy and Black
Duroc	Pietrain
Gloucester Old Spot	Poland China
Hampshire	Red Wattle
Hereford	Swabian-Hall Swine
Iberian	Tamworth
Iron Age	Vietnamese Potbelly
Kune Kune	Welsh
Lacombe	Wild Boar
Landrace	

Fig. P6 Pig breeds: British Saddleback (top left); Middle White (top right); Kune Kune (bottom left): Tamworth (bottom right).

P

pig influenza *See* **H1N1 VIRUS**

pig-dog hunting *See* **PIG-DOGGING**

pig-dogging, hog-dogging, pig-dog hunting A blood sport practised especially in parts of Australia and the USA, in which trained dogs wearing protective collars and breastplates corner and attack wild pigs until eventually a human handler kills the pig.

PiGIS *See* **PIG GRADING INFORMATION SYSTEM (PiGIS)**

piloerection The erection of the fur or hairs on the body of a mammal in response to a stimulus, e.g. cold.

pineal body *See* **PINEAL GLAND**

pineal gland, epiphysis, pineal body, third eye A small **ENDOCRINE GLAND** in the vertebrate brain which connects the **ENDOCRINE SYSTEM** with the **NERVOUS SYSTEM** and produces the hormone melatonin which affects sexual development and sleep–wake cycles. *See also* **CIRCADIAN RHYTHM**

pinioning *See* **FLIGHT RESTRAINT**

pinna

1. An extension of the external ear in mammals, supported by cartilage, which focuses sound on the ear. It is also important in **THERMOREGULATION** in some species (e.g. the African elephant (*Loxodonta africana*)), and in producing **SIGNALS** in species where it is mobile.

2. A fin, fin-like limb or similar appendage.

pinniped A member of the **PINNIPEDIA**.

Pinnipedia An order of aquatic **MAMMALS**: seals, sea lions, walrus.

pinyata A hollow papier mâché object containing food treats given to animals in zoos, e.g. primates, as a **FEEDING ENRICHMENT**. The animal must destroy the pinyata to reach the food.

PIR sensor Passive infrared sensor. Used to activate **CAMERA TRAPS**, alarms and other equipment by detecting **INFRARED RADIATION** produced by the bodies of humans or other animals.

Pisces An obsolete term for fishes: bony fishes, cartilaginous fishes, hagfishes, lampreys. The term is no longer used in modern classifications but is sometimes used in legislation. *See also* **FISH**

pit *See* **BEAR PIT**

pit bull terrier A type of dog bred especially for fighting, and consisting of several breeds. Possession of any of these breeds is banned in the UK by the **DANGEROUS DOGS ACT 1991**. *See also* **STATUS DOG, STATUS DOG UNIT**

pituitary gland, hypophysis A small **ENDOCRINE GLAND** in **VERTEBRATES** which is a protrusion of the **HYPOTHALAMUS**. *See also* **ANTERIOR PITUITARY, POSTERIOR PITUITARY**

placenta An organ that connects the developing mammalian **FOETUS** to the wall of the **UTERUS** during pregnancy, providing nutrients and oxygen, and allowing the removal of carbon dioxide and nitrogenous wastes. Many mammals consume the placenta after giving birth (placentophagy).

placental
1. Relating to the **PLACENTA**.
2. *See* **PLACENTAL MAMMAL**.

placental mammal, placental A mammal in which the female nurtures the foetus in the **UTERUS** via a **PLACENTA**; the majority of mammal species. *Compare* **MARSUPIAL, MONOTREME**

placental scar A mark left on the wall of the **UTERUS** where a **PLACENTA** was previously attached. Counting these scars during *POST-MORTEM* **(2)** examinations allows the identification of non-breeders, the calculation of the number of pregnancies a breeding female has had, and may be used to calculate **MEAN BIRTH INTERVAL**. Useful in field studies of **POPULATION DYNAMICS**.

placentophagy Consumption of the **PLACENTA** after giving birth.

plague
1. Any of several epidemic diseases, especially bubonic plague which is caused by the bacterium *Yersinia pestis* which affects **RODENTS** and is transmitted by fleas to humans.
2. An exceptional unwelcome increase in the population size of a species, which often does considerable economic damage to crops, e.g. locusts, rodents. Often caused by favourable environmental conditions, e.g. high rainfall causing an increase in food availability. House mouse (*Mus musculus*) plagues are a significant economic problem for grain farmers in Australia and western China. This is an **INVASIVE SPECIES** in Australia and has few competitors or natural predators.

plankton Small aquatic organisms which move with the water currents, consisting of phytoplankton (plants, mostly diatoms) and zooplankton (animals). Important as food for many larger aquatic organisms. *See also* **LIVE PHYTOPLANKTON, LIVE ZOO PLANKTON**

plantar Relating to the back (caudal) surface of the hindlimb below the tarsus or hock. *Compare* **PALMAR**

plantigrade Pertaining to walking on the sole of the foot with the heel in contact with the ground, e.g. as in humans, rabbits, bears etc. *Compare* **DIGITIGRADE, UNGULIGRADE**

plasma, blood plasma The liquid component of **BLOOD** in which the blood cells are suspended. It contains **HORMONES, GLUCOSE, MINERAL** ions, dissolved **PROTEINS** and carbon dioxide.

plasma membrane *See* **CELL MEMBRANE**

plasmalemma *See* **CELL MEMBRANE**

plastic song A type of **SONG** exhibited by birds during the development of the adult song, consisting of recognisable elements and vocalisations with distinct syllables.

plastic surgeon A person who is qualified in and licensed to practice **PLASTIC SURGERY**.

plastic surgery A specialist area of surgery concerned with the restoration, reconstruction and improvement of the form and function of the body, especially after damage caused by disease or injury. Plastic surgeons may sometimes treat animals that have been badly injured in fights or accidents.

platelets, thrombocytes Very small red bodies without a nucleus found in blood which are involved in **BLOOD CLOTTING** and release **SEROTONIN**.

platyhelminth (platyhelminths *pl.*) A member of the phylum **PLATYHELMINTHES**.

Platyhelminthes A **PHYLUM** of animals. The platyhelminths are free-living and parasitic flatworms. They include many economically important parasites such as tapeworms and liver flukes. They are characteristically flat and relatively featureless in form, however some have large eyes and legs. Platyhelminths possess a head where the sensory organs are concentrated, including simple eyes (ocelli), and there is a simple 'brain'. The mouth opens into a blind-ending sac which is often highly branched. There is no real circulatory system but a primitive excretory system is present. *See also* **FLUKE DISEASE**

platyrrhine A **NEW WORLD MONKEY**, which possesses widely separated nostrils that generally open to the side. *Compare* **CATARRHINE**

play A behaviour of many species, especially young mammals and birds, which allows them to develop and practise behaviours that are needed in adult life. Play in animals may be categorised as object play (interacting with objects), motor play (running, jumping etc.) and social play (interacting with conspecifics, e.g. play fighting, play mating etc.). It probably evolved as motor training and is now multifunctional in many species. Social play mimics adult agonistic competition. Zoos encourage object play as an enrichment to the lives of zoo animals. It is important that captive animals are provided with adequate opportunities for social play if they are to develop into competent adults. *See also* **SOCIAL FACILITATION, SPARRING**

Pleistocene megafauna A large animal fauna (mammals, birds and reptiles) that existed during the Pleistocene epoch and became extinct in a Quaternary extinction event.

Pleistocene Park A refuge in northeastern Siberia which was established in 1996 where horses, muskoxen, bison and other hervivores have been introduced and are converting the TUNDRA back to grassland. *See also* CONSERVATION GRAZING, PLEISTOCENE REWILDING

Pleistocene rewilding An idea that promotes the replacement of the extinct PLEISTOCENE MEGAFAUNA of North America by introducing large mammal species from Africa. The ecological communities of North America (especially the grasslands) evolved in the presence of large mammals and supporters of this idea believe that such REWILDING would improve the dynamics of these systems.

pleural effusion An abnormal accumulation of fluid between the two pleural membranes surrounding the lungs, which may impair breathing.

Pleuronectiformes An order of fishes: flatfishes.

PLOS/PLoS The Public Library of Science. A non-profit publisher which publishes open-access scientific research online.

plumage The feathers covering the body of a bird. It may have a different appearance at different ages and seasons, e.g. juvenile plumage, adult plumage, summer plumage, winter plumage.

Plumage League An organisation formed in Didsbury, Manchester (England) in 1886, by Emily Williamson and other Victorian ladies with the purpose of preventing the slaughter of birds by controlling the trade in the plumage of exotic birds for the fashion industry. Eventually the organisation joined with the FIN, FUR AND FEATHER FOLK to become the ROYAL SOCIETY FOR THE PROTECTION OF BIRDS (RSPB). *See also* SELBOURNE SOCIETY FOR THE PROTECTION OF BIRDS, PLANTS AND PLEASANT PLACES

plunge-diver A bird that feeds by hovering above water, especially the sea, and then diving for fish, e.g. common tern (*Sterna hirundo*), gannet (*Morus bassanus*).

pneumonia INFLAMMATION of the lungs caused by bacteria, fungi or chemicals. Signs include coughing, appetite loss, DEPRESSION and breathing difficulties.

PNS *See* PERIPHERAL NERVOUS SYSTEM (PNS)

poached out Relating to a species that has been removed from an area by POACHING.

poacher Someone who illegally captures animals. *See also* HUNT (1), POACHING

poaching The illegal taking of wildlife. Responsible for declines in the numbers of a very wide range of animals (e.g. ELEPHANTS, tigers).

pod A collective term for a group of cetaceans, e.g. a dolphin pod.

Podicipediformes An order of birds: grebes.

poikilotherm, ectotherm An animal that is not capable of regulating its body temperature physiologically. Internal temperature fluctuates considerably and is dependent upon temperature changes in the environment. However, it can be raised or lowered by moving into the sun or shade. All animal taxa apart from mammals and birds are poikilotherms. *Compare* HOMEOTHERM

poison A substance that may cause injury or death to an organism. Poisons may enter the body by being swallowed or inhaled, through a wound or sometimes through broken skin. A substance may act as a poison for one species and yet have no discernable effect on another species. This generally depends in part on whether or not the species possesses an enzyme capable of detoxifying the poison. *See also* VETERINARY POISONS INFORMATION SERVICE (VPIS)

poison avoidance The ability of an animal to learn to avoid poisonous substances. *See also* FOOD SELECTION

poisonous plant A plant whose parts (seeds, stems, leaves etc.) are toxic to animals or humans, e.g. rhododendrons, yew, lupins, laburnum and laurel. It is essential to exclude such plants from feedstuffs and from animal enclosures. Toxic chemicals produced by plants cannot be detoxified by certain animal species because they do not produce the appropriate ENZYMES.

poke box A box used for COGNITION experiments consisting of a tray from which an animal may select objects which are different colours, shapes, patterns etc.

polar Relating to either the north pole or the south pole.

Polar Bear Agreement *See* AGREEMENT ON THE CONSERVATION OF POLAR BEARS 1973

pole trap A small metal trap with spring-loaded jaws which are released when a central pressure plate is depressed. Mounted on fence posts to catch BIRDS OF PREY by their legs (Fig. P7). The use of pole traps was made illegal in the UK in 1904 by an amendment to the Wild Birds Protection Act 1902.

police dangerous dogs unit *See* STATUS DOG UNIT

police dog A dog trained to assist the police in controlling crowds, finding evidence, and locating and apprehending suspects; often a German Shepherd dog.

police horse A HORSE trained for crowd control, patrol in rural areas, ceremonial duties and other specialist roles. The Mounted Branch of the Metropolitan Police was founded in 1760. In 2011 it consisted of 140 officers and 120 horses based at eight operational stables across London. The Royal Canadian Mounted Police was founded in 1873 and originally called the North West Mounted Police.

Fig. P7 Pole trap. A kestrel (*Falco tinnunculus*) caught in a pole trap.

policy
1. An action plan, based on certain principles, determined by an organisation or individual. In the UK a **WHITE PAPER** is a document produced by the government which sets out its policy on a particular area of governance, e.g. wildlife conservation, agriculture, planning law. It has no legal status but is usually implemented by the passing of one or more **ACTS OF PARLIAMENT** or **STATUTORY INSTRUMENTS**. During World War II the Japanese government had a policy of destroying dangerous animals kept by zoos and circuses (**WARTIME ZOO POLICY**).
2. A code of conduct which employees, members of an organisation or others are expected to follow, or which is used to make decisions. Any organisation may produce a policy with which it requires its employees or members to comply. In Australia, the Director General of NSW Department of Primary Industries published a *Policy on the Management of Solitary Elephants in New South Wales* (pursuant to Clause 8(1) of the Exhibited Animals Protection Regulation, 2005).
See also **ANIMAL TRANSACTION POLICY, COMMON AGRICULTURAL POLICY (CAP), COMMON FISHERIES POLICY (CFP)**

poll
1. The top of the head. *See also* **POLL GUARD**
2. To remove the horns of cattle.

poll guard A cap worn by a horse which fits around the ears and covers the top of its head (poll) to protect it from bumps and scrapes, especially during transportation.

polled livestock Animals which do not have horns, either because they have been selectively bred to prevent them growing, or because they have been removed by **DISBUDDING** or **DEHORNING**. Livestock with horns may be dangerous, and are prone to catching their horns in fences and handling equipment.

polyandry A mating system in which one female animal mates with several males. *Compare* **POLYGYNY**

polydipsia Excessive drinking; drinking more than required by physiological needs. Often incorporated into sequences of **STEREOTYPIC BEHAVIOUR** such as **BAR-BITING** and **CHAIN-CHEWING**. Common in sows. It also occurs in disease, e.g. kidney failure, **DIABETES MELLITUS**, where it is in response to change in physiological need resulting from the disease.

polygamous Having more than one mate. *See also* **POLYANDRY, POLYGAMY, POLYGYNANDRY, POLYGYNY**

polygamy A mating system in which a single animal mates with more than one individual of the opposite sex. *See also* **POLYANDRY, POLYGYNANDRY, POLYGYNY**

polygene One of a group of genes which interact to control a continuously variable character such as height.

polygynandry A multi-male, multi-female **POLYGAMOUS** mating system in which females are usually more numerous than the males and mating occurs only within the group. Occurs in bonobos and lions.

polygyny A mating system in which one male mates with many females. In those species where a single male has a harem of females there may be excess males produced in captive populations. Some zoos keep **BACHELOR GROUPS** of these species and loan them to other zoos for breeding when required. *Compare* **POLYANDRY**

polymerase chain reaction (PCR) A molecular biology technique which amplifies small amounts of **DNA** and uses a polymerase **ENZYME** to assemble new strands of DNA for analysis. *See also* **DNA FINGERPRINTING, MICROSATELLITE MARKER**

Polymixiiformes An order of fishes: beardfishes.

polymorphic species A species which exists in a number of genetically distinct interbreeding forms. Some polymorphic species have previously been classified as several related species. Some polymorphisms can only be distinguished by genetic analysis. *Compare* **MONOMORPHIC SPECIES**

polyoestrous Having more than one **OESTROUS CYCLE** per year, e.g. most primates. *Compare* **MONOESTROUS**

polyparous Describing a female producing a large number of offspring. *See also* **MULTIPAROUS, NULLIPAROUS, PRIMIPAROUS, UNIPAROUS**

Polypteriformes An order of fishes: bichirs.

polysaccharide *See* **CARBOHYDRATE**

P

polytunnel A large plastic (polythene) structure with a semi-circular cross-section used on farms to protect small livestock (e.g. chickens) or plants from the weather (instead of a greenhouse) or to create a controlled micro-climate (e.g. for keeping and breeding butterflies and moths). Sometimes used in zoos for butterfly and moth exhibits.

polytypic A species is said to be polytypic if it exists as two or more distinctive geographical populations (SUBSPECIES). *See also* MONOTYPIC **(2)**

pony A HORSE TYPE consisting of animals with deeper bodies and shorter legs than other types.

Pony Club An international voluntary organisation for young people interested in ponies and riding. Founded in England in 1929, but now operates in 18 countries.

PopLink Software developed by LINCOLN PARK ZOO and used in the management and analysis of studbook databases. PopLink is shareware and is distributed free of charge by the zoo.

popsicle A block of ice made from fruit juice, and possibly containing pieces of fruit. Used as a FEEDING ENRICHMENT. *See also* BLOODSICLE

population

1. A group of organisms of the same species, living at the same time and in the same place, e.g. the zoo population of chimpanzees in 2009, the population of gannets on the Bass Rock in 1998.

2. In STATISTICS, the complete set of values of a particular variable in a given situation, e.g. the heights of all of the one-year-old ostriches in Kenya. Also called the parent population. *See also* SAMPLE

population control The process of preventing an increase in a population of animals possibly by culling or the use of CONTRACEPTION. *See also* CULL

population dynamics The branch of biology that is concerned with changes in the size and age structure of POPULATIONS and the environmental factors that influence them. *See also* POPULATION GROWTH, POPULATION SIZE

population estimate An estimate of the number of animals in a population made by using any one of a number of sampling techniques. *See also* AERIAL SURVEY, MARK–RECAPTURE TECHNIQUE, TOTAL COUNT

population genetics The study of inheritance at the POPULATION level, especially as it affects evolution. Important in the captive breeding of animals for farming or conservation purposes. *See also* GENE FREQUENCY, GENETIC DRIFT, HARDY–WEINBERG EQUILIBRIUM, SELECTION

population growth The increase in size of a population of organisms resulting from the positive effects of births and immigration and the negative effects of deaths and emigration. This may take a number of forms including EXPONENTIAL GROWTH, J-SHAPED GROWTH and LOGISTIC GROWTH (Fig. P8). *See also* ESSAY ON THE PRINCIPLE OF POPULATION

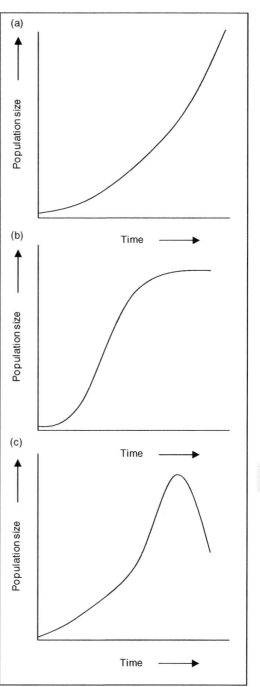

Fig. P8 Population growth: (a) Exponential growth; (b) Logistic growth; (c) J-shaped (boom-and-bust) growth.

P

Population Management 2000 (PM2000) Software used for the genetic and demographic analysis and management of pedigrees. It can be used as a stand-alone program but is most easily used in conjunction with the **SINGLE POPULATION ANALYSIS AND RECORD KEEPING SYSTEM (SPARKS)** studbook management software developed by the **INTERNATIONAL SPECIES INFORMATION SYSTEM (ISIS)**.

Population Management Plan (PMP) Programs A voluntary system whereby **ASSOCIATION OF ZOOS AND AQUARIUMS (AZA)** zoos manage and conserve captive populations of animals (as opposed to the Species Survival Plans in which the AZA requires cooperation). There are currently over 300 PMP Programs responsible for developing recommendations for the breeding and transfer of individual animals. Each PMP is managed by the corresponding **TAXON ADVISORY GROUP (TAG)** and administered by a PMP Manager who also acts as the **STUDBOOK KEEPER** for the species. *See also* **DISPLAY EDUCATION AND RESEARCH POPULATION, GOVERNMENT PROGRAM POPULATION, PHASE-IN POPULATION, PHASE-OUT POPULATION, SPECIES SURVIVAL PLAN® (SSP) PROGRAMS**

population size The number of individuals present in a **POPULATION** of organisms at a particular point in time (in the field or in captivity). In the field changes in population size are determined by the balance between **IMMIGRATION, EMIGRATION**, natality and **MORTALITY**. In a zoo, changes in population size are determined by the balance between births, deaths, animals received from other collections (or the wild) and animals transferred to other collections (or released into the wild). *See also* **MINIMUM VIABLE POPULATION (MVP)**

population supplementation The addition of individuals to a wild population of a species as a conservation measure to increase its abundance and potential for reproduction. *Compare* **REINTRODUCTION, REWILDING, TRANSLOCATION**

population viability analysis (PVA) A computer program that provides a quantifiable means of predicting the probability that a population will become extinct that can be used for prioritising conservation needs. It allows the calculation of a **MINIMUM VIABLE POPULATION (MVP)** for a species. The methodology takes into account both **DETERMINISTIC** factors (e.g. habitat loss or over-exploitation) and **STOCHASTIC** (random) factors (e.g. demographic, environmental and genetic factors). This type of analysis may be used to answer questions such as 'What is the risk of extinction of the Asian elephant in the next 100 years?'

porcine Of or relating to pigs.

Porifera A phylum of animals. The Porifera are sponges which are extremely simple aquatic organisms, almost all of which are marine. The body con-sists of a loose collection of cells arranged around a water-canal system. There are no body organs. Water is taken into the central cavity through tiny pores in special epidermal cells (porocytes) and expelled through a large osculum at the top of the body. Specialised flagellated cells remove particles of food from the current as the water passes over them. Sponges are immobile throughout most of the life cycle and attach themselves to the substratum with a structure called a holdfast. Sponges have the ability to reform the body if they are broken into smaller pieces.

porker A pig, especially one fattened to provide meat.

porpoising A behaviour which consists of leaping out of water and plunging back into it in the manner of a porpoise. Sometimes exhibited by penguins while swimming, especially when being pursued by predators. *See also* **PRONKING**

positive correlation *See* **CORRELATION**

positive reinforcement A term used in **OPERANT CONDITIONING** for the delivery of a **STIMULUS** (positive reinforcer – a reward) immediately or shortly after a response, that results in an increase in the probability that the response will occur in the future, or the future rate of response. The process of rewarding an animal for a correct response during training, e.g. food or praise. *See also* **LEARNING, REINFORCEMENT**. *Compare* **NEGATIVE REINFORCEMENT**

positive reinforcer *See* **POSITIVE REINFORCEMENT**

post hoc **study** A study of data after the experiment or study has been conducted, which looks for patterns which were not specified in advance.

post-anal tail A tail that extends beyond the anus. Characteristic of **CHORDATES**. It provides a method of locomotion in aquatic chordates and a moveable tail in most **MAMMALS**.

posterior Relating to the rear or hind end of a structure or the body; the caudal end of the body of quadrupeds (Fig. A8). *Compare* **ANTERIOR**

posterior pituitary, neurohypophysis The posterior lobe of the pituitary gland. Part of the **ENDOCRINE SYSTEM** which secretes **OXYTOCIN** and **VASOPRESSIN**.

post-mortem
  1. In Latin, literally, after death.
  2. Used as shorthand for a *post-mortem* examination (autopsy or necropsy) by a vet or scientist, which seeks to determine the cause of death.

postoccupancy evaluation (POE) An assessment of the performance of a new zoo exhibit (or other facility), once it has been commissioned and occupied, with respect to the improvement of animal welfare, the enhancement of the visitor experience and the improvement of the workplace for staff.

postpartum After **PARTURITION**.

postpartum oestrus The first **OESTROUS CYCLE** after giving birth.

post-release monitoring The monitoring of the survival, health, dispersal etc. of animals after they have been released from captivity following treatment for an injury or disease, or after being bred in captivity. Usually involves the **MARKING (1)** of individuals and often the use of **RADIO-TRACKING**. Important in assessing the success of a project. *See also* **POST-RELEASE SURVIVAL**

post-release survival The survival of animals that have been released into, or returned to, the wild as determined by **POST-RELEASE MONITORING**. May be measured as the number of days survived following release or the number of individuals sighted after a specific period of time following release.

post-surgery recovery pool *See* **UNDERWATER TREADMILL**

posture The position in which the body is held. This may send important signals to **CONSPECIFICS** and members of other species, e.g. lowering the body to the ground is a submissive behaviour (posture) in canids and some other species.

potassium A mineral which is important in the diet. It helps to control acid–base balance and osmotic pressure. Deficiencies are rare, but may occur in herbivores fed diets low in forage and high in concentrates (as grains are a poor source).

potassium permanganate An inorganic compound ($KMnO_4$) which is a strog oxidising agent with a wide range of uses including as a **DISINFECTANT**, for water treatment, and in diagnostic **PATHOLOGY**.

pouch

1. A 'pocket' located on the ventral surface of the lower **ABDOMEN (1)** in **MARSUPIALS** where **NEONATES** are kept during the early stages of development.

2. The pocket-like fold of tissue (cheek pouch) found in the mouths of many mammals (e.g. monkeys and many **RODENTS** including hamsters, gophers and squirrels) and used for carrying food or nesting materials.

poultice A moist, soft mass of material (such as clay) placed on a cloth or gauze, which is medicated and may have been heated, and is applied as a treatment to lesions, sores, boils etc.

poultice boot A waterproof boot used to contain a **POULTICE** for soaking an animal's foot, especially a horse's hoof.

poultry

1. Domesticated birds raised for meat or eggs, e.g. chickens, ducks, geese, turkeys. May be defined in the law.

2. Under s27(1) of the Wildlife and Countryside Act 1981, poultry is defined as domestic '*fowls, geese, ducks, guinea-fowls, pigeons and quails, and turkeys*'.

3. Meat from birds classed as poultry.

poultry ark Housing and an enclosed run for chickens and other poultry. Usually made of wood or plastic with framed wire mesh sides to the run, and with a semi-circular or V-shaped cross section. Suitable for small areas such as gardens. *See also* **CHICKEN COOP**.

poultry breeds *See* Fig. P9, **CHICKEN BREEDS** (Table C3). *See also* **GOOSE**, **TURKEY**

poultry register *See* **GB POULTRY REGISTER**

ppm Parts per million. A measure of concentration.

practice The repeated performance of an exercise which eventually accelerates the development of a behaviour. The improvement is the result of **ONTOGENY** rather than **MATURATION** or **LEARNING**.

prebiotic A food ingredient that provides a suitable environment for the multiplication of bacteria in the

P

Fig. P9 Poultry breeds: Dutch Crested Black Mottled or Polish chicken; Indian Runner ducks; Brahma chicken (left to right).

gut. Included in some **COLOSTRUM SUPPLE-MENTS**. *Compare* **PROBIOTIC**

precautionary principle A means of approaching policy, decision making and legislation in the absence of full scientific certainty. For example, the application of this principle would require policy makers or legislators to take into account the likely effects of climate change in the absence of scientific certainty about its existence and/or effects. It is important to invoke this principle when the consequences of doing otherwise could have disastrous effects on the ecosystem, human health or some other thing if the predicted adverse effects were to occur. The principle is an import feature of the **CONVENTION ON BIOLOGICAL DIVERSITY 1992 (CBD)**.

precipitation Any product of the condensation of atmospheric water vapour which falls due to gravity, e.g. rain, drizzle, mist, hail, sleet, snow etc. Rainfall is simulated in some animal houses, particularly **TROPICAL HOUSES**, using water sprays operated by a timer. Some zoos spray their animals with water on hot days to simulate the cooling effect of rain.

precocial Referring to young mammals or birds which are well developed when born or hatched, and can see, hear, walk and thermoregulate from a very early age, e.g. sheep, deer, cattle. In relation to bird chicks, ones which hatch in a well-developed form, can immediately move around unaided, and are independent of parents from an early age (Fig. P10). *Compare* **ALTRICIAL**

preconditioning In relation to the reintroduction of animals to the wild, the process of adapting individuals to their new environment in enclosures before release. This may include exposing predators to their natural prey organisms. Prior to release, captive-bred **BLACK-FOOTED FERRETS (***MUSTELA***

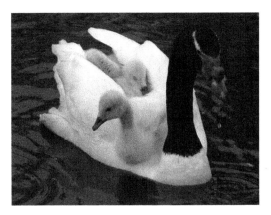

Fig. P10 Precocial chicks: a black-necked swan (*Cygnus melanocoryphus*) carrying cygnets on her back.

*NIGRIPES***)** are kept in pens containing naturalistic burrows and prairie dogs (*Cynomys*) which are their natural prey. *See also* **REWILDING (2)**

predator An individual or species of animal that feeds on other animals, usually of different species. *See also* **APEX PREDATOR**

predator-awareness training Training given to captive bred animals prior to release to the wild to aid predator recognition and improve **POST-RELEASE SURVIVAL** rates. Training of numbats (*Myrmecobius fasciatus*) at Perth Zoo involved exposure to bird warning calls, a hand-tethered live bird of prey and an overhead bird of prey silhouette on a wire and pulley system (Jose *et al.*, 2011).

preen gland, uropygial gland A gland which is located at the base of the tail in birds and which produces an oily substance called preen oil. The flow of oil is stimulated by the bird's bill and used to make the feathers water-resistant.

preen oil *See* **PREEN GLAND**

preening Grooming behaviour performed by birds for feather maintenance and as a **COMFORT BEHAVIOUR**. Consists of the arrangement and cleaning of feathers when dirty or wet. In some species water resistance is maintained with an oil substance from a **PREEN GLAND**. *See also* **ALLOPREENING**

preference In animal behaviour studies, a pattern of choosing. *See also* **CHOICE**

preference experiment An experiment used in **ANIMAL WELFARE** studies in which animals are able to express a preference for a particular item or condition, e.g. a choice of foods or a choice of environments. Expressing a preference may not necessarily equate to choosing good welfare. In a **CAFETERIA EXPERIMENT** an animal may express a preference for foods that promote obesity, thereby leading to poorer welfare.

pregnancy
1. In mammals, the period between **CONCEPTION** or **FERTILISATION** and birth, during which the **EMBRYO** develops in the uterus. Also called **GESTATION**. *See also* **PSEUDOPREGNANCY**
2. A particular instance of being pregnant. *Compare* **EMPTY**

pregnancy toxaemia, twin lamb disease A disease of ewes in late pregnancy when their energy requirements exceed their energy intake. The ewe diverts energy from her own body to the growing foetus, to her detriment, especially when carrying twins. A similar condition occurs in cattle.

prehensile Relating to an appendage or organ adapted for grasping or holding, e.g. the tail of geckos, chameleons and spider monkeys (*Ateles* spp.), the tongue of giraffes, the lips of rhinos and horses, the trunk of elephants and the nose of tapirs (*Tapirus* spp.).

prehensile tail *See* **PREHENSILE**

premolar Cheek teeth used for grinding and located between the **CANINES** and **MOLARS** in mammals.

prepartum Occurring before **PARTURITION**.

pre-seismic anticipatory behaviour Some animal species have been reported to exhibit a change in behaviour immediately prior to an earthquake, e.g. common toads (*Bufo bufo*) ceased spawning and only resumed normal behaviour after the event. The cue used may have related to changes in the ionosphere.

presenter, explainer, interpreter A person who presents, explains or interprets an exhibit, an animal, or a group of animals in a zoo to visitors. This may take the form of a one-to-one conversation but is often in the form of a short talk at the enclosure of a particular species, e.g. an 'elephant talk'. **KEEPERS (1)** may sometimes act as presenters.

pressure bandage A temporary tight bandage used on the extremities (e.g. limbs or tail) to reduce blood loss by applying pressure to blood vessels. Used while transporting patients to a surgery.

pressure point A location in the body where a major artery – usually supplying the extremities – passes near the body surface and over a bone, where the application of pressure may reduce or stop the flow of blood. May be used to stop blood flow to a wound for a few minutes.

pressurised carbon dioxide kit An apparatus which is used to add carbon dioxide to aquarium water from a pressurised gas cylinder through a pipe system. It is used to increase growth in aquatic plants by promoting **PHOTOSYNTHESIS**.

prevalence In **EPIDEMIOLOGY**, the ratio of the number of occurrences of a disease (or event) to the number of individuals at risk in the population, for a given period of time. *See also* **INDEX CASE**

preventative veterinary medicine The routine surveillance of the health of a population or collection of animals and the provision of health care (e.g. vaccinations, physical examinations, faecal examinations, treatment for parasites, **PEDICURES**, *POST-MORTEMS* (2), etc.) for the purpose of preventing disease and its transmission. This is particularly important in zoos because diagnostic procedures and treatments are less straightforward with zoo animals than domestic or farm species. *See also* **PROPHYLAXIS**

Prevention of Cruelty to Animals Act 1979 (NSW) A law in New South Wales, Australia, which protects animals from a very wide range of cruel treatment.

prey Any organism that is taken by another organism, a **PREDATOR**, as food.

pride A group of lions.

primary consumer A herbivore.

primary containment *See* **CONTAINMENT, PRIMARY**

primary feather One of the outer flight feathers, which originates from the manus ('hand') in birds. They are the principal source of thrust in flapping flight and may be individually rotated. *Compare* **SECONDARY FEATHER**

primary forest, ancient forest, ancient woodland, primeval forest, virgin forest A forest that has attained great age and possesses unique ecological features, including high **BIODIVERSITY**. It is a natural forest which is undisturbed or has not been disturbed by man for several hundred years and characteristically contains an abundance of mature trees. *See also* **SECONDARY FOREST**

primary host, definitive host An organism (host) in which a **PARASITE** reproduces sexually or where it becomes sexually mature. *See also* **FLUKE DISEASE**. *Compare* **INTERMEDIATE HOST**

primary producer *See* **AUTOTROPH**

primary production The total amount of organic matter synthesised by **AUTOTROPHS** (usually green plants) in a particular place over a particular period of time. *Compare* **SECONDARY PRODUCTION**

Primates An order of mammals: primates (lorises, pottos, galagos, lemurs, tarsiers, monkeys, apes, humans).

primestock Best quality livestock.

primeval forest *See* **PRIMARY FOREST**

primipara A **PRIMIPAROUS** female.

primiparous Describing a female who is bearing her first offspring or has borne only one offspring. *See also* **MULTIPAROUS, NULLIPAROUS, POLYPAROUS, UNIPAROUS**

prion An infective agent, consisting of a misfolded **PROTEIN**, which causes disease in animals and humans, e.g. **BOVINE SPONGIFORM ENCEPHALOPATHY (BSE), CHRONIC WASTING DISEASE (CWD), SCRAPIE**. It contains no **NUCLEIC ACIDS**.

Pristiformes An order of fishes: sawfishes.

Pristiophoriformes An order of fishes: saw sharks.

prize bull A bull that has won a prize in a competition because of its quality as judged by an acknowledged expert. Prize bulls are highly valued because of their capacity to father high-quality offspring either naturally or by **ARTIFICIAL INSEMINATION (AI)**. On 18 February 2012 a Limousin bull named *Dolcorsllwyn Fabio* was sold for 120 000 guineas (£126 000) at a livestock sale in Carlisle, Cumbria (UK), becoming the highest priced bovid in the UK and Europe.

probability In **STATISTICS**, the likelihood that an event will occur. If it is certain it is assigned a value of 1 and if it is impossible it is assigned a value of 0.

probiotic Live bacteria that support the useful and harmless bacteria already present in the gut against the harmful bacteria, and are beneficial to health. Added to some animal feed. *Compare* **PREBIOTIC**

problem animal control (PAC) The culling of animals that are raiding crops or killing people in a particular area. Generally used to refer to large dangerous and destructive mammals such as elephants.

problem solving A high-order cognitive ability which involves the ability to work through the elements of a problem to reach a solution. It requires the capacity to predict the consequences of particular alternative actions. Problem solving abilities have been demonstrated in a range of taxa including chimpanzees, orangutans, dolphins, jays and octopuses. *See also* **COGNITION, CREATIVITY, LEARNING**

Proboscidea An order of mammals: elephants.

procaryote *See* **PROKARYOTE**

procaryotic *See* **PROKARYOTIC**

Procellariiformes An order of birds: albatrosses, shearwaters, petrels.

production
1. The total biomass produced in an ecosystem over a particular period of time. *See also* **PRIMARY PRODUCTION, SECONDARY PRODUCTION**
2. Relating to the increase in mass of a particular organism or organisms, e.g. the production of a cow.

progeny Offspring

progesterone *See* **CORPUS LUTEUM**

prognosis A prediction of the probable outcome of a disease and the chance of recovery.

program animal A term used in the USA for an individual animal in a zoo who is not part of a breeding programme and is primarily used for educational purposes. *See also* **AMBASSADOR SPECIES, DISPLAY EDUCATION AND RESEARCH POPULATION, EDUCATION OUTREACH ANIMAL**

Project Nim *See NIM CHIMPSKY*

prokaryote, procaryote Any organism in which the genetic material is not enclosed within a **NUCLEUS**, and whose cells do not contain any other membrane-bound **ORGANELLES**, e.g. bacteria. *See also* **EUKARYOTE**

prokaryotic, procaryotic Relating to **PROKARYOTES**.

prolactin A hormone secreted by the **ANTERIOR PITUITARY** which promotes the secretion of **TESTOSTERONE** by the **CORPUS LUTEUM** in mammals and stimulates **LACTATION**.

prolapse The action of slipping or falling out of place. In anatomy, this may, for example, refer to the collapse of a structure such as the vagina or rectum. The uterus and vagina may prolapse due to weakening of the pelvic floor (the muscles and **LIGAMENTS (2)** which normally support them) after giving birth.

prolicide The killing of one's own offspring. *See also* **CANNIBALISM, FOETICIDE, FRATRICIDE, INFANTICIDE**

prone Lying on the belly (**VENTRAL** surface), with the front or face downwards. *Compare* **SUPINE**

pronging *See* **PRONKING**

pronking, pronging, stotting A gait adopted by some quadrupeds, especially some gazelles (e.g. springbok (*Antidorcas marsupialis*)), in which the animal jumps high into the air raising all four feet off the ground simultaneously. May be used when pursued by a predator or during play. This may be an adaptive behaviour as some biologists believe that it may signal the animal's comparative **FITNESS** to potential mates and potential predators. *See also* **PORPOISING**

prophylactic Relating to **PROPHYLAXIS**.

prophylaxis A drug, treatment, procedure or device used to prevent or reduce the risk of contracting a disease (e.g. postoperative prophylactic antibiotics) or protect against a particular unwanted event (e.g. **PREGNANCY**).

prosimian A member of the primate suborder Prosimii which includes lemurs, lorises, bushbabies and tarsiers.

prosocial behaviour Actions intended to benefit others. *See also* **ALTRUISM, RECIPROCAL ALTRUISM**

prostate gland An **EXOCRINE GLAND** in the reproductive system of male mammals of most species which provides some of the fluid component of **SEMEN** and whose smooth muscles aid in **EJACULATION**.

prosthesis An artificial limb or other body part fitted after an **AMPUTATION** or for some other reason, e.g. arm or leg.

protease *See* **PROTEOLYTIC ENZYME**

protected animal A type of animal which is given protection by a specific law. For example, in England and Wales, under section 2 of the **ANIMAL WELFARE ACT 2006**: *'An animal is a "protected animal" for the purposes of this Act if—*
(a) it is of a kind which is commonly domesticated in the British Islands,
(b) it is under the control of man whether on a permanent or temporary basis, or
(c) it is not living in a wild state.'
See also **ANIMALS (SCIENTIFIC PROCEDURES) ACT 1986**

protected area
1. An area of land which is given legal protection against damage from agriculture, development and other human activities and where it is unlawful to harm animals or plants and their habitats.
2. The UN **CONVENTION ON BIOLOGICAL DIVERSITY 1992 (CBD)** (Art. 2) defines a protected area as *'a geographically defined area which is designated or regulated and managed to achieve specific conservation objectives.'*
3. The **INTERNATIONAL UNION FOR THE CONSERVATION OF NATURE AND NATURAL RESOURCES (IUCN)** defines a protected area as *'a clearly defined geographical space, recognized, dedicated and managed, through legal or other effective means, to achieve the long term conservation of nature with associated ecosystem services and cultural values'* (Table P2).
See also **NATIONAL PARK (NP), SANCTUARY (1), SITE OF SPECIAL SCIENTIFIC INTEREST (SSSI), SPECIAL AREA OF CONSERVATION (SAC), SPECIAL PROTECTION AREA (SPA)**

protected contact, zero handling A method of handling animals that does not involve direct contact, and where the handler operates from behind a fence. Requires animal to be trained to move on command, present body parts for inspection, etc.

P

Table P2 Protected area: the IUCN Protected Areas Category System.

Ia Strict Nature Reserve
Strictly protected areas set aside to protect biodiversity and also possibly geological/geomorphical features. Human visitation, use and impacts are strictly controlled. These areas may serve as indispensable reference areas for scientific research and monitoring

Ib Wilderness Area
These areas are usually large unmodified or slightly modified areas, retaining their natural character and influence. There is no permanent or significant human habitation. Sites are protected and managed to preserve their natural condition

II National Park
Large natural or near natural areas set aside to protect large-scale ecological processes, along with the complement of species and ecosystems characteristic of the area, which also provide a foundation for environmentally and culturally compatible, spiritual, scientific, educational, recreational and visitor opportunities

III Natural Monument or Feature
These areas are set aside to protect a specific natural monument, which can be a landform, sea mount, submarine cavern, geological feature such as a cave or even a living feature such as an ancient grove. Generally quite small and often have high visitor value

IV Habitat/Species Management Area
These areas aim to protect particular species or habitats and management reflects this priority. Many will need regular, active interventions to address the requirements of particular species or to maintain habitats

V Protected Landscape/Seascape
Areas where the interaction of people and nature over time has produced an area of distinct character with significant, ecological, biological, cultural and scenic value. Safeguarding the integrity of this interaction is vital to protecting and sustaining the area

VI Protected Area with Sustainable Use of Natural Resources
These areas conserve ecosystems and habitats together with associated cultural values and traditional natural resource management systems. Generally large, with most of the area in a natural condition, where a proportion is under sustainable natural resource management and where low-level non-industrial use of natural resources compatible with nature conservation is seen as one of the main aims of the area

(Fig. F7). Used especially for elephants, particularly males in **MUSTH**. *See also* **TRAINING**. *Compare* **FREE CONTACT**

Protection Against Cruel Tethering Act 1988 An Act which added to the **PROTECTION OF ANIMALS ACT 1911** an offence of tethering *'any horse, ass or mule under such conditions or in such manner as to cause that animal unnecessary suffering'* in England and Wales. *See also* **TETHER**

Protection of Animals Act 1911 An Act of Parliament which applies to England and Wales that covers a wide range of cruelty offences which may be committed against domestic and **CAPTIVE ANIMALS**. It bans baiting and fighting of animals, the use of dogs as draught animals on public highways, attendance at animal fights and poisoning. It requires the provision of adequate food and water to animals kept in pounds and makes provision for the destruction of injured animals.

Protection of Badgers Act 1992 Legislation in Great Britain which makes taking, killing, selling, possessing and digging for badgers, and damage to their sets, illegal (Fig. P11). *See* **BADGER-BAITING**, **BADGER CULLING**

Fig. P11 Protection of Badgers Act 1992: a sign in the Lancashire countryside asking the public to report suspected crimes against wildlife to the police.

Protection of Wild Mammals (Scotland) Act 2002 An Act of the Scottish Parliament which banned hunting with dogs (**FOX HUNTING** and **HARE COURSING**). *See also* **HUNTING ACT 2004**

protective formation A behaviour exhibited by some herd animals whereby adults surround their young when attacked by predators. Musk oxen (*Ovibos moschatus*) exhibit this behaviour when attacked by wolves (*Canis lupus*). Some herd animals living in zoos may exhibit the behaviour when exposed to an unfamiliar noise or vehicle.

protein A complex biological molecule which consists of sub-units called amino acids. They are more complex than carbohydrates and lipids because their structure is determined by the precise sequence of amino acids. In the cell, proteins are constructed from instructions in the **DNA** contained in the chromosomes. These proteins perform a variety of functions. Some are **ENZYMES** and control the metabolic activity of the cell. Others act as **HORMONES** or **ANTIBODIES**. Proteins also form most of the structural components of cells, including collagen which strengthens skin, and actin and myosin, which are components of muscle. *See also* **ESSENTIAL AMINO ACID, PROTEIN SYNTHESIS**

protein concentrate A concentrated source of **PROTEIN** used for animal feed. It may be based on a number of protein sources, e.g. fish, rice, wheat, soya, and is fed to agricultural animals to increase the rate at which they gain weight. *See also* **CONCENTRATES, GROWER DIET**

protein skimmer, air-stripper, foam fractionator A device used in marine aquariums to remove organic compounds from water before they can break down into nitrogenous waste. Water flows through a chamber where it comes into contact with a column of fine bubbles. The bubbles carry proteins and other materials to the surface where the foam collects in a cup and is removed. A number of different designs exist.

protein synthesis The cellular process by which **PROTEINS** are assembled from **AMINO ACIDS** using information contained within **DNA**. Molecules of **TRANSFER RNA (TRNA)** collect amino acids and transport them to a **RIBOSOME** where they are added to a chain of amino acids whose sequence is determined by the **GENETIC CODE** in a molecule of **MESSENGER RNA (MRNA)**. *See also* **TRANSCRIPTION, TRANSLATION**

proteolytic enzyme, protease An **ENZYME** (e.g. pepsin) which breaks down large proteins into smaller parts (polypeptides and peptides), and ultimately into **AMINO ACIDS**. Important in the digestion of proteins in the gut.

protocol
1. A procedure or standard method for performing a particular task such as an experiment or veterinary procedure, e.g. a disinfection protocol.
2. A subsidiary agreement to an international treaty, e.g. **CARTAGENA PROTOCOL ON BIOSAFETY 2000**.

Protocol on Protection and Welfare of Animals 1997 A **PROTOCOL** to the Treaty of Amsterdam 1997 which amended the Treaty on European Union, the Treaties Establishing the European Communities and Related Acts and entered into force on 1 May 1999. In its revised form it states that:

'In formulating and implementing the Community's agriculture, fisheries, transport, internal market, research and technological and space policies, the Union and the Member States shall, since animals are sentient beings, pay full regard to the welfare requirements of animals, while respecting the legislative or administrative provisions and customs of the Member States relating in particular to religious rites, cultural traditions and regional heritage.' *See also* **EUROPEAN UNION (EU)**

Protoctista One of the five kingdoms of living things. They are aquatic eukaryotes which are not animals, plants or fungi. They include algae, diatoms, amoebae, ciliates and **PROTOZOA**, many of which are important in causing diseases.

Protozoa Within the **PROTOCTISTA**, a phylum of single-celled organisms (of plant-like and animal-like species) some of which are important in causing disease.

protozoan (protozoa, protozoans, protozoons *pl.*) A member of the phylum **PROTOZOA**.

proven Used to describe an animal which has demonstrated the capacity to reproduce, e.g. a proven bull.

proximal Relating to that part of a structure which is closest to its point of attachment to the body or to the centre of the body (Fig. A8). *Compare* **DISTAL**

proximate cause In relation to behaviour, an explanation that attempts to identify its immediate cause, e.g. the presence of a particular stimulus, a change in hormone levels etc. *Compare* **FUNCTIONAL CAUSE**

proximity logger A device (usually fitted to a collar, harness or ear tag) which incorporates a UHF transceiver and a VHF transmitter and which logs interactions between two or more animals when they are within a predefined distance of each other. It may be used to study contact between members of the same species or different species, presence at denning or nesting sites, how often an individual passes a fixed point and other behaviours. *See also* **RADIO COLLAR**

PRT Positive reinforcement training. *See* **REINFORCEMENT, TRAINING**

Przewalski's horse (*Equus ferus przewalskii*) A wild equid (Fig. P12) from the **STEPPES** of central Asia whose original range stretched from Germany to China and which was classified by the **INTERNATIONAL UNION FOR THE CONSERVATION OF NATURE AND NATURAL RESOURCES (IUCN)** as **EXTINCT IN THE WILD (EW)** until 2008. A captive breeding programme based on just 14 founders has resulted in the species being reintroduced into Mongolia and China. *See also* **INTERSPECIES EMBRYO TRANSFER**

pseudo-penis A structure found in some animals that superficially resembles a penis, e.g. the enlarged clitoris found in the spotted hyena (*Crocuta crocuta*).

Fig. P12 Przewalski's horse (*Equus ferus przewalskii*).

pseudopregnancy, false pregnancy The phenomenon whereby some mammals exhibit mammary gland enlargement, **LACTATION** and behavioural signs of **PREGNANCY** (e.g. nest building) when they are not pregnant. Believed to be caused by hormonal changes at the end of **DIOESTRUS**.

pseudo-sanctuary *See* **SCAMTUARY**. *Compare* **SANCTUARY (2)**

Psittaciformes An order of birds: cockatoos, parrots, macaws (Fig. M1).

psittacine A bird belonging to the order **PSITTACIFORMES**.

psittacosis, ornithosis, parrot disease, parrot fever A bacterial disease caused by *Chlamydophila psittaci* which results in severe respiratory illness in members of virtually all bird species, especially those of the parrot family (Psittaciformes), and in humans. Active infection is often triggered by **STRESS**. Infected birds may exhibit **LISTLESSNESS, DIARRHOEA, CONJUNCTIVITIS, SINUSITIS** and respiratory signs. The condition may be fatal. Treatment is with tetracycline or doxycycline in the feed.

psychological barrier, perceptual barrier A barrier used in a zoo or similar facility which would not physically contain the animal if it chose to escape, e.g. a gate or fence that is too low to prevent it from escaping by jumping over it; chains used to restrain an elephant which it could easily snap. Electric fences are psychological barriers because they carry insufficient current to do harm but animals nevertheless learn not to touch them.

psychology The scientific study of the mental processes and behaviour of animals and humans. Comparative psychologists study the mental life and behaviour of non-human animals. *See also* **ETHOLOGY**

puberty The onset of sexual maturity in animals, when the sex organs become functional, the **SECONDARY SEXUAL CHARACTERISTICS** appear and the individual becomes capable of reproduction.

public aquarium A place where aquatic organisms, especially fish, are exhibited to the public in large tanks. The first public aquarium was the **FISH HOUSE** at London Zoo.

public barrier, stand-off barrier, visitor barrier A barrier preventing the public from gaining access to the primary barrier of a zoo exhibit, e.g. a guard rail located in front of a chain-linked fence. In some exhibits the public barrier is the same as the primary barrier, e.g. where animals are held behind tempered glass windows.

P

public consultation The process of asking members of the public their opinions about a proposed project or activity with a view to taking these into account in deciding whether or not, or how, to proceed with a project. This is important in producing new laws and in **REINTRODUCTION** projects. *See also* **GREEN PAPER**

pullet A young hen, usually less than 1 year old.

pulmonary aspiration The taking in of foreign material into the trachea and lungs. *See also* **SUCTION DEVICE**

pulmonary circulation That part of the circulatory system which carries deoxygenated blood from the heart to the lungs and oxygenated blood back to the heart. *Compare* **SYSTEMIC CIRCULATION**

pulse A rhythmic beat that can be detected in an artery caused by the pumping of blood out of the heart. It may be felt by pressing an artery against a bone or it may be heard directly with a **STETHOSCOPE**. *See also* **PRESSURE POINT**

pulse oximeter A non-invasive device for measuring the **OXYGEN SATURATION** of blood by placing a sensor on the ear lobe, finger or some other suitable part of the body.

punishment **NEGATIVE REINFORCEMENT**. Sometimes used to promote **LEARNING**. *Compare* **POSITIVE REINFORCEMENT**

pupa (pupae *pl.***)** A stage in the life cycle of some insects. Called a chrysalis in Lepidoptera (butterflies and moths).

puppet mother A puppet which resembles the head of an adult bird and is used to feed **HAND-REARED** developing chicks to avoid **IMPRINTING** on humans. Used in the **CAPTIVE BREEDING PROGRAMME** for the **CALIFORNIA CONDOR** (*GYMNOGYPS CALIFORNIANUS*).

puppy milk Commercially availably replacement milk for puppies to assist with weaning.

puppy mill A breeding facility in which dogs are kept in inhumane conditions in small cages with mesh floors which is used to produce puppies to supply pet shops. The animals are often kept in unhygienic conditions, are deprived of proper veterinary treatment, and the bitches are killed when they are no longer able to reproduce.

puppy walker A person who trains young dogs as assistance animals, especially **GUIDE DOGS** for the blind. Volunteers foster the puppies in order to facilitate socialisation as well as training them in basic obedience and walking on a leash.

pure-bred Referring to an animal which has been bred from a recognised **BREED** or strain over many generations. *Compare* **CROSS, GRADE, HYBRID**

pus A yellowish-white viscous fluid formed in infected tissue containing white blood cells, cell debris and dead tissue. *See also* **ABSCESS**

puzzle box A box used in studies of animal intelligence, into which an animal, such as a cat, may be placed and where the only means of escape requires the animal to pull a string or push a button to open a door. Such boxes were used by Edward **THORNDIKE** to study **LEARNING**.

puzzle feeder An enrichment device which provides a small amount of food for an animal if it solves a puzzle, e.g. moves a series of levers in the correct order.

pygal Located on the rump or near the end of the backbone.

Pyrenean mountain dog A large powerful breed of dog used to protect sheep and other livestock from wolves in some areas of Europe, e.g. parts of Spain.

P

Fig. Q1 Q is for quarrel. Chimpanzees (*Pan troglodytes*).

Dictionary of Zoo Biology and Animal Management: A guide to terminology used in zoo biology, animal welfare, wildlife conservation and livestock production, First Edition. Paul A. Rees.
© 2013 John Wiley & Sons, Ltd. Published 2013 by John Wiley & Sons, Ltd.

Q fever A disease caused by the bacterium *Coxiella burnetii* which can infect humans, and farm and domestic animals. It may cause fever, abortion and infertility.

quad bike, all-terrain vehicle (ATV) A small three or four-wheeled vehicle weighing less than 550 kg (under UK law), designed to be used by a single person and steered by handlebars. These vehicles are used on difficult terrain and may be four-wheel or two-wheel drive and equipped with low-ratio gears and low-pressure tyres. They are used on farms and by some zoos to pull trailers carrying food, bedding and other materials within and between enclosures.

quadrat A square sample area used in field studies for sampling habitats or populations. It may be a small square frame – usually metal – which can be thrown to take RANDOM SAMPLES of vegetation or SESSILE organisms (e.g. snails or limpets) or a larger area marked out using surveying poles and rope (e.g. a 20 m × 20 m square). Useful for calculating the DENSITY of a species or calculating the percentage cover of plants.

quadruped A four-footed animal; an animal that uses four limbs for locomotion.

quadrupedal Relating to a QUADRUPED.

quagga A distinctive form of zebra which previously occurred in large numbers in the Cape Province of South Africa and the southern part of the Orange Free State, but became extinct in 1883 when the last known specimen died in Amsterdam Zoo. It was originally classified as *Equus quagga* but is now believed to be conspecific with the plains zebra (*Equus burchellii*). It was unusual because of the lack of distinct stripes on its hind quarters.

quarantine
1. The process of keeping an animal separate from other animals to prevent the possibility of it transmitting disease.
2. A building or other place where this process is carried out.

quarrel An angry dispute between two or more individuals (Fig. Q1). *See also* AGONISTIC BEHAVIOUR

queen
1. A female domestic cat.
2. The reproductive individual in a colonial species, e.g. bees, naked mole rats (*Heterocephalus glaber*).

quota *See also* HUNTING QUOTA

Q

Fig. R1 R is for rhea. Common rhea (*Rhea americana*).

Dictionary of Zoo Biology and Animal Management: A guide to terminology used in zoo biology, animal welfare, wildlife conservation and livestock production, First Edition. Paul A. Rees.

rabbit Any of a group of species belonging to the mammalian family Leporidae of the order **LAGO-MORPHA**. In the wild, rabbits live in systems of burrows called **RABBIT WARRENS**. They are hunted and farmed for their meat and fur. A large number of domesticated breeds of rabbit (*Oryctolagus cuniculus*) have been produced as pets. *See also* **HOUSE RABBIT, MYXOMATOSIS, RABBIT BREEDS**

rabbit breeds A wide range of breeds of different sizes, coat colour and hair length have been produced. Some major breeds are listed in Table R1.

rabbit warren
1. A system of connected underground tunnels produced and occupied by rabbits.
2. An area of land reserved for the breeding and preservation of rabbits. Formerly widely used for producing food and fur in medieval Britain and elsewhere. The archeological remains of these warrens have been well documented.

rabies An inoculable contagious disease caused by a lyssavirus. It affects virtually all mammals and occasionally occurs in birds. It is almost always fatal in humans. It is transmitted by bites or scratches and there is also a risk of infection from contamination of wounds or eyes by **SALIVA**. Rabies causes derangement of the nervous system, a change in temperament and, eventually, paralysis. Signs may occur as early as the 9th day after being bitten but may only appear after several months. The principal wild animal **VECTORS** are foxes, wolves, jackals, coyotes, badgers, martens, skunks, mongooses and bats. Foxes are highly susceptible to infection and their vaccination has been very successful at controlling rabies in western Europe. Rabies is a **NOTIFIABLE DISEASE** in most countries.

race
1. A narrow corridor in, or section of, an enclosure or cage consisting of parallel rails or panels just wide enough to accommodate a large animal. Used for controlling an animal's movement in order to conduct a veterinary examination or administer treatment, channelling a sheep to a **DIP** etc. *See also* **CRUSH**
2. An alternative name for **SUBSPECIES**.

raddle A pigment used to mark livestock, especially sheep. *See* **RAM HARNESS**

raddle crayon A crayon used to apply **RADDLE**.

raddle harness *See* **RAM HARNESS**

raddling The act of applying **RADDLE** to livestock. *See also* **RAM HARNESS**

Radford Report The report of the Circus Working Group established in Britain in 2006. Its report, *Wild Animals in Travelling Circuses* (2007), concluded that there was little scientific evidence available to inform the government when considering whether or not to ban the use of performing animals in **CIRCUSES (1)** and that there was no evidence that their welfare was either better or worse than that of animals kept in other captive environments.

radial symmetry Capable of bisection into two halves that are approximately mirror images of each other, in two or more planes. This type of symmetry is characteristic of adult coelenterates and echinoderms. Such organisms have no left or right side. *Compare* **BILATERAL SYMMETRY, PENTAMEROUS (RADIAL) SYMMETRY**

radio collar A transmitter fitted to a collar around the neck of an animal which transmits information about its location and, in some cases, physiological data such as pulse rate and respiration rate. *See also* **PROXIMITY LOGGER, RADIO TRACKING**

radio frequency identification (RFID) technology, electronic identification (EID) technology Technology that uses a **MICROCHIP**, that may be implanted under the skin or fitted to an animal's collar or an ear tag, which operates a device and/or provides information about the animal when it approaches an interrogator device (e.g. for identification purposes). May be used to allow access through a door or to an enrichment device in a zoo environment, or a cat flap in a house. This technology has also been used to track the movements of zoo visitors and to provide visitors with information about particular animals on **PERSONAL DIGITAL ASSISTANTS (PDAs)** as they approach an exhibit. **ROBOTIC MILKING MACHINES** allow trained dairy cows to be milked by voluntarily entering the

Table R1 Major rabbit breeds.

American	Harlequin
American Chinchilla	Havana
American Fuzzy Lop	Himalayan
American Sable	Holland Lop
Belgian Hare	Jersey Wooly
Beveren	Lilac
Blanc de Hotot	Mini Lop
Britannia Petite	Mini Rex
Californian	Mini Satin
Champagne d'Argent	Netherland Dwarf
Checkered Giant	New Zealand
Cinnamon	Palomino
Creme d'Argent	Polish
Dutch	Rex
Dwarf Hotot	Rhinelander
English Angora	Satin
English Lop	Satin Angora
English Spot	Silver
Flemish Giant	Silver Fox
Florida White	Silver Marten
French Angora	Standard Chinchilla
French Lop	Tan
Giant Angora	Thrianta
Giant Chinchilla	

machine, and record the amount of milk collected. This allows the farmer to identify any cows that have not been sufficiently milked. RFID technology may also be used on farms to monitor livestock movements thereby giving early warning of illness in which reduced mobility is a warning sign.

radio pill A small transducer which may be swallowed by, or implanted in, an animal that transmits physiological data (e.g. body temperature or movement) to a suitable receiver.

radio telemetry Wireless measurement made at a distance. *See also* **RADIO COLLAR, RADIO PILL, RADIO-TRACKING**

radiograph An image made using **X-RAYS**. *See also* **RADIOGRAPHY**

radiography The means of examining the inside of a body by recording images using **X-RAYS**.

radioimmunoassay A method of detecting or quantifying antibodies or antigens using radiolabelled substances, i.e. a substance that has had an atom replaced by a radioactive atom or substance so that it can be easily identified in tests.

radio-tracking A technique for locating and following animals. It involves attaching a battery-powered radio transmitter to the animal – usually on a collar – and locating it with a directional aerial. Used for determining home ranges, migratory routes etc. *See also* **RADIO COLLAR, RADIO PILL, RADIO TELEMETRY**

radius (radii *pl.*) A **LONG BONE** which, together with the **ULNA**, forms the lower forelimb in tetrapod vertebrates. It is the shorter and anterior of the two bones.

Raffles, Sir Thomas Stamford (1781–1826) Raffles was a British colonial administrator who was born in Jamaica and served in the British East India Company. He was at one time governor of Sumatra and founded Singapore in 1819. Raffles was a keen natural historian and became the president of the Batavian Society which studied the natural history of Java and adjacent islands. Raffles returned to England in 1824. He founded the **ZOOLOGICAL SOCIETY OF LONDON (ZSL)** in 1826 and became its first president. Raffles published descriptions of some 34 bird species and 13 species of mammals, mostly from Sumatra. He named many **NEW SPECIES** including the sun bear (*Ursus malayanus*).

rainforest A broad-leaved evergreen tropical forest which experiences very high rainfall. *See also* **CLOUD FOREST**

Rajiformes An order of fishes: skates.

ram harness, raddle harness A harness used to attach a crayon or chalk pad to the chest of a ram which leaves a coloured mark on the back of a ewe during mating. It is used to determine which rams have mated with particular ewes from the colour of the marks they bear. *See also* **RADDLE, RADDLING**

Ramsar Convention *See* **CONVENTION ON WETLANDS OF INTERNATIONAL IMPORTANCE ESPECIALLY AS WATERFOWL HABITAT 1971 (RAMSAR CONVENTION)**

Ramsar site A protected wetland area designated under the Ramsar Convention.

random sample In **STATISTICS**, a sample that has been taken in such a way that all possible individuals (or objects) in a **POPULATION** have an equal chance (**PROBABILITY**) of being selected.

range
1. The area over which a species or other taxon occurs. *See also* **HOME RANGE, RANGE STATE**
2. The distance from an observer to an animal or object in the field. May be measured using a **RANGEFINDER**.
3. In statistics, the difference between the highest and lowest values in a set of data, e.g. in the data 5, 8, 9, 11, 19, the range is 14 (19–5). *See also* **STANDARD DEVIATION, STANDARD ERROR OF THE MEAN, VARIANCE**

range state A state (country) where a particular species lives in the wild, e.g. Kenya is one of the range states of the African lion (*Panthera leo*).

rangefinder An optical/electronic device – similar to a telescope – used for determining distances in the field, usually using a laser. Useful for animal **CENSUSES**.

rank The position of an individual in a **DOMINANCE HIERARCHY**.

rapacious Referring to an animal that lives by catching prey.

rapid eye movements (REM) Involuntary quick movements made by the eyes of some species when closed while they are asleep, possibly indicating the occurrence of **DREAMING**.

raptor A **BIRD OF PREY**.

Rare Breeds Conservation Society of New Zealand An organisation formed to conserve, record and promote rare **LIVESTOCK** breeds found in New Zealand, with the aim of maintaining genetic diversity. It was founded in 1988 and has established a Rare Breeds Gene Bank. Breeds include horses, cattle, pigs, goats, poultry, rabbits, donkeys, deer, camelids, chinchillas, for example, the Auckland Island pigs, Australian Lowline cattle, Caspian horses, Enderby Island rabbits, Pitt Island sheep and Dexter cattle. *See also* **AMERICAN LIVESTOCK BREEDS CONSERVANCY (ALBC), RARE BREEDS SURVIVAL TRUST (RBST)**

Rare Breeds Survival Trust (RBST) A conservation charity, founded in 1973, whose aim is to prevent the extinction of UK native farm animal genetic resources. Since its formation no rare breed has become extinct. Rare breeds of animals are kept in approved conservation farm parks. They include Aberdeen Angus cattle, Boreray sheep, British Lop Middle White pigs, Bagot goats, Exmoor ponies and Old English Pheasant Fowl. The RBST maintains a 'Watchlist' which contains all native breeds of horses, cattle, pigs, sheep and goats (but not

R

Table R2 Selected examples of rare breeds on the RBST Watchlist 2012.

Category	Sheep	Cattle	Equine	Pigs	Goats
Critical	Boreray	The Chillingham Wild Cattle	Suffolk horse	–	–
Endangered	Leicester Longwool	–	Exmoor pony	–	–
Vulnerable	Hill Radnor	Lincoln Red (original population)	Clydesdale horse	Middle White	Bagot
At risk	Oxford Down	Gloucester	Shire horse	Berkshire	–
Minority	Dorset Horn	Hereford (original population)	–	British Saddleback	Golden Guernsey
Other native breeds	Badgerface Welsh	Dexter	Shetland pony	Large White	–

poultry), and it assigns each breed to a conservation category based on geographical concentration (Table R2). *See also* **AMERICAN LIVESTOCK BREEDS CONSERVANCY (ALBC)**, **RARE BREEDS CONSERVATION SOCIETY OF NEW ZEALAND**, **WATCHLIST CATEGORIES (RARE BREEDS SURVIVAL TRUST)**

rarity A species which is considered to be rare. Especially applied to birds by bird watchers.

rat-baiting A **BLOOD SPORT** which became popular in the UK after the **CRUELTY TO ANIMALS ACT 1835** banned the baiting of other animals, especially bulls and bears. *See also* **BEAR-BAITING**, **BULL-BAITING**

Ratel The journal of the **ASSOCIATION OF BRITISH AND IRISH WILD ANIMAL KEEPERS (ABWAK)**.

ratite A flightless bird that possesses no keel (**CARINA**) on its sternum, including ostrich, emu and rhea (Fig. R1).

RAVC *See* **ROYAL ARMY VETERINARY CORPS (RAVC)**

Raven's Cage A metal cage which originally housed ravens at **LONDON ZOO** and is now protected as a **LISTED BUILDING**.

RBST *See* **RARE BREEDS SURVIVAL TRUST (RBST)**

RCVS *See* **ROYAL COLLEGE OF VETERINARY SURGEONS (RCVS)**

reaction chain, chain reaction A sequence in which a behaviour exhibited by one individual acts as a stimulus eliciting a response in a second individual which then acts as a stimulus producing a response by the first individual, and so on until the sequence is complete. This occurs in the **COURTSHIP** and mating behaviour of the three-spine sticklebacks (*Gasterosteus aculeatus*) studied by **TINBERGEN**, in which the two sexes engage in a sequence initiated by the male when he appears and performs a zig-zag dance to attract her to his nest, and culminating when he fertilises her eggs.

realised niche The actual range of conditions that an organism occupies and the resources it can access as a result of limiting pressures from other species. *Compare* **FUNDAMENTAL NICHE**

rearing phase In agricultural animal production, the period of development of an animal (e.g. a cow) after birth when it feeds on milk until it is weaned. This phase is usually complete when a calf reaches 200 kg and is followed by the **GROWING PHASE**.

reasoning *See* **INTELLIGENCE**

receptor A sensory cell which is capable of receiving stimuli from the internal or external environment, e.g. in the eye and the ear, mechanoreceptors.

recessive allele The allele whose character is only expressed when two copies are present in the **GENOME** (one on each of a pair of **CHROMOSOMES**), in the absence of a **DOMINANT ALLELE** for the same character, except in the case of **SEX LINKAGE**.

reciprocal altruism A type of **ALTRUISM** in which an individual helps another at some cost to itself with the expectation of receiving help from that individual in the future. Such an arrangement is susceptible to **CHEATING** in which the recipient of help never helps others. Cheats would be at an advantage as they would receive a benefit at no cost. Cheating could be counteracted if individuals only helped those likely to reciprocate.

Recognition of Zoo Rules 2009 (India) A law in India, made under the **WILD LIFE (PROTECTION) ACT 1972 (INDIA)**, which requires the licensing (recognition) of zoos, regulates their staffing, lays down requirements for enclosures and veterinary facilities, the acquisition and breeding of animals, the conduct of educational and research activities, and other aspects of the operation of zoos.

recombinant DNA *See* **GENETIC ENGINEERING**

reconciliation Some species have evolved specific behaviours which allow the re-establishment of friendly relationships between individuals following an antagonistic encounter. Reconciliation is defined by Palagi *et al.* (2008) as '*the first postconflict affinitive contact between former opponents*'. This behaviour has been widely investigated in anthropoid primates. *See also* **APPEASEMENT BEHAVIOUR**. *Compare* **AGONISTIC BEHAVIOUR**

record keeping The recording of information about animals in zoos or farms, e.g. identification marks and numbers, date of birth or origin, identification of parents, size, weight, veterinary problems, vaccination status etc. Zoos and farms are required by law to keep animal records in most jurisdictions (e.g. *see* **ZOOS DIRECTIVE**). This is essential for tracing disease sources etc. *See also* **ANIMAL RECORD KEEPING SYSTEM (ARKS)**, **CATTLE TRACING SYSTEM**, **INTERNATIONAL SPECIES INFORMATION SYSTEM (ISIS)**, **MEDICAL ANIMAL RECORD KEEPING SYSTEM (MEDARKS)**, **MOVEMENT DOCUMENTS**, **REGISTRAR**, **ZOOLOGICAL INFORMATION MANAGEMENT SYSTEM (ZIMS)**

recovery plan, endangered species recovery plan A plan in the USA required by Section 4(f) of the **ENDANGERED SPECIES ACT OF 1973 (USA)** which must be completed by the **UNITED STATES FISH AND WILDLIFE SERVICE (USFWS)** or the **NATIONAL MARINE FISHERIES SERVICE (NMFS)** for the recovery of a population of a rare or endangered species to the point where it can be removed from the endangered list. It must outline the necessary goals, costs and estimated timeline for recovery. *See also* **BLACK-FOOTED FERRET (*MUSTELA NIGRIPES*)**, **CALIFORNIA CONDOR (*GYMNOGYPS CALIFORNIANUS*)**, **RED WOLF (*CANIS RUFUS*)**, *WYOMING FARM BUREAU FEDERATION V. BABBITT* **(1997)**

recovery position The position in which an unconscious mammal should be placed during first aid so that its airway is straight and its heart exposed for any emergency procedures that may be required. The animal should be on its right side with the head and neck straight and tongue pulled forward and behind the canine tooth to one side.

recruitment The addition of individuals to a population as a result of **REPRODUCTION** or **IMMIGRATION**.

rectum The lower part of the gut which stores faeces and ends in the anus.

recycling The process of reusing resources (e.g. water, wood, paper, plastic) in order to make activities more environmentally sustainable. Many modern zoos both engage in recycling themselves and encourage their visitors to recycle, by providing dedicated waste bins for plastic, paper etc. Some modern exhibits incorporate design features that recycle water from an animal pool, through a **REEDBED** filter to remove organic material, and then return it to the pool. Other exhibits have been constructed with recycled materials such as reclaimed wood. *See also* **GREEN EXHIBIT**

red blood cell *See* **ERYTHROCYTE**

Red Data books *See* **RED LIST**. *See also* **FISHER**

red leg A severe, usually acute, bacterial infection of amphibians which causes **HAEMORRHAGES** of the leg as a result of **SEPTICAEMIA**.

Red List A list of endangered and threatened species produced by the **INTERNATIONAL UNION FOR THE CONSERVATION OF NATURE AND**

NATURAL RESOURCES (IUCN). The information was originally published in the form of Red Data books, an idea conceived by Peter **SCOTT** in 1963. Species are divided into a number of red list categories based on their conservation status: **EXTINCT (EX)**, **EXTINCT IN THE WILD (EW)**, **CRITICALLY ENDANGERED (CR)**, **ENDANGERED (EN)**, **VULNERABLE (VU)**, **NEAR THREATENED (NT)**, **LEAST CONCERN (LC)**, **DATA DEFICIENT (DD)**, **NOT EVALUATED (NE)**. Some taxa are listed as 'data deficient' because too little is known about their wild populations or as 'not evaluated'. The status of a taxon may fluctuate with time and so it may move from one category to another. Red List criteria can be applied to any taxonomic unit at or below the level of species. There is a hierarchical alphanumeric numbering system of criteria and sub-criteria under the categories CR, EN and VU which indicates the reason for the classification, such as declining numbers or a reduced geographical range. The system is extremely complex. For example, a species may be categorised as EN B1ab(v); D. This means that the species is Endangered (EN) due to:

B – geographical range;

1 – extent of occurrence estimated to be less than 5000 km^3;

a – severely fragmented or known to exist at no more than five locations;

b – continuing decline, observed, inferred or projected in (v) the number of mature individuals;

D – population size estimated to number fewer than 250 mature individuals.

See also **FISHER**

Red List categories *See* **RED LIST**

red spot test, mark test, mirror test An experimental procedure used to establish whether or not an animal can recognise itself in a mirror. Devised by the American psychologist Gordon Gallup in 1970. A spot of red coloured dye is placed on the forehead or some other part of the face of the animal. The experimenter then records the extent to which the individual touches this spot compared with other (control) parts of its face. Animals that recognise their own image investigate the spot and appear to recognise their own image. The test has been used on primates, elephants and dolphins.

red wolf (*Canis rufus*) In 1967 in the USA the red wolf was listed as 'endangered' under the **ENDANGERED SPECIES PRESERVATION ACT OF 1966 (USA)**, and is now protected under the **ENDANGERED SPECIES ACT OF 1973 (USA)**. The **UNITED STATES FISH AND WILDLIFE SERVICE (USFWS)** declared it extinct in the wild in 1980. The genetic status of the red wolf has been the subject of considerable controversy. Some authorities claim that it is a distinct species and others claim that it is a hybrid between the grey wolf (*C. lupus*) and the coyote (*C. latrans*). However, the ESA does not cover hybrids. This distinction is important

R

because the USFWS has been involved in a multi-million dollar **RECOVERY PLAN** for the red wolf. *See also* **HYBRIDISATION**

redirected behaviour Behaviour that is related to one stimulus, but directed at something else. It usually involves aggression, e.g. a bird may attack an inanimate object instead of another animal that is the true target of its aggression. It is a type of **DISPLACEMENT ACTIVITY**.

rediscovered species A species that has been recorded in the wild after having previously been declared extinct, e.g. the ivory-billed woodpecker (*Campephilus principalis*) was believed to have been extinct since 1944 in the USA as a result of habitat loss and logging. Sightings were later reported in 1999, 2004 and 2005. *See also* **NEW SPECIES**

redox potential A measure (in volts) of the affinity of a substance for electrons (its electronegativity) compared with hydrogen (which is set at 0). Substances that are more strongly electronegative than hydrogen are capable of oxidising and have positive redox potentials. Substances that are less electronegative than hydrogen are capable of reducing and have negative redox potentials. It is important to monitor redox potential in aquariums as it is a useful indicator of water quality.

reedbed A habitat consisting of reeds. It may be natural or created as part of a **NATURALISTIC EXHIBIT** in a zoo, or for the purpose of filtering water so that it may be recycled within an exhibit.

reflex The simplest form of behaviour in which a stimulus is detected by a sense organ which sends a message to the **CENTRAL NERVOUS SYSTEM (CNS) (1)** and instructions are then sent to an effector organ (muscle or gland), e.g. the pupil constricts in response to exposure of the eye to a bright light. Reflexes are important in limb movements, posture and locomotion. Not all reflexes are completely automatic and in some cases the CNS may prevent them from occurring.

reflex arc A neural pathway which controls a reflex action. Most sensory neurones **SYNAPSE** in the **SPINAL CORD** in higher animals and do not pass directly into the brain, allowing them to function more quickly. Autonomic reflex arcs affect the inner organs while somatic reflex arcs affect the muscles.

reflex ovulator *See* **INDUCED OVULATOR**

REGASP *See* **REGIONAL ANIMAL SPECIES COLLECTION PLAN (REGASP)**

Regent's Park Zoo *See* **LONDON ZOO**

Regional Animal Species Collection Plan (REGASP) Collection planning software available from the **INTERNATIONAL SPECIES INFORMATION SYSTEM (ISIS)** under licence from the **ZOO AND AQUARIUM ASSOCIATION (ZAA)**. Institutions using REGASP have direct access to plans from other collections, allowing them to make contact to arrange animal movements.

regional association of zoos An association of zoos located in a particular geographical region, e.g. North America (**ASSOCIATION OF ZOOS AND AQUARIUMS (AZA)**) or Europe (**EUROPEAN ASSOCIATION OF ZOOS AND AQUARIA (EAZA)**). Promotes co-operation between members, and organises **CAPTIVE BREEDING PROGRAMMES**, conferences, and training. Facilitates involvement in *IN-SITU* **CONSERVATION** projects. *See also* **BRITISH AND IRISH ASSOCIATION OF ZOOS AND AQUARIUMS (BIAZA)**, **ZOO AND AQUARIUM ASSOCIATION (ZAA)**

regional collection plan (RCP) A plan made by a **REGIONAL ASSOCIATION OF ZOOS** for the exhibition and breeding of particular species; a set of recommended regional objectives for specific taxa based on conservation and other priorities.

Register of Veterinary Practice Premises The **ROYAL COLLEGE OF VETERINARY SURGEONS (RCVS)** register maintained to improve the traceability of **CONTROLLED DRUGS** and to ensure the compliance of veterinary surgeons with the **VETERINARY MEDICINES REGULATIONS 2011**.

registrar A person responsible for keeping records of individual animals in a zoo. Such records would include species, sex, age, **SIRE**, **DAM**, date and place of birth, and **ARKS NUMBER**.

regression analysis *See* **LINEAR REGRESSION ANALYSIS**

regulation
1. A rule drawn up by an organisation but which does not have the force of law, e.g. a university regulation.
2. In **EUROPEAN LAW**, an abbreviation for a **EUROPEAN REGULATION**.
3. In UK law, an alternative term for a **STATUTORY INSTRUMENT**.

regurgitation
1. Feeding regurgitation of certain species (e.g. penguins, hunting dogs (*Lycaon pictus*)) involves the passing up of partially digested food from the stomach to provide food for the young. In **RUMINANTS** partially digested food is regurgitated from the rumen to the buccal cavity as part of the process of rumination. *See also* **FOOD BEGGING**. *Compare* **VOMITING**
2. Passive regurgitation involves the passing up of undigested food from the oesophagus in pathological conditions of the oesophagus.

rehabilitation The process of preparing an animal for normal living after a period of illness, malnutrition or an accident, or after a period of captivity during which it has lost the capacity to survive in the wild. *See also* **PRECONDITIONING**

rehydration The process of restoring the fluid balance in a **DEHYDRATED** animal; to absorb water again after dehydration.

rehydration therapy The application of rehydration fluids to an animal that has become dehydrated.

reiki A Japanese holistic therapy that uses simple hands-on healing techniques to transfer energy ('life force') to an animal's body to promote self-healing. There is no scientific evidence that it has any beneficial effect. *See also* **HOLISTIC VETERINARY MEDICINE**

reincarnation The rebirth of an animal in a different form, generally a different species. Belief in this phenomenon protects some animal species from persecution because local people believe that certain animals are reincarnations of their dead ancestors, e.g. some lemur species in Madagascar. *See also* **FADY**

reinforcement In animal behaviour, the process by which the tendency to perform a particular behaviour becomes strengthened as a result of learning. A reward (positive reinforcer) or punishment (negative reinforcer) is used to encourage the performance of a desired behaviour. Although both types of reinforcer will modify behaviour, the use of negative reinforcers in animal training is generally considered unacceptable. *See also* **CLICKER TRAINING**

reintroduction The process by which a species is returned to an area where it once lived wild but is now absent (Table R3). *See also* **FIELD PROPAGATION AND RELEASE, POST-RELEASE MONITORING, PUBLIC CONSULTATION, RE-**

Table R3 Reintroduction: selected examples of species that have been reintroduced into the British Isles.

Species	Location
Beaver (*Castor fiber*)	Knapdale Forest, Argyll, Scotland
Red squirrel (*Sciurus vulgaris*)	Mynydd Llwydiarth forest and Newborough Forest Anglesey, Wales
Water vole (*Arvicola amphibius*)	Kennet and Avon Canal; Portbury Docks; The Trossachs, Scotland
White-tailed eagle (*Haliaeetus albicilla*)	Western Isles of Scotland and County Kerry, Ireland
Red kite (*Milvus milvus*)	The Chilterns; East Midlands; north-east England; Black Isle, Scotland; County Down, Northern Ireland
Corncrake (*Crex crex*)	Nene Washes, Cambridgeshire
Common crane (*Grus grus*)	West Sedgemoor, Somerset
Large blue butterfly (*Maculinea arion*)	Several sites in Somerset
Field cricket (*Gryllus campestris*)	Farnham Heath, Surrey; Pulborough Brooks RSPB reserve, West Sussex
Short-haired bumblebee (*Bombus subterraneus*)	Romney Marsh, Kent

INTRODUCTION GUIDELINES, REWILDING, *WYOMING FARM BUREAU FEDERATION V. BABBITT* (1997). *Compare* **INTRODUCTION, POPULATION SUPPLEMENTATION, TRANSLOCATION**

reintroduction guidelines A list of criteria that should be fulfilled before, during and after a species is reintroduced to the wild. Such guidelines have been published by the **INTERNATIONAL UNION FOR THE CONSERVATION OF NATURE AND NATURAL RESOURCES (IUCN), COUNCIL OF EUROPE, ASSOCIATION OF ZOOS AND AQUARIUMS (AZA)** and other organisations. Different organisations list different criteria and in some cases they contradict each other regarding best practice. Guidelines may include requirements for **PRECONDITIONING, PUBLIC CONSULTATION** and **POST-RELEASE MONITORING**. *See also* **REINTRODUCTION**

rejection The process by which a newborn animal is left by its mother to fend for itself shortly after birth when it is not capable of doing so. *See also* **HAND-REARED**

relative humidity (RH) *See* **HUMIDITY**.

relay neurone *See* **INTERMEDIATE NEURONE**

releaser *See* **SIGN STIMULUS**

relic population A **POPULATION** of a **SPECIES** which is isolated from other such populations as a result of **HABITAT FRAGMENTATION**, population decline or for some other reason, and is all that remains of a species which was formerly distributed over a wider area.

religion and animals *See* **ANIMALS AND RELIGION, RELIGIOUS SLAUGHTER**

religious slaughter The killing of an animal according to religious rules. In the UK, Jewish and Muslim communities are exempt under s22 of the **WELFARE OF ANIMALS (SLAUGHTER OR KILLING) REGULATIONS 1995** from the requirement to stun animals before slaughter. Animals may be slaughtered with a sharp knife by severing both carotid arteries and both jugular veins without stunning by a Jewish or a Muslim **SLAUGHTERMAN** who holds a licence issued under Schedule 1 of the Act.

REM *See* **RAPID EYE MOVEMENTS (REM)**

remote sensing

1. The process of gathering environmental data using cameras and other recording equipment, and especially using satellite imagery (Fig. R2).

2. *'The acquisition of data and derivative information about objects or materials (targets) located at the earth's surface or in its atmosphere by using sensors mounted on platforms located at a distance from the targets to make measurements (usually multispectral) of interactions between the targets and electromagnetic radiation'* (Short, 1982).

renal calculus, kidney stone, nephrolithiasis A condition caused by the accumulation of urinary salts bound together by a colloid matrix of organic materials. May be caused by infection.

R

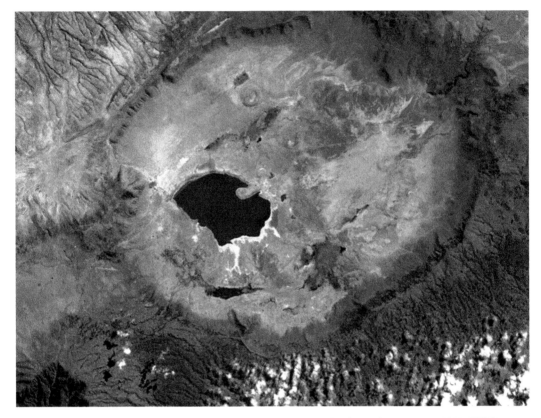

Fig. R2 Remote sensing: a satellite image of Ngorongoro Crater, Tanzania, taken by Landsat 7. (Source: NASA/USGS)

renderer A person (or organisation) that converts waste animal tissue and by-products (usually from slaughterhouses) into stable, useful materials, e.g. the conversion of animal fatty tissue into lard, production of bone meal etc.

renin An enzyme produced by the kidney which regulates blood pressure. *Compare* **RENNIN**

rennet A complex of digestive enzymes (including **RENNIN**) found in the stomach of mammals and important in the digestion of milk in young mammals.

rennin, chymosin An enzyme in mammalian stomachs which curdles milk thereby causing it to remain in the stomach for longer and improving absorption. *Compare* **RENIN**

reproduction The production of new individuals which are capable of living independently of their parents (or parent). *See also* **PARTHENOGENESIS**

reproductive strategy Reproductive behaviour which has been determined by the process of natural selection to maximise **FITNESS**, e.g. many parasites produce vast numbers of offspring because most will perish before finding a host, while mammals have relatively few young and provide parental care to ensure that as many survive as possible. *See also* **R-SELECTED SPECIES**, **K-SELECTED SPECIES**

reproductive synchrony The tendency of some animals to exhibit a particular stage of their reproductive cycle simultaneously, e.g. birth. This may act as an anti-predator swamping mechanism in some species, whereby too many young are available as prey at the same time, ensuring that many survive. In some species it ensures that births occur when environmental conditions are favourable. *See also* **OESTRUS SYNCHRONY**

reproductive system A system of organs which function together to produce new organisms by producing **GAMETES** and (usually) facilitating **FERTILISATION**, which consists essentially of the **GONADS** and the external genitalia.

reptile A member of the class **REPTILIA**.

Reptile and Exotic Pet Trade Association (REPTA) An organisation which was formed to represent the opinions of the reptile and exotic trade to ensure that no 'unreasonable' legislation was made in the UK regarding the keeping of and trading in exotic animals.

Reptilia A class of **CHORDATES** whose members are **POIKILOTHERMS**, the adults breathe air, and all have a skin covered in ectodermal **SCALES** (Figs. I1 and P1). In some taxa the body is supported by **SCUTES**. Reptile hearts are completely divided into two halves; they produce an **AMNIOTIC EGG** but **OVIVIVIPARITY** is common.

rescue(d) animal An unwanted (usually) domesticated animal (e.g. cat, dog, horse) who has been taken in by an animal rescue centre after being abandoned, confiscated or acquired for some other reason (Fig. I1).

research The process of acquiring new knowledge by experimentation, analysis of data, reinterpretation of existing knowledge or by some other rigorous means. Research in science is undertaken using the **SCIENTIFIC METHOD**.

research institute An organisation dedicated to **RESEARCH**, including its facilities and staff. It may be concerned with particular taxa (e.g. primates) or academic disciplines (e.g. genetics) and may be attached to, or part of, a university, zoo or other institution. For example *see* **INSTITUTE OF ZOOLOGY (IoZ)**, **JANE GOODALL INSTITUTE**, **ZOO RESEARCH INSTITUTE**.

resident species A species that spends its entire life cycle in a particular area, i.e. is non-migratory. *Compare* **INDIGENOUS SPECIES**, **MIGRATORY SPECIES**

resource In relation to an animal, something required for its survival, e.g. food, shelter.

resource partitioning, differential resource utilisation The complementary use of resources by similar **SYMPATRIC** species to avoid **INTERSPECIFIC COMPETITION**. *See also* **COMPETITIVE EXCLUSION PRINCIPLE**

respiration
1. The biochemical process by which organisms obtain energy by breaking down food. *See also* **AEROBIC RESPIRATION**, **ANAEROBIC RESPIRATION**
2. Breathing (ventilation)

respiratory chain *See* **ELECTRON TRANSPORT SYSTEM**

respiratory distress Difficulty breathing. May involve open mouth breathing and pronounced chest movements. May have a number of causes including asthma and pleural effusion.

respiratory failure Inability to adequately ventilate and/or oxygenate. *Compare* **DYSPNOEA**

respiratory rate, breathing frequency The rate at which an animal breathes in and out, usually calculated as breaths/minute. *See also* **BRADYPNOEA**, **DYSPNOEA, EUPNOEA, TACHYPNOEA**

respiratory system The complex of organs which facilitate the exchange of gases between an organism and the environment. In mammals this includes the trachea, bronchi, bronchioles and lungs and the respiratory muscles.

respiratory ventilator A machine which supports and monitors breathing, especially when the patient is anaesthetised during surgery and breathing is suppressed.

response A reaction to a particular **STIMULUS**.

responsiveness The extent to which an animal responds to environmental stimuli. A lack of response to external stimuli is considered an indicator of poor welfare in at least some species. *Compare* **APATHY, LETHARGY**

resting heart rate The **HEART RATE** when the body is at rest.

resting potential The electrical potential across the cell membrane of a **NEURONE** when it is at rest, i.e. when it is not conducting a **NERVE IMPULSE**. In this condition the inside is negatively charged with respect to the outside. *Compare* **ACTION POTENTIAL**

restraining tube A narrow transparent plastic tube used for restraining a snake during handling, e.g. for veterinary treatment. The head and anterior section of the body are placed inside the tube so that the snake may be handled safely.

restraint chute *See* **CRUSH**

restriction enzymes **ENZYMES** that catalyse the cleavage of **DNA** at specific sites producing discrete fragments. Used in **GENETIC ENGINEERING**.

reticulum The second stomach of a **RUMINANT**.

retina The light-sensitive surface of the back of the vertebrate eye which contains cells which detect colours (**CONES**) and cells which are sensitive to low light levels and create monochrome images (**RODS (1)**).

retinol *See* **VITAMIN A**

retirement In the context of working and performing animals (e.g. police horses, circus animals, elephants employed in forestry), the time at which they cease to work (usually due to old age) and for which provision needs to be made by their owners, e.g. by moving them to a **SANCTUARY (2)**.

return In fencing design, the top section of a vertical fence used as a containment barrier which leans inwards, usually at around 45 degrees (towards the animals) to prevent escape. It is often fitted with barbed wire or an **ELECTRIC FENCE**. A return is generally used to contain large, dangerous mammals capable of climbing, e.g. felids, canids, ursids.

reversal drug A drug used to reverse the effects of **SEDATIVE** and **ANAESTHETIC** drugs e.g. **REVIVON**.

Revivon A drug which reverses the effect of **IMMOBILON**.

reward band An identification band found on the legs of some wild birds. Used to collect information on migration, longevity etc. A reward is paid to anyone who finds a bird and returns its band to the organisation that owns it, e.g. such bands are used by the **UNITED STATES FISH AND WILDLIFE SERVICE (USFWS)**.

rewilding
1. The process of returning **ECOSYSTEMS** to their wild state by re-establishing animal and plant **COMMUNITIES** that are currently absent. *See also*

R

PLEISTOCENE REWILDING, REINTRODUCTION, TRANSLOCATION

2. The process of developing natural behaviours in animals that have been captive-bred, e.g. attempting to teach hunting behaviour to big cats that have not been taught to hunt by their mothers by exposing them to live prey. *See also* PRECONDITIONING

RFID technology *See* RADIO FREQUENCY IDENTIFICATION (RFID) TECHNOLOGY

rhesus monkey (*Macaca mulatta*) An OLD WORLD MONKEY which is widely used in medical research and was also used in space flights by the USA and Russia. The species was used in studies of maternal deprivation by HARLOW, and has given its name to the Rhesus blood group system. *See also* CHIMPONAUT

rhinal Relating to the nose.

rhinarium The moist, glandular, naked surface around the nostrils of the nose in most mammal species.

Rhinobatiformes An order of fishes: shovelnose rays.

Rhinoceros and Tiger Conservation Act of 1994 (USA) A US law which prohibits the import, export or sale of any product, item or substance containing anything derived from a tiger or rhinoceros species. These prohibitions also apply to anything labelled or advertised as containing any substance from these species.

Rhinos of Nepal exhibit *See* GREEN EXHIBIT

Rhynchocephalia An order of reptiles: tuataras.

rib cage *See* RIBS

ribonucleic acid *See* RNA

ribosomal RNA (rRNA) The type of **RNA** from which **RIBOSOMES** are partially constructed.

ribosome A very small cell ORGANELLE which is found in the cytoplasm of all living cells and attached to the ENDOPLASMIC RETICULUM (ER). It is made of protein and RIBOSOMAL RNA (RRNA), and assists in the assembly of PROTEIN molecules from AMINO ACIDS during the process of TRANSLATION.

ribs A series of slender bones which occur in pairs and surround and protect the THORAX (1) and articulate with the VERTEBRAL COLUMN forming a rib cage. Part of the AXIAL SKELETON of vertebrates.

rickets A disease of the skeletal system whereby the bones become soft and deformed as a result of VITAMIN D deficiency.

rickettsia Bacteria (*Rickettsia, Rochalimaea* and *Coxiella*) that infect mammals and arthropods. Diseases caused include epidemic typhus, endemic (murine) typhus, spotted fever.

rides Historically, animal rides have been popular in zoos. Elephants were once widely used to give rides to visitors, but this practice is now less common due to safety concerns. At LONDON ZOO in the past, rides were given by a Mongolian wild ass, camels, ponies and Asian elephants. Llamas were used to pull a carriage and the zoo used four zebras to pull

a cart advertising Mazawattee Tea around the city in an advertising campaign in 1914.

ridgling, rig A colloquial term used to describe either a MONORCHID or CRYPTORCHID, especially in horses.

Riding Establishments Act 1964 and 1970 Legislation in Great Britain which requires the licensing and inspection of premises which keep HORSES or DONKEYS to let out for hire or which provide riding instruction for payment. *See also* BEACH DONKEY

Rift Valley fever, enzootic hepatitis This disease is caused by a bunyavirus that is transmitted by mosquitoes. It causes necrosis of liver cells and abortion. The disease affects cattle, sheep, horses, donkeys, goats, buffaloes, camels and humans. It occurs mostly in Africa, but there is concern that it may spread through the Mediterranean countries and the Middle East. A live vaccine is available.

rig *See* RIDGLING

rights *See* ANIMAL RIGHTS, ANIMAL RIGHTS LEGAL CASES

rinderpest An infectious viral disease of cattle and some other even-toed ungulates that has been globally eradicated.

ring

1. A ring of metal or plastic fixed around the leg of an animal (often a bird) used for identification. A ring may bear an identification number or a series of coloured rings may be used to give an animal a unique mark.

2. *See* RINGING (2)

Ring of Bright Water A book and film, released in 1969, starring Bill Travers and Virginia McKenna , of the true story of the relationship between a man (Gavin Maxwell) living on the Scottish coast and a wild otter. *See also* BORN FREE

ringing

1. The process of attaching a RING (1) to the leg of an animal, especially a bird (BIRD RINGING). *See also* DOUBLE BANDING, REWARD BAND, TAIL RINGING.

2. The removal of a ring of bark from a tree, resulting in its eventual death as a result of the severing of the vascular tissue. It may occur when animals rub off or eat bark. Zoos often protect trees (especially from browsers) by fixing wire mesh, rope or some other material around vulnerable areas.

Ringling Bros. *See* BARNUM AND BAILEY

ringworm A contagious disease caused by fungi (e.g. *Trichophyton* spp., *Microsporum* spp., *Oidmella* spp.) which live on the surface of the skin or in the hairs of infected areas. It appears as patches of raised, dry, crusty skin, often more or less circular in form, where the hairs have fallen out, and scales and scabs have formed. It can affect a number of taxa of mammals and birds. Treatment is by oral administration of griseofulvin or topical application of, for example, natamycin or enilconazone.

R

riparian Relating to the banks of a natural water course or water body (a lake or pond) e.g. riparian woodland. *Compare* **RIVERINE**

riparium A **VIVARIUM** that simulates a **RIPARIAN** habitat: the watery edge of a pond, lake, river or stream; a **PALUDARIUM** with water circulating through pools located at different levels.

risk A measure of the likelihood (**PROBABILITY**) of a loss or injury occurring. It is a function of the probability of the event occurring and the seriousness of the potential injury or loss.

Risk = the probability of a hazard (an event

that *could* cause harm) resulting in an

adverse event × the severity of the event.

The risk of injury to the public is much higher if a lion escapes from an enclosure than if a penguin escapes from an enclosure. The probability of escape may be the same, but the severity if it occurs differs between the two species. *See also* **RISK ASSESSMENT**

risk assessment A process and/or a document that identifies and examines the health and safety **RISKS** associated with a particular activity and determines measures that should be taken to mitigate any dangerous elements, e.g. entering the enclosure of a dangerous animal to provide veterinary treatment.

ritualisation The process by which a behaviour becomes modified by evolution so that it comes to serve a purpose which is different from its original purpose, e.g. ritualised beak wiping has become part of the courtship behaviour of some finch species; ritualised **PREENING** movements form part of the courtship of some ducks; the collection of nesting material has become part of a courtship sequence in some bird species.

riverine Relating to rivers. *Compare* **RIPARIAN**

RNA Any of three types of ribonucleic acid: **MESSENGER RNA (mRNA)**, **TRANSFER RNA (tRNA)** and **RIBOSOMAL RNA (rRNA)**. It is constructed from a chain of **NUCLEOTIDES**.

roadside zoo An animal collection which is generally privately owned, run for profit and not accredited by a national or regional zoo organisation, where animals are primarily exhibited for the amusement of visitors, especially in the USA and Canada. Typically they are made up of barren cages and small enclosures surrounded by a chain-linked fence. Often visitors pay to have their photographs taken with animals such as tiger cubs and primates. Many of these zoos employ untrained staff and animals are deprived of veterinary care. Some are used to attract customers to other facilities, e.g. a petrol station. *See also* **SCAMTUARY**

Robert Jones bandage A padded bandage used as an external **SPLINT** for the temporary support of a limb after a fracture. It applies pressure which causes the tissues to reabsorb **INTERSTITIAL FLUID**, creates limb stability and protects from **TRAUMA**.

robotic behaviour A behaviour which is carried out automatically and does not require the animal to be aware of its actions.

robotic milking machine An automated milking machine which allows cows to decide when they want to be milked and allows them free access to milking equipment. The machine relies on **RADIO FREQUENCY IDENTIFICATION (RFID)** **TECHNOLOGY** to identify individual cows and monitor their milk output. *See also* **MILKING PARLOUR**

rocket-netting *See* **CANNON NETTING**

rocking A **STEREOTYPIC BEHAVIOUR** in which the animal moves its head or whole body backwards and forwards repeatedly in a rhythmical manner. Sometimes performed sitting with the legs hugged to the body in primates.

rod
1. A type of **PHOTORECEPTOR** cell found in the **RETINA** that is capable of detecting low-intensity light. Important in night vision. *Compare* **CONE**
2. A rod-shaped bacterium.

rodent A member of the order **RODENTIA**.

Rodentia An order of mammals: rodents (beavers, squirrels, chipmunks, prairie dogs, mice, rats, gophers, hamsters, lemmings, gerbils, voles, porcupines, springhare, mole-rats, cane rats, agoutis, capybaras, chinchillas, coypu, hutias).

rodenticide A chemical designed to kill rodents, e.g. **WARFARIN**.

Rolling Pen A device designed to move ewes and their lambs safely and reduce escapes. Consists of a floorless steel and weld mesh crate on wheels in which a ewe is penned while her lambs are carried in a box in front of her. As the shepherd pushes the pen forward the ewe follows her lambs.

Roman games Contests held in ancient Rome in which wild animals fought other animals and people for the entertainment of large audiences, resulting in large numbers of animal deaths. *See also* **CIRCUS MAXIMUS**, **COLOSSEUM**

roost
1. A place where birds or bats rest or sleep.
2. A group of birds or bats resting or sleeping.

Roslin Institute A research facility at the University of Edinburgh, Scotland, where an animal was cloned from an adult **SOMATIC CELL** for the first time: *DOLLY* **THE SHEEP**

rostral Located on the head, towards the nose.

rostrum *See* **BILL**

rotating cattle brush A large suspended rotary brush designed for **GROOMING** cattle (Fig. R3). It may be located in a cow shed or outside and is operated automatically by a switch which is activated when a cow comes into contact with the brush. May also be used in zoos for large mammals, e.g. rhinos.

rotational exhibit design, animal rotation exhibit A zoo exhibit which consists of a series of interlinked enclosures that may be used by different species (different groups of compatible species within a

R

Fig. R3 Rotating cattle brush. Inset: dairy cow activating the brush by rubbing against it.

MULTI-SPECIES EXHIBIT) at different times, on a 'time-share' basis. This gives each species (or multi-species group) access to a much larger area than if each was confined to a single enclosure. This design may be important in enriching the lives of the animals. At Louisville Zoo four display areas are shared by Sumatran tiger, tapir, babirusa, orangutan and siamang on a randomised basis. A rotation exhibit for fish could be created by linking several tanks and moving fishes between them in order to recreate seasonal movements.

Rothschild, Lionel Walter (1868–1937) A British banker, zoologist, animal collector and Member of Parliament. He started his first zoology museum at the age of 10 in a garden shed, and then built what became his Zoological Museum on land purchased by his father in Tring. It opened to the public in 1892 and was once one of the largest collections in the world. Most of his bird skin collection was sold to the **AMERICAN MUSEUM OF NATURAL HISTORY** in the early 1930s. Rothschild's museum is now the Natural History Museum at Tring, a division of the **NATURAL HISTORY MUSEUM (LONDON)**. Baron Rothschild was the first to describe the Rothschild giraffe (*Giraffa camelopardalis rothschildi*).

roughage *See* **FIBRE**. *See also* **BROWSE (1)**

Round House A gorilla house at **LONDON ZOO** designed by **LUBETKIN**. It was opened in 1933 and is now a **LISTED BUILDING**. It has a round structure which incorporates a rotating mechanism which allowed visitors to view the animals whether they were inside or outside. It is no longer used for gorillas.

roundworm *See* **NEMATODA**

Royal Army Veterinary Corps (RAVC) The Corps of the British Army which is responsible for the development of good husbandry, training practice, preventative medicine and care for service animals, especially horses and dogs.

Royal Australasian Ornithologists' Union *See* **BIRDS AUSTRALIA**

Royal College of Veterinary Surgeons (RCVS) The regulatory body for veterinary surgeons in the UK.

royal hunting grounds, royal park An area of land (generally forest) that was set aside for the exclusive use of royalty for hunting animals, e.g. many of the parks in London, the New Forest. Some former royal hunting grounds are now protected as nature reserves. *See also* **PARADEISOS, WOODSTOCK**

royal park *See* **ROYAL HUNTING GROUNDS**

Royal Society for the Prevention of Cruelty to Animals (RSPCA) The first society in the world dedicated to the protection of animals. It was established in the UK in 1824 by Richard Martin MP, the Reverend Arthur Broome and others. The Society operates **ANIMAL HOSPITALS** and employs inspectors who investigate cases of animal cruelty and work with the police to prosecute offenders. It campaigns against animal cruelty generally and the keeping of some animals in zoos, notably elephants, and it produces **RSPCA WELFARE STANDARDS**. In 2010 the RSPCA (in England and Wales) rescued and collected 130033 animals, investigated 159686 cruelty complaints and secured 2441 convictions to protect animals. *See also* **LIVE HARD, DIE YOUNG, RSPCA INSPECTOR**

Royal Society for the Protection of Birds (RSPB) The **SOCIETY FOR THE PROTECTION OF BIRDS** was founded in Manchester, England, in 1889 specifically to stop the slaughter of birds for their plumes. Branches were set up overseas and the branch in India secured the first measure against the plumage trade: an order from the Indian government in 1902 that banned the export of bird skins and feathers. In 1904 the society received a Royal Charter and became the Royal Society for the Protection of Birds (RSPB). The society raises funds for the conservation of birds and their habitats and manages 200 nature reserves in the UK covering 130000 hectares. It assists the police in the prosecution of people who commit crimes against birds and it is involved in programmes to reintroduce bird species in the UK, e.g. the Eurasian crane (*Grus grus*). The society publishes *LEGAL EAGLE*.

Royal Zoological Society of Scotland (RZSS) A conservation, education and research charity in Scotland (founded in 1909) which owns and operates

Edinburgh Zoo (opened in 1913) and the **HIGH-LAND WILDLIFE PARK** (opened in 1972). Edinburgh Zoo has the only giant pandas (*Ailuropoda melanoleuca*) in the UK and the Highland Wildlife Park is home to the UK's only polar bears (*Ursus maritimus*). In 2009 the society was involved in the trial reintroduction of beavers (*Castor fiber*) into Scotland.

rRNA *See* **RIBOSOMAL RNA (rRNA)**

r-selected species An r-selected species is an opportunist species. It exhibits rapid development, early reproduction, small body size and **SEMELPARITY**. It often exhibits type III survivorship (Fig. S16), has high colonising ability, density-independent mortality and poorly developed social behaviour (mostly schools, herds, **AGGREGATIONS**). An r-strategist typically lives in a variable, unpredictable environment. It is a species which produces many young, has a short life history and is capable of increasing the size of its population quickly when environmental conditions are favourable, e.g. voles and insects. *Compare* **K-SELECTED SPECIES**

RSPB *See* **ROYAL SOCIETY FOR THE PROTECTION OF BIRDS (RSPB)**

RSPB Investigations Unit A division of the **ROYAL SOCIETY FOR THE PROTECTION OF BIRDS (RSPB)** which investigates bird crime in the UK.

RSPCA *See* **ROYAL SOCIETY FOR THE PREVENTION OF CRUELTY TO ANIMALS (RSPCA)**

RSPCA inspector A person employed by the **ROYAL SOCIETY FOR THE PREVENTION OF CRUELTY TO ANIMALS (RSPCA)** to investigate cruelty to animals, rescue ill-treated and injured animals, and provide animal welfare advice to the public. An inspector has no special legal powers.

RSPCA Welfare Standards A series of documents published by the **ROYAL SOCIETY FOR THE PREVENTION OF CRUELTY TO ANIMALS (RSPCA)** detailing welfare standards for each of the major farm animal species.

Rules of Racing
1. Rules produced by the **BRITISH HORSERACING AUTHORITY (BHA)** which regulate horse races in Britain, including licensing, equipment, administration, appointment of officials, conduct, investigations, appeals etc.
2. Rules of the **GREYHOUND BOARD OF GREAT BRITAIN** which regulate greyhound racing as required by the **WELFARE OF RACING GREYHOUNDS REGULATIONS 2010**.

rumen The first stomach of a **RUMINANT**.

ruminant, cranial fermenter A mammal with a complex stomach. After chewing, food passes to the first stomach (rumen) where it is fermented by

microbes. It is then regurgitated to be chewed a second time and mixed with saliva. At this stage the food is referred to as 'cud' and ruminants are said to 'chew the cud'. Next, the food is swallowed a second time. It bypasses the rumen and passes directly to the second stomach chamber, the reticulum. Bacteria pass with the food to the omasum (third stomach) and then the abomasum (fourth stomach), where digestion is completed. Nutrients are absorbed in the small intestine and some additional fermentation and absorption occurs in the **CAECUM**. *Compare* **HINDGUT FERMENTER**

RumiWatch An automated health monitoring system for cows which consists of a halter that monitors feeding, rumination, water uptake, locomotion and rest, and transfers this information wirelessly to a computer for analysis.

run An enclosed area, usually outside, where small animals are able to move around relatively freely, often covered with wire mesh to prevent escape and exclude predators. Often provided for pets (e.g. rabbits, guinea pigs) or poultry, especially when their indoor accommodation is relatively small. *See also* **ANIMAL ENCLOSURE, POULTRY ARK**

running wheel A wheel inside which a small animal (e.g. a hamster, mouse or rat) is able to run, without moving forward, which may be used for measuring activity (Fig. H3).

rupture
1. A tearing apart of tissue or the bursting of a vessel.
2. A **HERNIA**.

Rural Payments Agency An executive agency of the **DEPARTMENT FOR ENVIRONMENT, FOOD AND RURAL AFFAIRS (DEFRA)** which encourages a thriving farming and food sector and strong rural communities. It makes **COMMON AGRICULTURAL POLICY (CAP)** support payments, traces **LIVESTOCK** (assigning **CPH NUMBERS**) and carries out inspections.

rut
1. A state of sexual excitement which occurs annually in some mammals, especially in male deer.
2. The period when rutting occurs (the rut).

Ryder, Richard (1940–) A British psychologist and philosopher who was important in the development of the **ANIMAL RIGHTS** movement and the originator, in 1970, of the concept of **SPECIESISM**. In 1975 he published *Victims of Science*, an attack on animal experimentation. Ryder is a former Chairman of the **ROYAL SOCIETY FOR THE PREVENTION OF CRUELTY TO ANIMALS (RSPCA)**. *See also* **ANIMAL TESTING**

R

Fig. S1 S is for Shire horses. The Shire horses of Thwaites Brewery (Lancashire, England).

Dictionary of Zoo Biology and Animal Management: A guide to terminology used in zoo biology, animal welfare, wildlife conservation and livestock production, First Edition. Paul A. Rees.
© 2013 John Wiley & Sons, Ltd. Published 2013 by John Wiley & Sons, Ltd.

sacred animal A species or type of animal which is considered to be sacred within a particular society or religion. *See also* **ANIMALS AND RELIGION, FADY, TOTEM**. *Compare* **UNCLEAN ANIMAL**

safari The Swahili word for journey, and generally used to refer to journeys on foot, or travelling on animals (e.g. horses or camels) or in vehicles to observe or hunt animals, especially **BIG GAME** in Africa. The term is used in the names of many zoos (including **SAFARI PARKS**) and zoo exhibits, e.g. *African Bird Safari* (**LONDON ZOO**), *TSAVO Bird Safari* (**CHESTER ZOO**), *NIGHT SAFARI* (Singapore Zoo).

Safari Club International An organisation based in the USA that supports hunting and wildlife conservation. it campaigns for the freedom to hunt animals nationally and internationally and to promote the image of hunters.

safari park, open range zoo A zoo in which animals are kept in large **ANIMAL ENCLOSURES** through which visitors drive on roads in their own cars, buses or other vehicles. Exhibits are often **MULTISPECIES EXHIBITS**, and early parks attempted to simulate the experience of visiting an African **NATIONAL PARK (NP)**. *See AFRICA USA, LIONS OF LONGLEAT. Compare* **ZOOLOGICAL PARK**

sagittal
1. Relating to the front-to-back suture on top of the skull where the parietal bones meet.
2. In a plane parallel to this suture.

sagittal crest A prominent ridge of bone which extends from the front to the back of the top of the skull of some mammals, reptiles and other taxa to which the jaw muscles are attached. It is particularly prominent in gorillas (*Gorilla gorilla*) and orangutans (*Pongo* spp.).

sahel A semi-arid transitional biogeographical zone located between the southern limit of the Sahara desert and the savannas located to the south.

saline solution A solution of sodium chloride in water used in intravenous drips and for washing wounds.

salinity (S) The total amount of dissolved material in saltwater expressed as grams per kilogram of sea water. Salinity is a dimensionless quantity and it has no units.

saliva A secretion of mucus and water produced by the **SALIVARY GLANDS**, which contains **ENZYMES** in some species, but in most it simply moistens food. Salivary **AMYLASE** is found in many bird and mammal species.

salivary glands **EXOCRINE GLANDS** that secrete **SALIVA** into the mouth in many terrestrial animals. Modified salivary glands produce venom in snakes and some other taxa.

salivation The release of **SALIVA** into the **BUCCAL CAVITY** by the **SALIVARY GLANDS**. This is an involuntary reflex response to the sight, smell or taste of food.

Salmonella A genus of enterobacteria whose members cause diseases, many of which are zoonotic, e.g. salmonellosis.

salmonellosis *See SALMONELLA*

salmonid A member of the family Salmonidae which includes salmon and trout.

Salmoniformes An order of fishes: salmons, smelts and their allies.

salt gland A gland in the head of some seabirds that secretes salt to maintain osmotic balance. *See also* **OSMOSIS**.

salt lick
1. A place (e.g. an area of soil) where animals go to obtain salt.
2. A block of salt given as a dietary supplement.

salt marsh An area of land which is periodically flooded with salt water. *See also* **MANGROVE FOREST, MARSH**

saltation
1. A single mutation which drastically alters the phenotype of an organism.
2. An evolutionary process in which there is sudden and dramatic change.
3. Saltatory locomotion. Moving by leaps or jumps, e.g. as in kangaroos, frogs and lemurs.

saltatory locomotion *See* **SALTATION (3)**

saltwater
1. Water which contains salt, especially that in the oceans; saline.
2. Marine; living in the sea; of or found in saltwater, e.g. a saltwater crocodile, saltwater fishing.

sample A small portion or small number of something whose characteristics (parameters) are taken to represent the whole. For example, a sample of crows taken from a **POPULATION** of crows; a sample of weights of mice taken from a population (of weights) of mice. *See also* **DESTRUCTIVE SAMPLING, RANDOM SAMPLE**

sampling The process of collecting or testing samples. In animal behaviour studies this may be achieved by, for example, collecting **FOCAL SAMPLES** or for instantaneous **SCAN SAMPLING**.

sampling error The difference between the true value of a **POPULATION** parameter and the value which has been estimated from a particular **SAMPLE**. For example, if the mean weight of a large population of birds is 100g and the mean weight estimated from a sample of ten birds taken from this population is 102g, the sampling error is $102 - 100 = 2g$.

San Diego Global *See* **ZOOLOGICAL SOCIETY OF SAN DIEGO (ZSSD)**

San Diego Zoo *See* **ZOOLOGICAL SOCIETY OF SAN DIEGO (ZSSD)**

San Diego Zoo Institute for Conservation Research *See* **INSTITUTE FOR CONSERVATION RESEARCH**

sanctuary
1. A place where wildlife is protected by law, e.g. a **WHALE SANCTUARY**, a **SHARK SANCTUARY**.

S

Fig. S2 Sand school.

2. A place where unwanted or rescued animals are kept, e.g. ELEPHANT SANCTUARY, monkey sanctuary, DONKEY SANCTUARY. *See also* SCAMTUARY

sand bath An area or tray of fine dry sand provided for captive birds or mammals (e.g. chinchilla (*Chinchilla* spp.)) to take a sand bath. This helps to keep the feathers/coat in good condition and remove ECTOPARASITES, e.g. lice. *See also* DUST BATHING

sand cat (*Felis margarita*) *See* INTERSPECIES EMBRYO TRANSFER

sand school, manage, manège An enclosed, surfaced area (usually rectangular) used for exercising horses and giving riding lessons. May be indoors (Fig. S2) or outdoors. *See also* MANÈGE for alternative meaning.

Sarcopterygii A subclass of bony fishes: lungfishes and coelacanth.

sarcoptic mange *See* MANGE

SARS *See* SEVERE ACUTE RESPIRATORY SYNDROME (SARS)

satellite image A photograph or other image taken of the Earth's surface from an orbiting or geostationary satellite. Such images may be digitised and analysed using GEOGRAPHICAL INFORMATION SYSTEMS (GIS) software and may be used in wildlife conservation, e.g. to analyse the distribution of different vegetation types (Fig. R2). *See also* AERIAL PHOTOGRAPHY

satellite tag An electronic tag attached to an animal which stores information about its location and transmits this to a satellite for analysis by scientists. Used for tracking the movements of animals ranging from large marine mammals to quite small birds. Battery powered and sometimes fitted with solar panels. Some tags are designed to drop off the animal after a predetermined time, especially those attached to marine animals which float to the water's surface after release. *See also* GEOLOCATOR

sausage pig A pig used specifically for sausages. They tend to be older sows and others that are no longer required.

savanna, savannah A dry tropical or subtropical grassland habitat characterised by drought-resistant vegetation dominated by grasses with scattered trees and inhabited by large grazing mammals. *See also* KOPJE, SERENGETI

savannah cat breed A hybrid cat produced by crossing a serval (*Leptailurus serval*) with a pedigree domestic cat (*Felis catus*). Popular as pets in the USA.

scale
1. A dermal or epidermal plate which occurs in large numbers typically covering the body of fishes, reptiles and some mammals.
2. One of the minute plate-like overlapping structures found on the surface of the wings of LEPIDOPTERA.

scamtuary, pseudo-sanctuary A derogatory term used by some animal welfare campaigners to refer to self-styled exotic animal 'sanctuaries' which breed animals (e.g. lions, tigers, TIGONS, LIGERS, bears and primates) and exhibit them for profit.

scan sampling A method of data collection used in animal BEHAVIOUR studies. All of the animals in a group are 'scanned' at regular intervals of time (e.g. every 5 minutes) and their behaviour at that instant in time is noted. *Compare* FOCAL SAMPLING

Scandentia An order of mammals: tree shrews.

scanning electron microscope A type of ELECTRON MICROSCOPE capable of producing a three-dimensional image of a very small structure.

scat Faeces.

scatter feed The provision of small pieces of food by scattering it around an enclosure. Usually not the main feed of the day but part of the overall diet. Sometimes used as a FEEDING ENRICHMENT for monkeys, apes, elephants, etc.

scavenger A species of animal (e.g. vultures, hyenas, foxes) that feeds on the dead bodies of other species, although some may also hunt.

SCBook (SPARKS Compatible studBook) A Microsoft Windows implementation of SINGLE POPULATION ANALYSIS AND RECORD KEEPING SYSTEM (SPARKS) which implements the functionality required by the EUROPEAN STUDBOOK FOUNDATION (ESB).

scent gland Glands found in many mammal species especially in proximity to the genitals, anus, eyes (preorbital glands) and elsewhere, which produce PHEROMONES that may indicate status, or sexual condition or are used to mark TERRITORY.

scent-marking The use of chemicals (usually urine or PHEROMONES) to communicate information to CONSPECIFICS, e.g. sexual condition, TERRITORY ownership etc. These chemicals may be released from specialised SCENT GLANDS. *See also* OVER-MARKING, TOTEM-TREE MARKING

Schaller, George B. (1933–) A zoologist and conservationist who is renowned for his detailed field studies of a number of iconic species including giant pandas, mountain gorillas and snow leopards. Dr. Schaller has held a number of posts at universities in the United States, including Stanford and Johns Hopkins, and was formerly Director of the New York Zoological Society's International Conservation Program. He is currently a Senior Conservationist at the **WILDLIFE CONSERVATION SOCIETY (WCS)** and has been instrumental in protecting important wildlife areas in the USA, China, Brazil, Pakistan and southeast Asia. Schaller's books include *The Year of the Gorilla* (1964), *The Deer and the Tiger* (1967), *The Serengeti Lion* (1972) and *The Giant Pandas of Wolong* (1985). *See also* **SERENGETI**

schedule In **LEGAL INSTRUMENTS**, a list that usually occurs at the end of a document, containing items to which the law applies, e.g. a list of protected birds, a list of banned chemicals.

schistosomiasis, bilharziasis A disease caused by infestation with a **PLATYHELMINTH** from the genus *Schistosoma*. They generally live in the portal and mesenteric veins and infect a wide range of animals including cattle, sheep, camels, water buffalo, horses, donkeys, dogs and humans. Infestation may be fatal. Transmission is via water and the **INTERMEDIATE HOSTS** are snails. **MOLLUSCI-CIDES** such as copper sulphate may be used to treat pasture and drugs such as praziquantel may be effective in treating infected animals.

Schmallenberg virus (SBV) A viral disease of sheep, cattle and goats which causes loss of condition in older animals, abortion and brain and limb defects in newborn animals.

Schomberg, Geoffrey A zoo design consultant and founder of the Federation of Zoological Gardens of Great Britain, in 1966, with the purpose of raising standards through regular inspection. This eventually became the **BRITISH AND IRISH ASSOCIATION OF ZOOS AND AQUARIUMS (BIAZA)**. He published a number of books on zoos including *British Zoos: a Study of Animals in Captivity* (1957), *Penguin Guide to British Zoos* (1970) and *General Principles of Zoo Design* (1972), and was editor of *INTERNATIONAL ZOO NEWS (IZN)*, 1974–1979.

Schönbrunn Zoo, Tiergarten Schönbrunn, Zoo Vienna The first modern zoo. Founded in 1752 by Franz Stephan – the husband of Empress Maria Theresa – in the grounds of the imperial palace of Schönbrunn. It was essentially a private collection, although the public was admitted occasionally. Enclosures were arranged around a central rococo pavilion which afforded the best views of the animals, which were kept behind high walls. Later, Josef II established a Society for the Acquisition of Animals and he financed collecting expeditions to Africa and the Americas. In its day Schönbrunn was the best animal collection in Europe and the zoo still exists as Tiergarten Schönbrunn, or Zoo Vienna.

school A collective term for a group of fish of the same species.

scientific method A method of investigation used by scientists by which they attempt to construct an accurate, reliable and consistent representation of the world. In simple terms use of the scientific method involves the following steps:

1. Make observations about some aspect of nature.

2. Propose a **HYPOTHESIS**: a guess about how the world works that is consistent with the observations.

3. Make predictions using the hypothesis.

4. Test the predictions by experimentation or additional observations.

5. Modify the hypothesis if necessary and then return to steps 3 and 4 until there are no inconsistencies between the hypothesis and experimental results or observations. Once consistency is achieved the hypothesis assumes the status of a theory.

See also **CONTROL, CROSS-SECTIONAL STUDY, LONGITUDINAL STUDY, META-ANALYSIS, NULL HYPOTHESIS (H_0), OCCAM'S RAZOR**, *POST HOC STUDY*, **SCIENTIFIC PAPER**

scientific name *See* **BINOMIAL NAME**

scientific paper A formal report of original scientific work published in an **ACADEMIC JOURNAL** by a scientist, vet, or other qualified person, after being subjected to **PEER REVIEW**. *Compare* **GREY LITERATURE**

scientific whaling The killing of whales for scientific purposes as permitted by Article VIII of the **INTERNATIONAL CONVENTION FOR THE REGULATION OF WHALING 1946**. This activity is widely believed to have been used by Japan and Iceland as a means of disguising commercial whaling.

Scorpaeniformes An order of fishes: scorpionfishes and their allies.

Scott, Sir Peter (1909–1989) Peter Scott was the son of the Antarctic explorer Robert Falcon Scott. He was a renowned wildlife artist and was responsible for the establishment of the Wildfowl Trust (now the **WILDFOWL AND WETLANDS TRUST (WWT)**). One of its early successes was in the captive breeding and reintroduction of the NĒ NĒ (*BRANTA SANDVICENSIS*). Scott was a past Chairman of the Survival Service Commission of the **INTERNATIONAL UNION FOR THE CONSERVATION OF NATURE AND NATURAL RESOURCES (IUCN)** and was largely responsible for establishing the concept of Red Data books. He was Chairman of the Council of the Fauna Preservation Society (now **FAUNA AND FLORA INTERNATIONAL (FFI)**) and also Chairman of the **WORLD WILDLIFE FUND (WWF)**, which he helped to found. *See also* **RED LIST**

Scottish Natural Heritage (SNH) The statutory nature conservation agency for Scotland.

scour *See* **DIARRHOEA**

scouring *See* **DIARRHOEA**

scrape A simple bird's **NEST** which amounts to little more than a disturbed area of ground, as made, for example, by some ground-nesting gull species.

S

scrapie A notifiable, fatal degenerative disease of the nervous system, caused by a **PRION**, which affects sheep and goats. Infected animals scrape their fleeces off against rocks and other objects.

screw picket A metal device, shaped like a corkscrew, used for fixing objects and tethering animals, to the ground. *See also* **PICKETING**

scute, scutum A horny, chitinous, or bony plate or scale such as that found in the shell of a turtle, on the underside of a snake, the skin of crocodiles or on the feet of some birds.

Sea Mammal Research Unit (SMRU) An organisation that conducts interdisciplinary research on marine mammals, at St Andrews University, Scotland. It is a collaborative centre of the **NATURAL ENVIRON-MENT RESEARCH COUNCIL (NERC)**.

Sea Watch Foundation A national marine conservation research charity dedicated to the protection of cetaceans around the UK. It monitors the status, numbers and distribution of species and changes in marine habitats.

seabird A vernacular term for birds associated with the sea, especially those that live far from the shore.

Seal, Ulysses (1929–2003) Dr. Ulysses Seal was a pioneer in the application of theoretical knowledge in genetics and population biology to practical conservation problems. He trained as a psychologist and then a biochemist. He worked as an endocrinologist at the Veteran's Administration Medical Center in Minneapolis, Minnesota, where he became interested in developing safe techniques for the anaesthesia of wildlife and contraception. In 1973 he founded the **INTERNATIONAL SPECIES INFORMATION SYSTEM (ISIS)** and in 1979 he became chairman of the Captive Breeding Specialist Group (now the **CONSERVATION BREEDING SPECIALIST GROUP (CBSG)**) of the **INTERNATIONAL UNION FOR THE CONSERVATION OF NATURE AND NATURAL RESOURCES (IUCN)**. Seal was instrumental in producing the first **SPECIES SURVIVAL PLAN® (SSP) PROGRAMS** and in saving the **BLACK-FOOTED FERRET (MUSTELA NIGRIPES)** from extinction.

sealing The hunting of seals for their fur, meat and other products.

seaquarium *See* **OCEANARIUM**

search and rescue dog A dog trained to locate humans by scent and used to find missing people and casualties in wilderness areas, natural disasters, large-scale accidents etc. (Fig. S3).

SEAZA South East Asian Zoo Association. *See also* **ZOO ORGANISATION**

sebaceous gland *See* **SEBUM**

Sebag studbook A Microsoft Windows based **SINGLE POPULATION ANALYSIS AND RECORD KEEPING SYSTEM (SPARKS)** compatible studbook administration program which is used by the **EUROPEAN STUDBOOK FOUNDATION (ESB)** and is available in English, German, Dutch and Spanish.

Fig. S3 Search and rescue dog: a lowland search and rescue dog.

sebum An oily substance produced by the sebaceous glands in mammalian skin which helps to prevent skin and hairs from drying out.

second filial generation, F_2 generation Offspring resulting from interbreeding F_1 hybrid individuals (or from self-fertilising them). *See also* **FIRST FILIAL GENERATION**

second nose *See* **VOMERONASAL ORGAN**

secondary bacterial infection An infection caused by bacteria which is not the primary cause of a disease or disorder.

secondary consumer A carnivore.

secondary containment *See* **CONTAINMENT, SECONDARY**

secondary feather An inner flight **FEATHER** located on the trailing edge of a bird's wing, between the bend of the wing and the body. *Compare* **PRIMARY FEATHER**

secondary forest A forest that has regenerated following a period of disturbance cause by humans (e.g. logging) or by a natural event such as a hurricane or fire. It exhibits differences in forest structure and species composition compared with **PRIMARY FOREST**.

secondary host *See* **INTERMEDIATE HOST, VECTOR**

secondary production
1. The total amount of **BIOMASS** produced by all **HETEROTROPHS (PRIMARY CONSUMERS** and **SECONDARY CONSUMERS)** in a particular place over a particular period of time. *Compare* **PRIMARY PRODUCTION**
2. The total amount of biomass produced by **PRIMARY CONSUMERS** only in a particular place over a particular period of time. *See also* **TERTIARY PRODUCTION**

secondary sexual characteristics Those features of an animal's body which are indicative of its sex, other than the **SEX ORGANS**, which generally appear at sexual maturity. They may be, for example, sex differences in hair growth, body size and muscular development, differences in particular features of the **PLUMAGE** in birds (in which males tend to be more colourful), differences in the size of **HORNS** in **ANTELOPES**, and differences in the **FINS** in some fish species.

Secretary of State's Standards of Modern Zoo Practice (SSSMZP) A series of standards issued in England under the **ZOO LICENSING ACT 1981**, which regulate the operation of zoos. They provide minimum standards in relation to the provision of food, water, a suitable environment; the provision of animal health care and the opportunity to express most normal behaviour; the provision of protection from fear and distress; the transportation and movement of animals; conservation and education; public safety, insurance, escapes; stock records; staff and training; public facilities, first aid, toilets, parking; and a requirement to display the **ZOO LICENCE**.

sedation
  1. The use of a drug to reduce **ANXIETY**, **STRESS** or excitement.
  2. The state induced by a **SEDATIVE**.

sedative A drug which reduces **ANXIETY** and awareness of the surroundings.

seeing-eye dog *See* **GUIDE DOG**

segmentation The division of the body, or a part of the body, into similar units as, for example, in the body of earthworms or the embryonic nervous and muscle systems of vertebrates.

seine net A fishing net which has a very fine mesh and is therefore capable of catching extremely small fish, many of which may not have had the opportunity to reproduce. Contributes to **OVERFISHING**.

seizure An episode of disturbed brain activity which results in changes in behaviour or attention. *See also* **EPILEPSY**

Selborne Society for the Protection of Birds, Plants and Pleasant Places The first national organisation in the UK specifically concerned with protecting wildlife. In 1885 the Selborne League was founded to perpetuate the name and interests of Gilbert **WHITE**. In the same year the **PLUMAGE LEAGUE** was also founded. They amalgamated in January 1886 as the Selborne Society for the Preservation of Birds, Plants and Pleasant Places. *See also* **NATURAL HISTORY AND ANTIQUITIES OF SELBORNE**

selection The process that determines which individuals survive and breed (Fig. S4). *See also* **ARTIFICIAL SELECTION, DIRECTIONAL SELECTION, DISRUPTIVE SELECTION, NATURAL SELECTION, STABILISING SELECTION**

selective breeding *See* **ARTIFICIAL SELECTION**

selective feeding When presented with a range of foods (especially **CONCENTRATES**) some animals will choose to eat only the foods they like and leave the less desirable items. In some situations this may lead to a dietary imbalance in captive or domesticated animals. *See also* **SELECTIVE GRAZING**

selective grazer An animal that practises **SELECTIVE GRAZING**, i.e. any animal that can aim for, and

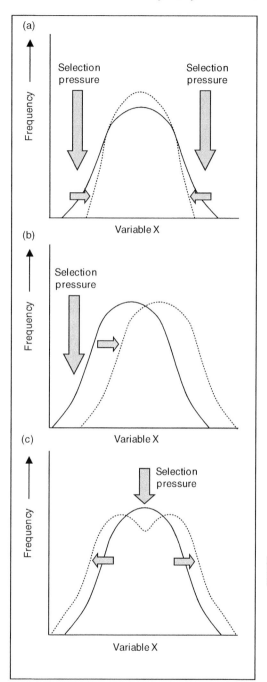

Fig. S4 Types of selection: (a) stabilising; (b) directional; (c) disruptive. Broken line shows the effect of selection.

intentionally select, a specific plant as food. *Compare* **BULK GRAZER**

selective grazing The preferential choosing of some plant species as food by grazers rather than others, when a choice is available. Many livestock species and wild herbivores practise selective grazing and this may affect the plant composition of the **SWARD**. Livestock are sometimes used to manage grasslands in a practice known as **CONSERVATION GRAZING**. *See also* **GRAZING FACILITATION**

selective pressure An environmental factor which affects the process of **EVOLUTION** by favouring some individuals in a population over others, thereby changing the population's genetic composition over time, e.g. very cold conditions may favour individuals with greater fat reserves. *See also* **SELECTION**

selenium A mineral nutrient. A component of glutathione peroxidase which protects cells from destruction by peroxides. Deficiency causes skeletal muscle degeneration, necrosis, calcification and liver pathologies. This may occur if animals are fed plants grown on selenium-deficient soils. Selenium toxicity may occur at relatively low levels.

self-awareness, self-recognition The capacity to be aware of one's own existence. *See also* **RED SPOT TEST**

self-directed behaviour Behaviours such as self-grooming, touching, or scratching. Their frequency may be used as an indicator of **ANXIETY** and **STRESS** in some species (especially non-human primates). A type of **DISPLACEMENT ACTIVITY**.

self-mutilation Self-inflicted harm, e.g. hitting the head against a wall, biting an arm, leg or tail. This sometimes occurs in captive animals, especially when kept in cages or enclosures which are too small or barren. *See also* **FEATHER PICKING**

self-narcotisation The act of inducing a condition in which the senses are dulled as a result of the release of **ENDORPHINS** from the brain. Some scientists have suggested that some **STEREOTYPIC BEHAVIOURS** may have this effect.

self-recognition *See* **SELF-AWARENESS**

Selous, Frederick Courteney (1851–1971) A British explorer, hunter and army officer who accompanied ex-president of the United States Theodore Roosevelt on his famous African **SAFARI** in 1909–1910. The Selous Game Reserve in southern Tanzania is named in his honour and is a UNESCO **WORLD HERITAGE SITE**. He wrote a number of books including *African Nature Notes and Reminiscences* (1908).

semantic communication The systematic use of **SIGNALS** by an animal to refer to different objects in the environment. *See also* **COMMUNICATION**, **INTERSPECIES SEMANTIC COMMUNICATION**, **SIGN LANGUAGE**

semelparity The condition of having a single reproductive episode during the lifetime of an animal. *Compare* **ITEROPARITY**

semen, seminal fluid A thick liquid that contains **SPERMATOZOA** produced by the **TESTES** and containing fluid from the **PROSTATE GLAND**. *See also* **SEX-SORTED SEMEN**

seminal fluid *See* **SEMEN**

sense organ A structure specifically evolved to detect stimuli in the environment, e.g. sound, light, chemicals etc.

sensitive period In animal development, an age-range when particular events are especially likely to affect the development of an individual, e.g. there is a sensitive period when young ducklings are particularly susceptible to **IMPRINTING** on other animals (normally the mother) or objects.

sensory enrichment The addition of sensory stimuli to an animal's enclosure as an enrichment: visual (e.g. television); auditory (recordings of **VOCALISATIONS** or music); olfactory (e.g. prey or predator faeces, **PHEROMONES**); taste and tactile stimuli. *See also* **AUDITORY ENRICHMENT**, **FELIWAY**

sensory neurone, sensory neuron A nerve cell which receives and carries information about stimuli detected by a sense organ to the **CENTRAL NERVOUS SYSTEM (CNS)**.

sensory receptor A sensory cell which responds to some change in an animal's internal or external environment by an alteration in its membrane voltage, thereby generating an **ACTION POTENTIAL**, e.g. a **CHEMORECEPTOR**, **MECHANORECEPTOR**.

sentience The capacity of an animal to have sensations, to perceive or feel.

sentinel *See* **SENTRY**. *See also* **ANIMAL SENTINELS**

sentry, sentinel An animal within a social group (e.g. meerkats (*Suricata suricatta*)) that is positioned at the edge of the group or on high ground and signals the presence of danger with an **ALARM CALL** (Fig. S5). *See also* **VIGILANCE BEHAVIOUR**

separation anxiety *See* **CANINE SEPARATION-RELATED BEHAVIOUR (SRB)**

separation area Part of an enclosure used to separate one or several individual animals from the others in a group. It may be used to isolate sick, injured or aggressive individuals, etc. *See also* **QUARANTINE**

sepsis *See* **SEPTICAEMIA**

septic Of a wound, contaminated with pathogenic bacteria. *See also* **SEPTICAEMIA**

septic shock, endotoxic shock A potentially fatal condition (possibly resulting in low blood pressure and multiple organ failure) caused by the combined effects of endotoxins produced by **BACTERIA** and the host's chemical mediators of **INFLAMMATION**. *See also* **TOXIC SHOCK SYNDROME**

septicaemia, sepsis, septicemia Blood poisoning. A serious condition caused by the presence of large numbers of bacteria or bacterial toxins in the blood. May be fatal. *See also* **SEPTIC SHOCK**

Serengeti A region of northern Tanzania, especially the Serengeti National Park, where the ecology of large mammals and **SAVANNA** ecosystems have been extensively studied. The area supports large numbers of lion, hyena, elephant, plains zebra, wildebeest, gazelles and other **LARGE MAMMALS**. *See also* **GRZIMEK**, **SCHALLER**

Serengeti Shall Not Die *See* **GRZIMEK**

Fig. S5 Sentry: a meerkat (*Suricata suricatta*) sentry.

Table S1 Sex determination in selected taxa.

Taxon	Sex chromosomes	
	Male	Female
Mammals	XY	XX
Duck-billed platypus (*Ornithorhynchus anatinus*)	XYXYXYXYXY	XXXXXXXXXX
Birds	ZZ	ZW
Some lizards	XY	XX
Other lizards and snakes	ZZ	ZW

seroconversion The development of **ANTIBODIES** in blood **SERUM** as a result of infection by disease organisms or **IMMUNISATION**. *See also* **SERONEGATIVE, SEROPOSITIVE**

seronegative Indicates the absence of specific **ANTIBODIES** that have been tested for in the blood **SERUM**. *Compare* **SEROPOSITIVE**

seropositive Indicates the presence of specific **ANTIBODIES** that have been tested for in the blood **SERUM**. *Compare* **SERONEGATIVE**

serotonin A **NEUROTRANSMITTER** which is involved in pain perception, mood, appetite, gastrointestinal function, sleep and male mating behaviour. It is found in **PLATELETS**, the gut and the **CENTRAL NERVOUS SYSTEM (CNS) (1)**.

serum, blood serum The clear fluid of the blood once the cells and clotting factors have been removed.

service dog A type of **ASSISTANCE ANIMAL** which has been trained to help people with visual or hearing impairment, a medical disorder (e.g. recurrent seizures) or a psychological disorder (e.g. post-traumatic stress disorder). In some cases it may carry medical equipment (e.g. an oxygen tank). *See also* **HELPER MONKEY**

service monkey *See* **HELPER MONKEY**

sessile Relating to animals, immobile, living attached to the substratum, e.g. corals, sponges, limpets. *Compare* **MOTILE**

seta (setae *pl.*) A hair-like structure found on the **EXOSKELETON** of **ARTHROPODS** and on the feet of geckos (Order: Squamata) which helps them to cling to smooth surfaces.

severe acute respiratory system (SARS) A new respiratory infection caused by a coronavirus known as SARS CoV, which is potentially fatal to humans. First discovered in November 2002 in the Guangdong Province of China and thought to have originated in animals. The disease has been found in Himalayan palm civets (*Paguma larvata*), a raccoon dog (*Nyctereutes procyonoides*) and a Chinese ferret badger (*Melogale moschata*). It has also been detected among people working in a live animal market in the area where the disease appears to have originated. High levels of antibody to the virus have been found in people trading in palm civets.

sex chromosomes The **CHROMOSOMES** that determine whether an individual is male or female and that carry the **GENES** associated with each sex. In most mammals males possess one X and one Y chromosome, while females possess two X chromosomes. *See also* **SEX DETERMINATION, SEX LINKAGE, SEX-SORTED SEMEN**

sex determination The establishment of the sex of an individual. Sex is determined by the types of **SEX CHROMOSOMES** present in the **SOMATIC CELLS**. The sex determination system varies between taxa (Table S1). In some turtles, lizards and crocodilians genes that control sex are temperature-dependent in a critical period during incubation. *See also* **SEXING ANIMALS, SEX-SORTED SEMEN**

sex hormone Any one of a number of hormones which is involved in reproduction and the development and functioning of the sex organs and secondary sexual characteristics. *See also* **OESTROGEN, TESTOSTERONE**

sex linkage The location of a gene on a **SEX CHROMOSOME**. In species where males have one X and one Y chromosome and the females have two X chromosomes, if a **RECESSIVE ALLELE** is located

S

on that part of the X chromosome which has no homologous section on the Y chromosome the male only needs to possess one copy of this allele in his GENOTYPE in order for the character it controls to be exhibited in his PHENOTYPE. The female would need to be HOMOZYGOUS recessive to exhibit the character. Such sex-linked recessive characters are consequently exhibited more in males than in females, e.g. ginger colour in domestic cats is sex-linked to the male.

sex organs The organs which make up the REPRODUCTIVE SYSTEM, e.g. OVARIES, FALLOPIAN TUBES, UTERUS, VAGINA, PENIS, TESTES

sex ratio The ratio of males to females in a POPULATION of animals.

sex skin The tissue around the genitals of female primates which swells under hormonal control when they are in OESTRUS.

sexed semen *See* SEX-SORTED SEMEN

sexing animals Determining whether an animal is male or female. In many species this can be achieved by observing the primary sexual characteristics (e.g. the presence of a penis or vagina). In sexually MONOMORPHIC SPECIES (3) sex may be determined by internal examination, e.g. using a LAPAROSCOPE or by DNA FINGERPRINTING. *See* SEX DETERMINATION, SEXUAL DIMORPHISM

sex-sorted semen, sexed semen Semen which contains a single type of sex chromosome. A chromosome-sorting technique is used in an attempt to control the sex of offspring produce by ARTIFICIAL INSEMINATION (AI). In mammals the male gametes determine the sex of the offspring. X chromosome-bearing and Y chromosome-bearing sperm from SEMEN can be separated to increase the probability of producing the desired sex.

sexual dichromatism Sex differences in the colour and markings on the fur, feathers etc. of a species.

sexual dimorphism The existence within a species of male and female individuals that have a distinctively different appearance due to the presence of distinguishing SECONDARY SEXUAL CHARACTERISTICS, e.g. different coloration, size, or the presence of TUSKS or ANTLERS in the male only. *See also* SEXUAL DICHROMATISM

sexual reproduction Reproduction which involves the production of HAPLOID gametes by MEIOSIS in both the male and the female of a SPECIES and the union of theses GAMETES (through FERTILISATION) to produce a ZYGOTE which is DIPLOID. *Compare* ASEXUAL REPRODUCTION, PARTHENOGENESIS

sexual selection The phenomenon whereby some individual animals are able to attract more mates than others by being more attractive to the opposite sex. This physical attractiveness may be linked to other attributes such as social dominance, health, etc. *See also* FITNESS, NATURAL SELECTION

sexual swelling *See* PERINEAL TUMESCENCE

sexually transmitted disease (STD) A disease which is passed from one animal to another as a result of sexual contact, e.g. bovine trichomoniasis.

shade An area or place that is not directly exposed to sunlight. *See* SUN SHADE

sham-chewing, vacuum-chewing Chewing with the mouth empty. A common STEREOTYPIC BEHAVIOUR of sows.

Shamu The first wild-captured killer whale (*Orcinus orca*) to survive more than 13 months, and the star of a show at *SeaWorld San Diego* in the mid 1960s. The name *Shamu* is now used in other killer whale shows.

Shape of Enrichment A journal and website which publishes articles on ENVIRONMENTAL ENRICHMENT and enrichment techniques for a variety of taxa.

shaping TRAINING an animal to perform a particular act by the progressive reinforcement of behaviours performed in the direction of the act. For example, to train a pigeon to push a button, it would first be rewarded for approaching the button, then rewarded for touching it, and then again for pressing it. *See also* OPERANT CONDITIONING

sharing The joint possession or apportioning of a resource between several individuals. Chimpanzees (*Pan troglodytes*) share tools and food. This behaviour may assist in maintaining group cohesion. Male chimps appear to share meat with females in return for sex. *See also* RECIPROCAL ALTRUISM. *Compare* CHEATING

shark finning The practice of removing the fins of a shark and discarding the rest of the fish at sea. Fins are used to make shark fin soup and in TRADITIONAL MEDICINES.

shark net A submerged net placed around the coastline near beaches to reduce the possibility of shark attacks on human swimmers. Important in protecting the tourism industry in some countries, e.g. South Africa.

shark sanctuary A protected area of marine water where it is illegal to fish for sharks, e.g. the waters around the island of Raja Ampat in Indonesia.

shearling
1. A sheep that has been sheared (shorn) once, usually a YEARLING.
2. The tanned skin from a sheep that has been sheared only once, with the wool left on.

Shedd Aquarium A major aquarium in Chicago named after businessman John G. Shedd. It opened in 1930 and was the first inland aquarium with a permanent saltwater collection.

sheep
1. An animal belonging to the genus *Ovis*.
2. A domesticated sheep (*Ovis aries*). A ruminant which is widely kept for its meat and wool.

sheep breeds A wide range of sheep breeds have been produced for their wool and meat (Fig. S6, Table S2).

Fig. S6 Sheep breeds: Greyface Dartmoor (top left); Oxford Down (top right); Wensleydale (middle left); Shetland (middle right); Hebridean (bottom left); Jacob (bottom right).

S

sheep dog trials A competition where owners of sheep dogs compete against each other in various activities, especially herding sheep into **CORRALS (1)**.

sheep shearing The removal of the fleece of a sheep, usually in one piece (Fig. S7). *See also* **SHEARLING**

sheep worrying *See* **WORRYING LIVESTOCK**

Sheldrick, Daphne (1934–) The wife of the late David Sheldrick (the first warden of **TSAVO** East National Park, Kenya) who established an orphanage for wild animals (especially elephants) and has returned many individuals to the wild. It was originally at Voi but is now located in Nairobi National Park. Dame Daphne perfected a milk formula for rhino and elephant calves. She has written several books, including *The Orphans of Tsavo* (1967).

shift cage A cage, which is either temporary or permanent, in which an animal may be held while moving it between one enclosure and another; sometimes used for night holding, e.g. in felids.

Shire horse A tall breed of draught horse, which was often used for pulling heavy carts, ploughs and other machinery, and particularly popular for pulling brewery wagons (Fig. S1).

shock, clinical shock A state of extreme collapse resulting from a serious failure of the circulatory system rendering it unable to supply vital organs, and characterised by low blood pressure and low temperature. It may occur as a result of HAEMOR-RHAGE, CORONARY THROMBOSIS, or extreme emotional disturbance. *See also* ANAPHYLACTIC SHOCK, CAPILLARY REFILL TIME, SEPTIC SHOCK, TOXIC SHOCK SYNDROME

shock collar *See* ELECTRONIC TRAINING COLLAR

shooting estate An area of land in the countryside – usually moorland – which is managed to encourage GAME BIRD species (e.g. grouse, pheasants) that are shot for sport. Shooting estates make an important contribution to the economy of some rural areas, e.g. the highlands of Scotland. Birds are

managed by a GAMEKEEPER who may supplement populations by CAPTIVE BREEDING.

shoulder girdle *See* PECTORAL GIRDLE

siblicide *See* FRATRICIDE

sibling A brother or sister. On average, such individuals have half of their genes in common. An individual may increase his or her FITNESS by helping siblings. *See also* ALTRUISM

sibling inhibition In relation to sexual behaviour, the tendency to avoid mating with a sibling. This can also occur between individual animals that have been reared together from a young age but are not related.

Sierra Club The largest and most influential grassroots environmental organisation in the United States, which works to protects communities, wild places and the planet. Founded in 1892 by John Muir.

sign
1. A clinical sign. In veterinary medicine, a behaviour or physical characteristic displayed by an animal which assists in the diagnosis of its condition, e.g. an abnormal gait, a discharge from the eye. *Compare* SYMPTOM
2. Signage: a displayed structure bearing symbols or written information used to indicate the location of exhibits in a zoo or other facility, exits, visitors services, etc., and to provide information about species and their conservation. *See also* INTERACTIVE SIGN, WAYFINDING
3. A gesture in SIGN LANGUAGE.

sign language A method of communication using hand movements and GESTURES. Some primates are capable of learning sign language devised by humans (e.g. American Sign Language). Wild bonobos (*Pan paniscus*) have their own system of hand signals.

sign stimulus, releaser An external stimulus (sign) from one animal to another which triggers a FIXED ACTION PATTERN. The red spot on the bill of an adult lesser black-backed gull (*Larus fuscus*) acts as a sign stimulus which elicits a fixed action pattern in

Table S2 Major sheep breeds.

Beulah Speckled-face	Herdwick
Black Welsh Mountain	Hill Radnor
Blackface	Jacob
Bluefaced Leicester	Kerry Hill
Border Leicester	Lleyn
Brecknock Hill Cheviot	Lonk
Cheviot	North Country Cheviot
Clun Forest	Romney
Dalesbred	Rough Fell
Derbyshire Gritstone	Shetland
Devon and Cornwall Longwool	South Wales Mountain
Devon Closewool	Southdown
Dorset Down	Suffolk
Dorset Horn	Swaledale
Exmoor Horn	Welsh Hill Speckled Face
Greyface Dartmoor	Welsh Mountain
Hampshire Down	White Face Dartmoor

S

Fig. S7 Sheep shearing.

Fig. S8 Sign stimulus: the red spot on the bill of a lesser black-backed gull (*Larus fuscus*) acts as a sign stimulus to its young.

the chick which consists of pecking the spot (Fig. S8). This stimulates the adult to regurgitate food.

signage *See* **SIGN (2)**

signal A message which may take the form of a **GESTURE**, **VOCALISATION**, scent or other form of communication between individuals. *See also* **FUNCTIONAL REFERENCE, SCENT-MARKING**

signalling The process of indicating one's status or intentions to another animal by the use of **SIGNALS**.

signature whistle A unique **VOCALISATION** made by an individual dolphin that broadcasts its identity to other dolphins in a **POD**. In effect the whistle functions as a name. When a particular dolphin calls out the signature whistle (name) of another dolphin the animal whose signature whistle is being called will call back. These signature whistles appear to help members of a pod to identify each other and stay together.

silage Green fodder made from grass crops and used as animal (**RUMINANT**) feed in winter, which has been compacted and stored in airtight conditions without being dried. Often stored in a **SILO**. *See also* **ALKALAGE**

Silent Spring A book published in 1962 by the American biologist Rachel **CARSON** which drew the attention of the public to the damage to human health and the environment being done by the overuse of **PESTICIDES**, especially **DDT**. The title is a reference to the disappearance of the sound of birdsong from the American countryside.

siliceous Consisting of silica. Some sponges (**PORIFERA**) have a siliceous skeleton.

silo A container used for the bulk storage of grain, animal feed and other materials. It usually consists of a large metal container suspended above the

Fig. S9 Silo: a food silo used to store animal feed.

ground with a funnel-shaped base through which the contents are removed (Fig. S9).

Siluriformes An order of fishes: catfishes and knifefishes.

silverback A large dominant male gorilla in a group, so named because of the silver-coloured hairs on his back (Fig. K3). *See also* **DOMINANT INDIVIDUAL**

simian
1. An **APE** or **MONKEY**.
2. Relating to an ape or monkey.

Singer, Peter (1946–) Peter Singer is an Australian philosopher and currently professor of Bioethics at Princeton University. He is the author of ***ANIMAL LIBERATION*** (1975) which is perhaps the most influential book on animal rights. Singer is a supporter of the **GREAT APE PROJECT** which is an international organisation of primatologists, psychologists, philosophers and others who advocate a UN Declaration of the Rights of Great Apes as a precursor to demanding the release of apes from captivity in research laboratories and elsewhere.

Single Population Analysis and Record Keeping System (SPARKS) Software which is used to manage studbook datasets. It calculates the genetic relationship between individuals in a population (**MEAN KINSHIP**). This helps zoos to decide which animals to use for future matings in order to avoid **INBREEDING** and maximise genetic diversity within the zoo population of a species.

singleton
1. A single animal.
2. An individual animal born on its own rather than as one of a multiple birth.

sinoatrial (SA) node The pacemaker of the heart. A group of modified cardiac muscle cells in the right atrial wall which initiate contraction of the atrial muscles when the heart beats. *See also* **ATRIOVENTRICULAR (AV) NODE**

sinus
1. A cavity or sac which naturally occurs in an organ or tissue in the body, e.g. the carotid sinus.
2. A similar structure caused by tissue destruction.
3. The paranasal sinuses: air cavities in the cranial bones, especially near the nose.

sinusitis **INFLAMMATION** of the paranasal **SINUSES** (3) which may result in a discharge. It may be due to bacterial infection, or secondary due to dental disease, facial trauma or other conditions.

sire
1. Father.
2. In relation to horses, according to the **AMERI-CAN STUD BOOK** (2008), *'A male horse that has produced, or is producing, foals.'*

Sirenia An order of mammals: dugong, sea cow, manatees.

site fidelity The tendency of an individual animal to remain in one locality.

Site of Special Scientific Interest (SSSI) A **PRO-TECTED AREA** in Great Britain designated under the **WILDLIFE AND COUNTRYSIDE ACT 1981** where there is a special conservation interest by virtue of the presence of rare animals, plants, geology etc.

sixth extinction A term sometimes used for the current **MASS EXTINCTION** of species.

skein A group of geese.

skeletal muscle, striated muscle, striped muscle, voluntary muscle Muscle which is striped in appearance under the microscope and is attached to the bones of the body by **TENDONS**. It is responsible for movement of the skeleton and is under the control of the **SOMATIC NERVOUS SYSTEM (SNS)**.

skeletal system The collective term for the bones of the body and the **TENDONS**, **LIGAMENTS (1)** and **CARTILAGE** that connect them. The teeth are also part of this system. *See also* **SKELETON**

skeleton A bony and/or cartilaginous framework which supports the body and protects the internal organs in vertebrates. *See also* **APPENDICULAR SKELETON, AXIAL SKELETON, EXOSKELETON**

skin The outer covering of soft tissue in vertebrates. Important in protecting the body from infection and preventing water loss in terrestrial species. *See also* **EPIDERMIS**. *Compare* **EXOSKELETON**

skin stapler, stapling device A machine used for the repair of surgical wounds in the lungs, intestine, liver, and also for stapling skin to close wounds. *See also* **SUTURE (1)**

Skinner box An experimental apparatus used in **LEARNING** experiments involving animals such as rats, mice and pigeons. Named after the experimental psychologist B. F. **SKINNER**. Skinner boxes provide the subject with an opportunity to learn a new behaviour (response), e.g. operating a lever or pressing a button, as a result of **REINFORCEMENT** (e.g. a food, water). The box offers a highly controlled environment where a researcher might address questions such as 'What schedule of reinforcement results in the highest rate of response?'

Skinner, Burrhus Frederic (B. F.) (1904–1990) An American experimental psychologist who conducted research on **OPERANT CONDITIONING** and invented the **SKINNER BOX**. He was one of the founders of **BEHAVIOURISM**.

Skippy the Bush Kangaroo An Australian 1960s children's television series starring a female Eastern Grey kangaroo called *Skippy* who was befriended by the son of the Head Ranger (played by Ed Devereaux) of a fictional national park near Sydney.

skittish Nervous, unpredictable, difficult to handle, lively, playful, moving quickly and lightly.

slat-chewing Nibbling at the edges of a wooden slat and ingesting small pieces of wood. An abnormal behaviour especially of sheep.

slaughterhouse A place where **LIVESTOCK** are slaughtered for food. *See also* **KNACKER, RELI-GIOUS SLAUGHTER**

slaughterman A person who kills animals for use as food, and for other animal products, in a slaughterhouse. *See also* **KNACKER, RELIGIOUS SLAUGH-TER, WELFARE OF ANIMALS (SLAUGHTER OR KILLING) REGULATIONS 1995**

sled dog racing Races in which dogsled drivers (mushers) compete for thousands of dollars and other prizes by racing sleds pulled by teams of 15 dogs. The Iditarod Trail Sled Dog Race is an annual race from Anchorage to Nome, Alaska. About 1500 dogs start the race each year but more than one third have to be flown out due to illness, injury or exhaustion. In extreme cases animals have died. Many suffer from lung damage.

sleep A reversible state of natural unconsciousness; a type of **DORMANCY** found in mammals and birds in which the brain exhibits characteristic patterns of electrical activity. *See also* **DREAMING, RAPID EYE MOVEMENTS (REM)**

Slimbridge The headquarters of the **WILDFOWL AND WETLANDS TRUST (WWT)**, located in Gloucestershire, UK (Fig. S10). Established by Peter **SCOTT**. *See also* **NĒ NĒ (***BRANTA SANDVICENSIS***)**

slipped disc *See* **INTERVERTEBRAL DISC**

sloughing The process that occurs when **NECROTIC** skin and other **TISSUE** becomes separated from the body, e.g. as in a wound or sore, the loss of fur, or the loss of tissue from the **ENDOMETRIUM** during **MENSTRUATION**.

slower-growing broiler breeds Breeds of **BROILER** chickens that are specifically bred to grow more

S

Fig. S10 Slimbridge, UK, headquarters of the Wildfowl and Wetlands Trust (WWT).

slowly than the fast-growing breeds favoured by the poultry industry. **SELECTION** for fast growth has contributed to welfare problems including chronic leg disorders, **ASCITES** and sudden death syndrome. Genetically slower-growing breeds include Hubbard JA57 and CobbSasso 150.

slum zoo A term used to refer to very poorly run zoos where animals are kept in barren cages and enclosures with little or no veterinary care and poor nutrition. The existence of many such zoos, particularly in eastern Europe, was a major impetus to the promulgation of the **ZOOS DIRECTIVE** by the EU.

small intestine The section of the gut located between the **STOMACH** and the **LARGE INTESTINE** where a considerable amount of digestion and absorption occurs. The inner surface is covered in **VILLI**. The small intestine is made up of the duodenum (which leads from the stomach), jejunum and ileum (which joins the **LARGE INTESTINE**).

small mammal A non-taxonomic term for mice, voles and other taxa of small (generally terrestrial) mammals. *Compare* **LARGE MAMMAL**

smart door A door that has been programmed to open only when approached by particular animals that are identified by **RADIO FREQUENCY IDEN-TIFICATION (RFID) TECHNOLOGY**. May be used in a zoo to allow access to certain areas to selected individual animals. *See also* **AUTOMATED CAT FLAP**

Smithsonian Institution, Smithsonian The world's largest museum and research complex which includes 19 museums and galleries (including the Natural History Museum) and the National Zoological Park, located in Washington DC, USA.

Smithsonian National Zoological Park A zoo in the USA which was created by an Act of Congress in 1889. It has two facilities: a 163 acre (66 ha) zoological park in northwest Washington DC, which is open to the public, and a non-public 3200 acre (1295 ha) Conservation and Research Center in Fort Royal, Virginia, which was established in 1975. The zoo

became part of the **SMITHSONIAN INSTITUTION** in 1890. Plans for the Zoo were drawn up by three men: William Temple **HORNADAY** (a conservationist and head of the Vertebrate Division at the Smithsonian Institution), Frederick Olmsted (the premier landscape architect of the day) and Samuel Langley – the third Secretary of the Smithsonian. In the early 1960s the zoo began to focus on the study and breeding of endangered species and a research division was created.

smolt A young salmon before it migrates from freshwater to the sea.

smooth muscle, involuntary muscle, non-striated muscle A type of muscle found in and responsible for the contractility of hollow organs, e.g. in **BLOOD VESSELS**, the **GASTROINTESTINAL TRACT**, the **BLADDER**, the **UTERUS** etc. It is primarily under the control of the **AUTONOMIC NERVOUS SYSTEM (ANS)**.

snails *See PARTULA* **SNAILS**

snake bag A bag designed for carrying snakes safely after capture.

snake farm A facility where snakes are kept and bred for research, the collection of **VENOM** and the development of **ANTIVENIN**. Often also functions as a tourist attraction. Snake farms occur in a number of countries including the USA, India, Thailand and many African states.

snake hook A hand-held device consisting of a pole terminating in a hook used to capture and safely handle snakes at a distance.

snare A loop of wire used to catch wild animals when they step into the loop and pull it tight around one of their legs, neck or other part of the body (Fig. S11). It is usually attached to a tree, post or other static object to prevent escape. Snares are widely used by **POACHERS**.

sneaky mating The mating of a low-ranking male with a female in a group to whom he would not normally gain access because of his **RANK**. May occur within the territory of another male. *See also* **DOMINANCE HIERARCHY**

sniffer dog A dog which has been trained to respond to the odour of a particular substance or organism. Such dogs are used in New Zealand by the Department of Conservation to locate kiwis (*Apteryx* spp.) during population surveys. Dogs have also been trained to find bumblebee nests (*Bombus* spp.) in the UK, track armadillos (Dasypodidae) in Brazil and cheetah (*Acinonyx jubatus*) in South Africa.

social behaviour The **BEHAVIOUR** that occurs between individuals within a social group. *See also* **AGONISTIC BEHAVIOUR, DOMINANCE HIER-ARCHY, SOCIAL ORGANISATION, SOCIOBIOLOGY**

social enrichment Contact between animals and humans or conspecifics which is either direct or indirect (auditory, olfactory or visual) and functions as an enrichment.

social facilitation The process by which an animal in a group is induced to perform a behaviour after

S

Fig. S11 Snare: a European lynx (*Felis lynx*) caught in a snare. The lynx is being held by veterinary surgeon Andrzej Fedaczyński DVM of the Rehabilitation Centre for Protected Animals in Przemysl, Poland. (Courtesy of Jakub Kotowicz.)

observing its performance by others. This often results in individuals within a group co-ordinating their behaviour so that they, for example, feed at the same time, with some degree of mutual stimulation.

social group A collection of individuals of the same species who live together. Some groups have little or no structure and are simple **AGGREGATIONS** (e.g. flocks of birds) while others exhibit a **DOMINANCE HIERARCHY** (e.g. primate societies). *See also* **BACHELOR GROUP**

social learning **LEARNING** which occurs in a social context, about the dominance status, abilities and **MOTIVATION** of others. This learning may be reinforced by the receipt of food or **GROOMING** and may result in emulation and **IMITATION**. *See also* **SOCIAL FACILITATION**

social organisation The social structure which results from the interactions between individuals of the same species and their social relationships. All species exhibit some degree of social organisation, from those in which individuals only come together to mate or merely form **AGGREGATIONS**, to colonial insects and those that form a complex **DOMINANCE HIERARCHY** such as many primates.

African elephants (*Loxodonta africana*) have a complex social system consisting of females and their offspring, family groups (Fig. S12), bond groups and clans (Fig. S13).

socialisation The process by which young animals learn social behaviour and how to function in a group.

society A group of related individuals of the same species which interact extensively with each other and exhibit **SOCIAL ORGANISATION**, which may include a **DOMINANCE HIERARCHY**. *See also* **COLONY**

Society and Animals Journal An academic journal concerned with human–animal studies which is published by the **ANIMALS AND SOCIETY INSTITUTE**.

Society for Companion Animal Studies A UK charity which promotes the study of human–companion animal interactions and raises awareness of the importance of pets in society. It was established in 1979 and works in partnership with the **BLUE CROSS**.

Society for Conservation Biology An international organisation which promotes the scientific study of the phenomena that affect the maintenance, loss and restoration of **BIODIVERSITY**. It publishes the journals *Conservation Biology* and *Conservation Letters*.

Society for the Acquisition of Animals *See* **SCHÖNBRUNN ZOO**.

Society for the Prevention of Cruelty to Animals An organisation founded in England in 1824 to promote the humane treatment of working animals, such as horses and cattle, and household pets, two years after the passing of **MARTIN'S ACT**. It later became the **ROYAL SOCIETY FOR THE PREVENTION OF CRUELTY TO ANIMALS (RSPCA)**.

Society for the Protection of Animals Liable to Vivisection The first anti-vivisection society in the world. Founded in England by Frances Power **COBBE** in 1875 and renamed the Anti-vivisection Society in 1897. It was instrumental in the passing of the **CRUELTY TO ANIMALS ACT 1876**.

Society for the Protection of Birds An organisation founded in Manchester, England, in 1889 specifically to stop the slaughter of thousands of birds every year for the plumes used in ladies' fashions, especially egrets, herons and birds of paradise. Branches were established overseas and the branch in India secured the first legal measure against the **PLUMAGE** trade in 1902, which banned the export of bird skins and feathers. The society became the **ROYAL SOCIETY FOR THE PROTECTION OF BIRDS (RSPB)** in 1904.

sociobiology '*The systematic study of the biological basis of all social behavior*' (**WILSON**, 1980).

sociogram A diagram, used in studies of animal behaviour, that indicates the relationships between individuals within a social group by drawing lines

S

Fig. S12 Social organisation: a family group (females and their young) of African elephants (*Loxodonta africana*) in Tarangire National Park, Tanzania.

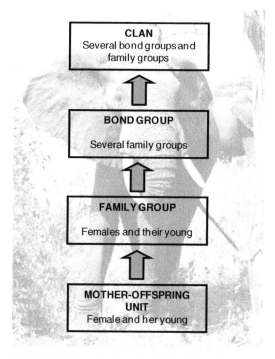

| CLAN |
| Several bond groups and family groups |

⬆

| BOND GROUP |
| Several family groups |

⬆

| FAMILY GROUP |
| Females and their young |

⬆

| MOTHER-OFFSPRING UNIT |
| Female and her young |

Fig. S13 Social organisation: levels of social organisation in the African elephant (*Loxodonta africana*). Based on information in Poole (1996).

connecting those individuals which associate with each other. The strength of the association may be calculated using an **ASSOCIATION INDEX** and indicated by the thickness of the line (Fig. S14).

sodium A mineral nutrient. It helps to maintain electrolyte balance and osmotic pressure. Deficiency may result in salt craving. It is most likely when lactating, growing or working. Sodium toxicity can occur if the animal is not given access to water.

soft release The release of an animal into a holding pen at a release site so that its progress may be monitored prior to its release into the wild. *See also* **PRECONDITIONING**. *Compare* **HARD RELEASE**

soft tissue injury Damage sustained to muscles, tendons and ligaments, often the result of a sprain, strain or blow which causes a **CONTUSION**.

Soil Association A UK charity, founded in 1946, which campaigns for healthy, humane and sustainable food, farming and land use. It promotes the use of **ORGANIC FARMING** methods.

solar bird scarer A rotating device powered by a solar panel which reflects sunlight to deter wild birds. Found attached to the top of aviaries. *See also* **BIRD SCARER**

somatic Of or relating to the body.

somatic cell A body **CELL**, i.e. one which has the **DIPLOID** number of **CHROMOSOMES**.

somatic cell nuclear transfer (SCNT) *See* **NUCLEAR TRANSFER**

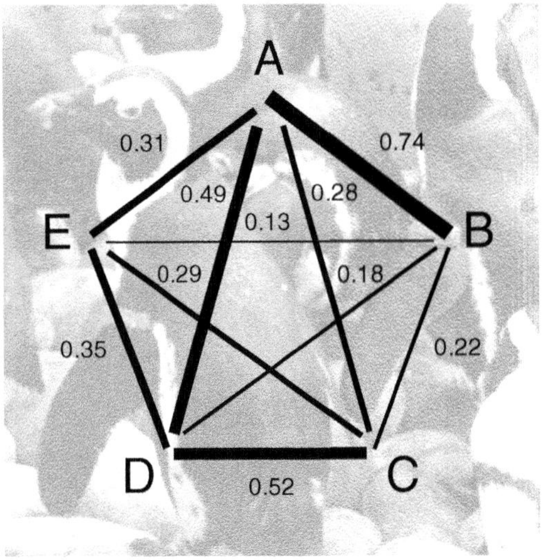

Fig. S14 Sociogram: hypothetical sociogram for five animals (A–E). Each number is the value of the association index for the dyad joined by the line.

somatic nervous system (SNS), voluntary nervous system That part of the **PERIPHERAL NERVOUS SYSTEM (PNS)** which regulates body movements via the control of **SKELETAL MUSCLES** and also receives information from the sense organs.

somatotrophin *See* **GROWTH HORMONE (GH)**

song A complex **VOCALISATION** made by an animal (e.g. whale, seal, bird, gibbon) as a form of **COMMUNICATION**. It may be analysed using a **SOUND SPECTROGRAM**. *See also* **SIGNATURE WHISTLE**

songbird A bird belonging to the suborder Passeri of the order **PASSERIFORMES**.

sonogram *See* **SOUND SPECTROGRAM**

sonograph *See* **SOUND SPECTROGRAM**

sound lure, tape lure A recording of an animal's **VOCALISATIONS** (e.g. bird song) used to attract it into the open, closer to the operator of the lure, or into a net. Used by some **BIRDERS** to attract birds. Especially useful in forest habitats. Used by some bird tour leaders to attract birds for customers.

sound pressure level meter An electronic device used for measuring sound pressure (in **DECIBELS**). It may be used in noise pollution studies and studies of communication in animals.

sound spectrogram, sonogram A visual representation of sound. A graph which shows the changes in spectral density with time, produced by an instrument called a sonograph. The horizontal axis represents time, the vertical axis represents frequency and colour may be used to represent the amplitude of a particular frequency at a particular time. Often used to study the different types of **VOCALISA-**

TIONS produced by a species, e.g. bats, whales. *See also* **SIGNATURE WHISTLE**, **SONG**

sow stall An individual stall for a single female pig. Banned by the EU **WELFARE OF PIGS DIRECTIVE**, which requires sows to be kept in social groups for welfare reasons, except for short periods during **FARROWING**.

SowCheck An automated **OESTRUS** detection system for pigs kept in group housing with electronic sow feeding stations and for sows in free-movement pens. Sows are analysed and automatically selected while visiting the feeding station.

sp. Abbreviation for a species. After the name of a **GENUS** it indicates a single unspecified species, e.g. *Panthera* sp. *Compare* **SPP.**

space blanket A thin sheet of metal used to cover animals to prevent heat loss.

SPARKS *See* **SINGLE POPULATION ANALYSIS AND RECORD KEEPING SYSTEM (SPARKS)**

sparring Practice fighting that occurs in some animal species, where there is no intention to cause harm, particularly between male **UNGULATES**, e.g. pronghorn antelope (*Antilocapra americana*). It may be used to test the strength of opponents and forms part of **PLAY** in some species.

spasm An involuntary **MUSCLE** contraction which is often painful.

spawn The eggs of aquatic animals, e.g. fishes, amphibians, molluscs etc.

spawning The process of producing **SPAWN**.

spaying Removal of the ovaries.

Special Area of Conservation (SAC) A **PROTECTED AREA** in the **EUROPEAN UNION (EU)**, designated under the **HABITATS DIRECTIVE**. *See also* **NATURA 2000**

Special Protection Area (SPA) A **PROTECTED AREA** for birds in the **EUROPEAN UNION (EU)**, designated under the **WILD BIRDS DIRECTIVES**. *See also* **NATURA 2000**

Specialist Group An association of experts within the **SPECIES SURVIVAL COMMISSION (SSC)** of the **INTERNATIONAL UNION FOR THE CONSERVATION OF NATURE AND NATURAL RESOURCES (IUCN)** who address conservation issues related to particular groups of plants or animals (e.g. ants, swans, Asian elephants, bears, canids) or focus on topical issues, such as the **REINTRODUCTION** of species into their former habitats or wildlife health. *See also* **CONSERVATION BREEDING SPECIALIST GROUP (CBSG)**

speciation The process by which new **SPECIES** are created during the course of **EVOLUTION**. This often occurs as a result of the geographical separation and isolation of two populations of the same species which subsequently evolve in different directions until they cannot produce viable offspring if interbred (genetic isolation). *See also* **ALLOPATRIC SPECIATION**, **PARAPATRIC SPECIATION**, **SYMPATRIC SPECIATION**

S

species
1. A definition of a species described by Ernst **MAYR** (1963) is *'groups of interbreeding natural populations that are reproductively isolated from other such groups.'* This is known as the biological species concept. *See also* **MORPHOLOGICAL SPECIES, NOMINATE SUBSPECIES, TYPE SPECIMEN**
2. In **TAXONOMY**, a subdivision of a **GENUS** (Table C5).
3. In legal instruments the term 'species' is often used in a general sense to include species, **SUBSPECIES** and **POPULATIONS** of a species.
4. In the United States the **ENDANGERED SPECIES ACT OF 1973 (USA),** section 3(16) states that, *'The term "species" includes any subspecies of fish or wildlife or plants, and any distinct population segment of any species of vertebrate fish or wildlife which interbreeds when mature'*.
5. In US Executive Order 13112 of February 3, 1999 a species is defined as *'. . . a group of organisms all of which have a high degree of physical and genetic similarity, generally interbreed only among themselves, and show persistent differences from members of allied groups of organisms'*.
6. In the **ANIMALS ACT 1971** (in relation to England and Wales), *'"species" includes sub-species and variety'*.
7. Council Regulation (EC) No 338/97 of 9 December 1996 on the protection of species of wild fauna and flora by regulating trade therein, Art. 2 defines the term *'species'* as meaning *'a species, subspecies or population thereof'*.

Species 2000 and ITIS Catalogue of Life A project whose goal is to create a validated checklist of all the world's species.

species diversity The number of different species present in a sample or geographical area, weighted by some measure of abundance (e.g. the number of individuals). Sometimes used synonymously with **SPECIES RICHNESS**. *See also* **ALPHA (α) DIVERSITY, BETA (β) DIVERSITY, GAMMA (γ) DIVERSITY**

species richness The number of different species present in a sample or geographical area. *Compare* **SPECIES DIVERSITY**

Species Survival Commission (SSC) The largest of the six volunteer commissions of the **INTERNATIONAL UNION FOR THE CONSERVATION OF NATURE AND NATURAL RESOURCES (IUCN)**, with a global membership of 7500 experts from almost every country in the world. It advises the IUCN and its members on the wide range of technical and scientific aspects of species conservation. Most members are deployed in more than 100 Specialist Groups and Task Forces which address conservation issues related to particular groups of animals or plants or topical issues such as **REINTRODUCTIONS** into the wild or wildlife health. The Commission produces a series of technical guidelines, e.g. *Guidelines for Reintroductions; Management of Ex-situ Populations for Conservation; Guidelines for the Prevention of Biodiversity Loss Cause by Alien Invasive Species; IUCN Red List Categories and Criteria.*

Species Survival Plan® (SSP) Populations An **ASSOCIATION OF ZOOS AND AQUARIUMS (AZA)** population management level which focuses on species that require the greatest conservation effort and requires members to co-operatively breed them. *See* **SPECIES SURVIVAL PLAN® (SSP) PROGRAMS**

Species Survival Plan® (SSP) Programs Co-operative breeding programmes in North America are managed by the **ASSOCIATION OF ZOOS AND AQUARIUMS (AZA)** and are called Species Survival Plan® (SSP) Programs and **POPULATION MANAGEMENT PLAN PROGRAMS**. These are managed by the AZA Wildlife Conservation Committee through the work of AZA **TAXON ADVISORY GROUPS (TAGS)**. The SSP species are often 'FLAGSHIP **SPECIES**'. Each SSP Program is responsible for producing a Master Plan. This identifies population management goals and recommendations to ensure the sustainability of a healthy, genetically and demographically diverse population (Table S3).

speciesism The view held by some animal rights supporters that discriminating against animals is wrong and the animal equivalent of racism in humans. The term was first coined by Richard **RYDER**.

species-specific Relating to one particular **SPECIES**, e.g. species-specific behaviour.

species-typical behaviour Behaviour which is normally exhibited by a particular species, especially behaviour exhibited in the natural environment.

specimen
1. An individual organism.
2. A part of an organism, e.g. a specimen of hair. The term is widely used in legal documents to mean any organism or part thereof.

spelt glumes The husks of spelt (a cereal) sometimes used as **LITTER (1)** for animal houses.

sperm *See* **SPERMATOZOON**

sperm bank A place where frozen sperm are stored for later use in **ARTIFICIAL INSEMINATION (AI)**. *See also* **FROZEN ZOO**

sperm competition The competition for fertilisation that occurs between **SPERMATOZOA** from different males in the female reproductive system when she mates with more than one individual. This may be reduced by **MATE GUARDING**. *See also* **SPERM PLUG**

sperm donor A male animal who provides sperm which is later used to inseminate a female of the same species (**ARTIFICIAL INSEMINATION (AI)**) possibly after freezing. The sperm may be collected by manual stimulation of the penis or massage, by the use of an **ELECTRO-EJACULATOR** or an **ARTIFICIAL VAGINA (AV)**. *See also* **FROZEN ZOO**

sperm plug, copulation plug, mucus plug, sphragis, vaginal plug A plug of gelatinous material that a male deposits in a female's **VAGINA, CLOACA (1)**

S

Table S3 Species Survival Plan® (SSP) Programs: examples of taxa included in the Species Survival Plan® (SSP) Programs of the Association of Zoos and Aquariums (AZA).

Koala, Queensland	Oryx, Scimitar-Horned
Leopard, Amur	Okapi
Leopard, Snow	Zebra, Grevy's
Tiger	Babirusa
Jaguar	Hippopotamus, Pygmy
Ocelot	Rhinoceros, Black
Cat, Fishing	Rhinoceros, Indian
Cat, Sand	Tapir, Malayan
Tamarin, Golden Lion	Bat, Rodrigues Fruit
Monkey, Goeldi's	Whale, Beluga
Lemur, Ring-Tailed	Kiwi, North Island Brown
Lemur, Ruffed	Mynah, Bali
Colobus	Ibis, Waldrapp
Mangabey	Condor, California
Monkey, Spider	Condor, Andean
Gibbon	Crane, White-Naped
Chimpanzee	Crane, Wattled
Bonobo	Parrot, Thick-Billed
Gorilla, Western	Peafowl, Congo
Elephant	Penguin, African
Anteater, Giant	Penguin, Humboldt
Wolf, Maned	Rail, Guam
Wolf, Red	Dragon, Komodo
Fox, Fennec	Iguana, Fiji Island Banded
Fox, Swift	Tortoise, Burmese Star
Dog, African Wild	Snake, Louisiana Pine
Otter, Asian Small-Clawed	Crocodile, Cuban
Bear, Polar	Alligator, Chinese
Bear, Andean	Toad, Puerto Rican Crested
Panda, Giant	Toad, Wyoming
Panda, Red	Cichlids, Lake Victoria
Bongo, Eastern	Snail, Partula
Gazelle, Speke's	Beetle, American Burying
Oryx, Arabian	

or other reproductive organ after mating to prevent other males from successfully mating with her, as a tactic in reproductive competition. Found in some mammals, reptiles, insects and spiders. *See also* ANTIAPHRODISIAC, SPERM COMPETITION

spermatogenesis The production of SPERMATOZOA in the TESTES by MEIOSIS followed by SPERMIOGENESIS.

spermatozoon, sperm (spermatozoa, sperm, sperms *pl.*) The male gamete. A motile haploid cell, which is flagellated in most species.

spermiogenesis The final stage in the production of sperm, when spermatids develop into mature, motile SPERMATOZOA.

Sphenisciformes An order of birds: penguins.

spheroidal joint *See* BALL AND SOCKET JOINT

sphragis *See* SPERM PLUG

SPIDER A model framework for assessing behavioural husbandry which may be used to assess the effectiveness of a TRAINING or enrichment programme. The acronym stands for: **S**etting goals, **P**lanning, **I**mplementing, **D**ocumenting, **E**valuating, **R**eadjusting.

spider-goat A TRANSGENIC ORGANISM consisting of a goat which contains spider genes and produces spider silk protein in its milk. This can be purified and spun into fibres to make artificial TENDONS, medical SUTURES (1) and a wide variety of other materials. The goat was developed by scientists at Utah State University.

spinal column *See* VERTEBRAL COLUMN

spinal cord Part of the central nervous system which consists of a tube of nerve fibres and NEURONES which is enclosed within the vertebral column in vertebrates. They connect with the brain and extend outwards to all parts of the body as the spinal nerves.

spinal disc herniation *See* INTERVERTEBRAL DISC

spindle cells, von Economo neurones Specialised NEURONES found in the brains of humans, great apes, elephants and cetaceans. Associated with high-level cognitive abilities such as the ability to experience EMOTION. *See also* COGNITION

spine *See* VERTEBRAL COLUMN

spiracle
  1. A small pore on the thoracic and abdominal segments of the body of an insect where the tracheae open to the outside. *See also* TRACHEA (2)
  2. An opening behind the eye of some fishes (rays, sharks and some bony fishes) which is a VESTIGIAL gill slit.

spleen An organ found in the abdomen of most vertebrates which is effectively a large LYMPH NODE. It removes old ERYTHROCYTES from the body, recycles iron, acts as a store of blood and is important in IMMUNITY as it synthesises ANTIBODIES. *See also* LYMPHOID TISSUE

splint A length of rigid material used to support and immobilise a broken limb. *See also* FIXATOR

spondylosis A degeneration of the VERTEBRAL COLUMN, in which spurs of bone grow on the VERTEBRAE and fuse them together, reducing mobility. The condition is common in old bears and big cats.

sponge *See* PORIFERA

spongy bone *See* BONE

spontaneous ovulator An animal in which OVULATION occurs spontaneously, without stimulation of the female by COPULATION with the male. This is the type of ovulation which occurs in most animals. *Compare* INDUCED OVULATOR

spoor The footprints, faeces and other signs of the presence of an animal. Useful in tracking animals and estimating their numbers in the wild.

spore
  1. A minute reproductive unit capable of giving rise to a new individual without sexual fusion, typically

Table S4 The cost of hunting a single elephant in Tanzania (based on Rees, 2007).

Expense/fee	Fee (US$)
21-day safari*	33 600
Conservation fee	2 100
Concession fee	2 100
Hunting permit	600
Trophy preparation	1 800
Government trophy handling fee	300
Elephant trophy fee**	14 000
Rifle import permit (per gun)	200
Staff gratuities	1 000
Total***	55 700

* The minimum length required under Tanzanian law for hunting an elephant.

** Based on one elephant with both tusks weighing more than 31.8 kg (70 pounds) each, and including 15% Community Development Programme Contribution.

*** Excludes air fares, hotel accommodation while not on safari, trophy shipping, handling and packing costs, ammunition import duty and trophy export fee.

consisting of a single cell. Occurs in many lower plants, fungi and protozoa.
2. A resistant form of a bacterium which is formed in adverse environmental conditions.

sport hunting, trophy hunting The killing of wild animals, usually large game animals (especially **LARGE MAMMALS**), for sport and to obtain trophies such as heads and skins. Sometimes whole bodies are mounted by a **TAXIDERMIST**. This activity is legal in a number of African countries, e.g. Zimbabwe and Tanzania, and an important source of tourist revenue (Table S4). The export of trophies is controlled by **CITES**. *See also* **BIG FIVE**, **CAMPFIRE**, **INTERNET HUNTING**, **SHOOTING ESTATE**

spotting scope A telescope used for counting and observing animals, particularly birds, often from a **HIDE (1)**.

spp. After the name of a **GENUS**, it indicates several members of that genus, e.g. *Canis* spp. *Compare* **SP.**

spread of participation index (SPI) A numerical measure of spatial behaviour which may be used to study the extent to which a captive animal utilises all of the areas within its enclosure.

$$SPI = \frac{M(n_b - n_a) + (F_a - F_b)}{2(N - M)}$$

where N = the total number of observations of the subject, M = the mean frequency of observations in all enclosure sites (i.e. N/number of sites), n_a = the number of sites with frequencies >M, n_b = the number of sites with frequencies <M, F_a = the total number of observations in sites with frequencies >M, F_b = the total number of observations in sites

with frequencies <M. Values of the index range from zero (all sites used equally) to 1 (only 1 site used).

SPSS A comprehensive package of statistical and analytical software produced by IBM and widely used within the scientific community, although the abbreviation stands for 'Statistical Package for the Social Sciences'.

Squaliformes An order of fishes: dogfish sharks and their allies.

Squamata An order of reptiles: iguanas (Fig. I1), lizards, monitor lizards, chameleons, geckos, lacertids, skinks, snakes (Fig. P1), adders, vipers.

Squatiniformes An order of fishes: angel sharks.

squeeze cage *See* **CRUSH**

squeeze chute *See* **CRUSH**

SSP *See* **SPECIES SURVIVAL PLAN® (SSP) PROGRAM**

SSSI *See* **SITE OF SPECIAL SCIENTIFIC INTEREST (SSSI)**

SSSMZP *See* **SECRETARY OF STATE'S STANDARDS OF MODERN ZOO PRACTICE (SSSMZP)**

stabilimeter A cage with a tilting floor which is used to measure activity in small animals.

stabilisation
1. The action of fixing a joint or broken limb in position. *See also* **FIXATOR**, **SPLINT**
2. The process of establishing normal respiration, heart rate, blood pressure, temperature and other physiological functions in a patient; the establishment of **HOMEOSTASIS**.

stabilising selection A type of **NATURAL SELECTION** in which variation in a character decreases as a result of the elimination of **PHENOTYPES** at both extremes of the distribution (Fig. S4). *Compare* **DIRECTIONAL SELECTION**, **DISRUPTIVE SELECTION**

stable floor matting Mats made of rubber or other material (e.g. ethylene vinyl acetate (EVA) foam) used to cover the floor of a stable to provide insulation, increased comfort and improved hygiene for horses.

stain A chemical used to colour a specific chemical or structure within a specimen prepared for viewing under a microscope, especially in **HISTOLOGY**. For example haematoxylin is used to stain cell nuclei; Sudan III may be used to stain fats.

stall A compartment in a stable, cowshed or other building where **LIVESTOCK** are kept, for housing a single individual.

stallion
1. An uncastrated male horse.
2. According to the **AMERICAN STUD BOOK** 2008, '*a male horse that is used to produce foals.*'

standard deviation A measure of the variation about the **MEAN** of a set of data. The square root of the **VARIANCE**. In a **NORMAL DISTRIBUTION** over 99% of all values fall between three standard deviations either side of the **MEAN**. *See also* **RANGE (3)**, **STANDARD ERROR OF THE MEAN**

S

standard error of the mean The STANDARD DEVIATION of the MEANS calculated from a series of SAMPLES taken from the same POPULATION. *See also* RANGE (3), VARIANCE

Standards for Exhibiting Circus Animals in NSW A set of standards issued pursuant to Clause 8(1) of the Exhibited Animals Protection Regulation 2005 in New South Wales, Australia. The standards relate to husbandry, housing, training, veterinary treatment, interaction with the public and the suitability of certain species for use in circuses.

standing stock *See* CRUSH

stand-off barrier *See* PUBLIC BARRIER

stapes *See* EAR OSSICLES

stapling device *See* SKIN STAPLER

stargazer A bird that exhibits STARGAZING.

stargazing A phenomenon in newly hatched chicks, in which the individual's head is permanently held back with the beak pointing upwards. Caused by a deficiency in thiamine.

startle response A reflex response to unexpected stimulation. In vertebrates, this is usually followed by arousal and an orienting response.

static life table *See* LIFE TABLE

statistical test A mathematical test which is used to determine whether the results of an EXPERIMENT (1) are likely to have occurred by chance or because of a real effect.

statistically significant Describing an event which is unlikely to have occurred by chance.

statistics
1. Numbers describing the properties of a population of measurements e.g. the MEAN, MEDIAN, MODE, STANDARD DEVIATION, VARIANCE etc.
2. The branch of mathematics concerned with drawing inferences from numerical data, based on probability theory.

statocyst A small balance organ found in many invertebrates.

status dog, weapon dog A dog used to intimidate and harass members of the public, often by young people who live on inner city housing estates, and those involved in criminal activity. The term is sometimes used synonymously with 'DANGEROUS DOG', but this term is defined in legislation in some countries. *See also* DANGEROUS DOGS ACT 1991, STATUS DOG UNIT

Status Dog Unit A unit of the Metropolitan Police Service launched in 2009 dedicated to the seizure of potentially dangerous dogs (STATUS DOGS), especially PIT BULL TERRIERS, from those who breed, sell and fight such dogs. These dogs are involved in gang-related crime, drug dealing and anti-social behaviour. *See also* DANGEROUS DOGS ACT 1991

status symbols, animals as Animals are widely kept by people, especially men, as status symbols. The animals kept range from big cats, wolves and birds of prey to particular breeds of domestic dogs. *See also* STATUS DOG

statutory instrument A type of delegated legislation in the UK and some other countries. Often used to add detail or make changes to an ACT OF PARLIAMENT.

STD *See* SEXUALLY TRANSMITTED DISEASE (STD)

steed A horse, especially one that is bold and lively; a high-spirited riding horse.

steer A young castrated bull or male ox, especially one reared for beef.

stem cell An undifferentiated cell which can divide to produce more stem cells and can give rise to a number of different types of specialised cells. There are embryonic stem cells (which are capable of differentiating into various cell types) and adult (somatic) stem cells (which are concerned with repair and tissue renewal). Animal experiments have been used to study the value of stem cells in disease treatment.

Stephanoberyciformes An order of fishes: princkle-fishes, whalefishes and their allies.

steppe A vast, semi-arid treeless grassland which extends across Asia from eastern Austria to China.

stereo microscope *See* DISSECTING MICROSCOPE

stereotyped behaviour Any repetitive behaviour which is relatively fixed in form, e.g. digging. It may be normal or abnormal (STEREOTYPIC BEHAVIOUR).

stereotyped route-tracing A STEREOTYPICAL BEHAVIOUR which consists of repeatedly walking the same route around an enclosure. Observed in some zoo animals (e.g. canids, felids, ursids, elephants) and farm animals (e.g. horses). In extreme cases the animal may step in its own footprints on each circuit. Part of the route often includes some of the boundary of the enclosure and in some cases the behaviour involves walking along a fence line repeatedly (Fig. S15).

stereotypic behaviour, stereotypical behaviour, stereotypy A repetitive behaviour which has no apparent purpose which often appears when an animal is under STRESS. Ödberg (1978) defined stereotypies as morphologically similar patterns or sequences of behaviour performed repeatedly and having no obvious function. The frequency and severity of stereotypical behaviour exhibited by a captive animal is often used as a measure of its WELFARE. Stereotypic behaviour may be associated with disorders of the BASAL GANGLIA. *See also* BAR-BITING, COPING HYPOTHESIS, PACING, ROCKING, STEREOTYPED ROUTE-TRACING, SWAYING, TONGUE-PLAYING

stereotypy *See* STEREOTYPIC BEHAVIOUR

sterile
1. Free from sources of infection (e.g. BACTERIA, VIRUSES).
2. Unable to reproduce.

sterile technique A series of methods used to prevent contamination of equipment, growth media etc.

S

Fig. S15 Stereotyped route-tracing: the path made by a spectacled or Andean bear (*Tremarctos ornatus*) in a zoo.

with unwanted microbes, especially such methods when used in **MICROBIOLOGY**.

sterilise

1. To treat with heat or chemicals to destroy **BACTERIA**, **VIRUSES** and other sources of infection.

2. To prevent an animal from breeding by, for example, removing its **TESTES** or **OVARIES**.

steriliser *See* **AUTOCLAVE**

sternal keel *See* **CARINA**

sternum

1. The breastbone. A long flat bone in most **VERTE-BRATES** (except fishes) that is situated along the ventral midline of the **THORAX (1)** and articulates with the **RIBS** and clavicles. *See also* **CARINA**

2. A thickened ventral plate on each segment of an arthropod exoskeleton.

steroid hormone *See* **STEROIDS**

steroids A large group of organic compounds with a characteristic molecular structure containing four rings of carbon atoms, including many hormones (e.g. **TESTOSTERONE**, **OESTROGENS**), choles-terol, **VITAMIN D**. *See also* **ANABOLIC STEROIDS**

stethoscope A device for listening to sounds made by the heart, lungs and other body systems to check for abnormalities and disease by pressing a diaphragm

against the surface of the body. Digital stethoscopes are more sensitive than acoustic stethoscopes and can be connected to computers and mobile phones to record and transmit sounds.

stewardship The concept that humans are responsible for the world and have an obligation to take care of it. The concept is used to emphasise humans' responsibility to protect wildlife and ecosystems. *See also* **EARTH JURISPRUDENCE, SUSTAINABILITY**

stimulus Any detectable change in an animal's environment. *See also* **RESPONSE**

stochastic In relation to an algorithm, **MODEL** or process, one where the outcome is unpredictable and determined by random processes, with no stable pattern or order. *Compare* **DETERMINISTIC**

stock fence A fence which is designed to control and restrict the movements of domestic livestock, e.g. cattle and sheep, to a defined area. *See also* **GAME FENCE, SHARK NET**

stock list A list of all of the species held by a zoo or all of the species of a particular **TAXON**, e.g. a bird stock list. In England the *SECRETARY OF STATE'S STANDARDS OF MODERN ZOO PRACTICE* (**SSSMZP**) require licensed zoos to produce an annual stock list which records the numbers of each species present on 1 January each year, and which

S

Table S5 An example of a stocktaking record for a species in a zoo. 2.4.0 indicates the presence of 2 male, 4 females and 0 of unknown sex.

Common name	Scientific name	Group at 1.1.2012	Arrive	Born	Death within 30 days of birth	Death	Depart	Group at 31.12.2012
Black rhinoceros	*Diceros bicornis*	2.4.0	0.1.0	1.1.0	0.1.0	0.0.0	0.1.0	3.4.0

accounts for any changes in number since the previous annual stock list was produced (births, deaths, arrived, departed, died within 30 days of birth). *See also* **STOCK RECORDS, STOCKTAKING**

stock prod *See* **CATTLE PROD**

stock records
1. Records of individual animals kept on a farm, in a zoo or other facility which keeps animals.
2. In relation to the requirements of the *SECRETARY OF STATE'S STANDARDS OF MODERN ZOO PRACTICE (SSSMZP)*, records kept and maintained of all individually recognisable animals and groups of animals in a zoo. Where possible, animals should be individually identifiable. Records should include information on the scientific name, identification, date of entry or disposal, date of birth or hatching, sex, distinctive markings or identification rings or other **MARKS**, clinical, behavioural and life history data, date of death and *POST-MORTEM* **(2)** results, details of any escapes and damage caused, food and diet (Table S5). The holdings of a particular species in a zoo are frequently abbreviated as follows: a.b.c, where a is the number of males, b is the number of females and c is the number of unknown sex (e.g. 1.4.2). *See also* **STOCK LIST, STOCKTAKING**

stocking density, animal density, animal loading A term used to describe the density of animals in an enclosure, i.e. the number of animals per unit area, e.g. 1 per hectare.

stocktaking The process of counting, sexing and recording all of the animals in a zoo in a list. Usually undertaken annually (Table S5). *See also* **STOCK LIST, STOCK RECORDS**

stomach A general term for a muscular bag used to store food in the **DIGESTIVE SYSTEM**; in mammals, a bag which receives food through the **OESOPHAGUS** and adds digestive **ENZYMES** to it before passing it by **PERISTALSIS** to the **DUODENUM**. *See also* **ABOMASUM, RUMINANT**

Stomiiformes An order of fishes: dragonfishes, lightfishes and their allies.

stool Faeces.

store cattle
1. Cattle purchased for rearing rather than for slaughter.
2. Cattle which have been grown slowly to avoid impairment of skeletal development. In these animals, muscle tissue is slightly below potential and fatty tissue is not developed.
Compare **FAT CATTLE**

stotting *See* **PRONKING**

stranding, beaching The washing up on a beach of marine mammals, usually cetaceans (sometimes seals), which subsequently die if not rescued and returned to the sea or taken into captivity. Mass strandings sometimes occur, involving the simultaneous beaching of a number of individuals of the same species, e.g. a pod of dolphins. Pilot whales (*Globicephala*) are particularly prone to stranding. They follow a single individual so if he becomes disorientated and becomes stranded the rest of the pod will follow. The species that mass strand navigate using **ECHOLOCATION** and it may be that animals become disorientated when this system fails, possibly as a result of acoustic pollution caused by ships. In the UK, deaths of stranded cetaceans must be reported to the UK Cetacean Strandings Co-ordinator at the **NATURAL HISTORY MUSEUM (LONDON)** who reports any trends to the **DEPARTMENT FOR ENVIRONMENT, FOOD AND RURAL AFFAIRS (DEFRA)**.

strangles, equine distemper A highly contagious disease of **EQUIDS** caused by the bacterium *Streptococcus equi equi*, characterised by **INFLAMMATION** of the **MUCOUS MEMBRANES** of the head and throat.

strategy In animal behaviour, a set of tactics used to achieve a particular goal which has evolved as a consequence of the effects of **NATURAL SELECTION** on alternative behaviours, taking into account their relative costs and benefits, e.g. **EVOLUTIONARILY STABLE STRATEGY (ESS), FORAGING STRATEGY, REPRODUCTIVE STRATEGY**.

straw The dry stalks of cereal plants which are used as bedding and fodder for livestock. *See also* **HAY**

stress A state of anxiety in an animal that results from specific internal or external stimuli (**STRESSORS**). The normal behavioural response is avoidance of the stimuli, indicating **AVERSION (2)**. In vertebrates stress is associated with changes in hormone levels, e.g. **CORTISOL**. Stress induces a syndrome of complex physiological changes known as **GENERAL ADAPTATION SYNDROME**. Stress may cause illness, inhibit reproduction and even result in death in some species. In captive animals it may cause **STEREOTYPIC BEHAVIOUR**. *See also* **COPING HYPOTHESIS, STEREOTYPIC BEHAVIOUR**

stressor A stimulus which is capable of causing **STRESS**, e.g. noise, excessive heat or cold, the presence of a dominant animal or predator, rough

S

handling by a caretaker, transportation, novelty, aggression, uncertainty, mother–infant separation. A stressor can only be identified by its effect on an animal's physiology.

striated muscle *See* SKELETAL MUSCLE

strict liability for the actions of animals *See* LIABILITY FOR THE ACTIONS OF ANIMALS

stridulation The process by which animals make sounds by rubbing together parts of their bodies. Insects such as grasshoppers and crickets produce sounds by rubbing a hind leg scraper against the adjacent forewing. In Madagascar, two species of streaked tenrecs (*Hemicentetes semispinosus* and *H. nigriceps*) communicate by rubbing together specialised quills on their backs which create ultrasonic signals.

Strigiformes An order of birds: owls (Fig. O1).

striped muscle *See* SKELETAL MUSCLE

stroke A sudden interruption of the supply of blood to the brain which may cause paralysis of certain parts of the body and death in extreme cases. It may be caused by the rupture of an artery, blockage of an artery by a blood clot etc.

stroke volume The volume of blood pumped out of the left VENTRICLE (2) of the heart with each HEARTBEAT.

structural enrichment, physical enrichment Alteration of the size or complexity of an animal enclosure, or the addition of FURNITURE, novel objects, vegetation or substrates (e.g. nestboxes, sleeping platforms, rocks).

Struthioniformes An order of birds: ostrich, tinamous, rheas, cassowaries, emus, kiwis.

stud farm A place where animals, especially horses, are bred.

studbook The official record of the PEDIGREE of a group of animals, especially valuable breeds of animals such as THOROUGHBRED horses, pedigree dogs etc., and also rare species.

studbook keeper A person who maintains a STUDBOOK for a particular species or breed of animal.

stunning In animal production, a process which renders an animal unconscious so that it cannot feel pain, prior to slaughter. It may be achieved by use of a CAPTIVE BOLT, an ELECTRIC STUNNER or a HUMANE SLAUGHTER PISTOL. Gas stunning is also used, causing unconsciousness or death through HYPOXIA or ASPHYXIA.

sty A type of housing for pigs. *See also* SOW STALL

styptic Relating to something which is astringent; causing contraction of the blood vessels and tissue thereby preventing bleeding.

styptic pencil A short stick of material which contains a STYPTIC drug used to stop bleeding.

subacute
1. In relation to a disease, intermediate in character between acute and chronic: one that develops more slowly than an ACUTE CONDITION but more quickly than a CHRONIC CONDITION.

2. Subclinical. Relating to a condition which is present in an animal that appears to be clinically well (i.e. is ASYMPTOMATIC) but which may be detected by laboratory or other tests.

subadult An animal which possesses some, but not all, adult characteristics and is not sexually mature. Some large birds take several years to attain their full adult plumage.

subalpine Relating to high upland slopes, especially the zone immediately below the tree line.

subclass A subdivision of a CLASS.

subclinical *See* SUBACUTE (2)

subcutaneous Located beneath the SKIN, e.g. subcutaneous fat.

submissive behaviour *See* APPEASEMENT BEHAVIOUR, RECONCILIATION

submontane Relating to the foothills or lower elevation slopes of a mountainous region.

suborder A subdivision of an ORDER.

subordinate individual An individual in an hierarchically organised SOCIAL GROUP whose status is lower than that of a more DOMINANT INDIVIDUAL. Subordinates often exhibit APPEASEMENT BEHAVIOUR towards dominant individuals during confrontations. *See also* DOMINANCE HIERARCHY

sub-Saharan Relating to the region south of the Sahara Desert.

subspecies A subdivision of a SPECIES. *See also* NOMINATE SUBSPECIES

substrate
1. The reactant which is consumed during an enzyme-catalysed reaction, e.g. when maltase (the enzyme) breaks down maltose (the substrate) to glucose.

2. *See* SUBSTRATUM

substratum, substrate
1. The material which makes up the floor of an enclosure or animal house, e.g. concrete, bark, sand, gravel, rock.

2. The material at the bottom of an aquarium, e.g. gravel.

subterranean Underground.

subtropical Relating to the regions immediately north and south of the tropical zone, which is bounded by the Tropic of Cancer and the Tropic of Capricorn.

succession *See* ECOLOGICAL SUCCESSION

suckler calf An unweaned calf that is still suckling.

suckling
1. The action of sucking milk from the mother.

2. A foal of any sex in its first year of life while it is still nursing (AMERICAN STUD BOOK 2008).

suction device, aspirator, suction equipment A device consisting of a tube connected to a pump and a collection jar which is used to remove liquid materials during an operation, to clear airways etc.

suction equipment *See* SUCTION DEVICE

suffering
1. An aversive aspect of MOTIVATION which may cause STRESS and may be associated with PAIN. It

S

is difficult to assess in animals because this can only be done by analogy with our concept of human suffering. Dawkins (1990) has defined suffering as occurring when unpleasant subjective feelings are acute or continue for a long time because an animal is unable to carry out the actions that would normally reduce risks to life and reproduction in those circumstances. It may occur when WELFARE is poor but poor welfare may occur in the absence of suffering, e.g. a injured animal suffers poor welfare but while sleeping is not suffering because it is not experiencing pain. *See also* BENTHAM

2. In England and Wales, under s62(1) of the ANIMAL WELFARE ACT 2006, ' *"suffering" means physical or mental suffering*'.

sulfur *See* SULPHUR

sulphur, sulfur A mineral nutrient which is important in some amino acids and in protein structure. Non-ruminants appear to have no dietary requirement for sulphur. Microbes in the gut of ruminants use dietary sulphur to synthesise some amino acids and B vitamins.

summer seasonal recurrent dermatitis, sweet itch A prevalent allergic disease of EQUIDS which is the result of an ALLERGIC REACTION to allergens in the saliva of biting midges (*Culicoides* spp.) and black flies (*Simulium* spp.). *See also* DERMATITIS

sun shade A structure that provides shade for animals in an outdoor enclosure. It is important for some species to provide a suitable environmental temperature. A shade may be small, e.g. for penguins, or very large, e.g. for gorillas or elephants. Sun shades are sometimes designed to appear as large parasols.

sunbathing, sunning The controlled, selective exposure of the body to the sun. This may be done in order to assist with THERMOREGULATION (e.g. in POIKILOTHERMS such as lizards), or to encourage VITAMIN D synthesis in the skin (as in some bird species). *See also* BASKING

sunning *See* SUNBATHING

super carnivore *See* APEX PREDATOR

super predator *See* APEX PREDATOR

superclass A TAXON above the level of CLASS.

superfetation *See* SUPERFOETATION

superfoetation, superfetation
1. The simultaneous presence of more than one stage of developing EMBRYO in the same animal.
2. In mammals, the formation of an embryo from an OESTROUS CYCLE while another embryo is already present in the UTERUS from a previous cycle.

supernormal stimulus A stimulus which is more effective in eliciting a response than the normal stimulus, and often preferred to the normal stimulus, e.g. herring gulls (*Larus argentatus*) prefer to retrieve large dummy eggs into their nests rather than normal-sized eggs.

superorder A TAXON above the level of ORDER.

super-organism An organism which consists of many individual organisms which exhibit a DIVISION OF LABOUR (2), e.g. a colony of ants, a colony of naked mole rats (*Heterocephalus glaber*). A social unit of a species that exhibits EUSOCIALITY.

supine Lying on the back (DORSAL surface), with the front or face upward. *Compare* PRONE

supplantation The replacement of one individual by another due to demonstrable superiority e.g. at a food source. *See also* DOMINANCE HIERARCHY

supplement *See* COLOSTRUM SUPPLEMENT, FOOD SUPPLEMENT, VITAMIN SUPPLEMENT

surface area:volume ratio The ratio of the surface area of a structure, or animal, to its volume. Very small organisms, for example protozoa, have a large surface area for each unit of volume. This means that they can absorb all of the oxygen they need for respiration and all the food they need over their body surface. Larger organisms have a much smaller body surface area per unit of body volume and consequently the surface is incapable of supplying sufficient oxygen to supply the needs of all of the cells. Many taxa have evolved lungs with a very large surface area in order to supply their tissues with oxygen, and complex digestive systems which provide a large surface area for the absorption of food. Imagine a theoretical organism which has a cubical form and sides of 2 units. Its surface area would be $2 \times 2 \times 6 = 24$ units, and its volume would be $2 \times 2 \times 2 = 8$ units; a surface area: volume ratio of 3:1. An organism with sides of 3 units would have a surface area of $3 \times 3 \times 6 = 54$ units, and a volume of $3 \times 3 \times 3 = 27$ units; a surface area to volume ratio of only 2:1.

surgical suture *See* SUTURE (1)

surplus animals In relation to a zoo, animals which are not required by a zoo for breeding or exhibition purposes. They may be transferred to other collections or euthanased if a suitable home cannot be found. Some zoos have been criticised for euthanasing healthy animals which they do not have room to house.

surrogate
1. A substitute parent who raises an animal after it is born or hatched. It may be a member of the same species, a different species or a human. *See also* ALLOPARENT, PUPPET MOTHER
2. A female mammal who carries and gives birth to an offspring to which she is not genetically related. *See also* CLONING, EMBRYO TRANSFER (ET), INTERSPECIES EMBRYO TRANSFER

survival of the fittest The concept in evolutionary biology whereby individual organisms that are best suited to the environment are most likely to survive and reproduce, passing their genes to the next generation. *See also* FITNESS, NATURAL SELECTION, SELECTION

survivorship curve A curve showing the pattern of mortality within a POPULATION with increasing

S

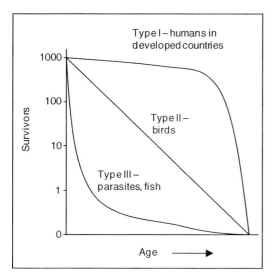

Fig. S16 Survivorship curves.

age, usually corrected so that the starting population is 1000 (Fig. S16). Graphs showing survivorship curves often use a **LOGARITHMIC SCALE** to indicate the number of survivors in each **AGE CLASS**. Survivorship curves have been used to compare the longevity of individuals in populations of zoo animals with that of individuals of the same species in the wild. *See also* **LIFE TABLE**

sustainability The ability to maintain the conditions under which people and ecosystems can exist in a harmonic relationship that permits the fulfilment of the social, economic and other requirements of present and future generations. *See also* **CARBON NEUTRAL, STEWARDSHIP, SUSTAINABLE USE**

sustainable use The UN **CONVENTION ON BIO-LOGICAL DIVERSITY 1992 (CBD)** (Art. 2) defines 'sustainable use' as *'the use of components of biological diversity in a way and at a rate that does not lead to the long-term decline of biological diversity, thereby maintaining its potential to meet the needs and aspirations of present and future generations.'*

suture
1. A surgical suture: a stitch, made of **COLLAGEN**, nylon, gut, wire or other material, used to hold body tissues together, e.g. close up a wound to aid healing.
2. The material used to make a surgical suture (e.g. gut, thread or wire).
3. The process of creating a surgical suture by sewing; the act of stitching body tissues together. *See also* **SKIN STAPLER**
4. A rigid joint between hard structures in animals, e.g. joints between bones in the skull, the spiral seam at the junction of the whorls of the shell of a gastropod.

swab
1. A piece of cotton wool, gauze or similar material used for cleaning a wound or taking a sample for examination or testing (e.g. a rectal swab).
2. The action of taking a sample or cleaning a wound with a swab.

sward An area of ground covered with grass. Management may be required to maintain an appropriate sward for particular species, e.g. it may be necessary to cut grass to maintain a short sward suitable for feeding for ducks and geese. *See also* **CONSERVATION GRAZING**

sway branch A tree branch to which food items are fixed before attaching it to an existing tree (possibly using a pulley system) as a moving enrichment feature for monkeys and other arboreal animals.

sway feeding pole A wooden pole which is resting on the ground at one end while the other end is raised using a pulley system. Food is attached to the raised end of the pole. Used as a moving enrichment for tigers and other large predators.

swaying Moving the body rhythmically from side to side: a type of **STEREOTYPIC BEHAVIOUR** seen particularly in elephants and bears.

sweat gland A gland in mammalian skin that secretes a dilute solution of essentially salty water to the outside which cools the surface by evaporation. This process assists in **THERMOREGULATION**.

sweat scraper A device, of various designs, which is rubbed across a horse's coat to remove excess water.

sweet itch *See* **SUMMER SEASONAL RECURRENT DERMATITIS**

swim bladder, air bladder, gas bladder, hydrostatic organ A gas-filled sac found in the abdomen of **ACTINOPTERYGII** which allows the fish to maintain **NEUTRAL BUOYANCY**. In some fishes it has a respiratory function, and it may also serve to create and receive sound.

swim bladder disease A common disease of aquarium fish. Generally caused by the fish taking in air with its food, thereby increasing its **BUOYANCY** and causing it to float to the surface.

swine A member of the mammalian family Suidae: **PIGS**, hogs and their relatives (Fig. W1).

swine fever *See* **CLASSICAL SWINE FEVER**

swine flu *See* **H1N1 VIRUS**

swine influenza *See* **H1N1 VIRUS**

symbiosis An association between two species from which both benefit, e.g. a black-throated honey guide (*Indicator indicator*) will direct a honey badger (*Mellivora capensis*) to a bee hive so that the honey badger can break into the hive and both can feed on the contents.

sympathetic nervous system (SNS) Part of the **AUTONOMIC NERVOUS SYSTEM (ANS)**. It acts antagonistically to the **PARASYMPATHETIC NERVOUS SYSTEM (PNS)** and tends to control the physiology in stressful conditions. For example, it dilates the pupils, inhibits salivation, relaxes the

S

airways, increases the **HEART RATE**, inhibits digestion, and stimulates the breakdown of glycogen and
the release of glucose into the blood.

sympatric Two **SPECIES** are said to be sympatric if
their distributions overlap in time and space. *Compare*
ALLOPATRIC, PARAPATRIC

sympatric speciation The formation of new species
from an ancestral species within the same geographical range. Although these new species are
capable of interbreeding they do not due to genetic
barriers (e.g. chromosomal mutation), differences
in their ecology, or for some other reason. *Compare* **ALLOPATRIC SPECIATION, PARAPATRIC
SPECIATION**

symptom A subjective feeling relating to illness or
injury which can only be described by a human
patient and which cannot be measured, e.g. feeling
sick. The term should not be used when discussing
animals as it is clearly impossible to determine how
an animal feels. *Compare* **SIGN (1)**

synapse The junction between two **NEURONES** into
which a **NEUROTRANSMITTER** is secreted to cause
a **NERVE IMPULSE** to pass from one neurone to
the other.

Synbranchiformes An order of fishes: swamp eels
and their allies.

synecology The ecology of entire communities of
species. *Compare* **AUTECOLOGY**

synergism The phenomenon whereby two events
occurring together have a greater effect than either
when it occurs alone, e.g. two drugs taken together
may act synergistically such that their combined
effect is greater than their individual effects.

synergistic Relating to or exhibiting **SYNERGISM**.

syngamiasis *See* **GAPEWORM**

syringe driver *See* **SYRINGE PUMP**

syringe pump, syringe driver A type of **INFUSION
PUMP** which is designed to administer a drug to a
patient gradually, either subcutaneously or intravenously, by pushing on the plunger of a syringe.

systematics
1. A synonym of **TAXONOMY**.
2. In relation to organisms, their **TAXONOMY** identification, **CLASSIFICATION** and nomenclature. *See
also* **BINOMIAL SYSTEM OF NOMENCLATURE**

systemic Affecting the entire body, e.g. a **SYSTEMIC
DRUG**.

systemic circulation That part of the circulatory
system in which the blood leaves the heart, services
the cells of the body (e.g. providing oxygen and
food, and removing carbon dioxide and nitrogenous
waste), and then returns to the heart. *Compare*
PULMONARY CIRCULATION

systemic drug A drug that acts by dispersing throughout the body. Drugs taken by mouth or injection
are generally systemic drugs. *Compare* **TOPICAL
DRUG**

systole The phase of the **CARDIAC CYCLE** when the
heart muscles contract, forcing blood out through
the arteries. *Compare* **DIASTOLE**

S

Fig. T1 T is for turkey (*Meleagris gallopavo*).

Dictionary of Zoo Biology and Animal Management: A guide to terminology used in zoo biology, animal welfare, wildlife conservation and livestock production, First Edition. Paul A. Rees.

T *See* THYMINE (T)

tachycardia A heart rate which is higher than the RESTING HEART RATE. *Compare* BRADYCARDIA

tachypnoea A higher than normal RESPIRATORY RATE. *Compare* BRADYPNOEA, EUPNOEA

TAG *See* TAXON ADVISORY GROUP (TAG)

taiga *See* BOREAL FOREST

tail An appendage which extends from the rear of the torso.

1. The flexible extension of the backbone in a vertebrate; that part of the body which is made up of the sacrum, coccyx and tail vertebrae in mammals, birds and reptiles. *See also* TAIL DOCKING, TAIL RINGING

2. Feathers at the hind end of a bird.

3. Some invertebrates also possess terminal structures which may be called tails, including some butterflies (e.g. swallowtails (Papilionidae)), scorpions and springtails.

Tails may be important in signalling (e.g. TAIL-FLAGGING), escape (*see also* AUTOTOMY), balance (especially in ARBOREAL species (*see also* PREHENSILE) and those that run at high speed) and in COURTSHIP DISPLAYS in some taxa (Fig. C9).

tail autotomy *See* AUTOTOMY

tail biting An abnormal behaviour, especially of pigs, whereby one animal bites the tail of another, often leading to serious damage. It may be caused by the need to perform exploratory and FORAGING BEHAVIOUR, barren accommodation etc. Controlled by some farmers by TAIL DOCKING.

tail docking, docking The removal of most of the tail in some dog breeds for cosmetic reasons. Docking may also be performed for hygiene reasons in LIVESTOCK, e.g. to prevent FLY STRIKE in sheep and TAIL BITING in pigs. *See also* TAIL RINGING

tail guard A device for covering and protecting a horse's tail, especially during transportation.

tail ringing A type of TAIL DOCKING used to remove most of the tail of a very young lamb (less than 3 days old) by applying a very tight rubber band around it, thereby cutting off the blood supply. The tail falls off after a few days. This practice is used to reduce the likelihood of FLY STRIKE.

tail-flagging, flagging

1. Swiping the tail from side to side. Occurs in a number of mammal species as a SIGNAL, e.g. California ground squirrels (*Spermophilus beecheyi*) use tail-flagging as a defensive signal to deter attacks by snakes. In some species tail-flagging may be indicative of the animal being in oestrus, e.g. female dogs hold their tails to the side when they are receptive to males.

2. A rhythmic motion of the tail which occurs in stallions when they ejaculate.

take, taking

1. In law, to seize or gain possession of. In relation to wild animals, this generally includes hunting, killing and capturing, and forms the basis of many offences.

2. In Great Britain, in the WILDLIFE AND COUNTRYSIDE ACT 1981, s.1(1) '. . . *if any person intentionally—(a) kills, injures or takes any wild bird; . . . he shall be guilty of an offence.*' The term 'takes' is not defined in the Act.

3. In the USA under s3(19) of the ENDANGERED SPECIES ACT OF 1973 (USA) '*The term "take" means to harass, harm, pursue, hunt, shoot, wound, kill, trap, capture, or collect, or to attempt to engage in any such conduct.*'

4. In the AGREEMENT ON THE CONSERVATION OF POLAR BEARS 1973, Art. I(2) '. . ."*taking*" *includes hunting, killing and capturing.*'

taking *See* TAKE

tally chart A method of recording the frequencies of objects or events in a table, e.g.

Age (yrs)	Frequency
1– 5	12
6–10	23
11–15	9
16–20	4

The information in a tally chart may be used to produce a bar chart, pie chart or other type of graphical representation. When data are being recorded each event or object being counted is recorded as a vertical stroke. For ease of counting every fifth item is recorded as a diagonal stroke across the previous four marks, e.g. 7 would be represented as: ⪢ ||

talon The claw of a bird of prey or similar claw of a predatory animal.

tameness A reduction in FLIGHT DISTANCE as a result of HABITUATION to human presence. Tameness may be defined as having no flight response with respect to man (Hediger, 1950). It is possible to selectively breed for tameness and this has been important in the process of DOMESTICATION.

tandem method A method of selective breeding in which a breeder selects for improvement in one characteristic at a time (e.g. milk yield). Once that character is adequately fixed in the strain he may then select for a different characteristic. *See also* ARTIFICIAL SELECTION. *Compare* TOTAL SCORE METHOD

tape lure *See* SOUND LURE

tapetum lucidum A layer of cells in the retina or choroid of the eye which contains zinc and a protein. It reflects light back through the retina thereby increasing light sensitivity, especially in nocturnal mammals. It causes EYE SHINE when illuminated by a bright light.

tapeworm Any of a number of species of platyhelminth endoparasite with a ribbon-like body and a head bearing hooks and suckers for attachment.

target A device used in the TARGET TRAINING of animals located in a fixed position, e.g. on a fence, or on the end of a pole held by the trainer. It may take a distinctive shape and/or be a particular colour or combination of colours.

target training A method of **TRAINING** animals by teaching them to approach or touch a **TARGET** for a reward. The target may be used to move the animal within an enclosure, move it into a position for a veterinary procedure or examination, or for some other reason. When elephants are trained to offer their feet for inspection, by a vet or keeper, from behind a **PROTECTED CONTACT** fence, two targets are used, each on a separate pole. The elephant is trained to stand with its head near one of the targets – so that it stands in the correct position – and then to lift its foot to the other target at an access point in the fence (Fig. F7). *See also* **POSITIVE REINFORCEMENT**

target weight In agricultural production, the weight at which an animal is deemed to be ready for slaughter. This will vary between species and between breeds.

tartar, calculus A hard deposit found on the teeth, usually yellowish-brown, made of organic secretions and food particles deposited in various salts, notably calcium carbonate.

Tarzan of the Apes A novel, written by Edgar Rice Burroughs and published in 1914, about a boy whose parents, Lord and Lady Greystoke, die in the African jungle and who is subsequently reared by an ape foster mother. The infant grew into a mighty warrior (*Tarzan*) and became leader of the ape tribe. The book and its sequels were made into films and TV series. The original book portrayed the African jungle as a dangerous place: '*From the dark shadows of the mighty forest came the wild calls of savage beasts – the deep roar of the lion, and, occasionally, the shrill scream of a panther.*' It reinforced the attitude that man should exert dominion over the animals while at the same time demonstrating an **EMPATHY** between *Tarzan* and his animal companions, especially his chimpanzee friend *CHEETAH*. *See also* **ANIMAL STARS**

taste A form of chemoreception in which chemoreceptors come into direct contact with chemical substances. Mammals can taste sweet, acid (sour), salt and bitter using taste buds located on the tongue. Food flavour depends upon taste and **OLFACTION**.

tattoo A permanent mark used to identify individual animals. Sometimes located in a visible place (e.g. inside the ear) but may be hidden from view.

taxidermist A person who practises **TAXIDERMY**.

taxidermy The practice of mounting the bodies of dead animals in naturalistic postures for exhibition in museums and elsewhere. Some well-known and unusual specimens from zoos have been preserved by museums. For example, the skin of the **TIGON** named *Maude* and the skeleton of the Asian elephant *Maharajah* formerly of **BELLE VUE ZOOLOGICAL GARDENS** in Manchester have been preserved by the Manchester Museum in the UK, and the **NATURAL HISTORY MUSEUM (LONDON)** has the preserved body of London Zoo's giant panda *Chi Chi* on display. *See also* **JUMBO**

taxis (taxes *pl.*) A type of orientation in which the animal moves directly towards or away from the source of stimulation. For example, movement towards light is positive phototaxis and movement away from light is negative phototaxis. A klinotaxis is a movement which involves a side-to-side motion of the head of an organism as it moves forward in response to a source of stimulation, which is caused by the alternating reaction of sensory receptors on either side of the body. *See also* **PHONOTAXIS**

taxon (taxa *pl.*) A group of organisms related by their evolutionary history, e.g. a **SPECIES**, **GENUS**, **FAMILY (1)**, **ORDER**, **CLASS**, **PHYLUM** etc.

Taxon Advisory Group (TAG) A group of individuals who work together to advance knowledge of the husbandry, nutrition, veterinary care and other aspects of the conservation of a particular **TAXON** and disseminate this to others. They are also involved in decisions about what species should be kept in zoos. *See also* **TAXON WORKING GROUP (TWG)**

Taxon Working Group (TWG) A group of experts on a particular **TAXON** established by the **BRITISH AND IRISH ASSOCIATION OF ZOOS AND AQUARIUMS (BIAZA)** as an alternative to a **TAXON ADVISORY GROUP (TAG)**.

taxonomy The scientific **CLASSIFICATION** of organisms. *See also* **CLADISTICS**

Taylor, David (1934–2013) An English veterinary surgeon who was the first to specialise in zoo and wildlife medicine and has worked with a very wide range of species in zoos around the world. He introduced the first dart gun for immobilising animals into the UK and co-founded the **INTERNATIONAL ZOO VETERINARY GROUP (IZVG)** in 1976. He has written a number of books including *Zoo Vet: World of a Wildlife Vet* (1976).

TCA cycle *See* **KREBS CYCLE**

teaching A form of behaviour (often exhibited by parents or older animals) in which younger animals learn by observing their elders. It involves the transfer of knowledge to the 'learner' with no immediate benefit to the 'teacher' and involves the 'teacher' modifying their behaviour in the presence of a naïve observer. It is not the same as **IMITATION** or **SOCIAL FACILITATION** or the cultural transmission of a tradition. Teaching is a controversial concept in non-human animals. Lions (*Panthera leo*) have been observed teaching their young to kill and primates have been seen teaching **TOOL USE**. *See also* **CULTURAL BEHAVIOUR**

teacup pup *See* **TOY BREED**

Tecton Group A group of architects responsible for a number of iconic zoo buildings, e.g. the Penguin Pool at London Zoo, and the bear pits at **DUDLEY ZOO**. *See also* **LUBETKIN**

temperate In **ZOOGEOGRAPHY**, referring to an area (temperate zone) which has a moderate climate, specifically those areas of the Earth lying between

T

Fig. T2 Territory: a black rhino (*Diceros bicornis*) marking his territory with dung at Chester Zoo.

the Arctic Circle and the Tropic of Cancer and between the Tropic of Capricorn and the Antarctic Circle.

temperature regulation *See* **THERMOREGULATION**

temporal gland A gland located on the side of the head (temple). In male elephants such glands secrete a strong-smelling liquid when they are in **MUSTH**.

temporary teeth, milk teeth The first of two sets of teeth that develop in mammals as they age. They are replaced by the adult (permanent) teeth. *Compare* **PERMANENT TEETH**

Ten Commandments In relation to animal welfare, ten rules that people should obey if we are to show respect to animals, proposed as a 'Bill of Rights' by Dr. Desmond **MORRIS** in *The Animal Contract* (1990). In relation to zoos, two are particularly important: '*No animal should be dominated or degraded to entertain us*', and '*No animal should be kept in captivity unless it can be provided with an adequate physical and social environment*'. *See also* **COMPARABLE LIFE TEST, FIVE FREEDOMS**

tendon In vertebrates, a connective tissue cord or band, made largely of collagen fibres, which attaches muscle to bone or some other structure.

terminal sire A sire (male) used for **CROSS-BREEDING** whose offspring will not be suitable for breeding themselves.

terrarium A **VIVARIUM** that simulates a dry habitat such as a **DESERT** or **SAVANNA**.

terrestrial Relating to dry land or the surface of the Earth, e.g. the tiger is a terrestrial species; tropical forest is a terrestrial ecosystem.

territory A defended area used by an individual or a group of animals. It may be marked by scent (in mammals) and defended by aggressive behaviour or displays (e.g. songs in birds) (Fig. T2). Territories may be separated spatially or temporally, i.e. different individuals use the same areas at different times, e.g. cheetahs (*Acinonyx jubatus*). Some species only hold territories during the breeding season, e.g. herring gulls (*Larus argentatus*). In some species holding a territory may be essential to breeding success. The territorial requirements of species should be taken into account in the housing of animals to reduce aggression and stress and to ensure successful breeding where this is desirable. *See also* **SCENT-MARKING**. *Compare* **HOME RANGE**

tertiary consumer An **APEX PREDATOR**.

tertiary production The total amount of biomass produced by SECONDARY CONSUMERS in a particular place over a particular period of time. *Compare* PRIMARY PRODUCTION, SECONDARY PRODUCTION

testis (testes *pl.***)** The male GONADS, which produce SPERMATOZOA and TESTOSTERONE.

testosterone The principal male sex hormone in mammals and birds. A STEROID hormone produced by the TESTES which is responsible for the development of the SECONDARY SEXUAL CHARACTERISTICS. In females it is an intermediate in the synthesis of OESTROGENS.

Testudinata An order of reptiles: turtles, terrapins, tortoises.

tether To tie or restrain an animal with a rope or chain fixed to the ground, a post or some other object, or confine to a particular spot. *See also* FLIGHT RESTRAINT, HOBBLE, JESSES, PICKETING, PROTECTION AGAINST CRUEL TETHERING ACT 1988, SCREW PICKET

Tetraodontiformes An order of fishes: triggerfishes and their allies.

tetrapod A VERTEBRATE with two pairs of limbs.

thalamus A part of the forebrain which lies above the HYPOTHALAMUS and is concerned with transmitting sensory information from the spinal cord, brainstem and CEREBELLUM to the cortex of the CEREBRUM (cerebral cortex).

themed exhibit A named EXHIBIT which showcases a particular species or habitat (MULTI-SPECIES EXHIBIT) which often has a sophisticated NATURALISTIC EXHIBIT design and usually includes a range of INTERPRETATION devices, e.g. at San Diego Zoo, *Polar Bear Plunge, Elephant Odyssey, Tiger River, Ituri Forest, Gorilla Tropics, Sun Bear Forest* and *Wings of Australia*; at Taronga Zoo (Sydney) *Wild Asia, Great Southern Oceans, African Waterhole, Australian Nightlife*. Themed exhibits are sometimes based on the habitat types found in particular NATIONAL PARKS (NP) and named after them, e.g. the *Tsavo Bird Safari* at Chester Zoo.

theory of mind The theory that some animals are capable of knowing what is in the mind of another. This is a controversial theory in relation to non-humans. *See also* COGNITION, DECEIT, MENTAL STATE ATTRIBUTION

Theria A subclass of the mammals which, among living forms, contains the METATHERIA (marsupials) and the EUTHERIA (placental mammals).

thermal imaging camera A device capable of producing images by converting INFRARED RADIATION to visible light (infrared thermography). Used for the veterinary examination of animals to identify unusual patterns of heat loss. For example, parts of the body affected by infection may appear warmer than normal due to INFLAMMATION, and unused muscles will appear colder than those that are used. These cameras may also be used to locate young animals within an enclosure which have been hidden by their mother, e.g. tiger cubs.

thermoreceptor A sensory receptor that is sensitive to temperature. Some taxa possess separate sensors which are sensitive to heat and cold. Important in thermoregulation in mammals and birds.

thermoregulation, temperature regulation The process of maintaining the internal body temperature (as part of HOMEOSTASIS) within limits which vary between species. In HOMEOTHERMS this is achieved physiologically while in POIKILOTHERMS it is achieved by behavioural means, e.g. by seeking shade when the body temperature is too high and moving into the sun when it is too low. In many mammals heat loss is achieved by sweating (panting in some species) while heat is generated by body organs (especially the liver), as a result of their high metabolic rate and by shivering, and retained by hairs (fur) and body fat. The capacity of mammals and birds to adapt to a range of temperatures has allowed many tropical taxa to be kept in zoos located far from their RANGE STATES without ill effect and without the need to keep them exclusively in heated enclosures. *See also* ADAPTIVE HETEROTHERMY, LETHAL TEMPERATURE

thermostat A device for regulating temperature. Important in controlling the temperature of an indoor enclosure which contains animals which need to be kept warmer or colder than the ambient temperature. Particularly important in regulating the temperature of enclosures which contain POIKILOTHERMS, e.g. an AQUARIUM (2) or a VIVARIUM. *See also* DIMMING THERMOSTAT

***Think Tank* exhibit** An orangutan exhibit at the SMITHSONIAN NATIONAL ZOOLOGICAL PARK which allows visitors to interact with the animals and demonstrates aspects of animal COGNITION. Visitors may engage in a tug-of-war with an orangutan, or watch an orangutan-controlled webcam, and the apes are able to spray visitors with water.

third eye *See* PINEAL GLAND

thirst A motivational state in animals caused primarily by dehydration.

thorax
1. The anterior part of the VERTEBRATE body which is protected by the rib cage, contains the heart and lungs, and is separated from the ABDOMEN (1) by the DIAPHRAGM. *See also* RIBS
2. In insects, the three body segments located between the head and the abdomen, each of which bears a pair of legs.

Thoreau, Henry David (1817–1862) An American naturalist who wrote *Walden; or, Life in the Woods* (1854), a book about two years he spent in a cabin he built in a forest near Walden Pond in Massachusetts. Here he reflected on the benefits of living a simple, self-sufficient life, close to nature.

Thorndike, Edward (1874–1949) An American psychologist who studied OPERANT CONDITIONING

T

and used **PUZZLE BOXES** to study learning in cats. *See also* **BEHAVIOURISM**

thoroughbred
1. Any breed of pure-bred horse.
2. A specific horse breed known as a Thoroughbred, used in horse racing.
3. In the United States, the **AMERICAN STUD BOOK** (2008) defines a thoroughbred as '*a horse that has satisfied the rules and requirements set forth herein and is registered in the American Stud Book or in a Foreign Stud Book approved by The JOCKEY CLUB and the INTERNATIONAL STUD BOOK COMMITTEE.*'

threat A form of communication which takes the form of a **DISPLAY** or **VOCALISATION**. Threats function in keeping rivals at a distance and reduce the need to expend energy on fighting or risk injury. Many threat displays are the result of **RITUALISATION**.

Threatened A **RED LIST** category used by the **INTERNATIONAL UNION FOR THE CONSERVATION OF NATURE AND NATURAL RESOURCES (IUCN)** which includes all of those taxa which are designated as **CRITICALLY ENDANGERED (CR)**, **ENDANGERED (EN)** or **VULNERABLE (VU)**.

Three Rs Guiding principles applied to **ANIMAL TESTING**: replacement, reduction, refinement. This refers to the replacement of animal experimentation with non-animal methods; the reduction in the use of animals in experiments; and the refinement of the methodologies used in animal experimentation to minimise animal suffering. These principles were first suggested by the zoologist Prof. William Russell and microbiologist Rex Birch in *The Principles of Humane Experimental Technique* (1959) while working with the **UNIVERSITIES FEDERATION FOR ANIMAL WELFARE (UFAW)**.

thrombocytes *See* **PLATELETS**

Thylacinus The Journal of the Australasian Society of Zoo Keeping.

thymine (T) A **NUCLEOTIDE** base that pairs with A (adenine) in **DNA**.

tibia (tibiae *pl.***)** A **LONG BONE** which, together with the fibula, forms the lower hindlimb in tetrapod vertebrates. It is the larger, stronger and anterior of the two bones.

Tiergarten Schönbrunn *See* **SCHÖNBRUNN ZOO**

Tierpark Hagenbeck A zoo opened by Carl **HAGENBECK** in Stellingen, Germany (now part of Hamburg) in 1907. This was the first zoo to exhibit animals in open enclosures retained by **MOATS**.

tiger farming The breeding of tigers in captivity in intensive conditions to supply the **TRADITIONAL CHINESE MEDICINE (TCM)** industry.

tigon A hybrid produced when a male tiger mates successfully with a female lion. *Compare* **LIGER**

time-lapse camera A video camera used to speed up viewing of events (e.g. behaviour) by recording images at a lower frequency than that at which they will be viewed, e.g. the movements of starfish or limpets. *Compare* **HIGH-SPEED CAMERA**

time-lapse photography A system of recording video images with a long interval between frames, in order to condense time. *See* **TIME-LAPSE CAMERA**

Tinbergen, Nikolaas (1907–1988) Niko Tinbergen was a Dutch zoologist who specialised in the study of instinctive behaviour. He was one of the founders of **ETHOLOGY** and shared the Noble Prize for Physiology or Medicine with Konrad **LORENZ** and Karl von **FRISCH** in 1973. Tinbergen is credited with originating the four questions that should be asked about any animal behaviour (*see* **TINBERGEN'S FOUR QUESTIONS**). He helped to found the Serengeti Research Institute in Tanzania and his published works include *The Study of Instinct* (1951) and *The Herring Gull's World* (1953).

Tinbergen's Four Questions The four questions devised by **TINBERGEN** that should be asked in order to explain any animal behaviour: how is it caused?; how does it develop?; how did it evolve?; and what is it for? These have been given the acronym ABCDEF: **A**nimal **B**ehaviour, **C**ause, **D**evelopment, **E**volution, **F**unction.

tissue A collection of **CELLS**, which may not all be of the same type, but have the same origin and work together to perform a particular function, e.g. heart muscle, hard bone, cartilage. A level of organisation between cells and **ORGANS**.

tissue fluid *See* **INTERSTITIAL FLUID**

TNR programme Trap–**NEUTER**–release (return) programme for **FERAL CATS**. Used as an alternative to eradication where feral cats are considered to be a problem. May also involve **VACCINATION** (trap–neuter–vaccinate–release).

tobogganing A term used to describe a method of locomotion which involves sliding on the ventral body surface across ice and snow, as practised by penguin species.

toilet claw, comb claw, grooming claw A specialised claw used for grooming found in some species, e.g. beavers and some primates. In beavers it is located on the second toe of each hind foot. *See also* **TOOTH COMB**

tolerance The ability to cope with environmental extremes, e.g. high or low temperatures. Some species are able to alter their range of tolerance by **ACCLIMATISATION**. This ability is particularly useful when animals are kept in zoos where the climatic conditions fall outside the range they might experience in the wild.

tom A male domestic cat.

tomogram *See* **TOMOGRAPHY**

tomograph *See* **TOMOGRAPHY**

tomography The study of the reconstruction of two- and three-dimensional objects from one-dimensional sectional images created by a penetrating wave (e.g. X-rays, gamma rays, **ULTRASOUND**) using computer software. The device used is a called a tomograph and the image produced is called a tomogram. Tomography is used in the diagnosis of disease.

T

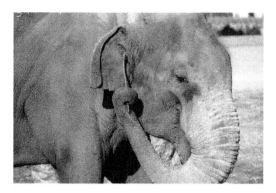

Fig. T3 Tool use: a bull Asian elephant (*Elephas maximus*) scratching his ear with a twig.

tongue-playing, tongue-rolling Swinging of the tongue outside the mouth from side to side or rolling the tongue inside the mouth repetitively. It is a common **STEREOTYPIC BEHAVIOUR** of cattle and horses.

tongue-rolling *See* **TONGUE-PLAYING**

tonic immobility A natural state of paralysis; an unlearned response triggered by physical restraint and characterised by a catatonic-like state of reduced **RESPONSIVENESS** to external stimuli. *See also* **DEATH FEIGNING**, **HYPNOSIS**

tool manufacture The making of a tool. Some species, e.g. chimpanzees (*Pan troglodytes*) and elephants (Elephantidae), are capable of making simple tools from materials present in their environment. They may, for example, break a small twig from a tree branch and use it for collecting food or scratching (Fig. T3). A mandrill (*Mandrillus sphinx*) at Chester Zoo was observed apparently making a tool to clean under its nails by stripping material from a small twig to make it narrower. *See also* **GREYBEARD**, **TOOL USE**

tool use The use of an object by an animal as an implement for performing or facilitating a mechanical operation, usually held in the hand, trunk or beak, e.g. chimpanzees (*Pan troglodytes*) use a stick to extract termites from a termite hill, elephants (*Elephas maximus*) use twigs to scratch their skin (Fig. T3). Wild gorillas (*Gorilla gorilla*) use tree branches to test the depth of water as they wade through it. Wild orangutans (*Pongo* spp.) use large plant leaves as 'umbrellas' to shelter from the rain. In captivity they put pieces of cardboard or old clothes on their heads (Fig. E3). Captive chimpanzees (*Pan troglodytes*) have been recorded using water as a tool to raise a peanut from the bottom of a deep container. One used his own urine to raise the peanut. In the wild the veined octopus (*Amphioctus marginatus*) has been observed using empty coconut shells as 'body armour'. If threatened it can hide itself in two half coconut shells held together with its tentacles. Zoos and other captive environments are useful for studying tool use. *See also* **ANVIL BEHAVIOUR**, **EGG-BREAKING BEHAVIOUR (2)**, **TEACHING**, **TOOL MANUFACTURE**

tooth (teeth *pl.*)
1. Enamel-coated structure in the jaw of most vertebrates used for biting, chewing and grinding food. Mammals possess **INCISORS**, **CANINES**, **MOLARS** and **PREMOLARS**. *See also* **DENTITION**, **EGG TOOTH**, **HETERODONT DENTITION**, **HOMODONT DENTITION**
2. A hard pointed structure found in some invertebrates, such as one of the pointed denticles or ridges on the **EXOSKELETON** of an **ARTHROPOD** or the shell of a **MOLLUSC**.

tooth comb, dental comb, dental tooth comb Any of a number of dental structures found in various mammalian taxa formed by spaces between the teeth (usually, but not always, **INCISORS**) and used for grooming. The presence of such structures in taxa as widely separated as lemurs and some African antelopes is the result of **CONVERGENT EVOLUTION**. *See also* **TOILET CLAW**

top carnivore *See* **APEX PREDATOR**

top predator *See* **APEX PREDATOR**

topical Of or relating to something (e.g. a substance or treatment) applied to a localised area of the external surface of the body.

topical drug A drug applied to a localised area of the external surface of the body, e.g. a topical **ANAESTHETIC (1)**, a topical antibiotic ointment. *Compare* **SYSTEMIC DRUG**

Torpediniformes An order of fishes: electric rays.

torpor A state of **HYPOTHERMIA** which is used by some **HOMEOTHERMS** as an adaptation to save energy. The body temperature may fall to within 1 °C of the environmental temperature and metabolic processes slow down to a small fraction of the normal rate. Animals enter torpor during **HIBERNATION**.

total count An animal **CENSUS** which counts all of the individuals of a species within a particular area. *Compare* **POPULATION ESTIMATE**

total protein *See* **CRUDE PROTEIN**

total score method A method of selective breeding in which the breeder selects for excellence in as many traits as possible at the same time. *Compare* **TANDEM METHOD**

totem An animal that serves among a particular traditional people as the emblem of a tribe, clan or family and is sometimes revered as its founder, ancestor or guardian. Some animals gain protection within a particular culture by virtue of being totems. *See also* **FADY**, **SACRED ANIMAL**

totem-tree marking The preferential repeated **SCENT-MARKING** of certain trees. Performed by some primate species, e.g. sifakas (*Propithecus* spp.).

touch pool An open-topped fish tank often found in aquariums where visitors may touch and/or feed the fish.

T

tourniquet A strap or similar device wrapped tightly around a limb to apply pressure to ruptured blood vessels to prevent loss of blood following a serious injury.

Tower of London Menagerie A MENAGERIE which was once located at the Tower of London. It was originally established at WOODSTOCK, England, by Henry I and was subsequently moved to the Tower, probably by Henry III.

toxaemia, toxemia
1. A condition caused by the presence of toxins in the blood.
2. A condition caused specifically by the presence of bacterial toxins in the blood.
See also PREGNANCY TOXAEMIA

toxemia *See* TOXAEMIA

toxic shock syndrome (TSS) A type of SEPTIC SHOCK. A potentially fatal condition which is the result of an infection caused by bacteria belonging to the genera *Streptococcus* or *Staphylococcus*, which cause SEPTICAEMIA. It may be a particular problem in dogs, especially puppies (canine streptococcal toxic shock syndrome).

toxicology The scientific study of the interactions between potentially harmful chemicals, living things and the environment.

toxicosis A pathological condition cause by exposure to a TOXIN or poison, e.g. canine zinc toxicosis.

toxin Any poison naturally produced by a plant, animal or microorganism.

toxocariasis An infection caused by roundworms from the genus *Toxocara*. It can occur in dogs, foxes, cattle and humans. Transmission is through resistant eggs in faeces which may survive in soil for long periods. Infection may cause poor growth, abdominal distension and diarrhoea and possibly impaction of the bowel and kidney damage. Infection in humans can cause granulomas to develop in the lung, liver, eyes and brain. Several ANTHELMINTICS are effective against the adult worms.

toxoplasmosis A disease of most HOMEOTHERMS including humans. It is caused by a coccidian parasite, *Toxoplasma gondii*. Cats are a particularly important VECTOR. Cats and other carnivores may become infected by ingesting cystozoites within cysts in the muscles of their prey, or from OOCYSTS present in feline faeces. Infection may cause a variety of signs, depending upon the species, including ABORTION, PERINATAL MORTALITY, coughing, distressed breathing, DIARRHOEA and ENCEPHALITIS. Pregnant women should avoid contact with infected faeces and other sources of infection.

toy breed, teacup pup, toy dog breed A very small breed of dog, e.g. the Chihuahua. *See also* HANDBAG DOG

toy dog breed *See* TOY BREED

toyger A type of DESIGNER CAT produced by the selective breeding of domestic shorthaired tabbies to increase the degree of striping in the coat, giving the appearance of a miniature tiger. *See also* ARTIFICIAL SELECTION

trace element An element required in very small quantities to maintain health, e.g. iodine. *See also* MICRONUTRIENT, VITAMIN. *Compare* MACRONUTRIENT

TRACES *See* TRADE CONTROL AND EXPERT SYSTEM (TRACES)

trachea (tracheae *pl.*)
1. In air breathing vertebrates, a tube of tissue connecting the BUCCAL CAVITY to the BRONCHI, supported by rings of CARTILAGE to keep it approximately circular in cross-section, forming an AIRWAY.
2. One of the cuticular tubes which make up the insect respiratory system. *See also* SPIRACLE (1)

tracheostomy, tracheotomy A surgical incision made in the ventral surface of the trachea to create an alternative airway when normal breathing is not possible. *See also* TRACHEOSTOMY TUBE

tracheostomy tube A tube inserted into a hole in the trachea (TRACHEOSTOMY) to create a clear temporary or permanent AIRWAY when the natural airway has been blocked.

trade *See* CITES, TRADITIONAL MEDICINE

Trade Control and Expert System (TRACES) A system used within the EU for notifying Member States of movements of live animals, germplasm and certain other commodities into or through their territories. It is a web-based service for the application for, and issuing of, Intra Trade Animal Health Certificates (ITAHCs) and Common Veterinary Entry Documents (CVEDs).

Trade Records Analysis of Flora and Fauna in Commerce (TRAFFIC) An organisation established in 1976 as the joint monitoring programme of the WORLD WILDLIFE FUND (WWF) and the INTERNATIONAL UNION FOR THE CONSERVATION OF NATURE AND NATURAL RESOURCES (IUCN) which works in co-operation with the CITES Secretariat to monitor trade in endangered species. It publishes the *TRAFFIC Bulletin* – which contains articles on many aspects of the global wildlife trade – and a number of identification guides on various taxa and wildlife products including ivory, bear gall bladders, crocodilians, turtles, tortoises, butterflies and seahorses. TRAFFIC International is based in Cambridge, UK, and the organisation has regional offices in many parts of the world. Many zoos exhibit animal products that have been seized at ports of entry (usually airports) in order to educate the public about the species protected by CITES.

trade sanctions Measures taken by one state against another often for the purpose of punishing it for a breach of INTERNATIONAL LAW. For example, in 1994 the USA imposed trade sanctions on Taiwan for failing properly to regulate trade in endangered species (especially tiger and rhino products).

T

Fig. T4 Traditional Chinese medicine.

Fig. T5 Trail monitor.

trade-off In animal behaviour, the balancing of competing priorities. For example, an individual must balance the gain in energy from finding food against the loss of energy sustained as a result of **FORAGING BEHAVIOUR**.

tradition *See* **CULTURAL BEHAVIOUR**

traditional Chinese medicine (TCM) **TRADITIONAL MEDICINES** used in Chinese culture, which may contain animal products such as tiger penis and rhinoceros horn (Fig. T4).

traditional medicine 'Cures' for a variety of illnesses and medical conditions which may or may not be effective and which contain animal and plant products taken from the wild or from species that have been bred in captivity. The use of some animal products threatens the survival of endangered animals e.g. bears, tigers, rhinoceroses, and drives the wildlife trade. Some species are farmed in cruel conditions. *See also* **BEAR FARMING, TIGER FARMING, TRADITIONAL CHINESE MEDICINE (TCM)**

TRAFFIC *See* **TRADE RECORDS ANALYSIS OF FLORA AND FAUNA IN COMMERCE (TRAFFIC)**

trail camera *See* **CAMERA TRAP**

trail monitor An electronic device for counting the number of times animals pass along a trail at different times of the day (Fig. T5). Used for monitoring general activity. Operated when an animal breaks an infrared beam set up across the trail or when it passes in front of a passive infrared sensor (**PIR SENSOR**). A trail monitor may be connected to a camera to act as a **CAMERA TRAP** and could be used to monitor enclosure use in a zoo.

trainer Someone who trains animals to perform particular tasks, e.g. **OBEDIENCE TRAINING** of dogs, training horses to jump etc. Some zoos employ trainers to train animals (especially primates, parrots, elephants and marine mammals) to submit to veterinary examination, receive injections, allow blood samples to be taken etc. This removes the need to capture and anaesthetise animals, thereby reducing **STRESS**. *See also* **TARGET TRAINING, TRAINING**

training The process by which an animal is taught a new behaviour or skill by a **TRAINER**. *See also* **OPERANT CONDITIONING, PROTECTED CONTACT, TARGET TRAINING**

tranquilliser A drug which acts on the central nervous system and has a calming effect on an animal, reducing anxiety or tension, without inducing sleep or drowsiness.

tranquilliser dart A **HYPODERMIC SYRINGE** containing a tranquilliser (e.g. **IMMOBILON**), fired from a rifle or a blowpipe and sometimes applied manually attached to the end of a pole. *See also* **ANAESTHETIC (1)**

transcription The cellular process by which the information carried on one strand of **DNA** is passed to a complementary strand of **MESSENGER RNA (mRNA)**. *See also* **PROTEIN SYNTHESIS**

transducer An electronic device that converts one form of energy into another. Used in telemetry

T

devices and **MICROCHIPS**. *See also* **RADIO FRE-QUENCY IDENTIFICATION (RFID) TECHNOLOGY, RADIO TELEMETRY**

transect A linear survey of an **ECOSYSTEM** or **HABITAT** that consists of the collection of data along, or either side of, a line (e.g. a footpath, a road, a path flown by an aircraft (**AERIAL SURVEY**)). Transects may be used to survey vegetation or to **CENSUS** animal populations.

transfer chute A channel through which animals must pass to move from one area of an enclosure to another, e.g. a suspended caged walkway connecting two cages. *See also* **CRUSH**

transfer RNA (tRNA) A group of small **RNA** molecules which bind with specific **AMINO ACIDS** and carry them to the **MESSENGER RNA (MRNA)** at the **RIBOSOMES** during **PROTEIN SYSTHESIS** so that they may be added to a polypeptide chain. *See also* **TRANSCRIPTION, TRANSLATION**

transgenic organism A genetically engineered organism created by inserting the DNA of one species into the **GENOME** of another. *See also* **CHIMERA, ONCO-MOUSE®, SPIDER-GOAT**

transhumance A traditional method of farming in mountainous areas in which herders move their livestock seasonally, to pastures at higher altitudes during the summer and to the lower valleys during the winter.

translation The process in **PROTEIN SYNTHESIS** in which the information in the **GENETIC CODE** contained within a strand of **MESSENGER RNA (MRNA)** is decoded to determine the specific sequence of **AMINO ACIDS** in a polypeptide chain.

translocation The movement of animals from one area to another in order to supplement the local population. This may be a useful conservation tool for some species. *See also* **INTRODUCTION, REINTRODUCTION, REWILDING**

transponder

1. A device which receives a radio signal and automatically transmits a different signal which can be used for detection.

2. An electronic tag implanted under the skin and used to identify individual animals by using a scanner. *See also* **MICROCHIP**

transport of live animals The movement of livestock between locations while still alive, as opposed to after slaughter (Fig. T6). A controversial practice opposed by many animal welfare organisations, e.g. **COMPASSION IN WORLD FARMING (CIWF)**.

transportation In relation to animals, the movement of individuals from one location to another. Special-

T

Fig. T6 Transport of live animals. Sheep being transported by road in the UK. (Courtesy of Clara Rees.)

ist companies ship domesticated, wild and zoo animals. *See also* **INTERNATIONAL AIR TRANS-PORT ASSOCIATION (IATA)**

trap A device for capturing an animal or animals, alive or dead, or for capturing evidence of their presence. Some types of traps are illegal in some jurisdictions, usually because they are considered inhumane. *See also* **AUTO-RESET PEST TRAP, CAMERA TRAP, GIN TRAP, HAIR TRAP, HELIGOLAND TRAP, INSECT TRAP, LONGWORTH TRAP, MOTH TRAP, POLE TRAP, SNARE**

trauma An acute physical injury, e.g. a bite from a large animal, a gunshot wound, the result of a vehicle collision.

Traveling Exotic Animal Protection Act (USA) A Bill presented to the US Congress in November 2011 to amend the **ANIMAL WELFARE ACT OF 1966 (USA)** to restrict the use of exotic and wild animals in travelling **CIRCUSES (1)** and travelling exhibitions in the USA by banning from use in any animal act any animal that was travelling in a mobile housing facility in the preceding 15-day period.

travelling menagerie, travelling zoo An itinerant animal exhibition which developed in the 18th and 19th centuries, often associated with fairgrounds and **CIRCUSES (1)**, although the larger ones travelled as independent entities. Animals travelled and were exhibited in simple cages carried on wagons pulled by horses and sometime elephants. The cages were located in large tents erected behind an impressive frontage. These **MENAGERIES** caused great excitement when they arrived in a town as they were the only opportunity most people had to see exotic animals. Many successful menageries operated in the UK and toured widely, including in Europe, and sometimes America. They often had elaborate and exotic names: *Polito's Grand Collection of Beasts*, *Miles's Grand Collection of Living Curiosities*, *Ballard's Grand Collection of Wild Beasts*, *William Mander's Grand National Star Menagerie*. Some changed names when they merged with others or when their proprietors adopted new names. *Anderton and Haslam's No. 1 Royal Menagerie* became *Anderton, Haslam and Forepaugh's Menagerie* and later *Professor Anderton and Captain Rowland's Combined Show*. In the late 19th century combined shows became common as menagerie owners added cinematograph shows to their businesses, e.g. *Crecraft's Wild Beast and Living Picture Show*, *Hancock's Living Pictures and Menagerie*. Travelling menageries also existed in America, e.g. the *Titus Menagerie* and *P.T. Barnum's Great Traveling Museum, Menagerie, Caravan and Hippodrome*. Some menagerie owners went on to establish zoos. *See also* **BARNUM AND BAILEY, BOSTOCK AND WOMBWELL'S ROYAL MENAGERIE, HAGEN-BECK, PIDCOCK'S WILD BEAST SHOW**

travelling stock route (TSR) A route which has been authorised for the movement of livestock (e.g. sheep and cattle) on foot in Australia. TSRs are long vegetated corridors of public land, often located alongside roadsides. They occur in all states of Australia, but predominantly in Queensland and New South Wales. The network of TSRs is collectively known as the 'Long Paddock'. Travelling stock reserves are fenced paddocks located at strategic positions along TSRs where livestock may be watered and rested overnight. TSRs act as reserves of natural vegetation and are considered to be important wildlife **CON-SERVATION CORRIDORS**.

travelling zoo *See* **TRAVELLING MENAGERIE**

Travers, Bill (William) (1922–1994) *See* **BORN FREE, RING OF BRIGHT WATER**

trawling A method of sea fishing which involves dragging a large bag-shaped net behind a boat called a trawler.

treadmill An electrically powered moving floor, similar to a conveyor belt, on which an animal walks or runs as a therapy after an injury, to study its **GAIT**, or while measuring some aspect of its physiology, e.g. respiratory rate. The position of the animal does not move in relation to the treadmill as the speed of the floor is regulated to match the animal's forward speed. *See also* **UNDERWATER TREADMILL**

treat dispenser A device that releases a small piece of food (a treat).

Treaty of Amsterdam 1997 *See* **PROTOCOL ON PRO-TECTION AND WELFARE OF ANIMALS 1997**.

tree-hugger A term used for an environmentalist, often in a derogatory manner. In northern India in the late 15th century, Jambeshwar, the son of a village leader, founded a Hindu sect called the Bishnois based on a religious duty to protect trees and wildlife. According to legend, in 1730, when the Maharajah of Jodhpur ordered that the few remaining trees in the area be cut down 363 women died hugging the trees while trying to protect them. *See also* **BUNNY-HUGGER**

tremors An involuntary quivering movement, e.g. head tremors in dogs.

triage The process in which cases are rapidly examined and classified by veterinary staff so that urgent cases can be prioritised for treatment. This is important in a busy veterinary hospital or some other situation where many animals need treatment at the same time.

trial and error learning *See* **OPERANT CONDITIONING**

tri-axial accelerometer *See* **ACCELEROMETER**

tribe A taxonomic rank between **FAMILY (1)** and **GENUS**.

tricarboxylic cycle *See* **KREBS CYCLE**

trichomoniasis A **SEXUALLY TRANSMITTED DISEASE (STD)** caused by a protozoan parasite (*Trichomonas* spp.).

trinomial name The scientific name of an organism which consists of the **BINOMIAL NAME** plus a third name which identifies the **SUBSPECIES**, e.g. *Panthera leo persica*.

T

tRNA *See* **TRANSFER RNA (TRNA)**

trocar A pointed metal instrument used to puncture the wall of a body cavity. Mainly used in **RUMINANTS** to allow the release of gas from the **RUMEN** in cases of bloat.

troglobite A small cave-dwelling animal that has lost many of its senses and much of its pigmentation, and cannot survive outside a cave environment. *See also* **TROGLOPHILE, TROGLOXENE**

troglophile A cave-dwelling animal that may complete its life cycle in a cave, but can also survive in above-ground habitats. Also known as a 'cave lover'. *See also* **TROGLOBITE, TROGLOXENE**

trogloxene An animal that uses caves for shelter but does not complete its life cycle in them, e.g. bats. Also known as a 'cave guest'. *See also* **TROGLOPHILE, TROGLOBITE**

Trogoniformes An order of birds: trogons.

trophic level The position that an organism occupies in a **FOOD CHAIN**: primary producer (*see* **AUTOTROPH**), **SECONDARY CONSUMER, TERTIARY CONSUMER, DECOMPOSER**.

trophy In **SPORT HUNTING**, an animal part or product (e.g. a head or skin) kept as a memento of a hunting expedition and a symbol of a hunter's prowess.

trophy fee An amount of money paid in return for being allowed to kill a particular species of wild animal as part of a **SPORT HUNTING** scheme.

trophy hunting *See* **SPORT HUNTING**

tropical Relating to the tropics: the area of the Earth that lies between the Tropic of Cancer and the Tropic of Capricorn.

tropical forest Any of a number of forest vegetation types which occurs in tropical areas, including **CLOUD FOREST, MANGROVE FOREST** and **RAINFOREST**.

tropical house A zoo building, open to the public, where **TROPICAL** animals are kept in warm humid conditions. Sometimes rainfall is simulated by spraying water into the atmosphere inside the building. It may contain animals within enclosures (e.g. crocodilians) and others in cages (e.g. toucans) or **VIVARIUMS**. Some species may be free-roaming (e.g. birds, chelonians) in areas planted with large tropical plants.

trypanosomiasis The name given to a group of diseases caused by flagellated protozoans of the genus *Trypanosoma* which are found in the bloodstream. One of these diseases, African trypanosomiasis, is transmitted by the tsetse fly (*Glossina*). The disease is usually chronic but acute cases occur and mortality rates may be high. Signs include intermittent fever, anaemia and loss of condition. **LYMPH NODES** are often enlarged. A **CHANCRE** (hard swelling) occurs at the site of the insect bite and is the first sign of infection. Drugs are used for both prophylaxis and treatment. American trypanosomiasis (Chagas disease) occurs in South and Central America and is transmitted to animals and people by blood-sucking triatomid bugs. Other **VECTORS** include rats, mice, foxes, ferrets and vampire bats.

trypsin A digestive enzyme which breaks down proteins. It is secreted into the small intestine by the **PANCREAS** as the inactive trypsinogen.

Tsavo An area of southern Kenya where Tsavo National Park is located. Used in the name of some themed African savanna zoo exhibits, e.g. *Tsavo Bird Safari* (Chester Zoo).

tsetse fly All of the species of the genus *Glossina*. Adults feed on the blood of vertebrates and are important vectors of several diseases including human sleeping sickness and animal **TRYPANOSOMIASIS**. Their bites typically produce a **CHANCRE**.

t-test A **PARAMETRIC** statistical test which is used to determine whether or not the **MEANS** calculated from two **SAMPLES** have come from the same **POPULATION**. A dependent (paired) t-test can be used to test the **NULL HYPOTHESIS (H_0)** that the means of two samples of paired data are equal, e.g. the scores obtained in two different cognition tests by 20 chimpanzees, where each animal has taken both tests. An independent (unpaired) t-test may be used to test the null hypothesis that the means of two independent samples are equal, e.g. the mean body length of beetle species X at location A compared with the mean length at location B.

tubal ligation The severing and sealing of the **FALLOPIAN TUBES** as a method of **CONTRACEPTION**.

tube feeding *See* **FORCE FEEDING**

tuberculosis (TB) A contagious disease caused by bacteria belonging to the genus *Mycobacterium*. It affects a wide range of mammals (including humans), birds, reptiles and fishes and is characterised by the formation of nodules or tubercles in almost any tissue or organ. Infection may be through the respiratory system, digestive tract, through a wound, contaminated feed, infected dung or by sexual contact. Treatment in domesticated animals is generally not attempted, but in zoo animals TB is sometimes treated with para-aminosalicylic acid. Control measures include good hygiene, good ventilation and good feeding. Bovine tuberculosis is transmitted between cattle and badgers (*Meles meles*) in the UK. Bovine TB is a notifiable disease. *See also* **KREBS REPORT**

Tubulidentata An order of mammals: **AARDVARK**.

tumescence A swollen or distended area caused by the normal engorgement with **BLOOD**, e.g. the erection of the mammalian **PENIS** is the result of penile tumescence. Many female primates exhibit **PERINEAL TUMESCENCE** (Fig. P4) as an indication of sexual receptivity. *See also* **DECEPTIVE SWELLING**

tumescent Relating to or exhibiting **TUMESCENCE**.

tumor *See* **TUMOUR**

T

tumour, neoplasm, tumor A swelling caused by the uncontrolled growth of cells which may be benign (harmless) or malignant (cancerous). *See also* CANCER

tundra A biome that consists of a treeless plain which experiences low temperatures and rainfall, and has low biological diversity and a simple vegetation structure, typically including lichens. It occurs in the Arctic and Antarctic. Arctic tundra encircles the north pole, north of the BOREAL FOREST, and is inhabited by polar bears, musk oxen, lemmings, reindeer, arctic foxes and arctic hares. Alpine tundra occurs on mountains above the tree line.

turkey A large pheasant-like bird (*Meleagris gallopavo*) which is native to Central and South America and is a popular poultry bird especially in the USA and the UK (Fig. T1).

turnstile A rotating gate through which visitors may pass when entering a zoo or other visitor attraction which is used to control entry and count visitors. Turnstiles used to be common in zoos but are now less so.

tush

1. A short narrow tusk, especially that sometimes found in female Asian elephants (*Elephas maximus*), that are normally tuskless.

2. The CANINE tooth of a horse.

tusk One of a pair of long, curved, pointed teeth which projects from the mouth in a number of species, e.g. elephant, walrus, wild boar, narwhal. *See also* IVORY, TUSH (1), TUSKER

tusker An animal bearing very well-developed TUSKS, especially an elephant, but also other species, e.g. a walrus or wild boar. *See also* TUSH (1)

TWG *See* TAXON WORKING GROUP (TWG)

twin lamb disease *See* PREGNANCY TOXAEMIA

twinning The production of twins (Fig. T7). They may be two individuals produced from the same ZYGOTE that split into two after FERTILISATION and therefore genetically identical (monozygotic twins), or produced when two sperm fertilise two different ova at the same time (dizygotic twins). The latter are not identical (fraternal twins) and are no more genetically similar than siblings born at different times. Monozygotic twins are CLONES (1). *See also* ARTIFICIAL EMBRYO TWINNING, PREGNANCY TOXAEMIA

twitch

1. An involuntary muscle spasm.

2. A device consisting of a noose or a hinged smooth metal ring used to grasp the upper lip of a horse to restrain it during veterinary or grooming procedures to prevent it from moving. This is thought to cause

Fig. T7 Twinning: twin lambs are common and may result in twin lamb disease in the mother.

the release of **ENDORPHINS** which calm the animal. Also called a humane twitch.

3. To use a twitch as described in (2).

4. To engage in **TWITCHING**.

twitcher A fanatical bird-watcher who 'collects' sightings of **BIRDS** and whose aim is to see as many different **SPECIES** as possible. *See also* **TWITCHING**. *Compare* **BIRDER**

twitching The act of searching for rare or infrequently seen birds. *See also* **TWITCHER**. *Compare* **BIRDING**

two year old In relation to horses, according to the **AMERICAN STUD BOOK** 2008 '*A colt, filly or gelding in its third calendar year of life (beginning January 1 of the year following its yearling year)*.'

tympanum, tympanic membrane The membrane of the ear which separates the outer ear from the middle ear and which vibrates in response to sound. The sound is then transmitted to the **COCHLEA** via the **EAR OSSICLES**.

type I conditioning *See* **CLASSICAL CONDITIONING**

type II conditioning *See* **OPERANT CONDITIONING**

type specimen, holotype The individual specimen upon which the original scientific description of a particular species was based. *See also* **NOMINATE SUBSPECIES**

typhoid *See* **FOWL TYPHOID**

T

Fig. U1 U is for up-ending.

Dictionary of Zoo Biology and Animal Management: A guide to terminology used in zoo biology, animal welfare, wildlife conservation and livestock production, First Edition. Paul A. Rees.
© 2013 John Wiley & Sons, Ltd. Published 2013 by John Wiley & Sons, Ltd.

U *See* URACIL (U)

UFAW *See* UNIVERSITIES FEDERATION FOR ANIMAL WELFARE (UFAW)

UFAW *Wild Animal Welfare Award* *See* UNIVERSITIES FEDERATION FOR ANIMAL WELFARE (UFAW)

ulcer A persistent sore on the surface of the SKIN or of a MUCOUS MEMBRANE lining a body cavity (e.g. the gut) which is often associated with INFLAMMATION.

ulceration The development or formation of an ULCER. *See also* CHANCRE

ulna (ulnae *pl.*) A **long bone** which, together with the radius, forms the lower forelimb in tetrapod vertebrates. It is the longer, larger and posterior of the two bones.

ultimate cause *See* FUNCTIONAL CAUSE

ultradian rhythm A pattern in time that has a period of considerably less than 24 hours. *Compare* CIRCADIAN RHYTHM, CIRCANNUAL RHYTHM, INFRADIAN RHYTHM, LUNAR RHYTHM

ultrasonic Relating to ULTRASOUND

ultrasonography A technique, involving an ultrasound scanner, in which high-frequency sound waves are bounced off internal organs and tissues and produce a pattern of echoes that are then used by a computer to create sonograms (images) of areas inside the body. Used for diagnosing disease and injury. *See also* SOUND SPECTROGRAM

ultrasound Sound frequencies higher than the limit of human hearing (above 20kHz), e.g. those produced by bats during ECHOLOCATION. *Compare* INFRASOUND

ultrasound scanner *See* ULTRASONOGRAPHY

ultraviolet light, UV light Electromagnetic radiation which lies in the ultraviolet range and has wavelengths shorter than visible light but longer than X-rays (10–400nm). It is made up of UVA, UVB and UVC light. Many species of reptiles require ultraviolet light, without which BONE RAREFACTION (weakening) may occur. Reptiles require both UVA, which affects activity cycles, and UVB, which is important in vitamin D_3 synthesis. Vivariums should provide a gradient of UV light and shade so that reptiles can self-regulate their exposure. UVB and some UVA light is blocked by glass.

umbilical cord A flexible tube of tissue which contains a bundle of blood vessels in a female mammal connecting the foetus to the uterus via the placenta and through which the developing foetus receives oxygen and food and gets rid of carbon dioxide and nitrogenous waste.

umbilical cord torsion Strangulation of the UMBILICAL CORD. This may cause foetal death and ABORTION if the cord wraps around the foetus, cutting off the blood flow in the cord and thereby reducing the transport of oxygen to the FOETUS.

UN *See* UNITED NATIONS (UN)

uncaged An international animal welfare organisation which campaigns against VIVISECTION and XENOTRANSPLANTATION, and for ANIMAL RIGHTS, and for democratic action on animal issues through the political system.

unclean animal, impure animal A species of animal that it is taboo to eat, or touch, in some religions, e.g. in Judaism (pigs, mice, camels), Islam (pigs, dogs). These taboos are based on the teachings of religious teachers and writings in religious books, e.g. the Bible and the Qur'an. *See also* ARK (1)

underfeeding The provision of insufficient food, especially for livestock. In sheep it may result in thin ewes, weak lambs and poor milk yield. *Compare* OVERFEEDING

undergrowth Low growing plants, shrubs and saplings on the forest floor.

understorey The lowest level in a forest: the area between the ground and a height of about 10m.

underwater treadmill, aquaciser, aquatred, hydrotherapy water walker, post-surgery recovery pool A TREADMILL submerged in a heated water tank used for HYDROTHERAPY, to treat animals (e.g. dogs) with conditions such as leg injuries, hip surgery, INTERVERTEBRAL DISC disease, hip or elbow DYSPLASIA, neurological conditions that affect walking, or as part of a weight loss or fitness programme. The water creates a resistance to the movement of the limbs thereby increasing the work done by the MUSCLES.

underwater tunnel A transparent tunnel made of acrylic material which passes through a large tank in an aquarium. It has a curved roof (which allows it to support the weight of water above) and is often equipped with a moving floorway to carry visitors continuously through the tank.

UNEP *See* UNITED NATIONS ENVIRONMENT PROGRAMME (UNEP)

UNESCO *See* UNITED NATIONS EDUCATIONAL, SCIENTIFIC AND CULTURAL ORGANIZATION (UNESCO)

ungulate A mammal with hooves; members of the ARTIODACTYLA and PERISSODACTYLA. The majority of large herbivores on Earth, consisting of some 260 species.

unguligrade Pertaining to walking on hooves, as in an UNGULATE. *Compare* DIGITIGRADE, PLANTIGRADE

Unimog A rugged, four-wheel drive Mercedes truck often used in difficult terrain and for fieldwork.

unipara A UNIPAROUS female.

uniparous

1. Describing a female who has had a single offspring.

2. Describing a female who is only capable of producing one offspring at a time (at each birth).

See also MULTIPAROUS, NULLIPAROUS, POLYPAROUS, PRIMIPAROUS

United Nations (UN) An international organisation which was established in 1945 (after the Second World War) whose primary objective is the achievement of world peace. It is based in New York, has 193 Member States which co-operate in the promulgation of international law, some of which is concerned with the protection of biodiversity, the sustainable use of resources and the prevention of environmental pollution.

United Nations Conference on Environment and Development (UNCED) *See* EARTH SUMMIT, RIO

United Nations Conference on the Human Environment 1972 A conference held in Stockholm which marked the beginning of a serious interest in environmental protection among the governments of many of the countries of the world.

United Nations Educational, Scientific and Cultural Organization (UNESCO) An organisation of the UNITED NATIONS (UN) whose mission is *'to contribute to the building of peace, the eradication of poverty, sustainable development and intercultural dialogue through education, the sciences, culture, communication and information.'* *See also* CONVENTION FOR THE PROTECTION OF THE WORLD CULTURAL AND NATURAL HERITAGE 1972

United Nations Environment Programme (UNEP) An organisation of the UNITED NATIONS (UN) whose mission is *'to provide leadership and encourage partnership in caring for the environment by inspiring, informing, and enabling nations and peoples to improve their quality of life without compromising that of future generations.'*

United States Department of Agriculture (USDA) The government department responsible for agriculture food and some aspects of natural resource management in the USA. USDA agencies include the ANIMAL AND PLANT HEALTH INSPECTION SERVICE (APHIS), the Natural Resources Conservation Service (NRCS), the Forest Service, the Farm Service Agency (FSA) and the National Institute of Food and Agriculture (NIFA).

United States Department of the Interior (DOI) The US Department of the Interior protects America's natural resources and heritage, and has responsibilities for Indian affairs, energy production and climate change. Agencies of the department include the UNITED STATES FISH AND WILDLIFE SERVICE (USFWS), the BUREAU OF LAND MANAGEMENT (BLM), the UNITED STATES NATIONAL PARK SERVICE and the US Geological Survey.

United States Fish and Wildlife Service (USFWS) The federal agency of the UNITED STATES DEPARTMENT OF THE INTERIOR (DOI) in the USA responsible for wildlife. The Service works to conserve, protect and enhance fish, wildlife, and plants and their habitats. *See also* USFWS FORENSICS LABORATORY, WILDLIFE INSPECTOR (2)

United States Food and Drug Administration (USFDA) The USFDA protects consumers by ensuring that all foods and drugs are safe and effective, and protects and advances public health. This includes the regulation of animal drugs, animal feed (including pet food) and animal cloning.

United States National Parks Service An agency of the UNITED STATES DEPARTMENT OF THE INTERIOR (DOI) which has responsibility for over 400 NATIONAL PARKS (NP). *See also* YELLOWSTONE NATIONAL PARK

Universal Declaration on Animal Welfare (UDAW) The WORLD SOCIETY FOR THE PROTECTION OF ANIMALS (WSPA) and other animal welfare groups are campaigning for a Universal Declaration on Animal Welfare (UDAW) to be endorsed by the UNITED NATIONS (UN). This would be an agreement among people and nations that: animals are sentient – they can suffer and feel pain; their welfare needs must be respected; and animal cruelty must end for good.

Universities Federation for Animal Welfare (UFAW) A charity that works to develop and promote improvements in the welfare of all animals through scientific and educational activity worldwide. UFAW organises conferences on welfare issues, publishes books, produces videos and technical reports, and provides advice to the UK government on animal welfare including legislation. It publishes the journal *ANIMAL WELFARE*, which contains, among other things, papers on farm and zoo animal welfare. UFAW funds research into animal welfare and makes an annual *Wild Animal Welfare Award*.

unzoo A concept developed by the zoo architect Jon COE who defined an unzoo as: *'A place where the public learns about wild animals, plants and ecosystems through interaction with and immersion in original or recreated natural habitats'* (Coe, no date). *Compare* LIVING MUSEUM

up-ending In wildfowl, the action of tipping the body so that the anterior end is submerged in water and the posterior end is held vertically as the bird collects food from the bottom of a pond or similar body of water. The position is maintained by paddling with the feet (Fig. U1).

uracil (U) A NUCLEOTIDE base which substitutes for T (thymine) in **RNA**, and pairs with A (adenine).

ureter One of a pair of tubes that drains the kidneys, carrying urine to the bladder.

urethra A tube which carries urine from the BLADDER to the outside of the body in mammals. In males it is joined by the VAS DEFERENS.

urine The waste liquid produced by the kidneys which consists of water, nitrogenous wastes and other waste materials. The loss of urine assists in maintaining water balance. *See also* OSMOREGULATION

urine washing A behaviour performed by some male primates whereby they urinate on their own hands and then rub the urine into their fur, e.g. capuchin monkeys (Cebidae). This appears to function as a signal by which males indicate their availability to females and make themselves attractive.

urolith A urinary stone.

uropygial gland *See* PREEN GLAND

ursid A member of the mammalian family URSIDAE.

Ursidae A family of mammals which consists of the bears, including the giant panda (*Ailuropoda melanoleuca*).

ursine Relating to URSIDS.

USDA *See* UNITED STATES DEPARTMENT OF AGRICULTURE (USDA)

USFDA *See* UNITED STATES FOOD AND DRUG ADMINISTRATION (USFDA)

USFWS *See* UNITED STATES FISH AND WILDLIFE SERVICE (USFWS)

USFWS Forensics Laboratory The laboratory is located in Oregon and is the only laboratory in the world devoted to crimes against wildlife. It consists of seven units: administration, chemistry, criminalistics, genetics, morphology, pathology and digital evidence. It is the official crime lab of the Wildlife Working Group of INTERPOL and CITES.

USGS National Wildlife Health Center (NWHC) A science centre of the Biological Resources Discipline of the United States Geological Survey, located in Madison. It was established in 1975 and assesses the impact of diseases on wildlife.

uterus, womb A muscular organ in which the embryo develops in mammals (except monotremes). It is paired in most taxa but singular in primates. It is connected to the FALLOPIAN TUBES through which fertilised eggs pass prior to implantation in the uterine wall, resulting in the formation of the PLACENTA. The uterus is connected to the outside of the body by the VAGINA, through which the embryo passes during PARTURITION as a result of contractions of the muscles in the uterine wall. *See also* ARTIFICIAL UTERUS

utilitarianism The proposition that the moral worth of an action is solely determined by its contribution to overall utility. In other words, the end justifies the means. It requires that one should act in such a way as to do the greatest amount of good for the largest number of individuals. This school of thought is generally credited to Jeremy BENTHAM. It is often used to justify the keeping of animals in zoos in order to conserve a species. *Compare* COUNTER-UTILITARIANISM

UV light *See* ULTRAVIOLET LIGHT

U

Fig. V1 V is for Victoria crowned pigeon (*Guora victoria*).

vaccination The administration of a **VACCINE**.

vaccine A material which produces an immune reaction (stimulates the production of **ANTIBODIES**) in an animal and subsequently confers an acquired **IMMUNITY** to a microorganism.

vacuum activity A behaviour performed without the presence of appropriate stimuli, e.g. nesting behaviour exhibited in the absence of any nesting material. *See also* **VACUUM DUST BATHING**

vacuum dust bathing **DUST BATHING** behaviour performed by **BATTERY CHICKENS** in the absence of 'dust'. *See also* **ENRICHED COLONY CAGES**

vacuum-chewing *See* **SHAM-CHEWING**

vagina A duct in mammals which receives the penis of the male during copulation and whose epithelium may undergo cyclic changes during the **OESTROUS CYCLE** under the influence of sex hormones. *See also* **ARTIFICIAL VAGINA**

vaginal plug *See* **SPERM PLUG**

vagrant An individual bird that has strayed from its normal geographical range. *Compare* **PASSAGE MIGRANT**

valve
1. A pocket or flap of tissue in the vertebrate heart wall that prevents the back flow of blood between chambers.
2. In the veins and lymphatics of birds and mammals, a pocket-like flap that prevents the back flow of blood.

variable *See* **CONFOUNDING VARIABLE, CONTINUOUS VARIABLE, DEPENDENT VARIABLE, DISCRETE VARIABLE, INDEPENDENT VARIABLE**

variance A measure of the dispersion of values around the **MEAN**. The square of the **STANDARD DEVIATION**. *See also* **RANGE (3), STANDARD ERROR OF THE MEAN**

variation *See* **ENVIRONMENTAL VARIATION, GENETIC VARIATION**

variety A taxon of organisms below the rank of species; a race, strain or breed. A formal category used in botany.

varmin *See* **VERMIN**

varmint *See* **VERMIN**

vas deferens (vasa deferentia *pl.*) One of a pair of muscular tubes which carries sperm from the testes (one from each) to the outside of the body. The vas deferens leads from the epididymis to the urethra in mammals or the cloaca in birds and reptiles. **VASECTOMY** is achieved by cutting and tying the vasa deferentia.

vascular Relating to the **CIRCULATORY SYSTEM** of an animal, e.g. vascular tissue.

vasectomy The severing of the **VASA DEFERENTIA**, to prevent sperm from entering **SEMEN**, as a means of **CONTRACEPTION**.

vasopressin, anti-diuretic hormone (ADH) A hormone produced by the **POSTERIOR PITUITARY** in mammals and released into the blood if blood water potential drops below normal. It causes vaso-constriction of the arterioles, thereby raising blood pressure, and also retention of water by the kidneys.

Vatican Menagerie A menagerie which was located in the Vatican. It thrived under Pope Leo X (1513–1523) and included monkeys, civets, lions, leopards, an elephant and a snow leopard. In addition to animals, one cardinal also kept people from a variety of ethnic origins. *See also* **HUMAN EXHIBITS**

veal Meat produced from a calf.

vector An agent that is responsible for the transmission of a disease or parasite from one organism to another. Often a biting insect. The vector may be affected by the disease or transmit it passively. The removal of vectors is an effective means of controlling some parasites, e.g. control of *Anopheles* mosquitoes reduces the incidence of **MALARIA**.

veggiesicle A block of ice containing vegetables. Used as **FEEDING ENRICHMENT**, e.g. for pandas.

vein A blood vessel that possesses a large lumen and a relatively thin, muscular wall, through which blood flows from the capillaries in the body towards the heart. In most cases it carries **DEOXYGENATED** blood.

veldt, veld Any area of open grassland with few or no trees, especially in southern Africa.

vending machine A coin-operated machine that sells food, drink or other small items. Used by some zoos to sell small bags of animal food or badges to raise money for conservation. A machine at the *Gorilla Kingdom* in London Zoo which dispenses gorilla badges asks users to 'vote' for how their money should be spent on conservation, e.g. on education, paying more rangers etc., and displays the total number of votes received for each category.

venom The poisonous fluid that some animals (e.g. some snakes, insects, etc.) produce and then inject into the bodies of their prey and enemies by biting, stinging or by some other means.

venomous Capable of producing **VENOM**.

vent *See* **CLOACA (1)**.

ventilation
1. The exchange of air between the lungs (or gills) and the environment.
2. The supply of air to the lungs by artificial means.
3. The exchange of air between a building and the outside environment. Draughts in livestock sheds can cause poor growth and encourage disease. *See also* **AIR FLOW, VENTILATION SHUTDOWN**

ventilation shutdown
1. A method of killing poultry by switching off ventilation systems in poultry buildings so that the birds die of **HYPOTHERMIA**, starvation, dehydration, lack of oxygen and possibly disease. Animal welfare groups consider this method of killing poultry to be inhumane.
2. In England, under the Welfare of Animals (Slaughter or Killing) (Amendment) (England) Regulations 2006 (SI 2006 No 1200) ventilation shutdown '*means the cessation of natural or mechanical*

ventilation of air in a building in which birds are housed with or without any action taken to raise the air temperature in the building'. This method of killing poultry may be authorised by the Secretary of State if he considers that no other method is practicable. It is intended as a measure to destroy birds in the case of an outbreak of disease such as **AVIAN INFLUENZA**.

ventral Of or relating to the front or anterior surface of the body; on or towards the lower abdominal plane. The opposite of **DORSAL** (Fig. A8).

ventricle
1. A general term for a chamber.
2. In the mammalian heart, one of two large muscular chambers which pumps blood to the lungs (right ventricle) and around the body (left ventricle).
3. In the brain, a chamber filled with **CEREBRO-SPINAL FLUID (CSF)**.

venule A narrow blood vessel which drains blood from a capillary. Venules combine to form veins.

vermicide *See* **ANTHELMINTIC**

vermiform Worm-like in appearance.

vermifuge *See* **ANTHELMINTIC**

vermin, varmin, varmint Wild mammals, birds and other animal **PESTS** that are harmful to crops, **FARM ANIMALS**, property, **GAME** or carry disease, and are often difficult to control. May also include insects such as cockroaches. There is no definition of vermin in UK law but various legislation concerned with pest control refers to moles, grey squirrels, rabbits, mink, stoats, weasels, rats and mice.

vernacular name The common name of a species, e.g. lion, blackbird, hippopotamus. Vernacular names vary between countries and languages, and over time. There may be several names used for a single species in a particular region. At various times and in various places in the UK the barn owl (*Tyto alba*) has been called the screech owl, silver owl, yellow owl, hobby owl, white owl, hissing owl, church owl, Jenny owl, ullat, oolert, willow owl and many other names. To avoid confusion scientists use the **BINOMIAL SYSTEM OF NOMENCLATURE** to assign a single Latin or scientific name **(BINOMIAL NAME)** to each species.

vernier caliper, vernier calliper *See* **CALIPER**

Versailles Menagerie A **MENAGERIE** built in 17th-century France by Louis XIV in the grounds of his palace at Versailles. The first animals were installed in 1665. When visitor numbers grew they did so much damage that the king had to restrict admittance to members of his court. The menagerie fell into disrepair and closed in 1792. The remaining animals were offered to the former Jardin du Roi in Paris, which was renamed the **JARDIN DES PLANTES**.

vertebra (vertebrae *pl.*) One of a series of bones which forms the vertebral column that protects the spinal cord in vertebrates.

vertebral column, backbone, spinal column, spine The series of bones (vertebrae) that runs along the long axis of the body of all vertebrates, surrounding and protecting the **SPINAL CORD** and providing attachment for muscles.

Vertebrata A subphylum of the phylum **CHORDATA** which contains the vertebrates. A vertebrate possesses a series of **VERTEBRAE** (segmented bones) surrounding the notochord and the nerve cord. The notochord is only present during embryonic development and its protective function is replaced by the vertebrae in the adult animal. Vertebrates possess complex, **CLOSED CIRCULATORY SYSTEMS** in which **BLOOD** containing **HAEMOGLOBIN** is pumped around the body by a **HEART**. This, and the development of a complex **RESPIRATORY SYSTEM**, has allowed the evolution of some very large species.

vertebrate A member of the **VERTEBRATA**.

vespertine Active at dusk or early evening. *Compare* **CATHEMERAL, CREPUSCULAR, DIURNAL, MATUTINAL, NOCTURNAL**

vestigial Relating to a structure which has become small and lost its function as a result of evolution, e.g. the pelvic bones found in some whale species.

vet *See* **VETERINARY SURGEON**

veterinarian *See* **VETERINARY SURGEON**

veterinary conservation medicine *See* **CONSERVATION MEDICINE**

Veterinary Laboratories Agency *See* **ANIMAL HEALTH AND VETERINARY LABORATORIES AGENCY (AHVLA)**

veterinary medicine The branch of medicine which is concerned with the cause, diagnosis and treatment of diseases and injuries in animals (companion, farm and wild) and the maintenance of their health. *See also* **CONSERVATION MEDICINE, PREVENTATIVE VETERINARY MEDICINE**

Veterinary Medicines Directorate An executive agency of the **DEPARTMENT FOR ENVIRONMENT, FOOD AND RURAL AFFAIRS (DEFRA)** which seeks to ensure the safe and effective use of veterinary medicinal products and aims to protect public health, animal health and the environment. It also promotes animal welfare by assuring the safety, quality and efficacy of veterinary medicines.

Veterinary Medicines Regulations 2011 A Statutory Instrument which lays down controls and procedures concerning the authorisation, manufacture, supply and use of veterinary medicines in the UK.

veterinary nurse A person who is qualified to administer nursing care to animals. In the UK, the **ROYAL COLLEGE OF VETERINARY SURGEONS (RCVS)** maintains a Non-statutory Register for Veterinary Nurses and persons listed on this register are know as Registered Veterinary Nurses (RVNs). An RVN is required to keep up-to-date and adopt a high standard of professional conduct.

V

Veterinary Poisons Information Service (VPIS) A 24-hour advice service for veterinary professionals in the UK for the diagnosis and management of poisoned animals.

veterinary public health The sum of all contributions to the physical, mental and social wellbeing of humans through an understanding and application of veterinary science (Anon., 1999). In this context veterinary science includes animal production.

Veterinary Record An academic journal of the BRITISH VETERINARY ASSOCIATION (BVA) which reports the results of research in veterinary science.

veterinary surgeon, vet, veterinarian A person who practises VETERINARY MEDICINE who must normally be registered with a professional body in order to practise, e.g. ROYAL COLLEGE OF VETERINARY SURGEONS (RCVS) in the UK. *See also* INTERNATIONAL ZOO VETERINARY GROUP (IZVG), ROYAL ARMY VETERINARY CORPS (RAVC), VETERINARY NURSE, VETERINARY SURGEONS ACT 1966

Veterinary Surgeons Act 1966 This Act makes provision for the management of the veterinary profession in Great Britain, for the registration of VETERINARY SURGEONS and veterinary practitioners, for regulating their professional education and professional conduct and for cancelling or suspending registration in cases of misconduct. Under the Act it is an offence for anyone who is not a registered vet to practise veterinary surgery, except in a small number of specific circumstances. For example, a medical practitioner or a dentist may treat or operate on an animal at the request of a registered vet.

vicarious liability The legal liability that an employer may have for the actions and omissions of an employee in certain circumstances when the employee commits a wrong while working on his employer's behalf. For example, if a visitor to a zoo or a farm is injured due to the negligence of a keeper or farm worker it may be possible for the injured person to sue the employer of the person that caused the injury. The employer would normally be in a better position to provide financial compensation to the injured party than the employee. Furthermore, the principle of vicarious liability provides an incentive for employers to supervise the behaviour of their employees closely and requires them to employ competent persons. *See also* LIABILITY FOR THE ACTIONS OF ANIMALS

Vienna Zoo *See* SCHÖNBRUNN ZOO

viewpoint A place from which something is viewed, e.g. an elevated position on a high-level walkway, a window, an UNDERWATER TUNNEL etc. The location of viewpoints is very important in enclosure design in zoos because they control the experience of the visitor.

vigilance behaviour A state of arousal in which an animal is attuned to detect specific events such as the presence of a predator; alertness or watchfulness; a state of readiness to predict uncertain events. This may take the form of looking up while feeding in, for example, birds and antelopes (Fig. V2). Animals may do this more often if they detect danger. Vigilance behaviour may be directed towards rivals or possible mates.

villus (villi *pl.***)** A finger-like projection which increases the surface area available for the movement of material across a surface. Found in the arachnoid membrane in the brain and the lining of the SMALL INTESTINE. *See also* MICROVILLUS

Vincennes Zoo *See* BOIS DE VINCENNES

viraemia, viremia The condition in which VIRUS particles are present in the blood.

viraemic, viremic Relating to VIRAEMIA.

viral Relating to a VIRUS.

viral disease A disease caused by a VIRUS, e.g. avian influenza.

viral haemorrhagic disease (VHD) An infectious disease of rabbits that is usually fatal.

viremia *See* VIRAEMIA

viremic *See* VIRAEMIC

virgin forest *See* PRIMARY FOREST

virology The scientific study of VIRUSES and viral diseases.

virtual zoo An on-line collection of profiles and photos of various animal species and zoo EXHIBITS. Often created by a zoo to encourage visits. The WORLD ASSOCIATION OF ZOOS AND AQUARIUMS (WAZA) has a virtual zoo as part of its website.

virus A non-cellular organism which consists of protein surrounding nucleic acid (DNA or RNA) with no metabolism of its own. It must infect another cell in order to reproduce using the host cell's metabolism to produce new virus particles. These are released from the host cells, often because they rupture, and infect new cells. Viruses infect all types of organisms and cause a wide range of diseases. *See also* PRION

viscera A collective term for the large internal ORGANS of the body, especially those of the ABDOMEN and THORAX.

vision The detection of light by the eyes and the behaviour response produced as a result of images formed in the brain. *See also* FIELD OF VIEW

visitor attendance The number of visitors that visit a zoo, usually expressed as number per year (annual visitor attendance). In most cases it is impossible to determine how many of these visits are repeat visits within the same year because records of individual visitors are not kept. The records are therefore of numbers of visits rather the numbers of visitors *per se*. Many factors affect annual attendance including the weather (traditional zoos see a fall in bad weather while aquariums see an increase), births of popular

V

Fig. V2 Vigilance behaviour: a herd of wildebeest (*Connochates taurinus*) in Ngorongoro Crater, Tanzania.

species, teachers' strikes (resulting in fewer school visits), opening of new exhibits. *See* **FOOTFALL,** *INTERNATIONAL ZOO YEARBOOK* **(IZYB)**

visitor barrier *See* **PUBLIC BARRIER**

visitor behaviour The actions of persons who attend visitor attractions such as zoos and museums, e.g. **VISITOR CIRCULATION,** the **DWELL TIME** at exhibits, especially the time they spend on particular activities such as reading signage and observing animals. An understanding of visitor behaviour is important when zoos and exhibits are being designed. *See also* **ATTRACTING POWER, FOOTFALL, HOLDING POWER, ORIENTATION (2), VISITOR CIRCULATION, WAYFINDING**

visitor circulation The pattern with which visitors move (circulate) around a zoo or other visitor attraction. *See also* **ORIENTATION (2), WAYFINDING**

visitor footfall *See* **FOOTFALL**

visitor numbers *See* **VISITOR ATTENDANCE**

visitor services Services provided by zoos and other visitor attractions for their visitors, e.g. toilets, café etc.

visitor studies The academic study of the behaviour of visitors to museums, art galleries, zoos, and other attractions. *See also* **VISITOR BEHAVIOUR**

vital signs monitor A device which monitors some combination of physiological parameters including respiration, blood pressure, blood oxygen saturation, temperature, heart function and other parameters. Often used in surgery.

vitamin An organic molecule that is essential in the diet. Vitamins serve a variety of functions. They are either lipid-soluble (A, D, E and K), and can be stored in body fat, or water-soluble (thiamine, riboflavin, niacin, B_6, pantothenic acid, biotin, folic acid, B_{12} and C) and must be ingested frequently because they cannot be stored. Vitamins are required in relatively small quantities and too much may be harmful. This is especially true for the lipid-soluble vitamins because they accumulate in the body. Animals kept in captivity often suffer from vitamin deficiencies. Some vitamins must be present in the food if the animal is to remain healthy. Others are synthesised within the animal's body. Different species require different vitamins. Any vitamin produced in the lower part of the alimentary canal only becomes available if the animal eats its own faeces (**COPROPHAGY**). Rabbits produce pellets that are rich in the B vitamins.

vitamin A, retinol This vitamin plays an important part in cellular metabolism, vision, bone development and epithelial cell integrity. Deficiency may cause ulceration of the cornea, blindness, xerophthalmia, and cellular changes in the trachea resulting in decreased resistance to infection. Vitamin A is only found in a preformed state in animal tissues,

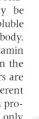

but some precursors, such as **CAROTENES**, are synthesised by plants. Carnivores may derive much of their requirement by consuming the livers of their prey where vitamin A is stored. This vitamin is toxic if consumed in excess and supplements should only be provided in small quantities. However, polar bears and pinnipeds contain very high levels in their livers, suggesting that they have a high degree of tolerance.

vitamin B complex This includes riboflavin, nicotinic acid, pantothenic acid, choline, biotin and thiamine. Most of the B vitamins are widely distributed in plant tissues or animal and microbial tissues. Many act as coenzymes that are important in a range of metabolic processes. Deficiency signs often involve central nervous system problems resulting in convulsions and lack of co-ordination. Other signs may also occur, such as diarrhoea, anaemia and impaired growth. Herbivores may obtain considerable amounts of B vitamins from the synthetic activity of gut microbes. Vitamin B_1 (thiamine) deficiency is rare but can occur in pinnipeds that consume raw fish as they contain enzymes that will destroy thiamine (thiaminases). Certain plants such as bracken, horsetails and sweet potato leaves contain similar anti-thiamine compounds and their consumption may cause a secondary deficiency. Deficiency signs include muscular spasms, a staggering gait and loss of appetite.

vitamin C, ascorbic acid This is important in many metabolic reactions and in the synthesis of the collagen found in cartilage. Some species can synthesise vitamin C, others need to absorb it from their diet. Vitamin C is a vitamin for primates, some birds, fishes and invertebrates, but not for amphibians, reptiles and many birds and mammals, which can synthesise it. Deficiency signs include sore joints, bones and muscles, listlessness, abnormal bone growth, anorexia and increased susceptibility to disease. Good sources of vitamin C include citrus fruits and the leafy parts of plants from the cabbage family.

vitamin D A vitamin concerned with the control of calcium and phosphorus in circulation and the absorption of calcium in the gut. It occurs in two forms, D_2 (ergocalciferol) and D_3 (cholecalciferol). Vitamin D_2 occurs in plant material after irradiation by sunlight, and vitamin D_3 is produced in the skin of animals as a result of exposure to ultraviolet light. Sun-dried hay contains more vitamin D_2 than hay that has been artificially dried. Vitamin D deficiency causes **RICKETS** in young animals and a softening of the bones in adults (osteomalacia). Species differences occur in the biological activity of the two forms of vitamin D, and there is evidence that **NEW WORLD MONKEYS** use D_3 more efficiently than D_2.

vitamin E, alpha-tocopherol A vitamin which acts as an **ANTIOXIDANT** and protects cells from oxidative changes caused by free radicals. Deficiency of this vitamin can result in cardiac and skeletal muscle myopathies, anaemia, **HAEMOLYSIS** and fat degeneration. Vitamin E is found widely in plants but little occurs in animal tissues.

vitamin K Vitamin K occurs naturally in two forms: K_1 which is synthesised by plants and K_2 which is synthesised by microbes. A synthetic form, K_3, is also called menadione. Vitamin K is important in blood clotting. Deficiency is highly unlikely in healthy animals that are fed natural foods or manufactured food products because it is widely distributed and it is synthesised by organisms in the gut.

vitamin supplement An addition to the diet which is given to increase the quantity of a single vitamin or a number of vitamins (multi-vitamin supplement).

vitamin-deficiency disease A disease caused by an inadequate quantity of a particular vitamin in the body, e.g. **RICKETS** is caused by a deficiency of vitamin D. Deficiency diseases vary between species and a variety of apparently unrelated conditions may be caused by a deficiency in a single vitamin.

vivarium An enclosed space designed to contain reptiles, amphibians or invertebrates, usually containing plants to create a **NATURALISTIC EXHIBIT**.

viviparity, vivipary Producing young which are active immediately after birth as a result of the embryo developing inside the mother. Sometimes described as giving birth to 'live young.' Occurs in placental mammals, some reptiles, amphibians, elasmobranchs and in some invertebrates, e.g. aphids. *Compare* **OVIPARITY**, **OVOVIVIPARITY**

viviparous Relating to **VIVIPARITY**.

vivipary *See* **VIVIPARITY**

vivisection Surgical procedures performed on live animals. Such activities require a licence and appropriate anaesthesia. *See also* **AMERICAN ANTI-VIVISECTION SOCIETY (AAVS)**, **ANIMAL TESTING**, *BAYLISS V. COLERIDGE* (1903), **DISSECTION**, **SOCIETY FOR THE PROTECTION OF ANIMALS LIABLE TO VIVISECTION**, **THREE RS**, **UNCAGED**

v-notching A method used by commercial lobstermen to mark egg-bearing female lobsters by clipping a small v-shaped notch from the tail flipper. After marking the lobster is returned to the sea and the mark protects her from harvest through additional moults, thereby preventing the removal of known breeders from the population.

vocalisation A sound made by an animal. Used for communication particularly in social species, e.g. an **ALARM CALL**, a mating call. Some species have a sophisticated range of vocalisations. Some monkey species have a primitive 'LANGUAGE' which includes different calls to warn **CONSPECIFICS** of the presence of different types of **PREDATORS**. Scientists give names to different vocalisations, e.g. chimpanzees make sounds known as 'pant hoots', 'pant grunts', 'food barks', 'waa barks' and 'copulation pants'. *See also* DIALECT, SIGNATURE WHISTLE, VOICE, VOICE RECOGNITION

voice In birds, a combination of the call and the SONG.

voice recognition The ability of some animals to identify CONSPECIFICS from their VOCALISATIONS alone. *See also* SIGNATURE WHISTLE

volar *See* PALMAR

voluntary muscle *See* SKELETAL MUSCLE

voluntary nervous system *See* SOMATIC NERVOUS SYSTEM (SNS)

vomeronasal organ, Jacobson's organ, second nose A specialised organ found in the nasal cavity of many animals (including primates and elephants) which is part of the OLFACTORY SYSTEM and is used to detect PHEROMONES and other chemical signals. *See also* FLEHMEN

vomiting The passing up of partially digested food from the stomach. May be a sign of stomach disease. Also occurs as an abnormal behaviour in some captive animals which vomit and then eat the vomit.

Vomiting can be prevented by using an ANTIEMETIC. *Compare* REGURGITATION

von Economo neurones *See* SPINDLE CELLS

VORTEX POPULATION VIABILITY ANALYSIS (PVA) software. It has been used extensively by the CONSERVATION BREEDING SPECIALIST GROUP (CBSG) of the INTERNATIONAL UNION FOR THE CONSERVATION OF NATURE AND NATURAL RESOURCES (IUCN).

Vulnerable (VU) A RED LIST category used by the INTERNATIONAL UNION FOR THE CONSERVATION OF NATURE AND NATURAL RESOURCES (IUCN): Vulnerable (VU) – best available evidence indicates that the taxon is facing a high risk of extinction in the wild.

Vulnerable Native Breeds British and Irish native dog breeds which have been identified as having annual registrations of 300 puppies or fewer by the KENNEL CLUB, e.g. Skye terrier, OTTERHOUND, Manchester terrier, Sussex spaniel.

V

Fig. W1 W is for warthog (*Phacochoerus africanus*).

Dictionary of Zoo Biology and Animal Management: A guide to terminology used in zoo biology, animal welfare, wildlife conservation and livestock production, First Edition. Paul A. Rees.
© 2013 John Wiley & Sons, Ltd. Published 2013 by John Wiley & Sons, Ltd.

wader A bird commonly found in shallow water on shores, estuaries etc., and belonging to the family Charadriidae (lapwings, plovers etc.) or Scolopacidae (curlews, sandpipers, stints, snipes etc.).

walk-over weighing system An automated weighing system used for livestock, especially cattle. Animals are weighed as they walk through a **RACE** and across a weighing platform. Individuals are identified during weighing by a smart reader which receives signals from a tag or collar worn by the animal using **RADIO FREQUENCY IDENTIFICATION (RFID) TECHNOLOGY**. *See also* **AUTO DRAFTER**

walk-through exhibit A zoo enclosure that visitors may enter and in which there may be few or no barriers between them and the animals, giving the visitor a feeling of immersion in the environment of the animals, e.g. a free-flight aviary, a bat cave, a lemur enclosure. Such exhibits must be carefully designed to prevent escape of the animals via the access points for visitors and may require the continuous presence of keepers to ensure the safety of both the animals and the visitors. *See also* **DOUBLE DOOR SYSTEM, IMMERSION EXHIBIT**

wall barrier A wall surrounding an enclosure which separates animals from the public. *Compare* **HA-HA**

Wallace, Alfred Russel (1823–1911) A British naturalist who studied the fauna and flora of the East Indies and who formulated a theory of **EVOLUTION** by **NATURAL SELECTION** independently of Charles **DARWIN**. Their ideas were presented jointly to the **LINNEAN SOCIETY OF LONDON** on 1 July 1858, although neither was present at the time. Wallace was also responsible for the concept of **WALLACE'S LINE**.

Wallace's line In **ZOOGEOGRAPHY**, an imaginary line which separates the fauna of the Oriental and Notogean regions (the latter consisting of Australia, Tasmania, New Zealand, Polynesia and the Hawaiian Islands). It was originally a line demarcated by Alfred Russel **WALLACE** between the Oriental and Australasian regions. It marks the western limit of Australian mammals and the eastern limit of the Oriental mammalian fauna. *See also* **FAUNAL REGIONS**

wallow
1. The action of immersing the body in mud.
2. A place where animals wallow.

war horse Any horse used in battle, especially those used by the Allied Forces during the First World War, either by cavalry or to transport guns, ammunition, food etc. Around 256 000 British horses died on the Western Front.

war memorials to animals A large number of animals have been used by the military forces of many nations during wartime including homing pigeons, horses, mules, donkeys, oxen, bullocks, reindeer, elephants, camels, dogs, cats and canaries. A number of memorials have been erected to honour these animals including one at Port Elizabeth, South Africa, to the horses that died in the Boer War, and the *Animals in War Memorial* in London, dedicated to animals who served, suffered and died alongside British, Commonwealth and Allied forces. *See also* **WAR HORSE, WARTIME ZOO POLICY**

warbles Warbles are swellings about the size of marbles on the backs of animals, caused by the larvae of various species of warble flies. They are found in cattle, horses, goats, reindeer, deer and other species. The condition commonly occurs in young animals, causing a loss of condition and sometimes even death. If a number of larvae are crushed death may be caused by **ANAPHYLACTIC SHOCK**. Some larvae may migrate through the body and damage the spinal cord. Larvae may be killed with a systemic insecticide.

warfarin A chemical used as a **PESTICIDE** (which kills rats and mice) and as an **ANTICOAGULANT**.

warmblood A **HORSE TYPE** consisting of sports and carriage horses.

warming unit A warm air blower for keeping veterinary patients warm.

warren *See* **RABBIT WARREN**

wartime zoo policy A government policy concerning the treatment and disposal of animals in time of war. During the Second World War the Japanese government systematically disposed of dangerous animals in circuses and zoos as a matter of national policy. Almost 200 'dangerous' zoo animals were disposed of between August 1943 and May 1945 in Japan. In addition 133 'dangerous' circus animals were destroyed in 1943. This was some time before US air strikes were anticipated. The policy is considered to have been an integral component of the government's military propaganda, aimed at mobilising the entire civilian population into total war, rather than a measure taken purely for public protection in anticipation of animals escaping if zoos were bombed (Itoh, 2010). *See also* **WAR MEMORIALS TO ANIMALS**

Washington Convention Colloquial name for the Convention on International Trade in Endangered Species of Wild Fauna and Flora 1973. *See* **CITES**

Washoe A female chimpanzee who was the first non-human to be taught to use American Sign Language (ASL), at the University of Nevada, Reno. She later taught ASL to her adopted son. *See also* **KANZI, KOKO, NIM CHIMPSKY**

Watchlist categories (Rare Breeds Survival Trust) Rare breeds of sheep, cattle, horses, pigs and goats (but not poultry) are listed by the **RARE BREEDS SURVIVAL TRUST (RBST)** in categories, based on the number of registered adult breeding females (Table R1). The definitions vary between species. For horses the categories are defined as:

1. Critical   <300
2. Endangered   300–500
3. Vulnerable   500–900

W

4. At risk	900–1500
5. Minority	1500–3000
6. Other native breeds	>3000

See also **RED LIST**

water conditioner A chemical additive which may be added to tap water to make it suitable for an aquarium by binding toxic heavy metals, neutralising chlorine and chloramines etc.

water drinker *See* **DRINKER**

water filter *See* **WATER TREATMENT SYSTEM**

water treatment system An apparatus for removing unwanted materials from water (organic matter, chemicals, particles etc.), e.g. in a pool within an animal enclosure or in an aquarium. *See also* **BIOLOGICAL FILTRATION, LIFE SUPPORT SYSTEM, MECHANICAL FILTRATION, WATER CONDITIONER**

waterbird *See* **WADER, WATERFOWL, WILDFOWL**

waterer *See* **DRINKER**

waterfowl

1. A term sometimes used to mean wildfowl: ducks, geese and swans, especially in North America.
2. All swimming waterbirds.
3. Article 1(2) of the **CONVENTION ON WETLANDS OF INTERNATIONAL IMPORTANCE ESPECIALLY AS WATERFOWL HABITAT 1971 (RAMSAR CONVENTION)** defines waterfowl as *'birds ecologically dependent on WETLANDS'*. *See also* **WADER**

Watson, James (1928–) and Francis Crick (1916–2004) James Watson is an American biologist. Watson and an English molecular biologist, Francis Crick, discovered that **DNA** consists of a double helix, made up of two parallel chains of alternate sugar and phosphate groups linked by pairs of organic bases, while working at the Cavendish Laboratory at the University of Cambridge, UK. This discovery – published in 1953 – earned Watson and Crick, together with Maurice Wilkins, the Nobel Prize for Physiology or Medicine in 1962. **DNA FINGERPRINTING** is now an essential tool for conservation biologists.

Watson, John Broadus (1878–1958) An American psychologist who was the founder of **BEHAVIOURISM**. He believed that the principles governing the behaviour of humans and other animals are essentially the same, and was greatly influenced by the work of **PAVLOV**. His books include *Behavior: An Introduction to Comparative Psychology* (1914).

wattle A fleshy caruncle or **DEWLAP** which hangs from the head or neck in several taxa, especially goats and birds. Some species of birds possess a brightly coloured wattle that is used in **COURTSHIP DISPLAYS**.

wave generator A mechanical device used in an aquarium or pool in a zoo or aquarium exhibit that agitates water to produce artificial waves, e.g. in the multi-species *Seal and Penguin Coasts* exhibit at Bristol Zoo (Fig. W2).

Fig. W2 Wave generator.

wayfinding In relation to a zoo visit, finding one's way around. *Compare* **ORIENTATION (2)**. *See also* **VISITOR CIRCULATION**

WAZA *See* **WORLD ASSOCIATION OF ZOOS AND AQUARIUMS (WAZA)**

weaner

1. A young mammal that has recently been weaned, especially a lamb, pig or calf. Also called a weanling.
2. A device that is placed over the mouth of an animal that is being weaned to prevent it from suckling.

See also **WEANING**

weaning The process, in mammals, of gradually withdrawing mother's milk and introducing the adult diet. A young mammal that has completed this process is said to be weaned. *See also* **WEANER, WEANLING**

weanling

1. *See* **WEANER (1)**.
2. Defined by the **AMERICAN STUD BOOK** (2008) as *'A foal of any sex in its first year of life after being separated from its dam.'*

See also **WEANING**

weapon dog *See* **STATUS DOG**

weapons, animals as A number of animal species have been used by the armed forces as weapons. In the Second World War the Russian army trained dogs with primed bombs strapped to their backs to run under German tanks. More recently, donkeys have been used to carry bombs in Afghanistan and the Middle East, and the Russian military has trained dolphins to attack enemy frogmen and warships.

weaving A type of stereotypic behaviour which consists of moving the head rhythmically from side to side, especially in elephants.

webcam An electronic camera used to record still images or video using a computer. It may be used to send images to the internet, e.g. a live video of animals in a zoo enclosure.

W

Fig. W3 Weigh bridge in a zoo.

Webster, A. John Emeritus Professor of Animal Husbandry at the University of Bristol. He was a founder member of the FARM ANIMAL WELFARE COUNCIL who developed the concept of the 'FIVE FREEDOMS' for farm animals in the 1960s.

weigh bridge A weighing machine located in the ground or the floor of a building. It may be used to weigh vehicles or large animals and is often built into ANIMAL ENCLOSURES for large mammals such as gorillas and elephants (Fig. W3). *See also* WALK-OVER WEIGHING SYSTEM

Weil's disease *See* LEPTOSPIROSIS

welfare *See* ANIMAL WELFARE

welfare epidemiology The evaluation of risk factors in ANIMAL WELFARE using an epidemiological (multi-institutional) approach and a broad spectrum of animal-based welfare assessments. *See also* EPIDEMIOLOGY

Welfare of Animals (Slaughter or Killing) Regulations 1995 A law in Great Britain which protects the welfare of animals prior to and during slaughter in slaughterhouses and elsewhere. It requires licensing of slaughtermen, makes provision for disease control, regulates the methods of killing, provides exemptions for RELIGIOUS SLAUGHTER etc.

Welfare of Farmed Animals (England) Regulations 2007 A law in England which regulates the keeping of animals on farms. It provides for their inspection, requires the keeping of records of veterinary treatment and mortality, appropriate staffing and suitable housing. It makes a variety of other provisions in relation to welfare including the provision of suitable feed and water.

Welfare of Pigs Directive EU legislation (Council Directive 2008/120/EC of 18 December 2008 laying down minimum standards for the protection of pigs) which makes provision for minimum standards of welfare for pigs, including space requirements, a requirement to keep sows in social groups and the strict regulation of the use of painful operations, e.g. TAIL DOCKING. *See also* SOW STALL

Welfare of Racing Greyhounds Regulations 2010 Regulations in England made under the ANIMAL WELFARE ACT 2006 which require the licensing of greyhound racing tracks, the attendance of a vet during every race, the provision of facilities for a vet and kennels for the dogs, and the keeping of records of dogs which race and any injuries they may sustain.

Welfare of Wild Animals in Travelling Circuses (England) Regulations 2012 The Welfare of Wild Animals in Travelling Circuses (England) Regulations 2012 (SI 2932) is a statutory instrument made pursuant to s13(1) of the ANIMAL WELFARE ACT 2006 and requires travelling CIRCUSES (1) which keep wild animals to be licensed and inspected, and requires the keeping of animal records.

welfare standards *See* RSPCA WELFARE STANDARDS

wellbeing *See* WELLNESS

wellness, wellbeing In relation to animals, the condition of good physical and mental health, especially when maintained by appropriate diet, adequate exercise etc. *See also* ANIMAL WELLNESS CENTRE

West Nile virus A viral infection found mainly in wild birds, especially corvids. Infected birds may die and literally fall out of the sky. The virus is related to yellow fever and Japanese encephalitis viruses, and is transmitted by mosquitoes. West Nile virus can infect primates, including humans.

wet feeding A method of feeding livestock (e.g. calves, WEANERS (1), BROILERS) with liquid food which is used to increase production. With automatic systems each animal may wear an electronic tag which controls how much food it receives. *See also* RADIO FREQUENCY IDENTIFICATION (RFID) TECHNOLOGY

wet moat A channel filled with water which is used as a barrier in some zoo enclosures. It may be shallow or deep and often varies in depth, being shallow on the inside of the enclosure (to reduce the risk of drowning to animals) and deep on the visitors' side (to prevent escape). Wet moats may be provided with ELECTRIC FENCES as a secondary barrier. *Compare* DRY MOAT

wetlands Defined by the CONVENTION ON WETLANDS OF INTERNATIONAL IMPORTANCE ESPECIALLY AS WATERFOWL HABITAT 1971 (RAMSAR CONVENTION) (Art. 1(1)) as, '. . . *areas of marsh, fen, peatland or water, whether natural or artificial, permanent or temporary, with water that is static or flowing, fresh, brackish or salt, including areas of marine water the depth of which at low tide does not exceed six metres.*'

whale sanctuary An area of marine water designated by the International Whaling Commission (IWC), or an individual state, in which commercial whaling is prohibited. The IWC established its first sanctuary in Antarctica in 1938. A second sanctuary was established in the Indian Ocean in 1979, and a third, the Southern Ocean Sanctuary, in 1994.

W

Fig. W4 Whale watching. A migrating humpback whale (*Megaptera novaeangliae*) observed by tourists (inset) in the Pacific Ocean.

Other sanctuaries exist which have been established under national laws, the Australian Whale Sanctuary (around Australia, its island dependencies and part of Antarctica), the Cook Islands Whale Sanctuary, and the Hawaiian Islands Humpback Whale National Marine Sanctuary. *See also* **INTERNATIONAL CONVENTION FOR THE REGULATION OF WHALING 1946**

whale stranding *See* **STRANDING**

whale watching The observation of whales as a leisure or tourist activity, usually from boats. It may occur in the open ocean or within **MARINE PARKS (1)** and is often focused at locations along migratory routes, e.g. the west coast of Costa Rica (Fig. W4). Regulations usually control the minimum distance boats must keep from whales to avoid unnecessary disturbance. Whale watching is important to the economies of some small communities and may help to support gift shops and other commercial activities.

whaling The occupation of catching and killing whales for commercial gain or as a subsistence activity. *See also* **INTERNATIONAL CONVENTION FOR THE REGULATION OF WHALING 1946, WHALING MORATORIUM**

whaling moratorium A ban on commercial whaling imposed by the International Whaling Commission (IWC) in October 1985. The ban was ignored by Iceland, South Korea, Norway, Japan and the former Soviet Union. *See also* **INTERNATIONAL CONVENTION FOR THE REGULATION OF WHALING 1946**

whip Implement used for striking an animal to urge it to move or move more quickly. It is often made of a strip of leather or cord fastened to a handle and may be used by a jockey to encourage a racehorse to run faster during a race. *See also* **WHIPPING REGULATION OF**

whipper-in A huntsman's assistant whose main job is to keep the pack of **FOXHOUNDS** together using a **WHIP**. *See also* **FOX HUNTING, KENNELMAN**

whipping, regulation of The number of times a jockey may strike a racehorse with a **WHIP** has been restricted by the **BRITISH HORSERACING AUTHORITY (BHA)** since September 2011. In its review – *Responsible Regulation: a Review of the use of the whip in Horseracing* – the BHA limited the use of the whip to up to seven times in flat racing and up to eight times in jump racing (and only five times in the last furlong/after the last obstacle).

W

Whipsnade Zoo, ZSL Whipsnade Zoo Whipsnade Zoo (formerly Whipsnade Wild Animal Park, now ZSL Whipsnade Zoo) was opened in 1931 by the **ZOOLOGICAL SOCIETY OF LONDON (ZSL)**, in Bedfordshire, in order to keep and study large animals in more natural surroundings than was possible at **LONDON ZOO**.

white adipose tissue *See* **ADIPOSE TISSUE**

white blood cell *See* **LEUCOCYTE**

White, Gilbert (1720–1793) Curate of the village of Selborne in Hampshire, England, who was one of the first English naturalists to make systematic recordings of nature. He published these in *The* **NATURAL HISTORY AND ANTIQUITIES OF SELBORNE** in 1789.

white hunter A generic term for a **BIG GAME** hunter, usually from Europe or North America, who hunted animals, especially **LARGE MAMMALS**, particularly in the first half of the 20th century. Some famous hunters have given their names to important national parks, e.g. 'Jim' **CORBETT**, Frederick Courteney **SELOUS**. *See also* **BIG FIVE**

white matter The component of the **CENTRAL NERVOUS SYSTEM (CNS) (1)** which contains glial cells and myelinated **AXONS** and occurs around the outside of the **SPINAL CORD** and the inside of the brain. *Compare* **GREY MATTER**

white paper In the UK, a document which sets out the details of government future policy in a particular area prior to the passing of legislation to implement this policy, e.g. Natural Environment White Paper. *Compare* **GREEN PAPER**

white spot A parasitic disease of fishes – particularly members of the carp family – in which white cysts occur all over the body. It is caused by the protozoan *Icthyophthirius multifiliis*. The organism lives at the bottom of ponds from where it releases the infective stage into the water. Treatment is by the application of zinc-free malachite green to the water.

white tiger A specimen of tiger (*Panthera tigris*) which is white in colour. These individuals are not **ALBINO** individuals. The white colour is conferred by a **RECESSIVE ALLELE** which can only be expressed if an individual possesses two copies of it: one inherited from its mother and the other from its father. All wild white tigers have been of the Bengal subspecies (*P. t. tigris*). Some white tigers are pink-eyed and pure white, while others have ice-blue eyes and black or brown stripes on a white, egg-shell white or cream background. In 1951 the Maharaja of Rewa began breeding white tigers in captivity and selling them to zoos.

WHO *See* **WORLD HEALTH ORGANIZATION (WHO)**

who A relative pronoun used (in place of a noun) in preference to 'that' or 'which' (which usually refer to inanimate objects) when referring to an animal, especially a particular individual, e.g. *Eric* was the chimpanzee *who* achieved the highest score in the test. It is generally considered to be grammatically correct to use 'that' in reference to an animal, but some **ACADEMIC JOURNALS** insist on the use of 'who' to emphasise their similarities to humans rather than their differences, and discourage the use of other derogatory terms. For example, contributors to the **JOURNAL OF APPLIED ANIMAL WELFARE SCIENCE (JAAWS)** are encouraged to '*use language that acknowledges the individuality and integrity of members of other species. For example, where possible use gender-specific personal pronouns ('he' or 'she') and personal forms of the relative pronouns ('who', not 'which'). Avoid terms such as 'it' and 'the organism,' and replace such phrases as 'laboratory, farm and zoo animals' with, for example, 'animals in the laboratory.'* The *Journal of Animal Ethics* requires that '*authors should avoid derogatory or colloquial language or nomenclature that denigrates animals (or humans by association), such as: beasts, brutes, bestial, beastly, dumb animals, sub-humans; companion animals should be used rather than pet animals, and free-living or free-ranging rather than (or in addition to) wild animals.*'

wholphin A **HYBRID (2)** produced when a whale breeds with a dolphin. A false killer whale (*Pseudorca crassidens*) mated with a bottlenose dolphin (*Tursiops* sp.) producing a 'wholphin' at the Sea Life Park on Oahu, Hawai'i in 1986.

Wilcoxon matched pairs signed-rank test A **NON-PARAMETRIC** statistical test used to compare the differences in the ranks of the pairs in a matched sample. Used as an alternative to a paired (dependent) **T-TEST**.

wild animal

1. An **ANIMAL** that normally lives wild and is not under human control. Under the law the term 'animal' may have a meaning other than the usual zoological meaning.

2. Under s27(1) of the **WILDLIFE AND COUNTRYSIDE ACT 1981** a wild animal is '*. . . any animal (other than a bird) which is or (before it was killed or taken) was living wild*'.

3. In the **ZOO LICENSING ACT 1981** (s.21(1)) '*. . ."wild animals" means animals not normally domesticated in Great Britain*'.

See also **ANIMAL**, *FERAE NATURAE*, **FREE-LIVING**, **WILD BIRD**, **WILDLIFE**

Wild Animals in Travelling Circuses *See* **RADFORD REPORT**. *See also* **TRAVELING EXOTIC ANIMAL PROTECTION ACT (USA)**

wild beast show *See* **BEAST SHOW**

wild bird

1. A bird that normally lives in its natural habitat and is not under the control of humans. It may be defined in legislation and may have a meaning other than the normal zoological meaning by excluding certain groups of birds.

2. A wild bird is defined by the **WILDLIFE AND COUNTRYSIDE ACT 1981** (s.27(1)) as '*any bird of a species which is ordinarily resident in or is a visitor to the European territory of any Member State in a wild state*

W

but does not include poultry or, . . . , any game bird.' In relation to most of the Act the definition excludes any bird bred in captivity unless it has been subsequently lawfully released into the wild as part of a repopulation or **REINTRODUCTION** programme.
3. The Protection of Birds Act 1954, s14(1) defines a wild bird as follows: ' *"wild bird" in sections five, ten and twelve of this Act means any wild bird, but in any other provision of this Act does not include pheasant, partridge, grouse (or moor game), black (or heath) game, or, in Scotland, ptarmigan.'*

Wild Bird Conservation Act of 1992 (USA) A US Act which promotes the conservation of exotic birds by encouraging wild bird conservation and management programmes in their countries of origin. In addition, it requires that all US trade in exotic birds is sustainable and of benefit to the species. Where necessary, the Act allows for the restriction or prohibition of imports of exotic birds.

Wild Birds Directives EU laws which protect birds and their habitats in the **EUROPEAN UNION (EU)**. The original Wild Birds Directive is Council Directive 79/409/EEC on the conservation of wild birds. The new Wild Birds Directive is Directive 2009/147/EC of the European Parliament and of the Council on the conservation of wild birds. The legislation requires Member States to protect wild birds and their habitats, including by establishing **SPECIAL PROTECTION AREAS (SPAs)**.

Wild Free-Roaming Horses and Burros Act of 1971 (USA) A law in the USA which designated wild horses and **BURROS** as *'living symbols of the historic pioneer spirit of the West'*, and protected them from capture, branding, harassment and death. The law is administered by the **BUREAU OF LAND MANAGEMENT (BLM)**. The BLM has removed large numbers of wild horses and burros from public lands and their management methods are controversial, partly because family groups are split as a result of mass roundups.

wild law Laws which are aimed at protecting the Earth's community, including animals, plants, rivers and ecosystems. *See also* **EARTH JURISPRUDENCE**

Wild Life (Protection) Act 1972 (India) The first comprehensive legislation designed to protect wildlife in India. It also created the **CENTRAL ZOO AUTHORITY (CZA)** and provided a power to regulate Indian zoos. *See also* **RECOGNITION OF ZOO RULES 2009 (INDIA)**

Wild Mammals (Protection) Act 1996 A law in Great Britain which makes it illegal to subject wild mammals to acts of cruelty such as drowning, asphyxiating, crushing, stabbing and burning. The Act did not ban **FOX HUNTING** and originally exempted any act done by a dog. Subsequent amendment (following the passing of the **HUNTING ACT 2004**) allows only cruel acts done by a dog if they are excluded by virtue of being classed as exempt hunting.

wild west show Travelling shows in the USA and Europe which performed to large audiences and presented a romanticised version of the American Old West, including the reconstruction of battles between native American Indians and the US cavalry. The most famous was *Buffalo Bill's Wild West Show* which operated from 1883 to 1913. These shows involved a large number of horses and cattle, and also a number of wild species including bears, elk and deer. *See also* **TRAVELLING MENAGERIES**

Wilderness Society A conservation organisation which was founded in 1935 and works to protect public lands in the USA.

wildfowl Ducks, geese and swans. *Compare* **WATERFOWL**

Wildfowl and Wetlands Trust (WWT) A conservation charity based in the UK which conserves wetlands and their wildlife. It was founded as the Wildfowl Trust in 1946 by Sir Peter **SCOTT** at **SLIMBRIDGE**. It operates nine wetland visitor centres in the UK and is also involved in conservation projects in other countries. *See also* **NĒ NĒ (*BRANTA SANDVICENSIS*)**

Wildfowl Trust *See* **WILDFOWL AND WETLANDS TRUST (WWT)**

wildfowler A person who hunts **WILDFOWL**, on foot or from a small boat (Fig. W5).

Fig. W5 Wildfowlers' gun punt. This boat was built in 1938–39 and was used by commercial wildfowlers on the Solway Firth. The long-barrelled gun was aimed by steering the boat and fired with a pull-cord.

wildlife

1. A collective term for **WILD ANIMALS** and plants which live in their natural habitat and are not under human control. The term is often taken to mean wild animals only. It generally applies to a specific locality, e.g. the wildlife of Kenya, the wildlife of Lundy Island. The term may be defined by law.

2. Under New York's Environmental Conservation Law Title 1, § 11-010s3 (6)(a), ' *"Wildlife" means wild game and all other animal life existing in a wild state, except fish, shellfish and crustacea.* '

3. In Ohio, the Ohio Revised Code, Title [15] XV, Conservation of Natural Resources, Chapter 1531.01, RR, ' *"Native wildlife" means any species of the animal kingdom indigenous to this state.* '

See also **WILD BIRD**

Wildlife and Countryside Act 1981 The most important English law protecting **WILDLIFE** in Great Britain. It protects animals (including birds' eggs) and plants and their shelters, including birds' nests. It lists species in **SCHEDULES** based on the degree and type of protection provided and prohibits the release or **REINTRODUCTION** into the wild of non-native species. It also makes provision for the designation and protection of **SITES OF SPECIAL SCIENTIFIC INTEREST (SSSIs)**. It allows certain acts under **LICENCE**, e.g. to photograph protected bird species at the nest, to take or keep protected species, reintroductions, and authorises the investigation of offences by **WILDLIFE INSPECTORS (1)**.

wildlife conservation *See* **CONSERVATION**

Wildlife Conservation Society (WCS) A conservation organisation in the USA which was founded in 1895 as the New York Zoological Society and was responsible for saving the American bison (*Bison bison*) from extinction. It now operates the Bronx Zoo, the New York Aquarium, Central Park Zoo, Prospect Park Zoo and Queens Zoo in New York City. In addition, it manages about 500 conservation projects in over 60 countries.

Wildlife Contraception Center A facility located at St. Louis Zoo, Missouri, which provides services to institutions that are members of the **ASSOCIATION OF ZOOS AND AQUARIUMS (AZA)**. It ensures the safety and effectiveness of **CONTRACEPTIVES** by organising monitoring programmes; tests new contraceptive methods; and assists animal managers and vets in the selection and administration of contraceptives.

wildlife corridor *See* **CONSERVATION CORRIDOR**

wildlife crime officer A police officer in the UK who deals with environmental crime, habitat damage and crimes involving wildlife including **BADGER-BAITING,** deer **POACHING, LAMPING (1), DOG-FIGHTING**, the shooting of birds, persecution of **BIRDS OF PREY, EGGING**, illegal trapping, bat crime and illegal wildlife trade (Fig. W6).

wildlife forensics A relatively new field of criminal investigation which uses scientific procedures to

Fig. W6 Wildlife crime officer: a confiscated bear skin displayed on a police vehicle used by wildlife crime officers in England.

examine evidence and crime scenes where offences against **WILDLIFE** have occurred. It includes identifying protected species from whole animal and plant specimens, their parts or products made from them. *See also* **USFWS FORENSICS LABORATORY**

wildlife hospital A specialist facility designed to care for injured, diseased and orphaned wild animals. In the UK a number of animal charities operate wildlife hospitals, including the **ROYAL SOCIETY FOR THE PREVENTION OF CRUELTY TO ANIMALS (RSPCA)**.

Wildlife Information Network (WIN) A veterinary science-based charity which provides information on the health, husbandry, diagnosis and treatment of wildlife and the control of emerging infectious diseases in free-ranging wildlife populations.

wildlife inspector

1. A person authorised (by the Secretary of State in England, the National Assembly for Wales in Wales, and Scottish Ministers in Scotland) to confiscate specimens, inspect documents and enter premises to collect evidence of some offences under the **WILDLIFE AND COUNTRYSIDE ACT 1981** (e.g. the taking of protected species).

2. A person employed by the **UNITED STATES FISH AND WILDLIFE SERVICE (USFWS)** who enforces a range of US and international laws, regulations, and treaties that protect wildlife and limit commercial traffic in endangered animals and plants. Inspectors work at the 14 designated ports through which commercial shipments of wildlife must pass and locations along the Canadian and Mexican borders.

wildlife law The body of law that protects animals and plants including **NATIONAL LAW, EUROPEAN LAW** and **INTERNATIONAL LAW**.

wildlife trade *See* **CITES, TRADITIONAL MEDICINE**

W

Wildlife Trusts, The An organisation made up of local Wildlife Trusts in the UK, Isle of Man and Alderney which manages around 2300 nature reserves under the umbrella of the Royal Society of Wildlife Trusts. The organisation has developed from the Society for the Promotion of Nature Reserves (SPNR) established by Charles Rothschild in 1912.

Wildlife Vets International A UK charity which provides veterinary services for conservation projects, training for staff working with endangered species, and a rapid response to conservation emergencies.

Wildscreen A global charity which promotes an appreciation of nature and **BIODIVERSITY** through wildlife imagery. It organises a biennial Wildscreen festival which showcases wildlife and environmental films and it maintains an archive of photographs and films in a digital library called ARKive.

Wilson, Edward O. (1929–) Professor Wilson is a world authority on the biology of ants, he established the study of **ISLAND BIOGEOGRAPHY** and he is an influential proponent of **BIODIVERSITY** conservation. He developed the controversial science of **SOCIOBIOLOGY**: the study of the biological basis for all social behaviour. Wilson is Professor of Entomology at Harvard University and has won the US National Medal of Science and many awards for his writing including the Pulitzer Prize twice for his books *The Ants* (1990) and *On Human Nature* (1978).

wind sucking An abnormal behaviour of equids which involves drawing in air (but not normally swallowing it), as in **CRIBBING**, but without grasping a surface.

wing loading (WL) The ratio of the mass of a bird's body to the total area of its wings.

$$WL = \frac{\text{body mass (g)}}{\text{total wing area (both wings)}\,(cm^3)}$$

A low value indicates light, buoyant flight; a high value, heavy laboured flight. *See also* **ASPECT RATIO (AR)**

wing rule A device used for measuring the wing length of a bird. *See also* **CALIPER**

winter lethargy A dormant state achieved by some mammals during winter when their metabolism is depressed and their temperature lowered but not to the same extent as in true **HIBERNATION**. Animals which exhibit this behaviour are able to wake easily during winter whereas true hibernators awaken slowly. Some authorities claim that black bears (*Ursus americanus*) and grizzly bears (*U. arctos*) exhibit winter lethargy rather than true hibernation.

wobble tree A **NATURALISTIC ENRICHMENT** device that is designed to look like a tree and moves when shaken, dispensing small food items. Useful for large mammals to encourage natural behaviour, especially bears.

wolf reintroductions *See* **WYOMING FARM BUREAU FEDERATION V. BABBITT (1997)**

wolf sanctuary An organisation which cares for wolves and **WOLFDOGS** that have been abandoned or rescued. Several exist the USA.

wolfdog, wolf–dog hybrid A hybrid created from a mating between a domestic dog and a wolf: usually a grey wolf (*Canis lupus*) with a wolf-like breed such as a German Shepherd, Siberian Husky or Alaskan Malamute. Very popular as pets in the United States.

womb *See* **UTERUS**

Woodland Park Zoo A zoo in Seattle, USA, which pioneered the development of naturalistic **IMMERSION EXHIBITS** in the mid 1970s, beginning with its gorilla exhibit designed by **JONES & JONES ARCHITECTS AND LANDSCAPE ARCHITECTS LTD.**, which opened in 1978.

Woods Hole Oceanographic Institution An institute in Massachusetts, USA, which conducts research on the oceans, including their wildlife, circulation, chemistry, pollution and interaction with the climate, along with polar research and underwater archaeology.

Woodstock The first zoo in England, created by Henry I. He built a seven-mile long wall around land at Woodstock in Oxfordshire creating the royal park of Woodstock. This was used for hunting and also contained his **MENAGERIE** which included lions, leopards, lynxes, camels, porcupines and a rare owl. The animals were later moved to the **TOWER OF LONDON MENAGERIE**.

wool biting Biting and ingesting a portion of the fleece. An abnormal behaviour common in sheep.

working animals Animals which are used by humans to perform work. *See also* **BEAST OF BURDEN, RETIREMENT, WAR HORSE, WORKING DOG**

working dog

1. A dog trained to perform a specific set of tasks e.g. protect livestock, search for people, guide a partially-sighted or blind person, etc.

2. In Australia, under s5 of the **COMPANION ANIMALS ACT 1998 (NSW)**, a working dog is '*a dog used primarily for the purpose of droving, tending, working or protecting stock, and includes a dog being trained as a working dog.*'

See also **FOXHOUND, GUIDE DOG, GUN DOG, HEARING DOG, OTTERHOUND, POLICE DOG, PYRENEAN MOUNTAIN DOG, SEARCH AND RESCUE DOG, YARRAMAN**

World Association of Zoos and Aquariums (WAZA) An umbrella organisation for the world zoo and aquarium community. Its members include leading zoos and aquariums, regional and national associations of zoos and aquariums and individual zoo professionals from around the world. WAZA began as the International Union of Directors of Zoological Gardens (IUDZG) in 1946. In 1950 the IUDZG became an international organisation member of the **INTERNATIONAL UNION FOR THE CONSERVATION OF NATURE AND NATURAL RESOURCES (IUCN)**. In 2000 the IUDZG was renamed WAZA.

W

Since 2001 its Executive Office has been located in the IUCN Conservation Centre in Gland, Switzerland. WAZA promotes co-operation between its member institutions with regard to conservation, the management and breeding of animals in captivity, and encourages the highest standards of animal welfare. It also represents zoos and aquariums in other international organisations, and promotes environmental education, wildlife conservation and research. The **INTERNATIONAL STUDBOOKS** for rare and endangered species are kept under the auspices of WAZA. In addition WAZA has produced a world conservation strategy for zoos and aquariums. *See also* **WORLD ZOO AND AQUARIUM CONSERVATION STRATEGY**

World Biodiversity Day Commemorated on 22 May each year. Also called the International Day for Biological Diversity. An international day for promoting biodiversity issues. Held since 1993 and commemorates the date on which the UN **CONVENTION ON BIOLOGICAL DIVERSITY 1992 (CBD)** was adopted at the Rio **EARTH SUMMIT** in 1992.

World Conservation Monitoring Centre (WCMC) An executive agency of the **UNITED NATIONS ENVIRONMENT PROGRAMME (UNEP)** which is based in Cambridge, UK, and is responsible for biodiversity assessment and supports the **CONVENTION ON BIOLOGICAL DIVERSITY 1992 (CBD)** and **CITES**. It was originally established as the Cambridge office of the **INTERNATIONAL UNION FOR THE CONSERVATION OF NATURE AND NATURAL RESOURCES (IUCN)** in 1979.

World Conservation Strategy (1980) A document produced by the **INTERNATIONAL UNION FOR THE CONSERVATION OF NATURE AND NATURAL RESOURCES (IUCN)** in co-operation with the **UNITED NATIONS ENVIRONMENT PROGRAMME (UNEP)**, **WORLD WILDLIFE FUND (WWF)**, **FOOD AND AGRICULTURE ORGANISATION (FAO)** and **UNITED NATIONS EDUCATIONAL, SCIENTIFIC AND CULTURAL ORGANIZATION (UNESCO)** in 1980. Its aim was to assist the achievement of sustainable development through the conservation of living resources by: maintaining essential ecological processes and life support systems; preserving genetic diversity; and ensuring the sustainable utilisation of species and ecosystems. *See also* **SUSTAINABILITY**

World Conservation Union Formerly an alternative name of the **INTERNATIONAL UNION FOR THE CONSERVATION OF NATURE AND NATURAL RESOURCES (IUCN)**.

World Environment Day Commemorated on 5 June each year and held annually since 1973. It is one of the principal means by which the **UNITED NATIONS (UN)** stimulates worldwide awareness of the environment and encourages political attention and action. World Environment Day commemorates the day on which the **UNITED NATIONS CONFER-**

ENCE ON THE HUMAN ENVIRONMENT 1972 began in Stockholm.

World Health Organization (WHO) WHO functions as the directing and co-ordinating authority for health within the **UNITED NATIONS (UN)** system. It was created in 1948 and its headquarters are in Geneva, Switzerland. Some of WHO's work concerns **VETERINARY PUBLIC HEALTH**.

World Heritage Convention *See* **CONVENTION FOR THE PROTECTION OF THE WORLD CULTURAL AND NATURAL HERITAGE 1972**

World Heritage Site A site protected by the **CONVENTION FOR THE PROTECTION OF THE WORLD CULTURAL AND NATURAL HERITAGE 1972**.

World Horse Welfare A UK charity concerned with the welfare of horses. Formerly known as The International League for the Protection of Horses. It operates rescue and rehoming centres for abused and neglected horses.

World Society for the Protection of Animals (WSPA) WSPA is the largest alliance of animal welfare societies in the world, with more than 1000 member organisations in 150 countries. It promotes responsible pet ownership, the proper treatment of farm animals, the care of animals during natural and man-made disasters and campaigns against the commercial exploitation of wildlife. WSPA's work is concentrated in regions of the world where there is little legislation protecting wildlife.

World Wetlands Day Held on 2 February each year, marking the date of the adoption of the **CONVENTION ON WETLANDS OF INTERNATIONAL IMPORTANCE ESPECIALLY AS WATERFOWL HABITAT 1971 (RAMSAR CONVENTION)** on 2 February, 1971. Used to raise awareness of wetland conservation.

World Wildlife Fund (WWF), Worldwide Fund for Nature The WWF was founded in 1961 as an international fundraising organisation for conservation, using the best scientific advice available from the **INTERNATIONAL UNION FOR THE CONSERVATION OF NATURE AND NATURAL RESOURCES (IUCN)** and other sources to channel funds. It has since evolved into a worldwide network of 30 national organisations. Current WWF conservation projects are concentrating on 'priority places' that must be saved in the next 50 years (e.g. the Amazon, the Arctic, Borneo and Sumatra, the Congo Basin, Galapagos, Madagascar), **FLAGSHIP SPECIES** and **FOOTPRINT-IMPACTED SPECIES**.

W

World Zoo and Aquarium Conservation Strategy The World Zoo Conservation Strategy was first published by the **WORLD ASSOCIATION OF ZOOS AND AQUARIUMS (AZA)** in 1993. This was superseded by *Building a Future for Wildlife – The World Zoo and Aquarium Conservation Strategy* in 2005. The purpose of the document is to provide a common set of goals for the zoo community and set out best practice in zoos in an attempt to allay the fears of those who are uncertain about the role of zoos and concerned about animal welfare.

World Zoo Conservation Strategy *See* WORLD ZOO AND AQUARIUM CONSERVATION STRATEGY

World's Zoological Trading Company Ltd. An animal dealer once based in London, which had agents in India, Canada, Australia and South America and provided wild-caught animals for zoos.

Worldwide Fund for Nature (WWF) Alternative name for the WORLD WILDLIFE FUND (WWF) in some countries.

worm A vernacular term for an organism with a long, thin form. The term is of no taxonomic value and is used to refer to various unrelated invertebrate taxa, e.g. annelid worms and NEMATODE worms. *See also* ANNELIDA, PLATYHELMINTHES

worming, deworming, drenching The application of an ANTHELMINTIC to an animal to remove intestinal parasites, e.g. TAPEWORMS.

worms A colloquial term for an infestation of parasitic worms (NEMATODES or PLATYHELMINTHS), especially in the gut.

worrying livestock In England and Wales, under the Dogs (Protection of Livestock) Act 1953 (s1(2)), 'worrying livestock' means 'attacking LIVESTOCK', or chasing livestock in a manner likely to cause injury, suffering or abortion, or loss of their produce, or being at large (i.e. not on a lead or under close control) in a field containing sheep. The Act makes worrying livestock illegal. Under the ANIMALS ACT 1971 a farmer may legally shoot a dog if it is worrying livestock (Fig. W7).

Fig. W7 Worrying livestock. A warning sign on farmland in North Wales: a dog found worrying livestock may be legally shot by a farmer under the Animals Act 1971.

wound management A series of procedures which control bleeding and aid the healing of wounds including the removal of foreign material and dead tissues, cleaning, closing and covering the wound. *See also* SKIN STAPLER, SUTURE (1)

Wright, Sewall Green (1889–1988) An American geneticist who was important in the development of population genetics and the modern theory of evolution, including the use of the INBREEDING COEFFICIENT in the study of pedigrees. *See also* MAXIMUM AVOIDANCE OF INBREEDING (MAI)

WWF *See* WORLD WILDLIFE FUND (WWF)

WWT *See* WILDFOWL AND WETLANDS TRUST (WWT)

***Wyoming Farm Bureau Federation v. Babbitt* (1997)** A legal case in the USA in which ranchers challenged the legality of grey WOLF (*Canis lupus*) REINTRODUCTIONS undertaken as part of a RECOVERY PLAN. The case was heard in the District Court for Wyoming and found the action of the UNITED STATES FISH AND WILDLIFE SERVICE (USFWS) in establishing a non-essential EXPERIMENTAL POPULATION of grey wolves in YELLOWSTONE NATIONAL PARK in Wyoming, Idaho, Montana, central Idaho and southwestern Montana to be unlawful. The wolves were reintroduced as a conservation measure but the programme was opposed by ranchers due to the threat posed to their livestock. At the time, under the ENDANGERED SPECIES ACT OF 1973 (USA) wolves could not be killed legally. However, conferring 'experimental status' on the translocated wolves under section 10(j) of the ESA allowed ranchers legally to kill any wolves found taking livestock on private land. The judge decided, however, that the effect of assigning these wolves 'experimental status' was to make the REINTRODUCTIONS illegal under the ESA because they were introduced within the range of non-experimental (native) populations. This meant that naturally migrating wolves could be mistaken for the 'experimental' wolves and killed by mistake, thereby threatening the wild (protected) individuals. The court ordered the reintroduced wolves to be removed but stayed the decision pending appeal. The Court of Appeal overturned the ruling and the wolves were allowed to stay. In August 2012 the Fish and Wildlife Service delisted grey wolves in Wyoming as it believed that they no longer needed the protection of the ESA. *See also* AMERICAN FARM BUREAU

Fig. Z1 Z is for zebra. Grevy's zebra (*Equus grevyi*).

X

Dictionary of Zoo Biology and Animal Management: A guide to terminology used in zoo biology, animal welfare, wildlife conservation and livestock production, First Edition. Paul A. Rees.
© 2013 John Wiley & Sons, Ltd. Published 2013 by John Wiley & Sons, Ltd.

X chromosome *See* SEX CHROMOSOMES

Xenarthra An order of mammals: sloths, anteaters, armadillos.

xenotransplantation The transplantation of animal ORGANS to humans. *See also* UNCAGED

X-ray

1. A form of electromagnetic radiation which is able to pass through some solids and liquids and may be used to produce images of broken bones, other internal tissue and organ damage, and some morphological signs of disease in animals.

2. A colloquial term for an image produced using X-rays; a radiograph.

Y chromosome *See* SEX CHROMOSOMES

yarraman

1. Australian slang term for a horse. *See also* MOKE (2)

2. A breed of Australian sheep and cattle dog.

yearling

1. An animal that is more than one but less than two years old. Often applied to animals such as antelope, deer, buffalo, horses, cattle.

2. A THOROUGHBRED racehorse between 1 January of the year after it was foaled and the following 1 January.

3. According to the AMERICAN STUD BOOK (2008), '*A colt, filly or gelding in its second calendar year of life (beginning January 1 of the year following its birth).*'

yellow body *See* CORPUS LUTEUM

Yellowstone National Park The first NATIONAL PARK (NP) in the world and now a WORLD HERITAGE SITE. It was founded in 1872 and is located in Wyoming, Montana and Idaho in the USA. It is important for its wildlife and volcanic features, and was the site of controversial wolf reintroductions in the late 1990s. *See also* WYOMING FARM BUREAU FEDERATION V. BABBITT (1997)

Yerkes National Primate Research Center A research centre at Emory University, Atlanta, Georgia. It houses almost 3400 non-human primates that are used for research in a wide range of biological disciplines including immunology, neuroscience and behaviour. Many studies of chimpanzee social behaviour have been conducted at the centre. *See also* DE WAAL, YERKES

Yerkes, Robert Mearns (1876–1956) An American primatologist, ethologist and comparative psychologist who studied human and primate intelligence and social behaviour in chimpanzees and gorillas. He became Professor of Psychobiology at Yale University where he founded the Yale University Laboratories of Primate Biology. This was renamed after his death and transferred to Emory University as the YERKES NATIONAL PRIMATE RESEARCH CENTER.

yoke gate A gate at the front end of a cattle CRUSH (RACE (1)) which holds the animal's head in place.

yolk A source of food used for growth and development in some developing embryos, consisting of a mixture of proteins and lipids. *See also* YOLK SAC

yolk sac A sac in the embryo of a vertebrate. It contains YOLK in reptiles, birds and monotremes which provides it with nourishment, but there is no yolk present in marsupials and eutherians.

Young, John Zachary (1907–1997) J. Z. Young was a British zoologist and neurophysiologist who was Professor of Anatomy at University College London until 1974. He worked on the nervous systems of cephalopods and discovered the squid giant AXON. Young was an influential zoologist who wrote many important textbooks including *The Life of Vertebrates* (1950), *The Life of Mammals* (1957) and *An Introduction to the Study of Man* (1971).

YouTHERIA A web portal containing data on the life history, ecology, taxonomy and geography of mammals, such as gestation length, activity cycle, body mass. The name is a combination of 'you' and 'THERIA'. *See also* PANTHERIA

ZAA *See* ZOO AND AQUARIUM ASSOCIATION (ZAA)

zebra (*Equus* spp.) Traditionally the animal that represents the letter Z in the alphabet. Any one of a number of species of equids from Africa south of the Sahara, which has a distinctive black and white striped appearance and inhabits savanna (Fig. Z1). An extinct form, the QUAGGA, formerly frequented parts of southern Africa but was hunted to extinction.

Zeiformes An order of fishes: dories and their allies.

zeitgeber Literally means 'time-giver'. A feature of the external environment that provides an animal with information about the passage of time, e.g. the movement of the sun or stars. *See also* BIOLOGICAL CLOCK, CIRCADIAN RHYTHM, CIRCANNUAL RHYTHM, INFRADIAN RHYTHM, LUNAR RHYTHM, ULTRADIAN RHYTHM

zero grazing A method of rearing livestock whereby they are fed indoors rather than allowed to graze in a field. This allows the farmer to control the nutrient inputs closely and increases productivity by reducing the loss of energy due to movement.

zero handling *See* PROTECTED CONTACT

ZIMS *See* ZOOLOGICAL INFORMATION MANAGEMENT SYSTEM (ZIMS)

zinc A mineral nutrient which functions as a cofactor in many metabolic reactions. It is involved in wound healing, PROTEIN SYNTHESIS and the functioning of the IMMUNE SYSTEM. Deficiency may result in growth retardation, anorexia and impaired reproduction (especially in males). The availability of zinc is low in some plants. Deficiency may occur if there is excess CALCIUM, cadmium or COPPER in the diet. Excess zinc may interfere with IRON and copper absorption and utilisation.

zoo

1. An abbreviation for zoological gardens. A place where wild animals are kept for exhibition, entertainment, breeding, research and conservation purposes. The legal definition varies between jurisdictions.

2. In the **Zoos Directive** (Council Directive 1999/22/EC of 29 March 1999 relating to the keeping of wild animals in zoos) Art. 2, *'"zoos" means all permanent establishments where animals of wild species are kept for exhibition to the public for 7 or more days a year, with the exception of circuses, pet shops and establishments which Member States exempt from the requirements of this Directive on the grounds that they do not exhibit a significant number of animals or species to the public and that the exemption will not jeopardise the objectives of this Directive.'*

3. In the USA, the **Animal and Plant Health Inspection Service (APHIS)** defines a zoo as: *'any park, building, cage, enclosure, or other structure or premise in which a live animal or animals are kept for public exhibition or viewing, regardless of compensation'* (9 Code of Federal Regulations, Ch.1 §1.1).

See also **Menagerie, public aquarium, safari park**

Zoo and Aquarium Association (ZAA) The regional association for zoos and aquariums in Australia, New Zealand and the South Pacific. It coordinates the Australasian Species Management Program, and has developed networks of specialists in wildlife conservation and environmental education. It was formerly known as the Australasian Regional Association of Zoological Parks and Aquaria (ARAZPA). *See also* **Regional Association of Zoos**

zoo architect A person who designs zoos or zoo exhibits. A number of architects have been influential in the development of zoos and zoo exhibits, including Sir Hugh **Casson**, Decimus **Burton**, Berthold **Lubetkin**, Jon **Coe**, and **Jones & Jones Architects and Landscape Architects Ltd**. *See also* **zoo architecture, ZooLex**

zoo architecture Many zoos contain historically important buildings and exhibits which have been designed by famous architects (*see* **zoo architect**) and which are now protected (*see* **listed buildings**). In the past some animal houses were very elaborate and made to look like Indian or Egyptian temples, manor houses or log cabins (*see also* **Buffalo House**). Aesthetic considerations were paramount and little regard was paid to the needs of the animals. Some of these buildings have now been converted for species other than those for which they were originally designed (e.g. the Elephant and Rhinoceros Pavilion at **London Zoo**) or they have been abandoned and remain unused by animals (e.g. the Penguin Pool at London Zoo, **bear pits** at **Dudley Zoo**). *See also* **ZooLex**

zoo art Posters, signage, sculptures, paintings and other works of art located and displayed in zoos. Some may be important in exhibit **interpretation**.

Zoo Bank The official registry of zoological nomenclature according to the **International Commission on Zoological Nomenclature (ICZN)**. It contains information about the original descriptions of new scientific names for animals,

publications containing these descriptions, their authors and information about the registration of **type specimens** (the specimen originally used to define a species). *See also* **binomial name**

zoo biology The scientific study of the biology of animals living in zoos, especially their behaviour, nutrition, reproduction, husbandry and captive breeding. See also **Hediger, zoo history, zoo sociology**

Zoo Biology A scientific journal that publishes research on animals living in zoos and other zoo-related studies which is published by the **Association of Zoos and Aquariums (AZA)**.

Zoo Branch The section of the **Department for Environment, Food and Rural Affairs (DEFRA)** which is responsible for the regulation of zoos.

Zoo Check Zoo Check was founded by the actors Bill Travers and Virginia McKenna after they had played the parts of George and Joy Adamson in the film *Born Free*. Zoo Check exposes suffering and exploitation in zoos and circuses, campaigns for better animal welfare legislation and challenges the education and conservation value of zoos. *See also* **Born Free Foundation, Captive Animals' Protection Society (CAPS)**

Zoo Dent International An organisation based in London and founded in 1985 by Peter **Kertesz** which provides specialist dental care for domestic animals and wildlife, especially animals kept in zoos.

zoo emergency response team A group of zoo staff trained to work together to resolve an emergency such as an animal escape. Members of the group should have access to radios, first aid equipment, weapons, vehicles, capture equipment, veterinary expertise etc. *See also* **emergency radio codes**

zoo enthusiasts Individuals who are interested in the activities and history of zoos. *See also* **Bartlett Society, zoo memorabilia**

Zoo Federation *See* **British and Irish Association of Zoos and Aquariums (BIAZA)**

zoo guest A zoo visitor.

zoo guide, zoo guidebook A guide to a particular zoo, usually containing a map and a description of the exhibits, the species kept and sometimes the conservation work of the zoo. Some zoos produce guidebooks designed like the **field guides** used to identify species in the wild. Some **travelling menageries** used to produce a catalogue of the species they exhibited with drawings of each and a short description (Fig. B7). Some Victorian zoo guide books had beautifully engraved covers and illustrations. Some **zoo enthusiasts** collect old zoo guides.

zoo guidebook *See* **zoo guide**

zoo history The history and development of zoos have become subjects worthy of academic study and are the subject of a number of academic texts including Bostock (1993) and Baratay and Hardouin-Fugier (2002) (Table Z1). *See also* **Exeter**

Z

Table Z1 Zoo history: some historical landmarks. (* Dates prior to 13th century are approximate.)

Date*	Zoo	Location
2097–2047 BC	King Shulgi (3rd Dynasty of Ur) probably owned the first zoo and kept large carnivores	Mesopotamia
1400 BC	Queen Hatshepsut kept exotic animals in her Garden of Ammon.	Egypt
1279–1213 BC	Per Ramesses (the 'House of Ramesses') was built by Ramesses II and appears to have contained a zoo where lions, elephants and possibly giraffes were kept	Egypt
c1100 BC	China, the Emperor Wen Wang created an 'Intelligence Park', which appears to have contained animals. Later Emperor Chi-Hang-Ti – of the Thsin dynasty – created a garden which was filled with animals and trees from all over his empire	China
879 BC	King Assurnasirpal II kept elephants in a 'zoo'	Assyria
283–246 BC	Ptolemy II founded a zoo at Alexandria	Egypt
c55 BC–1 AD	Large numbers of exotic animals were taken to Rome and killed in large-scale spectacles of slaughter	Rome
4th century	By this time most of the city states in Greece probably had their own animal collections	Greece
1086	In medieval Europe royalty and nobility often kept animals in deer parks. At this time 'deer' probably meant 'animal'. The Domesday Book records 35 deer parks around the year 1086	Europe
1100	Henry I established a menagerie at Woodstock. It was later moved to the Tower of London, probably by Henry III	England
13th century	Marco Polo saw lions and tigers wandering freely through a Chinese imperial palace. Around this time Kublai Khan, the fifth Great Khan of the Mongol Empire, had animal parks that were used for hunting, and he also kept tame cheetahs, tigers and falcons	China/Mongol Empire
1328–1350	Philip VI of France (reigned 1328–1350) kept lions and leopards at the Louvre	France
1368	A traveller to China reported that he had seen 3000 monkeys in the park of a Buddhist pagoda. A little later, a closed garden was reported from near Peking, which contained a high mountain inhabited by monkeys and other animals	China
1513–1523	The Vatican menagerie expanded under Pope Leo X (1513–1523) and included monkeys, civets, lions, leopards, an elephant and a snow leopard	Vatican
c1517–1521	A magnificent zoo was owned by the Aztec emperor Montezuma II, at his capital Tenochtitlan	Mexico
1552	Crown Prince Maximilian of Austria created a deer park and menagerie around the castle at Ebersdorf, near Vienna	Austria
1665	Louis XIV opened a menagerie in the grounds of his palace in Versailles	France

Dates of opening of selected zoos since 1752

1752	Schönbrunn Zoo, Vienna	Austria
1793	Jardin des Plantes, Paris	France
1828	London Zoological Gardens	England
1833	Dublin Zoological Gardens	Ireland
1836	Belle Vue Zoological Gardens, Manchester	England
1836	Bristol Zoological Gardens	England
1839	Royal Edinburgh Zoological Gardens	Scotland
1839	Amsterdam Royal Zoological Gardens	Netherlands
1843	Antwerp Zoological Gardens	Netherlands
1844	Berlin Zoological Gardens	Germany
1857	Rotterdam Zoological Gardens	Netherlands
1858	Frankfurt Zoological Gardens	Germany
1860	Jardin Zoologique d'Acclimatation, Paris	France
1860	Cologne Zoological Garden	Germany
1861	Dresden Zoological Gardens	Germany

Table Z1 (*Continued*)

Date*	Zoo	Location
1863	Hamburg Zoological Garden	Germany
1865	Breslau (now Wroclaw) Zoological Garden	Poland
1866	Budapest Zoological Garden	Hungary
1868	Lincoln Park Zoological Gardens	USA
1871	Stuttgart Zoological Garden	Germany
1872	Royal Melbourne Zoological Gardens	Australia
1873	New York Central Park Zoo	USA
1874	Basel Zoological Garden	Switzerland
1874	Philadelphia Zoological Garden	USA
1875	Cincinnati Zoo	USA
1876	Calcutta Zoological Gardens	India
1882	Ueno Zoological Gardens, Tokyo	Japan
1888	Cleveland Metroparks Zoological Park	USA
1888	Buenos Aires Zoo	Argentina
1888	Dallas Zoo	USA
1889	Atlanta Zoological Park	USA
1891	National Zoological Park, Washington DC	USA
1891	Giza Zoo, Cairo	Egypt
1892	St. Petersburg Zoological Garden	Russia
1895	Baltimore Zoo	USA
1896	Düsseldorf Zoological Garden	Germany
1896	Königsberg Zoological Gardens	Kaliningrad (now Russia)
1898	Pittsburgh Zoo	USA
1899	New York (Bronx) Zoological Park	USA
1899	Pretoria Zoo	South Africa
1899	Moscow Zoological Garden	Russia
1899	Toledo Zoological Gardens	USA
1907	Tierpark Hagenbeck (Stellingen)	Germany
1913	Edinburgh Zoological Gardens	Scotland
1916	San Diego Zoo	USA
1923	Paignton Zoological Gardens	England
1931	Chester Zoological Gardens	England
1938	Dudley Zoological Gardens	England
1952	Arizona–Sonora Desert Museum	USA
1959	Jersey Zoological Gardens	Channel Islands
1963	Welsh Mountain Zoo, Colwyn Bay	Wales
1966	*Lions of Longleat*, Wiltshire	England
1971	Knowsley Safari Park, Prescot	England
1994	South Lakes Wild Animal Park, Cumbria	England
1998	*Disney's Animal Kingdom*, Florida	USA
1998	*Blue Planet*, Ellesmere Port	England
2002	*The Deep*, Hull	England

EXCHANGE, HAGENBECK, LOISEL, MENAGERIE, MONTEZUMA II, SCHÖNBRUNN ZOO, TRAVELLING MENAGERIE, WOODSTOCK, WORLD'S ZOOLOGICAL TRADING COMPANY LTD., ZOO ARCHITECTURE, ZOO MEMORABILIA, ZOOLOGICAL SOCIETY OF LONDON (ZSL)

zoo inspector A person with experience of the activities of zoos (e.g. a vet or zoo curator) who makes regular visits to a zoo to determine whether or not it complies with the conditions of its ZOO LICENCE.

Zoo Inspectorate In the UK, a group of ZOO INSPECTORS employed by the DEPARTMENT FOR ENVIRONMENT, FOOD AND RURAL AFFAIRS (DEFRA) to inspect zoos.

zoo licence A legal document authorising the holder to operate a zoo; issued by a LICENSING AUTHORITY. In the UK this is the local authority (local government). *See also* ZOO INSPECTOR, ZOO LICENSING ACT 1981

Zoo Licensing Act 1981 A law requiring zoos in Great Britain to be licensed and setting out laws for the

Z

Table Z2 Zoo organisations.

African Association of Zoos and Aquaria (PAAZAB)
Alliance of Marine Mammal Parks and Aquariums (AMMPA)
Asociacion Espanola de Zoos y Acuarios (AEZA) – Spanish Association of Zoos and Aquariums
Asociación Ibérica de Zoos y Acuarios (AIZA) – Iberian Association of Zoos and Aquaria
Asociación Latinoamericana de Parques Zoológicos y Acuarios (ALPZA) – Latin American Zoo and Aquarium Association
Asociación Mesoamericana y del Caribe de Zoológico i Acuarios (AMACZOOA) – Association of Mesoamerican and Caribbean Zoos and Aquariums
Association Francaise des Parcs Zoologiques (AFdPZ) – Association of French Zoos
Association of Zoos and Aquariums (AZA) – North America
Asociación de Zoológicos, Criaderos y Acuarios de México AC (AZCARM) – Association of Mexican Zoos and Aquariums
British and Irish Association of Zoos and Aquariums (BIAZA)
Canadian Association of Zoos and Aquariums (CAZA)
Danske Zoologiske Haver & Akvarier – Danish Association of Zoological Gardens and Aquaria (DAZA)
Deutsche Tierpark-Gesellschaft (DTG) – German Animal Park Society
European Association of Zoos and Aquaria (EAZA)
European Union of Aquarium Curators (EUAC)
Verband deutscher Zoodirektoren e.V. (VdZ) German Federation of Zoo Directors.
Unione Italiana dei Giardini Zoologici ed Acquari (UIZA) – Italian Union of Zoos & Aquaria
Japanese Association of Zoos and Aquariums (JAZA)
Nederlandse Vereniging van Deirentuinen (NVD) – Dutch Zoo Federation
Österreichische Zoo Organisation (OZO) – Austrian Zoo Organisation
Sociedade de Zoológicos do Brasil (SZB) – Brazilian Zoo Society
South Asian Zoo Association for Regional Cooperation (SAZARC)
South East Asian Zoos Association (SEAZA)
Svenska Djurparksföreningen (SDF) – Swedish Association of Zoological Parks and Aquaria (SAZA)
Swiss Association of Scientific Zoos (ZOOSchweiz)
Syndicat National des Directeurs de Parcs Zoologiques (SNDPZ) – French zoo directors union
Union of Czech and Slovak Zoological Gardens (UCSZ)
World Association of Zoos and Aquariums (WAZA)
Zoo and Aquarium Association (formerly Australasian Regional Association of Zoological Parks and Aquaria (ARAZPA))
Zoological Association of America (ZAA)

regulation and inspection of zoos. The Act has been amended to implement the **Zoos Directive** and requires licensed zoos to have a conservation function. *See also* **Zoo Branch**, **Zoo Inspectorate**

zoo membership scheme A scheme run by a zoo which provides free entry in return for an annual subscription. Members may also be given additional benefits, e.g. a members' magazine, zoo shop discounts, lectures, members' trips etc.

zoo memorabilia Materials produced by or about zoos which are collected by **zoo enthusiasts**, including old zoo entrance tickets, **zoo guides**, postcards, photographs and souvenirs. Many zoos have archives of such material. Some museums and libraries hold zoo memorabilia and records for zoos that no longer exist, e.g. Chetham's Library in Manchester, UK, holds an archive of materials from **Belle Vue Zoo**, Manchester, which operated from 1836 until 1979.

zoo organisation An association of zoos which co-operate in captive breeding programmes, education, fund-raising for *in-situ* **conservation** projects

and other activities. They may be national, regional or international (Table Z2). Zoo organisations generally require members to comply with a set of standards relating to animal welfare, an **Animal Transaction Policy** etc., as a condition of membership.

Zoo Outreach Organisation (ZOO) A conservation, research, education and welfare NGO which was founded in 1985 with funds from the Government of India, originally to provide technical and educational support for zoos, and to enhance their public image. It also lobbies to improve zoo and animal welfare legislation and publishes *Zoos' Print*, *Zoos' Print Journal* and *Zoo Zen*.

zoo poo Dung sold to zoo visitors as garden fertiliser. Often elephant dung.

Zoo Quest A pioneering series of television programmes about wildlife made by Sir David **Attenborough** as a joint venture between the BBC and **London Zoo** and broadcast between 1954 and 1963. Attenborough travelled to a wide variety of locations around the world to capture animals for

Z

the zoo and filmed exotic species in the wild such at the Komodo dragon (*Varanus komodoensis*) which was the subject of the first episode.

zoo research institute A research institute dedicated to the study of zoo animals, conservation and wildlife. Important examples include the **INSTITUTE FOR CONSERVATION RESEARCH (ZOOLOGICAL SOCIETY OF SAN DIEGO (ZSSD))**, **INSTITUTE OF ZOOLOGY (IoZ)** (**ZOOLOGICAL SOCIETY OF LONDON (ZSL)**), the Leibniz Institute for Zoo and Wildlife Research (Berlin) and the Smithsonian Conservation Biology Institute.

zoo sociology The study of zoos as a reflection of changes in human society and attitudes to animals over time.

Zoo Time *See* **MORRIS, DESMOND** (1928–)

Zoo Zen A publication of the **ZOO OUTREACH ORGANISATION (ZOO)** which seeks to promote the 'zen' of zoo keeping and conservation biology, the art and science, discipline and practice of animal care and conservation. It publishes useful original material for zoo professionals and republishes material originally published elsewhere.

ZooChat An on-line community of animal conservation and **ZOO ENTHUSIASTS**. Its website hosts a number of special interest forums, a zoo photo gallery, information on zoo webcams and maps, and satellite images of many zoo locations.

zoochosis (zoochoses *pl.***)** A term coined by Bill Travers to describe abnormal **STEREOTYPIC BEHAVIOURS** in animals which are analogous to psychoses in humans. *See also* **ZOO CHECK**

zoochotic Relating to a **ZOOCHOSIS**.

zoogeographical regions *See* **FAUNAL REGIONS**

zoogeography

1. The scientific study of the distribution of animals, its causes and effects, and especially its relationship to evolution and the effects of continental drift.

2. A description of the distribution of a species or other taxon.

See also **FAUNAL REGIONS**

zookeeper *see* **KEEPER** (1)

ZooLex The ZooLex Zoo Design Organisation was established to help improve holding conditions for wild animals in captivity. It publishes and disseminates information related to zoo design and promotes appropriate holding conditions for wild animals in captivity. It also provides technical information and advice about zoo design and supports research related to zoo design and vocational training. *See also* **ZOO ARCHITECTURE**

zoological garden (s) *See* **ZOO**

Zoological Information Management System (ZIMS) A new global database of information on animal health and wellbeing, which is the first of its kind in the world. It contains pooled information from **INTERNATIONAL SPECIES INFORMATION SYSTEM (ISIS)** member institutions which can be accessed by other members with their permission. This data includes information on veterinary care, animal husbandry and behaviour. ZIMS will allow zoo professionals access to data from other institutions which was previously only available through personal contacts. The system also has the potential to track new and emerging animal diseases.

zoological park An alternative name for a **ZOO**, but the term generally refers to a more expansive facility, e.g. the **SMITHSONIAN NATIONAL ZOOLOGICAL PARK**, Washington DC. *Compare* **SAFARI PARK**

zoological society A group of people who have formed a society with the purpose of promoting interest in animals and, latterly, their conservation. Most societies own and operate a zoological garden, e.g. the **ZOOLOGICAL SOCIETY OF LONDON (ZSL)**, **ZOOLOGICAL SOCIETY OF SAN DIEGO (ZSSD)**, **WILDLIFE CONSERVATION SOCIETY (WCS)**

Zoological Society of London (ZSL) The Zoological Society of London was founded in 1826 by Sir Stamford **RAFFLES**. It operates **LONDON ZOO** and **WHIPSNADE ZOO**. In 1960 the Society established the **INSTITUTE OF ZOOLOGY (IoZ)** where scientists are employed to conduct zoological research.

Zoological Society of San Diego (ZSSD), San Diego Global The **ZOOLOGICAL SOCIETY** in California that operates San Diego Zoo, the San Diego Zoo Safari Park, and the San Diego Zoo **INSTITUTE FOR CONSERVATION RESEARCH**.

zoology The scientific study of animals.

zoonosis (zoonoses *pl.***)** A disease that is communicable from animals to humans or from humans to animals. Such diseases may be caused by **BACTERIA**, **VIRUSES** or other microbes, or by **PARASITES**. They include **RABIES**, **ANTHRAX** and other potentially fatal infections.

zoophilia *See* **BESTIALITY**

zooplankton *See* **PLANKTON**. *See also* **LIVE ZOOPLANKTON**

ZooRisk Software which is designed to help managers make scientifically based decisions about the management of captive populations by providing a quantitative assessment of a population's extinction risk due to demographic, genetic and management factors. This assessment is based on the history of the population, the biology of small populations and knowledge of our ability to manage captive populations. It was developed by **LINCOLN PARK ZOO** and is distributed free as shareware.

Zoos Directive Council Directive 1999/22/EC of 29 March 1999 relating to the keeping of wild animals in zoos. Its purpose is to improve conditions for animals in zoos within the **EUROPEAN UNION (EU)** while also requiring zoos to adopt a conservation role. Under Art. 2 . . . '*"zoos" means all permanent establishments where animals of wild species are kept for exhibition to the public for 7 or more days a year* . . .' All zoos are required to promote public education,

Z

provide adequate housing, prevent escapes and keep records. A zoo is required to engage in *one* of the following: research; training in conservation skills; information exchange; **CAPTIVE BREEDING**; repopulation or **REINTRODUCTION** to the wild (Art. 3). In addition to the requirements listed in Article 3, Member States are required to establish a licensing and inspection system for zoos in order to ensure that the requirements in Article 3 are met. If a zoo fails to meet these requirements the Directive makes provision for the closure of the zoo by a 'competent authority' which, in the UK would be the local authority which is responsible for issuing the licence. *See also* **ZOO LICENCE**, **ZOO LICENSING ACT 1981**

Zoos Expert Committee A committee that replaced the **ZOOS FORUM** in February 2011 as an independent advisor to the UK government (via the **DEPARTMENT FOR ENVIRONMENT, FOOD AND RURAL AFFAIRS (DEFRA)**) on zoo matters.

Zoos Forum An independent organisation that advised the UK government on zoo issues and policy until early 2011 when it was replaced by the **ZOOS EXPERT COMMITTEE**.

Zoos' Print A publication produced by the **ZOO OUTREACH ORGANISATION (ZOO)** which contains news of zoos and wildlife networks in south Asia.

Zoos' Print Journal An integral part of **ZOOS' PRINT** which contains scientific, peer-reviewed papers on the conservation, taxonomy, distribution, behaviour, welfare, veterinary care, trade, natural history and biology of south Asian and southeast Asian wild and captive fauna.

Zootrition Zootrition® is dietary management software that provides zoo managers with a means of comparing the nutritional content of specific food items and calculating the overall nutritional composition of diets. This allows the identification of potential nutrient deficiencies and toxicities. The software was developed by St. Louis Zoo with support from the **WORLD ASSOCIATION OF ZOOS AND AQUARIUMS (WAZA)**. It contains information about a wide range of foods and has a facility for zoos to enter information about others.

ZSL *See* **ZOOLOGICAL SOCIETY OF LONDON (ZSL)**

ZSL London Zoo *See* **LONDON ZOO**

ZSL Whipsnade Zoo *See* **WHIPSNADE ZOO**

Zuckerman, Sir Solly (1904–1993) A zoologist, academic and science advisor to the UK government who was the Secretary of the **ZOOLOGICAL SOCIETY OF LONDON (ZSL)** from 1955–1977 and as its President from 1977–1984. He was a pioneer in the study of primate behaviour and published *The Social Life of Monkeys and Apes* in 1931.

zygomatic arch A bone arch at the side of the skull to which the masseter (chewing) muscle is attached.

zygote The result of the **FERTILISATION** of an egg by a sperm, before it undergoes cell division, which normally restores the **DIPLOID** number of **CHROMOSOMES**. *See also* **EMBRYO**, **TWINNING**

Z

Acronyms and abbreviations

AAFC	Agriculture and Agri-Food Canada	APHIS	Animal and Plant Health Inspection Service
AArk	Amphibian Ark		
AAVS	American Anti-vivisection Society	APP	African Preservation Programme
AAZK	American Association of Zoo Keepers	AR	aspect ratio
		ARAZPA	Australasian Regional Association of Zoological Parks and Aquaria (now ZAA)
AAZPA	American Association of Zoological Parks and Aquariums (*now* AZA)		
AAZV	American Association of Zoo Veterinarians	ARBA	American Rare Breed Association
		ARKive	a digital library of film and photographs
ABA	American Birding Association		
ABS	Animal Behavior Society	ARKS	Animal Record Keeping System
ABWAK	Association of British and Irish Wild Animal Keepers	ART	assisted reproductive technology
		ARU	autonomous recording unit
ACCOBAMS	Agreement on the Conservation of Cetaceans in the Black Sea, Mediterranean Sea and Contiguous Atlantic Area 1996	ASAB	Association for the Study of Animal Behaviour
		ASCOBANS	Agreement on the Conservation of Small Cetaceans of the Baltic, North East Atlantic, Irish and North Seas 1991
ACh	acetylcholine		
ACRES	Audubon Nature Institute's Center for Research of Endangered Species		
		ASMP	Australasian Species Management Programme
ACTH	adrenocorticotrophic hormone	ASPCA	American Society for the Prevention of Cruelty to Animals
ad lib	*ad libitum*		
ADI	Animal Defenders International	ASZK	Australian Society of Zoo Keeping
ADMD	apparent dry matter digestibility	ATP	adenosine triphosphate
ADP	adenosine diphosphate	ATV	all-terrain vehicle
AEWA	African–Eurasian Migratory Waterbird Agreement (Agreement on the Conservation of African–Eurasian Migratory Waterbirds) 1995	AV	artificial vagina
		AWA	Animal Welfare Act of 1966 (USA)
		AWG	Aquarium Working Group (BIAZA)
		AZA	(American) Association of Zoos and Aquariums
AHVLA	Animal Health and Veterinary Laboratories Agency	AZK	Australasian Zoo Keeping Association
AI	artificial insemination		
AKAA	Animal Keepers Association of Africa	BAG	Behavioral Advisory Group (AZA)
		BAP	Biodiversity Action Plan
ALBC	American Livestock Breeds Conservancy	BARR	Breeds at Risk Register
		BAS (1)	British Alpaca Society
ALF	Animal Liberation Front	BAS (2)	British Antarctic Survey
ANI	animal needs index	BASC	British Association for Shooting and Conservation
ANOVA	analysis of variance		
ANS	autonomic nervous system	BBC	British Broadcasting Corporation
AONB	Area of Outstanding Natural Beauty	BBSRC	Biotechnology and Biological Sciences Research Council

BCEAW	Breeding Centre for Endangered Arabian Wildlife
BCI	bout criterion interval
BCMS	British Cattle Movement Service
BHA	British Horseracing Authority
BIAZA	British and Irish Association of Zoos and Aquariums
BIERZS	Bear Information Exchange for Rehabilitators, Zoos and Sanctuaries
BLM	Bureau of Land Management
BMI	body mass index
BMR	basal metabolic rate
BOD	biological or biochemical oxygen demand
BOU	British Ornithologists' Union
BP	blood pressure
BPA	British Pig Association
BRC	British Rabbit Council
BREEDPLAN	software used in livestock breeding
BSAVA	British Small Animal Veterinary Association
BSE	bovine spongiform encephalitis (encephalopathy)
BTO	British Trust for Ornithology
BUAV	British Union for the Abolition of Vivisection
BVA	British Veterinary Association
BVD	bovine viral diarrhoea
BWG	Bird Working Group (BIAZA)
C	Celsius
CAGR	Canadian Animal Genetic Resources (Program)
cal	calorie
Cal	1000 calories
CAMPFIRE	Communal Areas Management Programme for Indigenous Resources
CAP (1)	Common Agricultural Policy
CAP (2)	Global Captive Action Plan
CAPACITY	software used in conservation breeding programmes
CAPS	Captive Animals' Protection Society
CAT	computer-assisted tomography/ computed axial tomography
CAWC	Companion Animal Welfare Council
CBD	Convention on Biological Diversity 1992
CBSG	Conservation Breeding Specialist Group (formerly Captive Breeding Specialist Group)
CCAMLR	Convention on the Conservation of Antarctic Marine Living Resources 1980
CCIA	Canadian Cattle Identification Agency
CCTV	closed-circuit television
CDC	Centers for Disease Control and Prevention

CFP	Common Fisheries Policy
CFR	Code of Federal Regulations
CG	chorionic gonadotrophin
CheCS	Cattle Health Certification Standards
CITES	Convention on International Trade in Endangered Species of Wild Fauna and Flora 1973
CIWF	Compassion in World Farming
CLTS	Canadian Livestock Tracking System
CMS	Convention on the Conservation of Migratory Species of Wild Animals 1979
CNS	central nervous system
COSHH	Control of Substances Hazardous to Health
COTES	Control of Trade in Endangered Species (Enforcement) Regulations 1997
CPH	County Parish Holding
CR	Critically Endangered (Red List category)
CRES	Conservation and Research for Endangered Species (formerly Center for Reproduction of Endangered Species) (ZSSD)
CRoW	Countryside and Rights of Way Act 2000
CSF (1)	cerebrospinal fluid
CSF (2)	classical swine fever
CSF (3)	cafeteria-style feeding
CSIRO	Commonwealth Scientific and Industrial Research Organisation
CT	computed tomography
CVED	Common Veterinary Entry Document
CWD	chronic wasting disease/cervid chronic wasting disease
CZA	Central Zoo Authority (India)
DBI	DanBred International
DD	Data Deficient (Red List category)
DDT	dichlorodiphenyltrichloroethane
DEFRA (defra)	Department for Environment, Food and Rural Affairs
DM	dry matter
DMI	dry matter intake
DNA	deoxyribonucleic acid
DNS	did not survive
DOI	United States Department of the Interior
EAZA	European Association of Zoos and Aquaria
EAZWV	European Association of Zoo and Wildlife Veterinarians
EBV	estimated breeding value
ECG (1)	electrocardiogram
ECG (2)	electrocardiograph
ECJ	European Court of Justice

ECOLEX	a database of natural resources law and policy		GAA	Global Amphibian Assessment
EDGE	Evolutionarily Distinct and Globally Endangered		GAE	gross assimilation efficiency
			GAHMU	Great Ape Health Monitoring Unit
			GASP	Global Animal Survival Plan
EEG (1)	electroencephalogram		GBGB	Greyhound Board of Great Britain
EEG (2)	electroencephalograph		GBIF	Global Biodiversity Information Facility
EEKMA	European Elephant Keeper and Manager Association			
			GBO-3	Global Biodiversity Outlook 3
EEP	European Endangered Species Programme		GEF	Global Environment Facility
			GENES	software used in conservation breeding programmes
EGGS	software used in conservation breeding programmes			
			GH	growth hormone
EIA (1)	Environmental Impact Assessment		GI	gastrointestinal
EIA (2)	Environmental Investigation Agency		GIS	Geographical (Geographic) Information Systems
EID	electronic identification			
EKG (1)	electrocardiogram		GM	genetically modified
EKG (2)	electrocardiograph		GMO	genetically modified organism
EMA	Elephant Managers Association		GMT	Greenwich mean time
EN	Endangered (Red List category)		GnRH	gonadotrophin-releasing hormone
ENFAP	European Network for Farm Animal Protection		GPS	Global Positioning System
			H_0	null hypothesis
EOL	Encyclopedia of Life		H_1	alternative hypothesis
EPA	Environmental Protection Agency		H1N1	swine flu virus
ER	endoplasmic reticulum		H5N1	bird flu virus
ESA	Endangered Species Act of 1973		ha	hectare
ESB	European Studbook		HAR	human–animal relationships
ESF	European Studbook Foundation		HBOT	hyperbaric oxygen therapy
ESU (1)	evolutionarily significant unit		HLLE	head and lateral line erosion
ESU (2)	electrosurgical unit		HPA	hypothalamic–pituitary–adrenal
ET	embryo transfer		HSUS	Humane Society of the United States
ETFE	ethylene tetrafluoroethylene			
EU	European Union		IAH	Institute for Animal Health
EUAC	European Union of Aquarium Curators		IATA	International Air Transport Association
EUROBATS	Agreement on the Conservation of Bats in Europe 1991		ICP	institutional collection plan
			ICSH	interstitial cell-stimulating hormone
EW	Extinct in the Wild (Red List category)		ICU	intensive care unit
			ICZ	International Congress of Zookeepers
EX	Extinct (Red List category)			
F	Fahrenheit		ICZN	International Commission on Zoological Nomenclature
F_1	first filial generation			
F_2	second filial generation		IDA	In Defense of Animals
FAO	Food and Agriculture Organisation		IFAW	International Fund for Animal Welfare
FAP	fixed action pattern			
FBI	Federal Bureau of Investigation		IGFA	International Game Fish Association
FCI	Federation Cynologique Internationale			
			INTERPOL	an international police organisation
FDA	Food and Drug Administration (*See also* USFDA)		IoZ	Institute of Zoology (ZSL)
			IPPL	International Primate Protection League
FERA	Food and Environment Research Agency			
			IR	infrared
FFI	Fauna and Flora International		ISAZ	International Society for Anthrozoology
FMD	foot and mouth disease			
FONZ	Friends of the National Zoo (USA)		ISIS	International Species Information System
FSA (1)	Farm Service Agency			
FSA (2)	Food Standards Agency		ITAHC	Intra-trade Animal Health Certificate
FSH	follicle-stimulating hormone			
FWAG	Farming and Wildlife Advisory Group		ITIS	Interagency Taxonomic Information System

IUCN	International Union for the Conservation of Nature and Natural Resources
IUDZG	International Union of Directors of Zoological Gardens
IUPN	International Union for the Protection of Nature (now IUCN)
IUU	illegal, unreported and unregulated (fishing)
IVF	*in-vitro* fertilisation
IWC	International Whaling Commission
IWRC	International Wildlife Rehabilitation Council
IZE	International Zoo Educators Association
IZES	Independent Zoo Enthusiasts Society
IZN	*International Zoo News*
IZVG	International Zoo Veterinary Group
IZYB	*International Zoo Yearbook*
J	joule
JAZA	Japanese Association of Zoos and Aquariums
JMSC	Joint Management of Species Committee
JMSP	Joint Management of Species Programme
JNCC	Joint Nature Conservation Committee
kHz	kilohertz
LaCONES	Laboratory for the Conservation of Endangered Species (India)
LC	Least Concern (Red List category)
LD_{50}	lethal dose 50%
LH	luteinising hormone
LHRH	luteinising hormone-releasing hormone
LMO	living modified organism
Ltd	limited company
M99	Immobilon
MAB	Man and the Biosphere Programme
MAI	Maximum Avoidance of Inbreeding
MateRx	software used in conservation breeding programmes
MCZ	Marine Conservation Zone
MDI	metered dose inhaler
MedARKS	Medical Animal Record Keeping System
MLA	Meat and Livestock Australia
MOC	multiple ocular coloboma
MPA	Marine Protected Area
MPI	maintenance of proximity index
MRCVS	Member of the Royal College of Veterinary Surgeons
MRI	magnetic resonance imaging
mRNA	messenger RNA
MSI	mate suitability index
MSY	maximum sustainable yield
mtDNA	mitochondrial DNA
MTT	mental time travel
MVE	Murray Valley encephalitis
MVP	minimum viable population
MWG	Mammal Working Group (BIAZA)
NAIS	National Animal Identification System
NAIT	National Animal Identification and Tracing
NASA	National Aeronautics and Space Administration
NC3RS	National Centre for the Replacement, Refinement and Reduction of Animals in Research
NCA	Ngorongoro Conservation Area
N_e	effective population size
NE	Not Evaluated (Red List category)
NERC	Natural Environment Research Council
NFSCo	National Fallen Stock Company
NFU	National Farmers' Union
NGO	non-governmental organisation
NIFA	National Institute of Food and Agriculture
NLIS	National Livestock Identification System
nm	nanometre
NMFS	National Marine Fisheries Service
NNR	National Nature Reserve
NP	national park
NRCS	National Resources Conservation Service
NSW	New South Wales
NSWG	Native Species Working Group (BIAZA)
NT (1)	National Trust
NT (2)	Near Threatened (Red List category)
NWHC	National Wildlife Health Center (USGS)
NZ	New Zealand
OIE	Office International des Epizooties (World Organisation for Animal Health)
OUTBREAK	software used in conservation breeding programmes
p	frequency of the dominant allele
PAAZAB	African Association of Zoos and Aquaria
PAC	problem animal control
PanTHERIA	a database of information on extinct mammals
PARTINBR	software used in conservation breeding programmes
PAW	Partnership for Action Against Wildlife Crime (UK)
PAWS	Performing Animal Welfare Society
PCR	polymerase chain reaction
PDA	personal digital assistant
PDSA	People's Dispensary for Sick Animals

PETA	People for the Ethical Treatment of Animals Foundation	SPIDER	Setting goals, Planning, Implementing, Documenting, Evaluating and Readjusting
PIG	Pig Improvement Company		
PiGIS	Pig Grading Information System	SPSS	Statistical Package for the Social Sciences
PIR	passive infrared		
PLOS/PLoS	*Public Library of Science*	SR	Statutory Rule
PM2000	Population Management 2000	SRB	separation-related behaviour
PMP	Population Management Plan (AZA)	SSC	Species Survival Commission
		SSI	Scottish Statutory Instrument
PNS (1)	parasympathetic nervous system	SSP	Species Survival Plan®
PNS (2)	peripheral nervous system	SSSI	Site of Special Scientific Interest
POE	postoccupancy evaluation	SSSMZP	*Secretary of State's Standards of Modern Zoo Practice*
PRT	positive reinforcement training		
PVA	population viability analysis	STD	sexually transmitted disease
PWG	Plant Working Group (BIAZA)	TAG	Taxon Advisory Group
q	frequency of the recessive allele	TB	tuberculosis
RAVC	Royal Army Veterinary Corps	TCA	tricarboxylic acid
RAWG	Reptile and Amphibian Working Group (BIAZA)	TCM	traditional Chinese medicine
		TIWG	Terrestrial Invertebrate Working Group (BIAZA)
RBST	Rare Breeds Survival Trust		
RCP	regional collection plan	TNR	trap–neuter–release
RCVS	Royal College of Veterinary Surgeons	TRACES	Trade Control and Expert System
		TRAFFIC	Trade Records Analysis of Flora and Fauna in Commerce
REGASP	Regional Animal Species Collection Plan		
		tRNA	transfer DNA
REM	rapid eye movements	TSE	transmissible spongiform encephalopathy
REPTA	Reptile and Exotic Pet Trade Association		
		TSS	toxic shock syndrome
RFID	radio frequency identification	TWG	Taxon Working Group
RH	relative humidity	UDAW	Universal Declaration on Animal Welfare
RNA	ribonucleic acid		
rRNA	ribosomal RNA	UFAW	Universities Federation for Animal Welfare
RSPB	Royal Society for the Protection of Birds		
		UHF	ultra-high frequency
RSPCA	Royal Society for the Prevention of Cruelty to Animals	UK	United Kingdom of Great Britain and Northern Ireland
		UN	United Nations
RVN	Registered Veterinary Nurse	UNCED	United Nations Conference on Environment and Development (UNCED)
RZSS	Royal Zoological Society of Scotland		
S	salinity	UNCLOS	United Nations Convention on the Law of the Sea 1982
SAC	Special Area of Conservation		
SARS	severe acute respiratory syndrome	UNEP	United Nations Environment Programme
SAZARC	South Asian Zoo Association for Regional Co-operation		
		UNESCO	United Nations Educational, Scientific and Cultural Organization
SBV	Schmallenberg virus		
SCNT	somatic cell nuclear transfer	US	United States (of America)
SE	standard error	USA	United States of America
SEAZA	South East Asian Zoo Association	USDA	United States Department of Agriculture
SI (1)	Statutory Instrument		
SI (2)	Système International d'Unités (International System of Units)	USFDA	United States Food and Drug Administration
SMRU	Sea Mammals Research Unit	USFWS	United States Fish and Wildlife Service
SNH	Scottish Natural Heritage		
SNS(1)	sympathetic nervous system	USGS	United States Geological Survey
SNS(2)	somatic nervous system	UV	ultraviolet
SPA	Special Protection Area	VHF	very high frequency
SPARKS	Single Population Analysis and Record Keeping System	VORTEX	software used in conservation breeding programmes.
SPI	spread of participation index		

VPIS	Veterinary Poisons Information Service	WSPA	World Society for the Protection of Animals
VU	Vulnerable (Red List category)	WWF	World Wide Fund for Nature/ World Wildlife Fund
WAZA	World Association of Zoos and Aquariums	WWT	Wildfowl and Wetlands Trust
WCA	Wildlife and Countryside Act 1981	YouTHERIA	a database of information on living mammals
WCMC	World Conservation Monitoring Centre	ZAA	Zoo and Aquarium Association (formerly ARAZPA)
WHO	World Health Organization		
WIN	Wildlife Information Network	ZIMS	Zoological Information Management System
WL	wing loading		
WoRMS	World Register of Marine Species	ZOO	Zoo Outreach Organisation
WSC	Wildlife Conservation Society	ZSL	Zoological Society of London
WSI	Welsh Statutory Instrument	ZSSD	Zoological Society of San Diego

References

Anon. (1987) *Handling Geographic Information*. Department of the Environment, HMSO, London.

Anon. (1999) *Future trends in veterinary public health: report of a WHO study group* (WHO technical report series: 907). WHO Study Group on Future Trends in Veterinary Public Health, Teramo, Italy.

Anon. (2001) *Climate Change 2001: The Scientific Basis. IPCC. IPCC Third Assessment Report*. Cambridge University Press, Cambridge.

Anon. (2011a) *IUCN Red List version 2011.1*: Table 1 http://www.iucnredlist.org/documents/summarystatistics/2011_1_RL_Stats_Table_1.pdf. Accessed 29.3.12

Anon. (2011b) *Crop Areas, Livestock Populations and Agricultural Workforce 2011. England – Final Results*. DEFRA, London.

Anon. (2012) *Living Planet Report 2012*. WWF International, Gland, Switzerland.

Ascione, F.R. (1998) Battered women's reports of their partners and their children's cruelty to animals. *Journal of Emotional Abuse* 1(1): 199–133.

Baker, R.R. (1978) *The Evolutionary Ecology of Animal Migration*. Hodder and Stoughton, London.

Baratay, E. and Hardouin-Fugier, E. (2002) *Zoo. A History of Zoological Gardens in the West*. Reaktion Books Ltd, London, UK.

Bekoff, M. and Pierce, J. (2009) *Wild Justice: The Moral Lives of Animals*. University of Chicago Press, Chicago.

Bostock, S. St.C. (1993) *Zoos and Animal Rights. The ethics of keeping animals*. Routledge, London and New York.

Briffa, M. and Greenaway, J. (2011) High *in situ* repeatability of behaviour indicates animal personality in the beadlet anemone *Actinia equina* (Cnidaria). *PLoS ONE* 6(7): e21963. doi:10.1371/journal.pone.0021963

Broom, D. (1986) Indicators of poor welfare. *British Veterinary Journal* 142: 524–526.

Broom, D.M. (1991) Animal welfare: concepts and measurement. *Journal of Animal Science* 69: 4167–4175.

Burt, W.H. (1943) Territoriality and home range concepts as applied to mammals. *Journal of Mammalogy* 24: 346–352.

Coe, J. (no date) *The Unzoo Alternative* http://www.zoolex.org/publication/coe/Unzoo150805.pdf. Accessed 29.3.12

Dawkins, M. (1990) From an animal's point of view: motivation, fitness and animal welfare. *Behavioral and Brain Sciences* 13: 1–9.

Hambler, C. (2004) *Conservation*. Cambridge University Press, Cambridge.

Hediger, H. (1950) *Wild Animals in Captivity*. Butterworth Scientific Publications Ltd., London.

Hirst, K.K. (no date) *Animal Domestication Table of Dates and Places. When and Where Animal Domestication Occurred*. http://archaeology.about.com/od/dterms/a/domestication.htm. Accessed 21.6.12.

Ironmonger, J. (1992) *The Good Zoo Guide*. Harper Collins Publishers, London.

Itoh, M. (2010) *Japanese Wartime Zoo Policy. The Silent Victims of World War II*. Palgrave Macmillan, New York.

Jose, D., Bradfield, K., Power, V. and Lambert, C. (2011) Predator awareness training at Perth Zoo – a review. *Thylacinus* 35(3): 2–7.

Lovejoy, T.E. (1980). Changes in Biological Diversity. In: Barney, G.O. (Ed). *The Global 2000 Report to the President of the US, Vol. 2 (The Technical Report)*. Penguin Books, Harmondsworth, pp 327–332.

Manning, A. (1972) *An Introduction to Animal Behaviour* (2nd ed.). Edward Arnold (Publishers) Ltd., London.

May, R.M. (2011) Why Worry about How Many Species and Their Loss? *PLoS Biology* 9(8): 1–2.

Mayr, E. (1963) *Populations, Species, and Evolution. An Abridgement of Animal Species and Evolution*. The Belknap Press of Harvard University Press, Cambridge, Mass.

Neesham, C. (1990) All the world's a zoo. *New Scientist* 127: 31–35.

Norse, E.A. and McManus, R.E. (1980) *Environmental Quality 1980: The Eleventh Annual Report of the Council on Environmental Quality*. Council on Environmental Quality, U.S. Government Printing Office, Washington, DC, pp 31–80.

Nowak, R.M. (1999) *Walker's Mammals of the World* (6th ed). Johns Hopkins University Press, Baltimore and London.

Ödberg, F.O. (1978) Abnormal behavior: stereotypies. *Proceedings of the first World Congress of Ethology Applied to Zootechnics*, Madrid, Industrias Graficas Espana, pp 475–480.

Paine, R.T. (1969) A note on trophic complexity and community stability. *American Naturalist* 103: 91–93.

Palagi, E., Antonacci, D. and Norscia, I. (2008) Peacemaking on treetops: first evidence of reconciliation from a wild prosimian (*Propithecus verreauxi*). *Animal Behaviour* 76(3): 737–747.

Poole, J. (1996) The African elephant. In: Kangwana, K. (ed.) *Studying Elephants. AWF Technical Handbook Series No. 7*. African Wildlife Foundation Nairobi, Kenya, pp 1–8.

Rees, P.A. (2007) Sport hunting and game viewing: two faces of ecotourism in Tanzania. In: Hosetti, B.B. *Ecotourism, Development and Management*. Pointer Publishers, Jaipur, India, pp 151–171.

Rees, P.A. (2009) Activity budgets and the relationship between feeding and stereotypic behavior in Asian elephants (*Elephas maximus*) in a zoo. *Zoo Biology* 28: 79–97.

Shepherdson, D.J. (1998) Introduction. Tracing the path of environmental enrichment in zoos. In: Shepherdson, D.J.,

Dictionary of Zoo Biology and Animal Management: A guide to terminology used in zoo biology, animal welfare, wildlife conservation and livestock production, First Edition. Paul A. Rees.
© 2013 John Wiley & Sons, Ltd. Published 2013 by John Wiley & Sons, Ltd.

Mellen, J.D. and Hutchins, M. (eds.) *Second Nature. Environmental Enrichment for Captive Animals*. Smithsonian Institution Press, Washington DC and London, pp 1–12.

Shivik, J.A., Palmer, G.L., Gese, E.M. and Osthaus, B. (2009) Captive coyotes compared to their counterparts in the wild: does environmental enrichment help? *Journal of Applied Animal Welfare Science* 12: 223–235.

Short, N. (1982) *The Landsat Tutorial Workbook*. National Aeronautics and Space Administration, Washington, DC.

Watts, P.C., Buley, K.R., Sanderson, S., Boardman, W., Ciofi, C. and Gibson, R. (2006) Parthenogenesis in komodo dragons. *Nature* 444: 1021–1022.

Wilson, E.O. (1980) *Sociobiology. The abridged edition*. The Belknap Press of Harvard University Press, Cambridge, Mass.

Wilson, E.O. (1993) Biophilia and the conservation ethic. In Kellert, S.R. and Wilson, E.O. (eds). *The Biophilia Hypothesis*. Island Press, Washington, DC, pp 31–41.

Printed and bound by CPI Group (UK) Ltd, Croydon, CR0 4YY

27/10/2024